Engineering
Cost
Estimating

THIRD EDITION

Engineering Cost Estimating

Phillip F. Ostwald

University of Colorado
Boulder, Colorado

Prentice Hall, Englewood Cliffs, New Jersey 07632

Library of Congress Cataloging-in-Publication Data

OSTWALD, PHILLIP F.
   Engineering cost estimating/Phillip F. Ostwald. — 3rd ed.
     p. cm.
   Rev. ed. of: Cost estimating. 2nd ed. c1984.
   Includes bibliographical references and index.
   ISBN 0-13-276627-2
   1. Engineering—Estimates. I. Ostwald, Phillip F. Cost
estimating. II Title.
TA183.O83 1992
658.1′55 — dc20   91-26490
             CIP

Acquisitions editor: Elizabeth Kaster
Editorial/production supervision
  and interior design: Irwin Zucker
Copy editor: Michael Schwartz
Cover design: Patricia McGowan
Prepress buyer: Linda Behrens
Manufacturing buyer: David Dickey
Supplements editor: Alice Dworkin
Editorial assistant: Jaime Zampino
Series logo design: Judith Winthrop

© 1992, 1984, 1974 by Prentice-Hall, Inc.
A Paramount Communications Company
Englewood Cliffs, New Jersey 07632

Printed in the United States of America
10 9 8 7

ISBN   0-13-276627-2

Prentice-Hall International (UK) Limited, *London*
Prentice-Hall of Australia Pty. Limited, *Sydney*
Prentice-Hall Canada Inc., *Toronto*
Prentice-Hall Hispanoamericana. S.A., *Mexico*
Prentice-Hall of India Private Limited, *New Delhi*
Prentice-Hall of Japan, Inc., *Tokyo*
Simon & Schuster Asia Pte. Ltd., *Singapore*
Editora Prentice-Hall do Brasil, Ltda., *Rio de Janeiro*

To Doris, Mark, Phillip, and Lynne

Contents

Preface   xiii

1 Introduction   1

1.1   Profit Is Necessary for Business Survival   1
1.2   Stewardship Necessary for Economic Survival   2
1.3   Competition and Failure   2
1.4   Motivation for Engineering Cost Analysis   3
1.5   Design   3
1.6   Classification of Estimates   7
1.7   Information   10
1.8   Units and Money   11
1.9   A Look at the Book   16
        Questions   17
        Problems   18
        More Difficult Problems   20
        Case Study   21

2 Labor Analysis   22

2.1   Labor   22
2.2   Measured Time   23
2.3   Wage and Fringe Rates   41
2.4   Joint Labor Cost   46
2.5   Learning   49
        Summary   50

Questions   51
Problems   51
More Difficult Problems   58
Case Study   60

3   *Material Analysis*   *61*

3.1   Material   61
3.2   Shape   64
3.3   Material Cost Policies   69
3.4   Joint Material Cost   75
          Summary   81
          Questions   81
          Problems   82
          More Difficult Problems   88
          Case Study   92

4   *Accounting Analysis*   *93*

4.1   Business Transactions   93
4.2   Fundamentals   95
4.3   Chart of Accounts   97
4.4   Structure of Accounts   100
4.5   Balance Sheet Statement   104
4.6   Profit and Loss Statement   104
4.7   Budgeting   107
4.8   Depreciation   111
4.9   Overhead   118
4.10   Job Order and Process Cost Procedures   127
          Summary   127
          Questions   128
          Problems   129
          More Difficult Problems   136
          Case Study   138

5   *Forecasting*   *140*

5.1   Graphic Analysis of Data   141
5.2   Least Squares and Regression   146
5.3   Moving Averages and Smoothing   163
5.4   Cost Indexes   170
          Summary   176
          Questions   177
          Problems   178
          More Difficult Problems   186
          Case Study   190

6   Preliminary and Detail Methods   191

6.1   Design and Evaluation   191
6.2   Opinion   192
6.3   Conference   193
6.4   Comparison   193
6.5   Unit   195
6.6   Cost and Time Estimating Relationships   197
6.7   Probability Approaches   209
6.8   Standard Time Data   215
6.9   Factor   224
6.10 Post Estimate Analysis   228
      Summary   238
      Questions   238
      Problems   240
      More Difficult Problems   247
      Case Study   249

7   Operation Estimating   250

7.1   Operation Cost   251
7.2   Manufacturing Work   252
7.3   Metal Cutting Operations   253
7.4   Metal Part Case Study   261
7.5   Nonrecurring Initial Fixed Costs   272
7.6   Operation Cost   277
7.7   Make-Versus-Buy Analysis   278
      Summary   279
      Questions   279
      Problems   281
      More Difficult Problems   284
      Case Study   286

8   Product Estimating   288

8.1   Price   288
8.2   Estimating Engineering Costs   292
8.3   Information Required for Product Estimating   295
8.4   Financial Documents Required for Product
      Decision   298
8.5   Learning Applications   300
8.6   Methods   308
8.7   Pricing Methods   312
8.8   Spare Parts   318
8.9   Design to Cost   320

Summary   322
Questions   322
Problems   323
More Difficult Problems   330
Case Study   332

9   Project Estimating   334

9.1   Project Bid   335
9.2   Work Package   339
9.3   Estimating   348
9.4   Cost and Bid Analysis   352
9.5   Engineering Economy   367
9.6   Example   377
Summary   380
Questions   381
Problems   382
More Difficult Problems   391
Case Study   392

10   System Estimating   394

10.1   Effectiveness   395
10.2   Definitions   398
10.3   Analytical Aids   402
10.4   Methods   404
Summary   419
Questions   419
Problems   420
More Difficult Problems   425
Case Study   427

11   Estimate Assurance   428

11.1   Analysis of Estimates   428
11.2   Behavioral Considerations   433
11.3   Operation Estimate Assurance   435
11.4   Product Estimate Assurance   441
11.5   Project Estimate Assurance   444
11.6   System Estimate Assurance   448
Summary   449
Questions   449
Problems   450
More Difficult Problems   454
Case Study   455

12 *Contract Considerations* *457*

12.1   Importance   457
12.2   Basic Contract Types   458
12.3   Fixed Price Arrangements   460
12.4   Cost Reimbursement Arrangements   463
12.5   Contract Clauses   465
12.6   Negotiation, Award, and Audit   467
12.7   Ethics   470
        Questions   471
        Problems   471
        Case Study   473

Appendices *474*

Appendix I: Values of the Standard Normal
               Distribution Function   475
Appendix II: Values of the Student *t* Distribution   476
Appendix III: Learning Tables   477

Selected Answers *482*

References *473*

Index *495*

Preface

This third edition of *Engineering Cost Estimating* provides students and professionals the latest principles and techniques for the cost evaluation of design. Certainly, the principles are more accepted and identified, and the techniques, widely practiced by professionals, are introduced in an academic setting. Professional engineering examples, if printed with all their minutia, would be too voluminous for a textbook. Instead, we have abridged some to highlight certain applications.

This book is adaptable to many courses that relate engineering design to economic evaluation. Engineering cost estimating courses in many universities are the same as cost analysis or economic evaluation in other schools. Typical course titles are Cost Analysis, Cost Engineering, Cost Estimating, Engineering Economy, Industrial Analysis, Manufacturing Estimating, and Technology Planning. At the University of Colorado, where I teach, our course Design Estimating is a co-requisite to the capstone design course. The number of courses that study economic appraisal of technology is growing. In rewriting the third edition we harmonized those differences that are so necessary for economic analysis of design.

This book is suitable for engineering and technology students when they reach their first level of specialization. Some programs restrict early opportunity for experimentation and broadening, and design evaluation is deferred. Although a diversity of occupational problems is included, instructors may wish to supplement the problems and design studies by using their own experiences.

Courses that have student teams or multifunctional discipline teaching of those topics are becoming more common because the profession of engineering cost estimating is becoming interdisciplinary.

Instructors teaching engineering economy will find this text useful. Traditionally, engineering economy texts concentrate on compound interest equations. This book condenses those principles to ten pages and, instead, discusses the broader matters of estimating costs and time that are crucial for the design, and eventually the time value of money equations. This is the routine followed by the engineering community.

Many who use engineering cost estimating techniques are from the practical ranks of industry, construction, business, and government. Often, self-study is necessary to supplement an intimate grasp of practice with an appreciation of academic topics. This book will give a taste of the principles of a special kind of topic.

Though designs certainly differ, principles and practices used for design appraisal are remarkably similar. We state, without proof, that all technology undergoes a business appraisal. This book, then, covers those subjects that contribute positively to the successful economic attainment of the design.

Design is given a broad and liberal interpretation. Every design is a new combination of pre-existing knowledge that satisfies an economic want. This three-part definition includes airplanes, bridges, buildings, cars, chemical plants, computers, highways, machined parts, machine tools, mining development, production lines, rockets, semiconductors, ships, systems of machines and people, and toys. With technology as the focal point, engineering cost estimating is the body of theory and business practice that provides an economic value for the design.

The arrangement of the chapters and topics allows for a variety of teaching and self-study approaches. Basically, the text is divided into four areas: design and business environment, methods, estimating, and assurance. Design customarily precedes its cost calculation. No reference is made to any specific designing. However, practices of engineering cost estimating are built on the design base. The estimating portion, which is the largest of the four areas, discusses the kinds of information and estimates for four categories of design. Operation, product, project, and system design contexts are constructed. Techniques pertinent to each are associated with that kind of design. Processes of assurance and contractual consideration are presented after various designs are cost estimated.

Cost is a nebulous term that has no standard definition and, when used in some contexts, may imply a meaning that is clearly not cost. To appreciate those distinctions, we must be prepared to understand the particular context in which the word is used. Surprisingly, the word "cost" could mean price, bid, or effectiveness, and dollars are one dimension for those measures. For management and many engineers, dollars are a more important dimension than amperes, foot pounds, or mass flow. Subtle variations of cost are understood.

More than 250,000 business firms in the United States provide a product. More than 700,000 firms are in the construction industry. Perhaps when we consider professional estimating activities there may be over one million firms in the United States alone. The field is not small. Each of those firms must use modern engineering cost estimating practices. The days of "guesstimating" are past.

An experienced engineer, after a few times "looking down the barrel" defending an estimate (management calls it probing for softness; the engineer calls it picking on professionalism), can misunderstand management's interest in cost estimating. Cost overruns for weapons systems and public work projects testify to the embarrassment that engineering cost estimating has faced and the importance of this specialization for management. Indeed, the well being of firms and our country may rest, in part, on economic evaluation of design.

Businesses have realized that computer management information systems are helpless in overcoming a lack of trust in the truth of estimated data in cost forecasting systems. International trade and foreign competition with past and future trading partners present a challenge, and cost and cost estimating will be important. Productivity needs an index, and engineering cost estimating plays a prominent role in its measurement. Technology's impact on changing times, employment, growth and development, pricing efficiency, income, gold and foreign trade, and the blessings of our kind of democratic society are topics of today.

Engineering cost estimating discloses the strengths of a company, country, or trading group for executive management and should never hide weakness from management. Decisions, both great and small, depend in part on estimates. "Looking down the barrel" need not be an embarrassment for the engineer if newer techniques, professional staffing, and a greater awareness are assigned to the engineering cost estimating function.

This book has more material than can be covered in a single semester course. Instructors can make selections based on the needs and interests of the class. Considerable material exists on manufacturing and construction. Those classes having a manufacturing bent should be sure to include chapters 7 and 8. Construction estimating courses may use chapters 9 and 11. It is a feature of the book that those differences are harmonized, and engineering cost estimating principles relevant to both can benefit student understanding.

In a great measure the usefulness of this book has been enhanced by my association with many engineers in industry and government and in clinics and seminars for over 25 years. I hope this text does justice to their practice. Several classes of students have been patient and understanding of the many poor drafts. The names used in various problems and case histories are real students, professional engineers, and friends who were helpful.

And most of all, I wish to thank my wife Doris for her help and encouragement without which this book would have never been completed. And, ἀ γ α π α ω

Phillip F. Ostwald
Boulder, Colorado

1

Introduction

1.1 PROFIT IS NECESSARY FOR BUSINESS SURVIVAL

Profit, as the Winston Dictionary defines it, is the amount by which income exceeds expense in a given time. This notion about profit leads to unfortunate conclusions. First, profit is necessary for taxes, dividends, and capital reinvestment. *Taxes,* whether national, state, or local, are the inescapable reward for successful operation, a vital contribution to continue a democratic society. If *dividends,* the rent paid for invested capital and money, were not paid, then investors would be less likely to supply money for growth. After taxes and dividends are removed from gross profit, a portion, referred to as *plow back,* is necessary for new equipment and other modernization needs. Successful managements do not ignore debt repayment, research funding, maintenance cost, salaries, or other expenses, but it is surprising that profit is sometimes overlooked. Is profit less vital than anticipated costs? It is important that profit becomes a planned expense.

A new approach can be suggested: Everything is going to be spent. Thus, it is a question of partitioning income and expense. To use a simple illustration, assume that sales revenue is going to be $1000. You expect to realize a net profit of $50. On the basis of your calculation, the "net profit dollars after taxes" is $50 and all other costs, including income taxes, must be found within the $950. It is common at this point to hear the following excuse: "You can't tell until afterward." What about prediction of sales income? Can it be safely approximated? Sales forecasts are surprisingly accurate and provide a foundation for profit estimating. The planning recognizes that what counts are current costs and not those of the previous quarter.

How successfully can expenses be held to 95%? In controlling costs versus targets, experience indicates that management can react to unplanned events. The assurance of profit remains a primary goal for long-term survival.

1.2 STEWARDSHIP NECESSARY FOR ECONOMIC SURVIVAL

Business, whether large or small, is not alone in its quest for survival. The pursuit of this objective includes government and the governed. A democratic government with authority to impose economic laws is not a wealth producer and has no inexhaustible source of wealth. Government activities, such as public works, welfare, and the military, use the resources of the country. Despite the nobility of cause and honest-meaning goals, governments suffer from financial bankruptcy. Curtailment of welfare programs, deevaluation, and heavy tax loads are symptoms of failure. Politics do not shield against economic ruin because the accounting ledger between nations is a reminder for long-term fiscal sobriety.

Even churches, charitable organizations, and not-for-profit trusts must have positive balances between short- and long-term income and debt. Individuals need no economic reminders. Despite credit or loans, bankruptcy or poverty is not uncommon. Unfortunately, no inviolate equation exists that avoids financial failure. The notion that receipts and expenses must maintain a positive cash flow is an oversimplification. Benefit-cost ratios, where social goals are evaluated in monetary terms, provide a narrow solution. Legally imposed restrictions on credit and spending are imperfect. Knowing the profound nature of this problem, a general objective for any steward is to simply husband resources.

1.3 COMPETITION AND FAILURE

It is generally assumed that competition of all kinds is increasing. This statement can be examined on pragmatic grounds. A monopolist's product must be indispensable and there must be no opportunity for substitution, competition, and control by the government. Those conditions are practically impossible to find, although they are sometimes approximated. Pure competition is present when many firms provide a standard product to numerous purchasers. A single supplier or purchaser is unable to affect the price significantly. Pure competition does not prevail either. Rather, imperfect competition is the usual marketplace.

Evidence of financial failure is found by examining companies, products, governments and their programs, and individuals. The profit squeeze on companies may result in public disclosure of bankruptcy. Mergers or sales of assets of the company, changes in title, and interdivisional failures within a larger corporation disguise the more subtle failures. A good deal of empirical evidence exists that products fail. New products that enter the market but are withdrawn within a short time is a case in point. Curtailment or complete abandonment of various governmental programs, although politically inspired, is asserted to be a result of increasing social competition. If poverty may be accepted as evidence of failure for the individual, then its popularity is well known.

Although evidence of all types of failure is clear, the factors causing failure are not. With production exceeding public demand, particularly true in the Western world, a temperamental society cannot guarantee long-term stability in the marketplace. Shifting consumer preferences, pliable and elusive, illustrate the short- and long-term effects of increasing competition. Consequences of government legislation are an obvious business factor. Increasing costs of production, inflation or recession, rising policy costs, new inventions, and improving technology are reasons for business failure.

Inasmuch as we are concerned with engineering, our attention is naturally directed to the matters of invention and technology. This requires application of engineering, science, business, and mathematics. Thus, we deal with inventions and technology as factors of increasing competition within the firm and government.

1.4 MOTIVATION FOR ENGINEERING COST ESTIMATING

Motivation for professional engineering cost estimating results from the necessity for profits, stewardship of resources, and competition. The alternative to failure is clear: Rational cost analysis is necessary to reflect the economic advantages of the firm or government or individual. It does not always follow that an engineer dominates those economic decisions, but the act of cost estimating requires knowledge of engineering, science, business, and mathematics.

The importance of the engineering cost specialty is now an accepted fact, and increasing activity in industry and education is foreseeable. Many professional societies regularly present papers, hold meetings, and sponsor clinics devoted to the ramifications of engineering cost analysis. Some groups recognize that the engineer and business professional, unlike a scientist, has an economic interest in the success of the design. A scientist, on the other hand, is more concerned with the idea or principle and less interested in an economic justification for the principle.

1.5 DESIGN

At this point we shall define several terms. Engineering cost estimating is concerned with evaluation of engineering design. The term *engineering cost estimating* could well be the term *profit engineering* because the "making sure" of profits is a higher priority for many. The estimate is the result of engineering work. When used as a noun, the word *estimate* implies an evaluation of a design expressed as cost, amount, or value. When used as a verb, *estimating* means to appraise or to determine. There are four types of estimates: operation, product, project, and system. Their ordering does not suggest any ranking of difficulty.

The word *design* is given the broadest possible definition. It does not mean "computer aided work." Design requires creative engineers and is defined as follows: Every *design* is a new combination of pre-existing knowledge that satisfies an economic want. The phrase, "is a new combination," emphasizes novelty and suggests the unusual characteristics of the designer or design team and the circumstances. The designer must possess ability or be the recipient of serendipity and its fortunate concentration of forces. "Pre-existing knowledge" relates to a design's (not designer) intellectual past and to the industry for which it has been built. For "satisfies an economic want," the engineering activity ultimately fulfills economic satisfaction.

The driving forces for design are knowledge and wants because each alone is insufficient. Without wants, no problems exist; without knowledge, no problems can be solved. This is the chicken and the egg riddle. How can a want-knowledge milieu be created? Simple answers are not possible, and, to avoid reader-author disputes, let us assume that demand and intellectual curiosity or sheer happenstance spurs on the design.

Elaboration of the design procedure is shown in Fig. 1.1. We should realize that a precise sequential process is not intended. The design procedure is actually a hodgepodge of simultaneous continuous actions. The elements are

1. Problem
2. Concepts
3. Engineering models
4. Evaluation
5. Design

Problem.   The initial description of a problem is a vague statement satisfying some want. It is necessary to transform this vague statement into an enlarged picture. By presuming the existence of a primitive problem, information (technical or nontechnical), costs, and other data are superficially gathered to give form to the problem. Suppliers, customers, competitors, standardization groups, safety and patent releases, and laboratories are sources for ideas. You may wonder why all the fuss over a simple problem statement. To give heed to a raw and imperfect problem is not unknown; students are not the only ones guilty. But a thoughtful and reasoned statement specifying the problem may lead to a more efficient result. Questions may guide the formation of a problem statement such as, "Does this fit the company's needs, interests, and abilities? Are the people connected with the problem capable of carrying it to completion, or can suitable people be hired?"

Concept.   The stage is set for the concept search after a problem with subsidiary restrictions has been defined. The quest may start with idealism such as the perfect gas laws, frictionless rolling, or perpetual motion. It is searching, learning, and recognizing; it is not application, because that comes later. A timely and fortunate search may uncover unapplied principles. When we consider that about a million scientific and technical articles are published annually, it should be clear that a

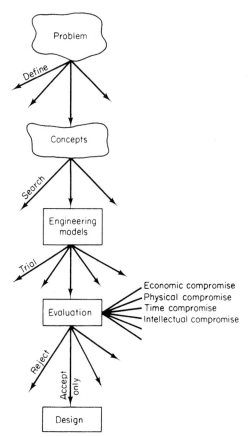

Figure 1.1. Engineering design process.

listing (ignoring study for the moment) of all information even within a narrow field is a hopeless cause. Here is a recognized defeat in the face of overwhelming odds. Nonetheless, the chance of finding the basic idea for a new development may be found through patent disclosures, new texts, or journals in the field. Instead of the unobtainable goal of completeness, there are other goals capable of being achieved in the search. Knowing where to start and when to stop in the search for concepts are lessons of experience.

Engineering models. The engineering model involves application of creditable concepts uncovered earlier. The formation of this model may range from a casual back-of-envelope model to a complicated physical shape. The formation of a model is an engineering trait and distinguishes the engineering pattern of thought. Models, whether experimental or rational, permit manipulation for theories or testing. Laboratory testing provides numerical answers by using physical mockups. Data are obtained, results are noted, and conclusions are stated. An engineer chooses the cheapest of the methods to state and understand a model.

Model has many meanings. We define model as a representation to explain some aspect. We are seldom able to manipulate reality because it may be either impossible or uneconomical. The market in a free economy is an illustration of the former, and a nuclear reactor for electrical generation illustrates the latter. Prediction of reality through mathematical abstraction is the reason for models. Engineers apply those ideas to scale larger problems. For example, an analyst may be unable to comprehend an actual system, and, with limited powers of perspective, a model may be a satisfactory substitute. Discovery of pertinent variables, rejection or confirmation of prototypes, and comparison to a standard are reasons to use modeling.

We segregate models according to physical, schematic, or mathematical notions. All three are used in the design process. The scale of abstraction proceeds from physical models to mathematical formulations.

Physical models involve change of scale. A globe looks like the earth, for instance.

A schematic model results when one set of properties represents a second set of properties. The model may or may not have a look-alike appearance to the real-world situation. Coding processes may be used, such as the chalkboard demonstration of a football play appearing as crosses and circles. Hydraulic systems are beneficial in understanding electrical systems and vice versa. Organization flow-process charts are other examples of schematic models. Schematic models capture the critical feature and ignore the unimportant.

Mathematical models operate with numbers and symbols in their imitation of relationships. Although those models are more difficult to comprehend, they are the most general.

In an approximate way, mathematical models explain the real situation. It is customary to manipulate mathematical models according to the conventional rules of mathematics. Mathematical models are desirable in engineering because they are easy to manipulate. The unit cost formula is an example of a model found in operation estimating. Whenever an engineer uses a recapitulation sheet, where labor, material, and overhead are summed, a mathematical model is used in a procedural sense. The discounted cash flow model used to calculate a rate of return of an asset is one way for dealing with project estimates.

There are precautions in modeling. Models should be flexible to permit repeated applications. Mathematical manipulation should be simple. Arithmetic is preferred over algebra. Algebra is preferred over calculus. Calculus is preferred over vectors. Computerized spreadsheets may be sufficient. Simple models are sometimes preferred because of saleability to management, and increased confidence results from their use.

Evaluation.   Ultimately, the engineering model reaches the evaluation point. A compromise forced on the engineer by economics, physical laws, social mores, ignorance, and the human fault of stupidity discolors evaluation.

Even moral questions may be debated. The wisest person cannot foresee all the future effects of a design, but it is bold to ask. Others may cooperate at this point: The stylist may abridge and direct the progress, and the manager or marketing person may foresee other problems. But here is an appropriate place to stop, pause, and evaluate.

The cost analysis proceeds along lines similar to those that have been discussed. After the problem is wisely stated (couched in the design engineering model), estimating ideas are considered, an engineering cost model is selected, evaluation trials are started (shortcomings with time, money, staff, programs, or information impede the model), and, finally, the cost estimate is completed.

Actually, the optimum evaluation of engineering models is a continuous process. Experience suggests that it takes a long time to pass from the idea stage to the design stage.

Questions exist when performing evaluations: What is the total cost of developing this design up to and including the sales promotion? What is the profit of the total investment during the first few years of production and sales? How long will it take for the initial investment to be returned? Does the new product coincide with the abilities and experience of the company? A number of factors are noted when considering those questions. Experience, study, and a questioning attitude are required traits for evaluation.

Design.   *Design* is the execution of the plan into being and shape. In fulfilling functional requirements, design involves, for example, computing, drafting, checking, and specifying, and answers, "How shall it be built?" rather than "How will it work?" By emphasizing design as a term in the design process we do not intend to overinflate its importance. Nonetheless, a greater proportion of the designer's time is tied to designing.

As a new combination of pre-existing knowledge, design satisfies an economic want. Engineering cost analysis provides the measure for this compliance. In view of this relationship, design is considered on the basis of an operation design, product design, project design, and system design. This artificial classification is used to show compatibility with engineering cost analysis and only suggests a method for teaching of engineering cost analysis. It does not describe a new classification for design.

1.6 CLASSIFICATION OF ESTIMATES

Initiation of engineering cost analysis arises from design. Practical examples of designs include bridges, cars, chemical plants, computers, highways, machine tools, radios, piece parts, service work, and systems of machines and people. With design as the focal point, engineering cost estimating is the body of theory and practice that provides a measure of the economic want of the design. Note that the engineering cost estimating activity is not involved with the "satisfaction" part of the design. Satisfaction is determined by either the marketplace or by the firm's management who may stop an undesirable design. Politics may also approve or abrogate the satisfaction requirement of designs that deal with society. Thus, the engineer finds an economic measure for the design.

There are four kinds of estimates associated with the operation, product, project, or system design. Table 1.1 is a tabular description.

TABLE 1.1. CLASSIFICATION OF ESTIMATES

Design	Fundamental Characteristic	Symbolic Measure of Economic Want	Examples
Operation	Worker and tool	Cost	Assembler and hand tools, secretary in office, crew work, driver and transportation vehicle
Product	Quantity	Price	Toys, radios, houses, typewriters, bridges, computer-controlled machine tools, transportation vehicles
Project	Single end item	Bid	Bridge, plant addition, refinery, 500 kV transmission line, capital tooling for product, right-of-way structure for transportation vehicles
System	Configuration	Effectiveness	Weapon system, hospital, rapid transit system

The "way of working" establishes the content of an *operation design*. An *operation estimate* is a forecast of labor and material required for an operation design. The design may be a toy or computer or building, and work, a worker, and a tool are involved. The tool may be simple or complicated. An operation estimate includes one worker with one tool, one worker with multiple machines, or crew work with a machine and is appropriate for the factory, construction site, office, service station, maintenance yard, hospital, or government. The worker can be categorized as skilled, unskilled, craftsman, apprentice, journeyman, or professional. Work is either classified as direct if it can be clearly traceable to the function of the design or indirect. The work that engineers perform on a weapons system can be classified as direct or indirect, an optional management choice. An ordinary example of a direct classification is a numerical-controlled lathe operator making parts that are identified clearly in a product or an electrician wiring a commercial building under construction. Methods used for operation estimating are different from those required for other types of estimates.

An entire product, rather than an operation, is estimated for a *product design*. In product estimating there is the fundamental characteristic of design change and quantity. For instance, model II is similar to model I. The engineer would not have estimated model II from scratch. Instead, by several methods, the engineer adds and subtracts for the replicated distinctions in design. A product estimate depends on production quantity, a few too many. There is a mistaken belief that large quantities of consumer goods are necessary for a product estimate, and that the production time should be reasonably brief in relationship to the quantity produced. Bridges, turbines, airplanes, toys, computers, and other consumer products are products that are similar in the general sense. Homes and apartments constructed in a single development have different facades, interiors, and so on, but there are some similarities. The engineer when dealing with product designs uses many similar methods whether the product estimate is for $2 toys or $200 million turbines. Product estimating methods are different from those required for other types of designs.

A *project design,* whether a plant, equipment, capital tooling for a product, or a prototype, is a single end item. In a project design the emphasis is on the end item. The design is custom, perhaps, and there will be only one manufactured or constructed. Usually, the dollar amount is considered capital, or large, rather than expense, or minor. Examples include refinery, plant, turbine, bridge, prototype, and airplane. While some of those examples were listed in the product estimating category, project estimating is for singular rather than plural goods. Methods used in project estimating are essentially different from those used for other types of estimates. However, the methods may require operation and product estimates as input information. Project designs usually require a significant period of time for manufacture or construction.

Another distinction between product and project estimating is found in a buyer-seller context. For example, if a factory producing numerically controlled (NC) machine tools were manufacturing several or more units, then the factory engineer would associate the problem with product design and use methods of product estimating. The engineer in evaluating the NC machine for purchase of one piece of equipment would approach the problem as a project design because of the single object and then apply methods of project estimating.

System design involves designs of operations, products, and projects in any arrangement. The fundamental characteristic of a system is configuration. Thus, we define system design especially for the purpose of engineering cost analysis. A system design is complex, and the elements of the estimate include operation, product, and project estimates. In our terminology, system design deals in the public, government, or not-for-profit domain of enterprise. In those areas factors exist that are political or altruistic and provide for general needs and goals of society. Profit is not determined explicitly for the system estimate. One example of a system design is a public rapid-transit system and the engineer represents the public authority spending the money. This would involve, for example, initial nonrecurring costs such as securing right-of-way and constructing the road structure for the vehicles and would involve project estimates. Transit vehicles, not all identical, would be necessary, and product estimates would be required. A driver and the vehicle, or the worker and tool, represent the grist for an operation estimate and become necessary information. Thus, a system design is a configuration of operations, products, and projects and uses those estimates in arriving at a value for the measure of the economic want.

The task facing the engineer is to provide a fact representative of the *economic want of the design.* A *want* is a value exchanged between competing and self-interested parties: the price a buyer is willing to pay for a toy, a contractor-owner agreement on the bid value of a building project, or the fiscal-year budget value for a rapid transit system that the public authority proposes and elected representatives of public voters approve. (Table 1.1 provides some symbolic measure of wants.) While dollar cost is usual for operation estimates, so are man-hours or man-days. Price is the measure of want for products. In projects, we are interested in the value of the bid. In system designs, public effectiveness may imply a measure such as benefit-cost or a budget fiscal-period total that sums cost streams throughout the system life cycle. However desirable various dimensions for the measurement of economic want may be, the engineer deals principally with money and dollars.

The dollar magnitude of an estimate varies with the design. A $0.10/unit price for a product and 10^8 bid for a project are not unknown. We assert that the intellectual requirements for either estimate are the same. The range of a minimum to maximum of 10^9 times underscores the variability of the engineering cost analysis field.

1.7 INFORMATION

It is necessary to have information of various kinds before the engineer is able to estimate the economic want of a design. Two extremes in the amount of information exist. Visualize the case where virtually no information is available. Then we presume that it is unlikely that an estimate can be made. At the other extreme, assume that all data are available, which implies that the money has been spent for the design and an estimate is unnecessary. Whenever the data are all available, the process of after-the-fact cost analysis is likely to be accounting and not engineering cost analysis. The engineer works in a cost data and design environment where the information is not fully disclosed and thus the reason for estimating unknown information. We separate the accounting from engineering cost analysis field on the principle that the accountant deals with cost quantities that have been spent and consistently recorded. The engineer reckons in cost quantities that have not been spent for a design. Cost accounting and engineering cost analysis are specializations that work together and have much in common.

Some cost data are *historical*. Accounting reports are historical because they emphasize the transactions recorded through cost-controlling accounts that may be kept in a ledger system. Money is expended, and materials, labor, services, and expenses (such as power or heat) are received. Specific accounting procedures must be provided for recording the acquisition and disposition of materials, for the recording and use of labor, and for their distribution. The internal function of cost accounting as it relates to our interest is discussed in chapter 4. Cost accountants are primarily responsible for historical information.

Some data are *measured*. Engineers may find that work measurement methods give information that is amenable to estimation, either in time or dollar dimensions. Material quantities calculated from drawings and specifications are a form of measured data.

Finally, some data are *policy* and have the property of being fixed for engineering purposes. The origins of policy data are varied and are accepted as factual and often unchallengeable by the engineer. Union-management wage settlements or union hall hiring of construction labor where predetermined policies dictate the wages and types of labor on equipment to be operated are examples of policy data. Budgets and legislative restrictions for municipal to national laws dictate codes of conduct and cost. The federal government requires a social security tax from the employer for the purpose of providing old-age benefits. An unemployment compensation tax, sometimes called Federal Unemployment Insurance Cost, is collected by the states to provide funds to compensate workers during periods of unemployment. Those data may emanate from internal departments within the organization, official government sources, international agencies, trade associations, trade unions, sampling organizations, or any office that gathers and divulges design and economic information.

Accounting is a major source of information, but there are other departments internal to the organization that provide information. The personnel department, charged with the handling of employees, interprets the union contract (where unions exist), conducts labor contract negotiations, and keeps personnel records regarding wages and fringe costs.

Operating departments, whether, for example, in construction, manufacturing, or crafts, are the producing organizations and are concerned with doing. The foreman or department manager knows the operating details at that moment. Frequently, he or she is a direct source for information. The foreman may often assist in obtaining data on special forms that report extraordinary costs of process equipment, manning, efficiency, scrap, repairs, or down time. Sometimes he or she is the oracle for a "guesstimate" on operations with which he or she is familiar.

The purchasing department in many organizations is responsible for spending money for materials. Some companies believe that the purchasing department is responsible for the outside manufacturer. The purchasing department knows about purchasing and shipping regulations and can be a frequent source of information.

The contribution of sales and marketing is apparent in the pricing of products: for instance, market demand, sales, consumer analysis, advertising, brand loyalty, and market testing.

A variety of basic economic facts and trends is available from the U.S. government. The Bureau of Labor Statistics (BLS) provides elements of cost on the prices of materials and labor.

Data are found from manufacturers' agents and jobbers, who, although they promote special interests, are willing to release information given to them by their clients. Trade associations, subsidized by groups of businesses sharing a common need, are typical organizations that publish data for engineering cost analysis.

1.8 UNITS AND MONEY

The student of engineering can no longer be uninformed about worldwide systems of units and money. Understanding in international trade and dimensional and monetary conversions is required. U.S. customary and metric units and the dollar, pound, deutsche mark, peso, and so on are used throughout this book to provide a familiarity.

Le Systeme International d'Unites, known officially worldwide as SI, is a modernized metric system and incorporates many advanced unit concepts. With SI it is possible to have a simplified, coherent, decimalized, and absolute system of measuring units. Because it will be many decades before the United States deals exclusively with SI units in engineering, we use a mixture of SI and U.S. customary units. Sometimes the U.S. customary value is given first and then followed by SI units in parentheses: for example, 5.008 in. (127.20 mm).

The SI system has seven base units (meter, kilogram, second, ampere, Kelvin, candela, and mole), two supplementary units (radian, steradian), and additional derived units. The list of derived units within the SI system is extensive. Basic to SI is the following definition: 1 newton is the force required to accelerate a mass of 1 kilogram at the rate of 1 meter per second squared; 1 joule is the energy involved when a force of 1 newton moves a distance 1 meter along its line of action; and 1 watt is the power that in 1 second gives rise to the energy of 1 joule.

The SI units for force, energy, and power are the same regardless if the design is mechanical, electrical, hydraulic, or chemical. Confusion is often found in the U.S. customary system of using both pounds force and pounds mass, but this is avoided in SI. The SI system has a series of approved prefixes and symbols for decimal multiples and is shown in Table 1.2.

TABLE 1.2. SI PREFIXES

Multiplication Factor	Prefix	SI Symbol
$1,000,000 = 10^6$	mega	M
$1,000 = 10^3$	kilo	k
$10 = 10^1$	decka	da
$0.1 = 10^{-1}$	deci	d
$0.01 = 10^{-2}$	centi	c
$0.001 = 10^{-3}$	milli	m
$0.000001 = 10^{-6}$	micro	μ

Be careful with capitalization of SI symbols to avoid confusion: for example, K is Kelvin but k is kilo. Further, we use the U.S. practice of commas to separate multiples of 1000. SI uses a space instead of a comma.

Data for units and money are from a variety of sources and have been recorded with varying degrees of refinement. Specific rules are observed when engineering and cost data are added, subtracted, multiplied, or divided. Consider the example of adding three numbers for engineering analysis, where the first is reported data in millions, the second in thousands, and the third in units, such as in example (a):

(a)	163,000,000	(b)	163,000,000
	217,885,000		217,900,000
	96,432,768		96,400,000
	477,317,768		477,300,000

If those numbers were pure engineering data, then the numbers should first be rounded to one significant digit farther to the right than that of the least accurate number and the sum given as in example (b) and then rounded to 477,000,000. If the numbers are pure cost data, then the overriding choice depends on final disclosure of the information. The rule adopts example (a) as preferred but would accept example (b) if an approximation is all that is required.

The rule for multiplication and division of engineering data follows: The product or quotient must contain no more significant digits than are contained in the number with the fewest significant digits used in multiplication or division. The difference between this rule for addition and subtraction should be noted. The last rule requires rounding digits that lie to the right of the last significant digit in the least accurate number.

Multiplication: $113.2 \times 1.43 = 161.876$ rounds to 162

Division: $113.2 \div 1.43 = 79.16$ rounds to 79.2

Addition: $113.2 + 1.43 = 114.63$ rounds to 114.6

Subtraction: $113.2 - 1.43 = 111.77$ rounds to 111.8

The product and quotient are limited to three significant digits because 1.43 contains only three significant digits. The rounded answers in the addition and subtraction examples contain four significant digits.

Numbers that are exact counts are treated as though they consist of an infinite number of significant digits. When a count is used in computation with a measurement, the number of significant digits is the same as the number of significant digits in the measurement. If a count of 113 is multiplied by a value of 1.43, then the product is $161.59. However, if 1.43 were a rough value accurate only to the nearest 10 and, hence, contained only two significant digits, then the product would be 160.

Rules for cost engineering are similar, although there are exceptions for dealing with the requirement for exaggerated precision. Those are given in later chapters. Essentially, the requirements for exaggerated precision come from large-quantity considerations. Rules for rounding are the same as for estimating and engineering practice. If 3.46325 is rounded to four digits, then it would be 3.463. If it is rounded to three digits, then 3.46. If 8.37652 is rounded to four digits, then it would be 8.377. If it is rounded to three digits, then 8.38. If the digit discarded is exactly 5, then the last digit retained should be rounded upward if it is odd or not if it is even. For example, 4.365 when rounded to three digits becomes 4.36, and 4.355 would be rounded to 4.36 if rounded to three digits.

This book makes no exception to the practice of avoiding cents as a dimension. For instance, a unit of a product would be expressed as $1.43 and never as 143¢ because of potential confusion between the two dimensions.

Conversions from U.S. customary units to SI, and vice versa, are made by using Table 1.3. Consider the following examples. Convert 12.52 ft to meters, 17.2 ft^3 to cubic meters, 5.15 lbm to kilograms, 2.005 in. to millimeters, and 2.4637 in. to millimeters.

TABLE 1.3. APPROXIMATE CONVERSION FACTORS FOR U.S. CUSTOMARY UNITS TO SI

To convert from:	To:	Multiply by:
Area		
square foot	square meter (m²)	9.290×10^{-2}
square inch	square meter (m²)	6.451×10^{-4}
Energy		
Btu	joule (J)	1.055×10^{3}
kilowatt hour	joule (J)	3.600×10^{6}
Force		
pound-force	newton (N)	4.448
Length		
inch	millimeter (mm)	2.54×10^{1}
foot	meter (m)	3.048×10^{-1}
mile	kilometer (km)	1.609
Mass		
ounce	kilogram (kg)	2.834×10^{-2}
pound	kilogram (kg)	4.535×10^{-1}
ton (short, 2000 lbm)	kilogram (kg)	9.071×10^{2}
Power, horsepower (550 ft lb/sec)	watt (W)	7.456×10^{2}
Pressure, pound-force per square inch	pascal (Pa)	6.894×10^{3}
Temperature, degree Fahrenheit	degree Celsius (C)	$(t_f - 32)/1.8$
Velocity		
foot per second	meter per second (m/s)	3.048×10^{-1}
mile per hour	kilometer per hour (km/h)	1.609
Volume		
cubic foot	cubic meter (m³)	2.831×10^{-2}
cubic inch	cubic meter (m³)	1.638×10^{-5}
cubic yard	cubic meter (m³)	7.645×10^{-1}
board foot	cubic meter (m³)	2.359×10^{-3}
gallon	liter	3.785

Here are the answers: $12.52 \times 3.048 \times 10^{-1} = 3.8161$, which, using the rule of precision of the original measurements, becomes 3.82 m; $17.2 \times 2.831 \times 10^{-2} = 0.48693$ becomes 0.487 m³; $5.15 \times 4.53 \times 10^{-1} = 2.3330$ becomes 2.33 kg; $2.005 \times 2.54 \times 10^{1} = 50.927$ becomes 50.93 mm, because a two-place decimal for the millimeters has the same degree of accuracy as contained in the original inch measurement; and $2.4637 \times 2.54 \times 10^{1} = 62.5780$, which when rounded to three decimal places becomes 62.578 mm because of similar degrees of accuracy implied by a measurement.

For many companies, export opportunity spreads risk among several countries and markets. But cross-frontier trade inevitably causes foreign exchange exposure for either one or both of the trading partners. Consider the example of a company that will purchase machinery from Germany. We assume that the purchaser is in the United States and has the choice of currency for buying and paying for the machine.

Three options are available: The company may agree to buy in deutsche marks, dollars, or some third-country currency. If the purchase is in dollars, then it may appear that the foreign exchange has avoided a problem, but do not forget that the problem exists for the seller of the machine. The seller will be in receipt of dollars and will sell them for deutsche marks (DM). On the other hand, if the machine is invoiced in deutsche marks, then the buyer will buy deutsche marks before buying the machine. A foreign exchange transaction is involved either way. If the goods are invoiced in a third country, for instance British sterling, then the U.S. buyer arranges payment in British sterling to the German supplier, who in turn sells the sterling and converts into deutsche marks. In this event we have two foreign exchange transactions, whereas previously it was only one. It is axiomatic that movement of goods or services across a frontier causes a foreign transaction. Exporters and importers think in terms of their own national currency.

The prevailing *foreign exchange rates* between any two countries are the prices at which bills of exchange on one of those countries will sell in the currency of the other. The bills of exchange are expressed in terms of the price or rate at which the currency of one country is exchanged for that of another. Table 1.4 is a sample of foreign exchange rates. For instance, it takes $1.986 U.S. to exchange 1£, or $0.6573 U.S. for 1 DM. But free and uncontrolled foreign exchange rates fluctuate daily. Low exchange (meaning a lower cost of purchasing foreign currencies) normally indicates a strong demand for foreign currency or a heavy offering of U.S. dollars. A distinction is also seen in Table 1.4 between spot and future exchange, perhaps 1 year from the date of the spot value. When an importer purchases spot exchange, delivery is actually taken for a definite amount of foreign exchange at the time of the purchase for which he or she pays the rate then quoted for his or her particular bill of exchange. When the importer purchases a future exchange contract, the importer agrees to purchase a given amount of exchange on a fixed date in the future or within a fixed period to pay for it at the rate specified in the future contract. This future rate may be higher or lower than the spot rate. Consider the following examples.

TABLE 1.4.   TYPICAL EXCHANGE RATES IN U.S. DOLLARS

	German Mark	Swiss Franc	Japanese Yen	Canadian Dollar	British Pound	Italian Lira	French Franc	Mexican Peso
Spot	0.6573	0.7987	0.00791	0.8900	1.986	0.000895	0.1972	0.000374
One year	0.6629	0.8074	0.00713	0.8930	2.006	0.000910	0.1984	0.000327

A material is priced currently at $3.65/lb. Find the equivalent spot market in British pounds and SI.

$$\left(3.65\frac{\$}{\text{lbm}}\right)\left(\frac{1}{1.986}\frac{£}{\$}\right)\left(\frac{1}{0.4535}\frac{\text{lbm}}{\text{kg}}\right) = £4.0526/\text{kg}$$

German material is currently priced at 485 DM/m^3. Determine equivalent values 1 year hence in U.S. dollars and customary dimensional units.

$$\left(485 \frac{DM}{m^3}\right)\left(0.6629 \frac{\$}{DM}\right)\left(0.02831 \frac{m^3}{ft^3}\right) = \$9.1018/ft^3$$

Observe that the trailing decimals on the right side of the equations are increased because of the exact nature of currency conversion.

1.9 A LOOK AT THE BOOK

This book provides the kinds of thinking that are found in engineering cost work. The circular riddle—do problems provide the stimulus in finding solution methods or do techniques that have been unused discover and solve problems—is really never answered. An engineer would not seriously consider redesign of the wheel or feel any guilt in exploiting its theory and practice.

An effort has been made to assimilate theories and practices that are broadly attractive to all engineering students, whether they are (or are to be) employed in research, development, design, production, construction, sales, or management. Though engineering cost practices vary among the several fields of technology, the principles do not. This becomes clearer by noting the organization of the book.

In the first chapter we couple engineering design to its business environment. In this text we assume that design neither leads nor lags its economic shadow. Vital to the design is the cost information on which decisions must be based. With the design and cost data at hand, the engineer builds a corresponding cost-design structure. In the past this consisted of columnar and recapitulation sheets. Now, however, the cost model is too involved for those simple maneuvers.

Chapters 2, 3, and 4 are concerned with labor, material, and accounting analysis. Labor and material costs items are major contributions for the design estimate. When accounting costs are considered, the cost is more complete. But before those costs can be estimated for a design, there must be analysis. Data are unfortunately out of date, demand is history, and budgets need review. Those experiences demonstrate the need for forecasting. In this book, forecasting implies numerical analysis of information, and chapter 5 suggests several popular approaches. Methods are classed as preliminary or detail. Chapter 6 provides general methods suitable to various designs.

How might the act of estimating be classified? For engineering cost analysis, we follow the design in a logical manner to provide a scheme of estimating. We could concoct an estimate classification according to purpose, accuracy, time, type of commitment, or design. If we were to classify estimates based on purpose, as many are, we would find estimates for the verification of a vendor quotation, appropriation, budgeting and funding, and evaluation studies or design feasibility in addition to cost or price. If accuracy were the determining factor, then we could imagine an estimate classification as order of magnitude, say $\pm 50\%$; as ball park, say $\pm 20\%$; and as accurate, $\pm 5\%$. Initial and final are other possibilities. The classification scheme adopted in this book is associated with design. The designs, whether operation, product, project, or system, provide the identifying feature. Formats, procedures, and ramifications vary for those types of designs. Methods of estimating do not. Those design estimates are covered by operation (chapter 7), product (chapter 8), project (chapter 9), and systems (chapter 10).

The use of analysis, judgment, and experience will forever remain a vital factor in the engineering field. Certainly, without judgment chaos results. However, estimates will not be successful unless there are activities that ensure their success. This follow through is called estimate assurance and is described in chapter 11.

Contractual practices are presented in chapter 12. The thrust of the last chapter is dominated more by a discussion of goals than by rigid rules.

With a liberal interpretation, cost may mean cost, profit, income, expense, or any economical measure of want and is an important dimension. Many believe that a car, computer, or rocket design has economic value as the first and last requirement. The thought is this: Given a design (aerospace, agricultural, chemical, civil, electrical and electronic, industrial, manufacturing, marine, mechanical, metallurgical, mining, petroleum), physical and real-world restrictions are its companion. Engineering, production, marketing, sales, and finance conform to the engineering design, because the drawings and specifications are the authority for construction and operation. The sales person sells the design, the service engineer maintains this design, the accountant classifies costing points about this design, the manager plans production schedules to build the design, and the construction engineer selects processes and equipment to construct the design. Design, used in its broadest context, causes a long chain reaction. The engineering student who brings to his or her job an understanding of the economic consequence of design is valued in industry, business, and research and becomes an engineer, designer, project engineer, supervisor of engineering activities, or a business person who works closely with design, development, and the research team.

QUESTIONS

1.1. Define the following terms:

Profit	Administrative practices
Estimator	Operation estimate
Estimating	Product estimate
Design	Project estimate
Economic want	System estimate
Engineering model	SI
Historical information	Exchange transaction
Policy information	Spot exchange
Measured information	Bills of exchange

1.2. Prepare a list of career opportunities in estimating from the classified want ads of a newspaper or professional trade magazines.

1.3. How does competition and failure of the enterprise interact with principles of cost estimating? Discuss.

1.4. How would you define profit? Discuss fully. What are the consequences of negative profit (loss)? How can profit be made appealing to the individual?

1.5. List positive and negative results of failure in business. Should governments prevent business failure? Does company size and political power affect your answer?

1.6. How many not-for-profit organizations can you name? How do governments protect their interests?

1.7. Distinguish between product and technology competition. Which affects your life more?

1.8. What distinguishes the act of estimating? Will estimating become a science?

1.9. How do economic laws and physical laws differ? Are the well-ordered cause-and-effect relationships separable in business and engineering fields?

1.10. Describe how computers have improved on back-of-envelope techniques in estimating.

1.11. Relate the role of cost estimating to design engineering. Contrast these roles.

1.12. Distinguish among an operation, product, project, and system estimate. Cover mutually exclusive descriptions applicable to those four estimates. How are they similar? How are they different?

1.13. What do exchange rates between countries reflect? Why do they fluctuate?

1.14. Consider another rationale for the classification of the estimate. Would labor estimating, material estimating, and tools and machine estimating be complete?

1.15. Do you agree that the dimension "dollar" is as important as other engineering dimensions?

PROBLEMS

Convert the following 11 problems from U.S. customary units to SI units or from SI to U.S. units. Show correct abbreviations.

1.1. (a) 17 ft^2, 2.4 ft^2, 450 in.^{2}, 5000 ft^2 to meters^2 (m^2)
   (b) 0.15 in.^{2}, 0.035 in.^{2}, 20.61 in.^{2} to millimeters ^{2} (mm^2)

1.2. (a) 280,000 British thermal units (Btu) to joules (J)
   (b) 7,500,000 kilowatt hours (kWh) to J

1.3. (a) 18 pounds force (lbf) to newtons (N)
   (b) 180,000 lbf, 1.8 $\times$ 10^6 lbf to newtons with appropriate prefix

1.4. (a) 18 ft, 2.0 ft to meters (m)
   (b) 10 in., 0.01 in., 100 in., 0.00015 in. to millimeters (mm)

1.5. (a) 15 ounce mass (ozm) to kilogram (kg)
   (b) 25 tons to kilograms (kg)
   (c) 1400 kg to pounds (lb)

1.6. (a) 15 pounds-mass/foot^3 (lbm/ft^3 = density) to kilogram/meter^3 (kg/m^3)
   (b) 180 kg/m^3 to lbm/ft^3

1.7. (a) 2500 pounds-force/foot^2 (lbf/ft^2) to megapascals (MPa)
   (b) 180 pounds per in.^{2} (psi), 1750 psi to Pa
   (c) 17 MPa to psi

1.8. (a) 1500 Celsius (C), 200 C, 1000 C to Fahrenheit (F)
  (b) 200°F, 1000°F to C (Note: Do not use degree symbol with Celsius.)
1.9. (a) 180 feet/minute (fpm), 500 fpm to meters/second (m/s)
  (b) 180 inches/second (in./s), 1855 in./s to meters/min (m/min)
1.10. (a) 0.37 ft^3, 125 ft^3, 700 ft^3 to meters^3 (m^3)
  (b) 0.01 in.^{3}, 12 in.^{3}, 150 in.^{3} to millimeters^3 (mm^3)
  (c) 1000 yards^3 to m^3
  (d) 250 mm^3, 80 mm^3, 1500 mm^3 to in.^{3}
1.11. (a) 800 ft^3/min, 65 ft^3/sec to m^3/s
  (b) 1000 m^3/s, 75 m^3/s to ft^3/min

Use Table 1.4 for Problems 1.12 to 1.15.

1.12. A casting costs 17.50 U.S. dollars per unit. What is the spot value of the casting
  (a) in German marks?
  (b) in Swiss francs?
  (c) in Canadian dollars?

1.13. If a catalyst is worth \$85 per gallon in the United States, then what is an equivalent value
  (a) in German marks and metric units?
  (b) in Italian lira and SI?

1.14. International export opportunities may send materials from one country to another and back to the originating country because of labor cost or a technology advantage. An electronic product is transported to Japan, and a value of 12,432 yen is added.
  (a) What is the U.S. value?
  (b) If a value of 58.65 U.S. dollars is estimated for equivalent U.S. work, then what is the exchange rate that is indifferent to the decision?
  (c) For the work to remain in the United States, must the exchange rate increase or decrease relative to the indifference rate?

1.15. An American business woman travels from New York to four countries. She will start with U.S. currency and exchange her dollars in each country by using that country's prevailing exchange rate. In each country she will buy the next air fare and will incur business expenses. Travel and expense budget expressed in the currency of the country are as follows:

U.S. to Canada	Canada to England	England to France	France to Germany	Germany to New York
\$875 (U.S. dollars)	\$2200 (Canadian dollars)	£1600 (United Kingdom pound sterling)	12,000 F (French francs)	12,700 DM (deutsche marks)

What minimum amount of U.S. cash will she need in New York?

1.16. A small country is preparing for its anniversary, and a politician wants 2500 busts of the president for distribution. Copper alloy will be used, but the weight of the bust depends on the copper alloy used. When the master of the bust is submerged in water, it displaces 0.00098 m^3 of water. An 80–20% copper-zinc alloy requires \$2.50 per kilogram for copper and \$2.07 per kilogram for zinc. The densities of copper and zinc are 8906 and 7144 kg/m³. Your billing price is figured at seven times the metal cost. The country's currency is renolas, figured at a current exchange rate of 14.3 units to 1 U.S. dollar.

(a) Find the price for 2500 units in renolas and U.S. dollars.

(b) Convert the values to U.S. customary units and repeat.

MORE DIFFICULT PROBLEMS

1.17. A U.S. contractor will design, fabricate, and supervise a project in country X. Designing and fabricating are performed in the United States, while erection and assembly are done in country X by using the national labor of that country. The deal is agreed to at base time zero and will be adjusted for inflation within country X and exchange rate. The contract at time zero is as follows:

Time	U.S. Material	U.S. Cost for Erection in Country X
0	$6 million	$2 million

The engineer believes that the rate of inflation will be 10% and 25% annually in the United States and country X. The exchange rate of country X to the U.S. base is 1:1, 2:1, and 3:1 for years 1, 2, and 3. U.S. material cash flow is spread evenly over the first 2 years while erection will occur during year 3 in country X. The contract requires that the full lump sum is paid at year-end 3, adjusted for inflation and exchange rate. Assume that the inflation and exchange rates are independent.

(a) If the U.S. material costs are spread uniformly over the first 2 years, then roughly what total amount will be spent because of inflation? (Hint: Consider inflation to increase at a compound amount, and that money spent at the start of the year will inflate for the entire year while money spent at the end of the year will not inflate for that year. Thus, consider average inflation and the time period.)

(b) If the contractor will pay the national labor in that country's currency during year 3, then how many country X value units are roughly expected? How many U.S. dollars?

(c) What approximate sum of U. S. money is due to the contractor at year-end 3? What country X currency?

1.18. A reactor vessel can be built in Germany, England, or Italy and shipped to country Y for integration in a refinery. Quotes are received from contractors and ratioed to the U.S. estimate in terms of U.S. currency by using Table 1.4 and spot exchange rates. According to the contract, payment is transacted between the two parties on the day the reactor reaches the port of entry of country Y. The exchange rate for the scheduled arrival is, however, different from that listed by the table and is shown as a relative change to the table.

	Germany	England	Italy
Ratio of subcontract bid to U.S. estimate	0.9992	1.0065	1.0062
Exchange rate on arrival compared to Table 1.4	+0.3%	−0.5%	0.01%

The U.S. estimate for the reactor is $100 million.

(a) In which country should the vessel be built, and what is that cost in U.S. currency?

(b) What is the dollar penalty if the next-lowest country is selected?

(c) Where does nonnumerical judgment enter the decision process?

CASE STUDY:
THE VACUUM CHAMBER ENVIRONMENTAL TEST UNIT

A new vacuum chamber environmental test unit is required by an aerospace firm to help existing units because of an increased work load. The units were already scheduled to capacity during the contract time period. The test units are necessary because spacecraft components require testing in a vacuum to simulate the altitude at which they will be orbiting. Existing vacuum units are a small 42-in. diameter chamber with a high pumping capacity and fast vacuum pull down time and a large 10-ft diameter chamber for large-component testing. A satisfactory commercial system cannot be purchased, and a number of other solutions are proposed:

1. Repeat the existing design and construction of the small chamber, thereby saving the engineering design costs. This design sacrifices inside dimensions.

2. Design and build an intermediate-sized chamber with a larger 32-in. diffusion pump. This should relieve some of the load on the larger chamber.

3. Design and build an intermediate-sized chamber with the same pump stack as the small chamber, sacrificing some pumping capacity and pump time but saving the cost of the larger pump stack and its engineering.

4. Design and build an intermediate-sized chamber that is the same as described in solution 3, but add an additional port for another pump in the future, if required.

A single solution to this case problem cannot be provided because pertinent information remains undisclosed, but consider the initial design process facing the engineer. What steps of the (1) problem, (2) concepts, (3) engineering models, (4) evaluation, and (5) design are most significant here? Why? What information is necessary to estimate these designs? What elements of information do you classify as available or unavailable? List typical measures of economic want that will allow selection of the preferred design. Prepare your answers in the form of a report.

2

Labor Analysis

Labor comprises one of the most important items of an operation. Before we are able to estimate labor, there is an almost unquestioned dependence on an objective measurement or historical value of time. Methods to find labor time include time study, man-hour reports, and work sampling and are studied in this chapter. Once time values are known for a design, the time values are multiplied by the wage or the wage and fringe costs. Sometimes labor time or cost are joint and must be separated to find unit cost. Thus, analysis is required before labor estimating can begin.

2.1 LABOR

Labor has received intensive study, and many recording, measuring, and controlling schemes exist in an effort to manage labor. Labor can be classified in a number of ways: for instance, direct–indirect, recurring–nonrecurring, designated–nondesignated, exempt–nonexempt, wage–salary, blue collar–management, and union–nonunion. Social, political, educational, and type of work are other divisions that classify labor. Payment for wages may be based on attendance or performance. We select the direct-indirect classification for cost analysis.

In operation designs there is the simple qualitative formula

$$\text{labor cost} = \text{time} \times \text{wage} \tag{2.1}$$

The selection of time matches the requirements of the operation design. Time is expressed relative to a unit of measure and is denominated by terms such as piece, bag, bundle, container, 100 units, 1000 board feet (2.4 m^3), and so on. The unit of time may be second, minute, hour, man-day, man-week, month, or year. Thus, we may use 15.025 hours per 100 units in manufacturing or 14 man hours per frame for building construction or 15 man-months per tank for chemical processing construction. Neither the application nor the magnitude of the time affects the generality of Eq. (2.1). In early estimates the design may be roughly known, and larger chunks of time are used. But, as the design becomes detailed so does the refinement of time. In some situations the estimate of time is a guesstimate and may be unrelated to any measured, referenced, or analyzed data. A *guesstimate* is based on the engineer's observational or rough experiences. Circumstances exist where those personal judgmental values are unavoidable.

The second part of Eq. (2.1), wage, is defined in the context of the operation design. The design may be for one worker and one machine or a crew with one machine or a crew with several machines or processes. In the simplest case, one on one, the job description and job design are available to the engineer. The number used for the wage corresponds to the time period of work.

Units for the wage are dimensionally compatible to the time estimate. If the time estimate is in hours per unit, then the wage is expressed in dollars per hour. The wage may be the amount that the worker sees in the paycheck or it may include all or part of the fringe costs. The choice is coordinated with accounting analysis. Labor estimating and analysis is concerned with direct and not indirect labor. Direct labor time for work is *directly* related to the design. An example of direct labor is an assembler in manufacturing or the carpenter, mason, roofer, and so on, in construction. Engineers are classified as direct labor in some project designs. Indirect labor in manufacturing is shop labor other than direct labor, such as stockroom clerks, foremen, and material handlers. Indirect labor could be timekeepers or foremen in construction projects. Stated differently, direct labor "touches" the design, and indirect labor is supportive of that effort. Indirect labor and its costs are covered by factory overhead, a topic discussed in chapter 4. Table 2.1 gives many of the definitions used in this chapter.

2.2 MEASURED TIME

A school of thought exists that asserts that "time is the measure of cost." Although the slogan is debatable, there is an element of truth when applied to the needs of detail cost estimating. For this activity there is an almost unquestioned dependence on an objective measure of time.

Historical records provide costs about labor, supervision, methods, and a host of endeavors but are subject to greater error. The measurement of time, on the other hand, is limited mostly to direct labor. Two categories of work are direct and indirect. Work can be segregated into one or the other category inasmuch as the direct-indirect category is determined by practice and definition. Discussed in this section are techniques of measuring labor. Though time is measured, it is cost that is ultimately required.

TABLE 2.1.   DEFINITIONS FOR LABOR ANALYSIS

Actual time	The time reported for work, which may include delays, idle time, and inefficiency as well as efficient effort
Allowance	An adjustment to the normal time providing for personal needs, fatigue, and special delays inherent in the work
Allowed time	Normal time increased by appropriate allowances (see standard time); also, reported man-hours adjusted because of work differences or judgment
Avoidable delay	Interruption in the work that caused additional time that could have been avoided or minimized by the worker by using better skill or judgment
Constant element	An element whose normal time is constant with respect to various independent effects on the element
Continuous timing	A method of time study where the total elasped time from the start of the study is recorded at the end of each element
Cycle	The total time of elements from start to finish in a repetitive operation
Delay allowance	One part of the allowance included in the standard time for interruptions or delays beyond the operator's ability for prevention
Element	A subpart of an operation separated for timing and analysis; beginning and ending points are described, and the element is the smallest part of an operation observed by time study
Elemental breakdown	The description of the elements of an operation in a measurable sequence
Fatigue allowance	One part of the allowance caused by physiological reduction in ability to do work, sometimes included in the standard time
Foreign element	Unrelated to the operation and removed from the time study
Frequency of occurrence	The number of times an element occurs per operation or cycle
Idle time	A time interval in which the operator, equipment, or both are not performing useful work
Incentive	Financial methods motivating a worker to exceed a standard
Machine interference	Idle machine time occurring as a result of operator attention on another machine in multiple-machine work
Machine or process time	Time required by a machine or process
Man-hour	A unit of measure representing one person working for 1 hour. See also person-hour
Normal time	An element or operation time found by multiplying the average observed time by a rating factor
Observed time	The time observed on the stopwatch/electronic clock and recorded on the time study sheet during the measurement process
Operation	Designated and described work subject to work measurement, estimating, and reporting
Personal allowance	One part of the allowance included in the standard time for personal needs that occur throughout the working day
Person-hour	A unit of measure representing one person working for 1 hour.

TABLE 2.1.   DEFINITIONS FOR LABOR ANALYSIS (*cont.*)

Rating factor	Involves comparing the performance of the operator under observation by using experience or other bench marks; additionally, a numerical factor is noted for the elements or cycle; 100% is normal, and rating factors less than or greater than normal indicate slower or faster performance
Regular element	Elements that occur once in every cycle
Snap back	A method of timing that records the elemental time at the end point of the element
Snap observation	An observation made virtually instantaneous as to the state of the operation (i.e., idle, working, or the nature of the element)
Standard time	Sums of rated elements that have been increased for allowances
Variable element	An element whose normal time varies or depends on one or more dependent effects

Although engineers may not be concerned personally with the measurement of labor, they do depend on the results from work measurement. Engineers are satisfied if such labor measurements are objective, as far as possible, and are willing to use the information provided that professional techniques are used in the finding of time for an operation. Three methods for the measurement of time are useful: time study, work sampling, and man-hour reports. Although we may argue in favor of a particular method, each is suitable and necessary for different occasions.

2.2.1. Fundamentals of Time Study

For our purpose it is not necessary to delve into the historical background of time study. Suffice it to say that Frederick W. Taylor was the founder of time and motion study and that Frank and Lillian Gilbreth were two leading pioneers.

Time study is the analysis of an operation to eliminate unnecessary elements and to determine the better and cheaper method of performing the operation; to standardize methods, equipment, and conditions; and then, and only then, to determine by measurement the number of standard hours required for an average worker to do the job.

Time study has many advantages and is useful in industry. Because of competition in any industry, it is necessary to know various cost factors and how much a proposed change decreases or increases cost. The determination of cost comparisons is possible by time study, which is the backbone of many cost systems.

The stopwatch procedure goes something like this: (1) conduct analysis of methods and improve if necessary; (2) record significant data; (3) separate the operation into elements; (4) record the time consumed by each element as it occurs each time; (5) rate the pace or tempo at which various elements of work are performed; (6) determine the allowances; and (7) convert rated elements into normal time, include allowances, and express the standard in common units of production. Although the stopwatch procedure is criticized, its role in gathering time for production remains significant. Many businesses in the United States depend on time study for a substantial amount of the data used in cost analysis.

The equipment for taking a time study is simple: a clipboard and an electronic timer or decimal minute or hour stopwatch.

The first and most important phase of taking a time study is its preparation. Is the job ready for timing? The time study technician resolves this question by answering the following questions: Is the proper tooling being used, and is it laid out correctly? Is the material laid out in an economical manner? Are proper machines or tools being used? Is the quality of the finished part or operation up to the inspection standards? Is the motion pattern employed the most economical that can be devised at this time?

The second phase of the time study is to record the information on the time study form. After the title block information is recorded, the elements of work must be identified and written down in sequence. The technician keeps in mind when breaking down the operation that the elements should be as short as possible but be long enough for timing accuracy. Wherever possible, manual operator time is separated from machine time. Wherever possible, elements that are constant (or nearly so) are separated from variable elements. Elemental start and stop time should be easy to identify.

Consider the example of the assembly of a common electrical receptacle. A present method where assembly fixtures are not used is examined for possible improvements. The step to record significant information starts with a plan view sketch of the bench layout. Observe in Fig. 2.1 that only the barest of detail is shown because the precise location of the screwdriver or stack bins is not critical.

A sketch of the parts for assembly is given in Fig. 2.2. Certainly, drawings are available if additional information becomes necessary. Note that the title block lists other information that may be used for additional detail.

At this point the observer watches the details of the operation and separates the operation into elements. Trial and error is expected because short-time elements that have clearly defined start and stop points are important. Observe that element 1 is "Right hand reaches for back plate at bin A and grasps." Bin A has been previously located on the layout sketch. In a similar way the other seven elements are described. There may be some experimental timing to get a feeling for the length of the time for the elements.

Observe Figure 2.3, where the elements are abbreviated. Each line number is for a particular cycle. A cycle for the receptacle would be eight elements. Space on the form is available for foreign elements that are unrelated, and a letter from A to I is used for coding if foreign elements occur unintentionally during a cycle. Foreign elements are removed from time studies later.

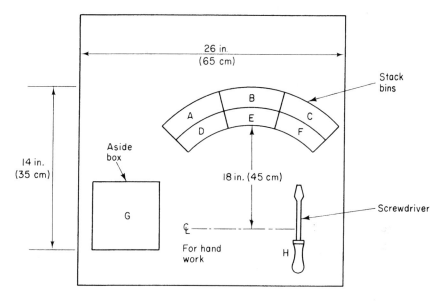

Figure 2.1.   Layout of receptacle bench assembly: present method.

Actual timing is continuous or snap back. If the time study were continuous, then the entry for line 1, element 1, would be 0830 (8:30 a.m.) under column R, and each element would be registered consecutively. At the conclusion an approximate time of 0945 would be entered in line 15, element under column R. The R designates a time-study reading. A snap-back method records the time for each element individually. On mechanical clocks this involves depressing the crown, which resets the clock hands to zero. A similar "zeroing" button exists for electronic clocks. But observe that under element R, column 8, the times 0.60, . . . , 0.48 are the cycle times for the operations. The time is reset to zero following the end of the cycle, at which time the next line starts with zero. Thus, the elements of this time study are continuously timed but have a cycle snap back.

Time studies for repetitive operations may be recorded in seconds, minutes, or hours. The time study of Fig. 2.3 is in decimal minutes. Once time recording is under way, the observer time-marks the end of each element for each line under the column R. The reading of a moving hand of a clock or electronic digital number and marking the value on the time-study form, which corresponds to the conclusion of the element, requires training. The number of readings necessary for a good average is a matter of judgment and depends on the consistency of the cycle time.

During the time study the observer judges the effort of the operator in performing the elements and operation. This is called *rating*. The observer watches the performance or speed as motions, elements, and operations are made and compares this mentally to a 100% speed. The operator's effort, which is reflected in the rating factor, includes intangibles such as health, interest in work, skill, and speed. If the operator's effort is considered in excess of normal (100%), then the rating factor is

Timed by R. Van Jours   Checked by Teo Davies   Workplace or mach 0738   Mach. no. —

Operators name Ronald Jeller   Clock No. 303 - 9709   Material See Sketch   R.P.M. —

Time study no.   Dept No. Electrical assembly   Lubricant —   Strokes per min —

Special tools used None   Feed

Part name Electrical receptacle   Part No 1050   Operation No. 40

Remarks

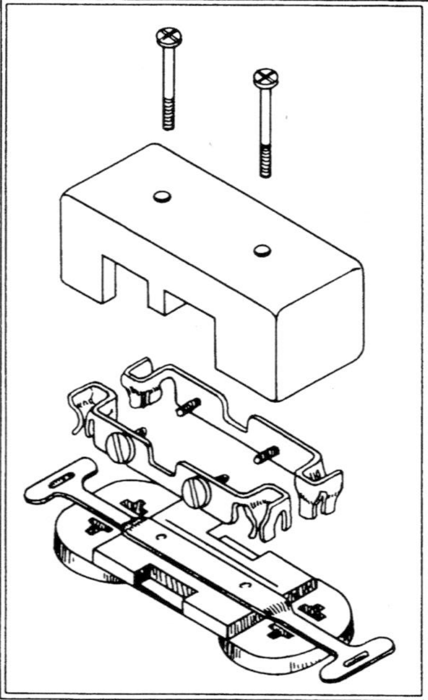

DETAILED DESCRIPTIONS OF ELEMENTS

1. Right hand reaches for back plate at bin. A and grasps

2. Moves to left hand, reaches for mounting ear at B and grasps, mounting ear and inserts. Left hand holds.

3. Right hand reaches for contact and inserts (from C)

4. Right hand reaches for contact and inserts (from C)

5. Right hand reaches, grasps, transports, and places free plate over back plate. Left hand holds.

6. Reaches for A screw at F, grasps. Transports, positions with right hand. Left hand holds.

7. Same as 6

8. Right hand reaches for screwdriver at H, grasps and transports to back plate, and tightens screw. Left hand assembly. Right hand returns screwdriver.

Figure 2.2. Sketch and elements of time study.

Time study observation

Date: Sept. 23.
Time start: 8^30 AM
Time stop: 9^45 AM
Elapsed time: 1^15

Sheet no. 1
No. sheets 1

Elements	Pick up back plate		Mount ear to back plate		Contacts in back plate		Contacts in back plate		Face plate over back plate		Place screw in back plate		Place screw in back plate		Tighten screw and aside			
Number	1		2		3		4		5		6		7		8			
Line	T	R	T	R	T	R	T	R	T	R	T	R	T	R	T	R	T	R
1	.03	.03	.06	.08	.08	.16	.05	.21	.05	.26	.09	.35	.04	.39	.21	.60		
2	.03	.03	.03	.06	.05	.11	.03	.14	.06	.20	.04	.24	.05	.29	.17	.46		
3	.02	.02	.04	.06	.04	.10	.04	.14	.08	.22	.04	.26	.05	.31	.16	.47		
4	.02	.02	.04	.06	.04	.10	.04	.14	.06	.20	.04	.24	.05	.29	.16	.45		
5	.02	.02	.05	.07	.05	.12	.03	.15	.05	.20	.04	.34	.05	.29	.15	.44		
6	.02	.02	.03	.05	.05	.10	.04	.14	.04	.18	.06	.24	.04	.28	.15	.43		
7	.02	.02	.03	.05	.05	.10	.05	.15	.06	.21	.04	.25	.06	.31	.16	.47		
8	.04	.04	.03	.07	.05	.12	.05	.17	.04	.21	.06	.27	.08	.35	.14	.49		
9	.02	.02	.04	.06	.05	.11	.04	.15	.05	.20	.04	.24	.08	.32	.15	.47		
10	.04	.04	.04	.08	.04	.12	.04	.16	.05	.21	.06	.27	.05	.32	.16	.48		
11	.02	.02	.05	.07	.07	.14	.07	.21	.05	.26	.04	.30	.09	.39	.14	.53		
12	.03	.03	.04	.07	.08	.15	.07	.22	.05	.27	.04	.31	.04	.35	.21	.56		
13	.02	.02	.03	.05	.06	.11	.03	.14	.08	.22	.05	.27	.05	.32	.15	.47		
14	.03	.03	.04	.07	.02	.09	.06	.15	.12	.27	.06	.33	.05	.38	.18	.56		
15	.03	.03	.02	.05	.06	.11	.03	.14	.08	.22	.05	.27	.05	.32	.16	.48		
16													Sum =			7.36		

Summary

	1	2	3	4	5	6	7	8	
Total time	.39	.52	.79	.67	.92	.75	.83	2.45	
No. of readings	15	15	15	15	15	15	15	15	
Average of readings	.026	.035	.053	.045	.061	.050	.055	.163	
Frequency	One	out	of	one					
Average time	.026	.035	.053	.045	.061	.050	.055	.163	
Element rating factor	1.05	1.00	1.00	.95	.90	1.00	1.00	1.05	
Element normal time	.027	.035	.053	.043	.055	.050	.055	.171	

Sum = 0.489

Foreign elements

SYM	R	T	Description
A			
B			
C			
D			
E			
F			
G			
H			
I			

Av. cycle time 7.36/15 = .491
Cycle rating factor 1.05
Normal cycle time 0.515

Percent allowances

Pers.	Fat.	Delay	Total
5%	5%	5%	15%

Std. time per unit 0.606
Pieces per hour 99
Std. hours per 100 1.010

Allowances in minutes

Pers.	Fat.	Delay	Total
.008	.008	.008	.024

Avail. prod. min. per hr. ______
Pieces per hour ______
Std. hrs. per 100 ______

Figure 2.3. Time study.

greater than 1. On the other hand, if the effort is considered less than 100%, then the rating factor is less than 1. The rating factor is entered for each element in the row, "Element Rating Factor." The rating factor for the elements and the operation are not necessarily equally weighted. This rating may be at the conclusion of the study. Though the rating process may be criticized as arbitrary, observers can be trained. Despite the criticism it is evident that rating is necessary for effective cost analysis. The shop portion of the time study is concluded when the rating factor is posted, and the observer then returns to his or her desk to complete the analysis.

The next step analyzes the time study. Subtractions are made for each element and posted under column T, which stands for time. For instance, line 1, element 4, has $0.21 - 0.16 = 0.05$, which is written under column T4. This subtraction is made for all T columns and cycles. If the timing had been snap back, then the subtraction would not be required.

Next we total column T for each element and divide by the number of observations to obtain the average of the readings. In this time study, the elements are regular because they occur once for each cycle. It is possible that an element may occur several times or fractionally for each element and be irregular. This fraction of the noncyclic elements is entered in the frequency row. Once the element is multiplied by the element rating factor, we have *normal time,* which is the time for an average qualified person working at a normal pace. This entry for each column T is made for the bottom row, "Element Normal Time," and the row sum equals 0.489.

Observe the sum of 7.36 arising from the total under 8R. The average cycle time gives $7.36/15 = 0.491$, which is marked on the form. When multiplied by the 1.05 cycle rating factor, the normal cycle time is 0.515 ($= 0.491 \times 1.05$), which compares to the 0.489. That those two values are not identical is not surprising because the rating factor is a trained but arbitrary value.

After determining the normal cycle time, we add for the job *allowances.* During a time study all irregular elements or interruptions for personal and unavoidable delays are purged. Thus, a time study is a picture of regular work elements only. But other legitimate needs are necessary for an operator to sustain work throughout the shift. *Personal* and functional body needs are added, and about 4 to 5% is the U.S. customary value.

Operator interruptions beyond his or her control is required for tool breakage, out of parts, foreman instructions, and so on, and is considered. The usual range is 2 to 8%. Those are added to the allowances and are called *delays.*

Fatigue is another factor included within allowances. Fatigue is physiological body wear reducing the ability to do work. It is not "tiredness," because that is expected. Fatigue is not subject to precise measurement. For instance, hot, heavy, dirty work such as forging of steel billets has a higher allowance than light assembly work in a controlled, air-conditioned factory. Some practitioners believe that fatigue does not exist in most factory environments. But fatigue allowance percentages are determined by practice or union negotiation or common sense. Fatigue may vary from 0 to 25%, depending on average weight handled, percent of time under load, repetitive or nonrepetitive work, sitting or standing, and cycle time.

These allowances—personal, fatigue, and delay—are sometimes referred to as PF&D allowances. A typical allowance is 15%, but it can vary from 8 to 35%. The allocation of the allowance in cost analysis procedure is as a percentage of 480 minutes. Because productive time in the workday is inversely proportional to the amount of PF&D allowance, the allowance should be expressed as a percentage of the total work day. The three portions of the allowance are added as a percent of the workday. Next we divide the total workday by the productive day expressed as a percent of the workday,

$$F_a = \frac{100\%}{100\% - PF\&D\%} \tag{2.2}$$

where F_a = allowance multiplier for PF&D

PF&D = personal, fatigue, and delay allowance expressed as a percentage

Assume that allowances total 15%, which is 72 minutes of the 480-minute workday. Converting this allowance to a percentage of the 408-minute productive day results in a multiplier of

$$F_a = \frac{100\%}{100\% - 15\%} = 1.176$$

We apply the allowance by multiplying the normal time by the allowance factor. If the rated productive time is 408 minutes, then the job standard is 408 × 1.176 = 480 minutes.

The job *standard* is computed as

$$T_s = T_n \times F_a \tag{2.3}$$

where T_s = standard time for a job per unit

T_n = normal cycle time per unit

An alternative common expression is $T_s = T_n(1 + PF\&D)$, where PF&D allowance is expressed as a decimal. It is not as accurate in definition but is often used for simplicity. Observe that for the electrical receptacle and Eq. (2.3) we have 0.515 × 1.176 = 0.606 standard minute per unit.

Equation (2.3) did not state whether the time was minutes, hours, or mandays. That depends on the dimension of the time measurement (i.e., week, day, hour, minute, or second). Though the basic approach is unaffected by the magnitude of the time units, it is customary in production work to use minutes or hours. If minutes are the dimension of the measurement, then minutes are converted to pieces or units of production per time period by a reciprocal relationship. If pieces per hour are desired, then

$$\text{pieces per hour} = \frac{60}{T_s} \tag{2.4}$$

Various firms prefer expressions such as pieces per hour, units per minute, or packages per week. In some businesses it is common to express the rate per dozen or gross. However, the expression may be written, the dimension units per time is not as preferred as hour per unit or units for cost analysis work. The conversion is handled by

$$H_s = T_s \times \frac{N}{60} \tag{2.5}$$

where T_s = standard minutes per unit

N = standard quantity of units (i.e., 1, 10, 100, 1000, or 10,000)

H_s = hours per quantity of units

The value N may be 1, 10, 100, 1000, or 10,000, depending on the nature of the production (i.e., one or several to large volume). A standard is always implied when hours are used, meaning that the rated time is adjusted by an allowance multiplier. Because N may range from 1 to 10,000 (and even 10^6 quantities are common in cigarette manufacture), a principle called *exaggeration of precision* is stated for purposes of cost engineering sensitivity. We recommend the following number of trailing decimals:

Standard Quantity	Required Trailing Decimals	Time-Study Example (Fig. 2.3)
1	0.x	—*
10	0.xx	0.10/10 units
100	0.xxx	1.010/100 units
1000	0.xxxx	10.1020/1000 units
10,000	0.xxxxx	101.01961/10,000 units

*Scientific notation (1.01×10^{-2}) is not used in cost engineering work.

If H_s is expressed as "hours per 100 units," then we would have for the time study (Fig. 2.3) $H_s = 1.010$ $(= 0.606 \times 100/60)$. Pieces per hour are 99 $(= 60/0.606)$. Industrial practice will round down and drop the fractional part. Pieces per hour is a rough shop value. The more accurate H_s is preferred for calculations. Once H_s is known for a job, we are ready to calculate operation labor cost after we have determined the wage.

Nonrepetitive time study.   It should be understood that what we examined previously dealt with repetitive work found in production industries. Frequently, we are unable to study more than one cycle. Nonrepetitive time study, although not as accurate and free from error as other methods, does provide information.

Sometimes called production study or all-day time study, nonrepetitive time study has been found useful for construction operations, indirect labor, production setups for highly repetitive work, allowances for repetitive work, audits, and direct-labor long-cycle type of work. This class of work is called undesignated because it is difficult to preplan.

The nonrepetitive time study differs from the repetitive motion and time study in a number of ways. The job study is not as complete, rating is frequently ignored, the preliminary investigation may be less, and preknowledge of the elements is unknown. The study uses continuous timing with a stopwatch, and the length of the elements is based on judgment. Examples of the types of work for which this is best suited include carpentry, railroad line work, material handling, maintenance, and secretarial activities.

Predetermined motion-time data.   These data systems are an organized body of information used in evaluating manual work. The data are expressed in time units, such as hours or minutes, for human motions. The analysis breaks down the elements of an operation into motions and assigns a time value that has been previously determined and tabulated. These time values are openly published and generally available. Observe that the electrical receptacle time study for element 1 is described as "Right hand reaches for back plate at bin A and grasps." Reach and grasp are typical motions too brief in terms of time to observe and record by using stopwatch practices. One system, such as Time Measurements Methods, will provide time for motions. Predetermined motion-time data systems require care in application, and, in the common form of tabular presentation, are not significant in cost engineering (except in high-volume industries). Additional information may be studied by examining the references.

2.2.2. Man Hour Reports

For certain types of work, such as construction, crew work, long-cycle production, job shops, or professional work, methods of obtaining time information for analysis and for later estimating are different from those found within short-cycle production work. Man hour reports are the usual way to deal with nonrepetitive work, because nonrepetitive work is difficult to measure. In some cases there may be company policies prohibiting the practice of time study.

Man hour, man-month, or man-year are the measures of time eventually desired from those reports. The basic unit of measure represents one worker working for 1 hour. Though the number of working days will vary in a month or year owing to holidays, vacation, sick leave, and so on, for cost analysis the *man-year* is defined as work of 52 weeks with 40 hours in each week. A *man-month* is 173.3 hours per month. Man-minutes or man-weeks are uncommon units and are discouraged in cost analysis application. Some cost analysis activities plan their calendar as 13 months of 4 weeks to avoid the unequal months. In practice, qualifications can be added such as the (1) time spent as actual clock hours or (2) adjustment of work effort to allowed time. In this first qualification the individual is assumed to have spent time in idleness, relaxation, or nonproductive effort, and in the second those conditions are removed and an allowance is contrived to give the standard requirement of the PF&D for the estimated work. Man-hours are used for estimating if the man hours are relatively constant, preknown, and unaffected by changes in wage rates, overtime, bonuses, and so forth. Examples of man-hour units follow:

Number of man hours per inch (2.54 cm) of weld

Number of man hours to erect 100 ft^2 (9.3 m^2) of framework

Number of man-months per mile (1850 m) of road construction

The term "person hour" is sometimes used to replace man hour. Information for man-hour analysis is usually obtained from the job ticket or the foreman report. Workers usually complete their own job ticket, which is verified by a time clock with computer terminal entry or collected by the owner, foreman, or manager. The foreman's report, on the other hand, will have similar information and content but is compiled by first-level management or a shop manager. In either situation the information can be computerized, but electronic reporting and analysis does not alter the fundamental approach. The job ticket or foreman's report is used initially to find the number of man hours. The same reporting document is later used for cost and time control.

Consider an example in a small contract welding shop where job tickets are used. Observe from Table 2.2 that the items of interest are operation, elapsed time, and units completed. Note that this job ticket has no foreman's approval, timekeeper's mark, or any external verification. From the job tickets, and with the aid of other instruction sheets or engineering drawings, the engineer will examine the data for consistency, completeness, and accuracy. The engineer may check back with the operator or foreman on questions that arise. In a contract job shop there is little repetition of similar work, so the engineer may choose to purge or alter information. This reworking of data is risky, but, in analyzing historical work data, the first step

TABLE 2.2.   JOB TICKET USEFUL FOR MAN-HOUR ANALYSIS

TRITON JOB TICKET

Name _Michael Blakely_   Employee no. _505 30 9709_

Date _May 13_   Order no. _101_

Part no. _6682_   Part name _Bracket_

Time started _9:15_   Time stopped _11:43_ Quantity _3_

Operation _Weld 40 x 80 x 3/4 in. (100 x 200 x 1.9 cm)_

plate. Submerged arc weld. Butt weld two fillets, 240 in.

_(600 cm)_

Special instructions _Automatic weld_

is to clean up the information as rationally as possible. If raw data were used as received, then other problems may arise later.

Job tickets are collected for similar work, and a spread sheet is devised for the data. A spread sheet brings together the facts and prepares the data for subsequent mathematical analysis. (No single best way exists to devise a spread sheet.) At this point judgment is necessary. From Table 2.3 we can see that the welding operations are varied and that the engineer chooses to analyze the general operation welding rather than divide the process into one of the many welding methods. Observe that several thicknesses, welding methods, and two-part numbers have been posted on the spread sheet. The mixing of processes and handling time into one lump time value is not an immaterial detail. But the engineer in this small job shop may choose to reduce, leave as is, or increase the three job-ticket times to give *allowed time*. The allowed times will later be considered for man-hour estimating data, but in this chapter our concern is measurement, collection, and data refinement.

TABLE 2.3.   SPREAD-SHEET EXAMPLE OF A MAN-HOUR REPORT FOR WELDING

Part no.	Material	Thickness x length	Welding method	Job ticket actual man – hours	Remarks	Allowed time man – hours
6682 (3 units)	Steel	3/4 (1.9) x 240 in. (600 cm)	Submerged arc butt weld 2 fillets	2.47	Fillet 2 sides reduce by 1/2 for one side and 2 extra units	1.24
7216	Steel	1/2 (12.7) x 84 in. (213 cm)	Shielded metal arc, 1 fillet	1.06	No change	1.06
8313	Steel	3/16 (4.8), 1/4 (6.4), 1/2 (12.7), x 344 in. (874 cm)	Shielded metal arc	1.18	Increase by 1/4 for various thicknesses	1.48

If *man hour spread sheets* are to be of value as a permanent measured record, their backup data must include a variety of information to permit the engineer to weigh deviations from the observations. For instance, in construction work the weather conditions, skill of the crews (experienced or green, native or imported), equipment, hazards, location of work, material condition, and type of construction are vital factors that need to be known. Those remarks appear on the foreman's report, and the engineer makes office adjustments for those effects. Those adjustments lead to allowed time, which is an intermediate value from observed hours to standard man hours. The adjustments may be derived from engineering instructions, other information, or, as a last resort, be based on the engineer's experience or hunch. Sometimes the foreman's report lumps together elements of work unsuitable for this analysis.

In summary, man hour reports are derivative time values adjusted by the engineer from job tickets or foreman reports. That the data are nonhomogeneous is evident, and judgment and skill are inescapable to match the work to allowed man hours.

2.2.3. Work Sampling

Work sampling is a technique for gathering information about large segments of a work force population. It is a counting method for quantitative analysis in terms of time or percent of the activities of workers, machines, or any observable state of an operation. It is useful in analysis of undesignated or nonrepetitive work activities and allowances. Work sampling is relatively inexpensive to obtain and convenient to perform and can be conducted without recourse to a stopwatch or the necessity for historical reports.

A work-sampling study consists of a number of observations taken that pertain to the specific activities of the person(s) or machine(s) at random intervals. Those observations are classified into predefined categories directly related to the work situation. Tally marks are made by the technician, such as "work," "idle," or "absent," during the course of the work-sampling study. The key to accuracy is the number of observations, which may vary according to the requirements. One survey may require very broad areas to be investigated, in which case relatively few observations will be necessary to obtain meaningful results. On the other hand, many thousands of observations may be needed to establish production standards. To determine the number of observations necessary, the technician predetermines the accuracy of his or her results. Four thousand observations will provide more reliable results than 400. However, if accuracy is unimportant, then 400 observations may be ample.

Because work sampling is a statistical technique, the laws of probability must be followed to obtain accuracy of the sampling estimate. In this type of observation, an event such as equipment working or idle is instantly tallied. For this selective choice mathematicians define a binomial expression where the mean of the binomial distribution is equal to Np_i, with N equaling the number of observations and p_i, the probability or relative frequency of event i occurring. The variance of this binomial distribution is equal to $Np_i(1 - p_i)$.

As N becomes large the binomial distribution approaches the normal distribution. Because work-sampling studies involve large sample sizes, the normal distribution is an adequate approximation to the binomial. In work sampling we take a sample of size N observations in an attempt to estimate p_i or

$$p_i' = \frac{N_i}{N} \tag{2.6}$$

where p_i' = observed proportion of occurrence of an event i expressed as a decimal

   N_i = number of instantaneous observations of event i

   N = total number of random observations

Equation (2.6) is related to the binomial, and this random variable is called a proportion. As is shown in textbooks on probability, the standard error of a sample proportion for a binomial distribution may be expressed by

$$\sigma_{p'} = \left[\frac{p'(1 - p')}{N}\right]^{1/2} \tag{2.7}$$

where $\sigma_{p'}$ is the standard deviation of the proportion of the binomial sampling distribution. Bias and errors may occur in any sampling procedure and may result in a deviation between p' and p or the true value. A tolerable maximum sampling error in terms of a confidence interval I commensurate with the nature and importance of the study is pre-established. For instance, if $p' = 62\%$, and if a maximum interval of 4% is desired, then $I = 4\%$ for 60 to 64%, or $62\% \pm 2\%$ where $I/2 = 2\%$. The relative accuracy is $0.02/0.62 \times 100 = 3.2\%$. This confidence and interval may be viewed by examining Fig. 2.4. The factor 1.645 is obtained from a table of probabilities (see appendix 1 for example) and is the normal distribution for a confidence of 90%, which is usual for work sampling studies. The total area under a normal curve

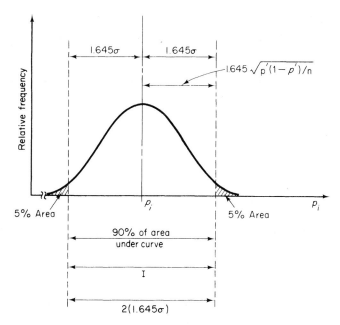

Figure 2.4. Analogous curve from "normal" relationship to binomial work-sampling practices.

is 100%, and the opportunity for a sampling value p' to fall within the tails is given in appendix 1 and is equal to $\pm Z = 2 \times$ (probability from the table for a given Z). For instance, if Z ranges from -1.645 to $+1.645$, then the probability that the true value p will be between the limits is $2 \times 0.45 = 0.90$, while the probability that the true value will be outside the limits is $1 - 0.90$ or 10%. The value 0.4500 is determined from appendix 1. Some values of Z corresponding to confidence areas follow:

Area Between Limits(%)	$-Z$ to $+Z$	Area Outside Limits(%)
68	± 1.000	32
90	± 1.645	10
95	± 1.960	5
99	± 2.576	.1

The sampling interval is given by

$$I = 2Z\left[\frac{p_i'(1 - p_i')}{N}\right]^{1/2} \tag{2.8}$$

where I is the desired interval expressed as a decimal and Z is the factor from normal tables for a chosen confidence.

We expect that the true value of p falls within the range $p' \pm 1.645\sigma_{p'}$ approximately 90% of the time. In other words, if p is the true percentage of work estimate, then the estimate will fall outside $p' \pm 1.645\sigma_{p'}$ only about 10 times in 100 owing to chance or to sampling errors alone. Equation (2.9) may be solved for the sample size when the other factors are either assumed or known, as

$$N = \frac{4Z^2 p_i'(1 - p_i')}{I^2} \tag{2.9}$$

That value of N that is maximum from the events i is chosen as N for the work-sampling study. Relative accuracy is found as $I/2p'$.

As an example of work sampling in construction, assume a job where carpenter crews are forming lay up walls on a dam. Those concrete retaining walls are numerous and standard. In addition, the carpenters are doing other work, but their principal activity relates to the lay up walls. Once the work is understood, we define the job elements as

1. Form lay up walls
2. Inspection waits
3. Set up for form work
4. Crane waits
5. Miscellaneous

A preliminary study such as Table 2.4 helps to uncover problems before the major study is started. The study sells the idea, and, importantly, a percentage of observations falling into each activity gives a useful, though rough estimate of the universe percentages. Note in Table 2.4 that we might decide that inspection waits are not an important enough element to consider separately, especially if we have little control over them, so we decide to add inspection waits to the "miscellaneous" activity. But we found that material waits in miscellaneous deserved a separate category, and we may have some management control over those waits. Rearranging the data we would have (1) form layup walls, 62%; (2) set up for form work, 16%; (3) crane waits, 12%; (4) material waits, 4%; and (5) miscellaneous, 6%.

TABLE 2.4.   EXAMPLE OF STICK CHART FOR PRELIMINARY WORK SAMPLING STUDY

Element	Observations	p'
1. Form layup walls	THL THL THL THL THL THL I	62 %
2. Inspections	III	6 %
3. Setup for form work	THL III	16 %
4. Crane waits	THL I	12 %
5. Miscellaneous	II	4 %
		100 %

With the job elements now defined, the next step deals with calculating the number of observations. The number of observations for the study depends on the percentage of observations in each element and the size of the desired confidence interval. For example, we might estimate that an element will occur somewhere close to 60 or 64%, and a 90% confidence interval on a plus-minus tolerance of 2% or less is desirable. Our first estimate of sample size in our example is based on preliminary percentages, and Table 2.5 indicates the appropriate tolerances. The largest sample size will control the study and in this case we need about 2,858 observations.

TABLE 2.5.   FINDING SAMPLE SIZE FOR CONSTRUCTION STUDY BY USING 90% CONFIDENCE INTERVAL

Element	Rough, p'	Desired Interval, I	N_i	Relative Accuracy, $\frac{I}{2p'} \times 100$ (%)	90% Confidence Interval
1	0.62	0.04	1594	±3.2	0.60–0.64
2	0.16	0.03	1616	9.4	0.145–0.175
3	0.12	0.02	2858	8.3	0.11–0.13
4	0.04	0.02	1042	25	0.03–0.05
5	0.06	0.02	1527	17	0.05–0.07

Note that it is element 3, which is not the largest proportion. Our next step is to spread the observations equally among the days and then randomly within the working day, excluding break and lunch periods.

In retrospect it becomes possible to determine the magnitude of the sampling error after the test is underway or concluded. Assume that for element 2, and partially through the study where $N = 875$, a total of $N_2 = 184$ tally marks are indicated, and $p' = 0.21$. This would give an interval of

$$I = 2(1.645)\left[\frac{0.21 \times 0.79}{875}\right]^{1/2} = 0.0453$$

and we can say that the true value lies within 0.21 ± 0.023 with a probability of 90%. Observe that the lower value $0.187 (= 0.21 - 0.023)$ does not lie in the interval of element 2 (0.145, 0.175) as given in Table 2.5. The student may want to explain this observation in terms of sampling errors, interval, and probability.

With work-sampling information about the event or job and a percentage fact for each, it is possible to compute labor cost. A model using these data is

$$H_s = \frac{(N_i/N)HR(1 + PF\&D)}{N_p} \tag{2.10}$$

where H_s = standard man-hours per job element i
     N_i = number of event i observations
     H = total man-hours worked during study
     R = rating factor
    $PF\&D$ = allowance decimal
     N_p = work units accomplished during period of observing this event

Assume a construction job where carpenters form lay up walls on a dam. The study runs for 3 weeks for 14 carpenters. Ratings similar to time-study practices are made during the course of random work sampling, and $R = 0.96$ is determined. A total of 17 frames are finished, and work sampling indicates that this event $N_i/N = 62\%$ of the total effort of the gang. For an allowance of 20% the time per frame may be found as

$$H_s = \frac{0.62(14 \times 3 \times 40) \times 0.96 \times 1.20}{17} = 70.6 \text{ man hours per frame}$$

Work sampling is a tool that has broad ramifications for the engineer. It allows the work sampler to get the facts in an easy and fast way. In summary, the following considerations should be kept in mind for work sampling:

1. Explain and sell the work-sampling method before putting it to use.
2. Isolate individual studies to similar groups of machines, operations, or activities.
3. Use as large a sample size as is practical, economical, and timely.

4. Observe the data at random times.

5. Take the observations over reasonably long periods, (e.g., 2 weeks or more), although rigid rules must bend with the situation and design of the study.

2.3 WAGE AND FRINGE RATES

Labor costs, which are dollars paid for wages or salaries for work performed, are a major ingredient of an estimate. In heavy construction, for example, labor costs constitute 40% or more of the bid on a building.

A *wage* is paid or received for work by the hour, day, or week, and, once denominated by a period of time, it becomes a wage rate (e.g., $12.75 per hour). A salaried employee is paid for a period of time, say a week, month, or year.

Payrolls cover two classes of workers: first, management and general administrative employees, who may or may not be on a salary basis; and, second, hourly labor. General administrative employees may be management, engineers, foremen, inspectors, and clerk-typists.

Hourly labor is sometimes classified as direct or indirect. *Direct* refers to employees who can be associated with a product directly, such as milling machine operators or truck drivers; that is, any work that can be preplanned or designated. *Indirect* refers to workers who are generally performing undesignated work, such as clerk-typists, janitors, industrial engineers, and superintendents. In an allocation sense of cost, their work and effort is usually for a variety of tasks, making it difficult to designate precisely what portion of their work contributes to the particular operation, product, project, or system. In the main the indirect operator is not clearly identifiable to a particular task.

Fringe benefits are one of the costs of labor. In the past, engineers concluded that wages or salaries received directly constituted the total sum of labor costs. This is not so. *Fringe benefits,* which are related to wages and salaries, constitute as much as 30% of the actual cost incurred for labor.

Variety exists in the handling of wages and fringes. But if a deduction occurs from the employee alone, then it is not explicitly identified for estimating because it is included in the employee wage rate. There are labor costs not included in the wage rate that are known loosely as fringes, and it is necessary to understand their content and value.

Once the design is known, the work is described and job descriptions of employees are matched to the work. Those job descriptions indicate the skill, knowledge, and responsibility required of the worker. Each company will have occupational descriptions. Table 2.6 is an example of an assembler. The occupations are graded, and a company will have a pay scale that increases from the very simple occupation to the most difficult. Those job descriptions and their wage rates are policy information available to the engineer, especially for operation and product designs. However, unions or labor-management contracts may be the source of information for wages in construction.

TABLE 2.6. JOB DESCRIPTION

Assembler (Bench assembler; floor assembler; jig assembler; line assembler; subassembler)

Assembles and/or fits together parts to form complete units or subassemblies at a bench, conveyor line, or on the floor, depending on the size of the units and the organization of the production process. Work may include processing operations requiring the use of handtools in scraping, chipping, and filing of parts to obtain a desired fit as well as power tools and special equipment when punching, riveting, soldering, or welding of parts is necessary. Workers who perform any of these processing operations exclusively as part of specialized assembling operations are excluded.

Class A.   Assembles parts into complete units or subassemblies that require fitting of parts and decisions regarding proper performance of any component part or the assembled unit. Work involves any combination of the following: Assembling from drawings, blueprints, or other written specifications; assembling units composed of a variety of parts and/or subassemblies; assembling large units requiring careful fitting and adjusting of parts to obtain specified clearances; using a variety of hand and powered tools and precision measuring instruments.

Class B.   Assembles parts into units or subassemblies in accordance with standard and prescribed procedures. Work involves any combination of the following: Assembling a limited range of standard and familiar products composed of a number of small-or medium-sized parts requiring some fitting or adjusting; assembling large units that require little or no fitting of component parts; working under conditions where accurate performance and completion of work within set time limits are essential for subsequent assembling operations; using a limited variety of hand or powered tools.

Class C.   Performs short-cycle, repetitive assembling operations. Work does not involve any fitting or making decisions regarding proper performance of the component parts or assembling procedures.

Wages and fringe effects are considered by one of two methods: (1) wage only and (2) wage and fringe combined, or the gross hourly cost. In the first method the fringe effects are collected in overhead, which will be described in chapter 4.

2.3.1. Wage-Only Method

Wage payment can be classified into general groupings: those that pay for *attendance* and those that pay on *performance*. In time attendance wage plans, gross wages are figured easily. The time in attendance is multiplied by the rate. An engineer who earns $54,000 per year earns $4,500 per month. If he or she starts or leaves within the month, the pay is prorated to the number of calendar or working days, depending on company policy. The qualitative formula given by Eq. (2. 1) can be more formally expressed as

$$C_{dl} = H_a \times R_h \tag{2.11}$$

where C_{dl} = time-rate cost for direct labor effort, dollars

H_a = actual hours

R_h = rate per hour in dollars

Naturally, R_h can be found from annual, monthly, or weekly pay scales. For a $4,500 monthly scale, the weekly scale is $4500 \times 12/52 = \$1038.46$. There are $173\frac{1}{3}$ hours per month ($= 52 \times 5 \times 8 \times 1/12$), and the hourly scale is $25.96. Those calculations use the popular 40-hour week. In attendance-based plans (sometimes called day work), the worker is paid on the amount of time spent on the job.

Of course, the worker is interested in as large a wage as possible. The employer, on the other hand, is interested in the labor cost reduction. If the employer is able to encourage increased output from the worker, the employer might be willing to pay higher wages. In these performance plans, the worker's earnings are related to productive output and are called incentive or piece-rate plans.

$$C_{dl} = N_p R_p \qquad (2.12)$$

where C_{dl} = piece rate cost for labor, dollars

N_p = number of pieces produced

R_p = standard rate per piece in dollars

If $R_p = \$0.1466$ per unit and $N_p = 128$ units, then $C_{dl} = \$18.76$, which is paid to the worker. (Subsidy or bonus pay schemes have much in common with incentive plans, but will not be discussed here.) Most incentive plans do not pay additional wages until 100% standard is reached. Below that level a guaranteed wage or day work is pledged. A formula expressing this relationship is given as

$$C_{dl} = H_a R_h + R_h(H_s N_p - H_a) \qquad (2.13)$$

subject to

$$C_{dl} \geq H_a R_h$$

where C_{dl} = total cost per pay period for labor, dollars

H_s = standard hour per unit

Assume that $H_a = 0.8$ hour, $R_h = 14.50$, $N_p = 128$, and $H_s = 1.010$ hour per 100 units. Then,

$$C_{dl} = 0.8 \times 14.50 + 14.50\left(\frac{1.010}{100} \times 128 - 0.8\right) = \$18.75$$

If only 70 units had been produced under the piece-rate plan during 0.8 hour, then earnings would be $10.26. Under the guaranteed 100% plan, earnings would be $11.60 because $C_{dl} \geq H_a R_h$.

Earned hours are recorded for pay purposes when an incentive plan exists. Thus, a job is set in standard hours, and the worker may earn more dollars by producing the job in fewer actual or clock hours. The advantage of this approach is that the standard hour base is the same for each operation unit, whereas the actual hours for an operation unit vary with the worker's efficiency on that job. The ratio of the task amount of work, or hours standard to hours actually taken in performing the work, is

$$E = N_p \frac{H_s}{H_a} \times 100 \qquad (2.14)$$

where E is the labor efficiency in percent. For the case where $N_p = 128$ units, $E = (128 \times 1.010/100)/0.8 \times 100 = 162\%$, and for 70 units $E = 88\%$. Other measures of efficiency are discussed in chapter 11.

2.3.2. Gross Hourly Cost

In some cases the fringe costs are not included in overhead and are determined explicitly for estimating. Combined wages and fringes are called *gross hourly cost* and may be determined for production workers.

Federal and state laws regulate wages and salaries paid by employers. Two are prominent: the wage-hour law and the Walsh-Healy Act. Manufacturers engaged in interstate commerce are covered by the wage-hour law. The Walsh-Healy Act covers only companies having federal contracts. Both require the payment of wages at time-and-a-half rates for more than 40 hours in 1 week. Walsh-Healy adds the same requirement for more than 8 hours in 1 day. Some groups are exempted from the wage-hour law, such as management and engineers. A worker with a 40-hour base rate of $12 must be paid $18 per hour for his or her overtime hours. The wage-hour law specifies a minimum per hour as the lowest paid wage. For example, the wage minimums as established by Congress constitute a floor for employed labor in most categories. Other labor laws exist of course, such as: discrimination, limiting work hours for women and children, safety and sanitation laws, and unemployment insurance, taxes, and social security taxes.

Contractual agreements may specify additional requirements, such as number of holidays, time off, sick leave, and uniforms. In addition to wages earned by employees, the employer must pay the appropriate government agency or insurance carrier an additional amount.

A partial listing of *fringe costs* may include (1) legally required payments, such as payroll taxes and workmen's compensation; (2) voluntary or required payments, such as group insurance and pension plans; (3) wash-up time, paid rest periods, and travel time; (4) payment for time not worked, holidays, vacations, and sick pay; and (5) profit-sharing payments, service awards, and payment to union stewards. Fringe costs depend on local situations and must be determined individually for each case.

The Social Security Act requires that businesses pay a tax for old-age and medical benefits, called social security tax, or FICA (Federal Income Contribution Act). This amounts to a percentage of a fixed sum of gross earnings of an employee. The employee contributes an amount equal to the employer. This rate and the base on which it is levied are subject to change by Congress and have marched steadily upward since the first law back in 1935. Initially, the employee paid 1% of the first $3000 earned and the employer paid an equal amount. Recently, though, the employee deduction was calculated as 6.20% of eligible income to a ceiling of $55,500 for FICA and 1.45% to a ceiling of $130,200 for meditax. These deductions are changed whenever Congress finds it necessary. This employer percentage amount is added to the cost of business. The employee's share of the FICA tax is not a business cost.

Workmen's compensation is a levy against employers for continuation of income to employees in periods when they cannot work because of accidents occurring on the job. Payments may be made to the state compensation insurance fund or to state-approved insurance carriers. Employees are grouped into work types, and rates based on experience of risk incidence are established for each type. In the event of injury or death, the insurance carrier provides financial assistance to the injured person or to the survivors in the event of death.

Another federal-state tax on wages is for unemployment insurance and was established by the Social Security Act of 1935. Setting up the employment services was delegated to the states, which collect the major portion of the tax to operate their own unemployment offices. State legislation had to conform to the requirements of the federal Act. Built into the system is a merit-rating procedure to reward "good" employers (i.e., employers who operate their labor pool to prevent repetitive hirings and firings and who terminate employees only for sound economic reasons). When a business is established or enters the system, it pays the maximum state rate. If its employment record becomes stable, then the rate drops until the minimum rate is reached. The employer is rewarded for good employment practices. Not all employees, such as farm workers or domestic help, are covered. If an employee is laid off, then he or she is entitled to unemployment "pay" for a limited period. The unemployment insurance pays for this involuntary employee cost.

Fringe benefits include expenditures, other than payroll taxes and workmen's compensation insurance, that benefit employees individually or as a group. Because those benefits are not legally required, they may vary from employer to employer and be based on labor-market competitiveness, industry practice, management's attitude toward employees, management's social consciousness, or labor-management contract.

One such benefit is the portion of medical and dental insurance paid for by the employer. Supplemental medical insurance covers the employee and family for all or part of ordinary illness, or could apply only to catastrophic illness. Usually, a schedule of payments by the employee and the employer is established, a percentage of present employees are required to initiate the plan, and all new personnel are included automatically. The company's portion of the premium needs to be considered for estimating. Employee contributions are not a business cost.

Another supplemental benefit is life insurance. A schedule is established showing the employee's and employer's portion of the premium. In initiating this plan the personal histories of present employees are secured, and a stated percentage of the work force must participate. All new personnel may be automatically included. The company's portion of the premium needs to be considered by estimating.

Vacation pay is also considered a fringe benefit. Vacation policy varies as to length. When the employee becomes eligible to participate, the amount of vacation earned but yet unpaid is a company cost.

Sick pay is similar to vacation pay. It may not be formally set up on the accounting books as such but is charged to overhead when taken. It can, however, be estimated on experience.

Another fringe benefit is an employee supplemental pension plan, which provides retirement benefits in addition to social security. The plan may be self-funded, in which case the company agrees to invest contributions, or the plan may be funded by an outside agency such as an insurance company or a mutual fund company. In the latter case, the company makes the payments required under the plan. In a third arrangement, the plan may be fully funded by the company or partially by employees.

Another fringe benefit is the stock-purchase plan, by which the company encourages employees to become shareholders in the company by allowing the employees to purchase stock at less than market value. The immediate benefit the employee receives is the difference between market price and the price paid.

Other miscellaneous fringe-benefit expenses include

1. Cost of operating the cafeteria less receipts for sales made. The benefit to the employees is the difference between what they would pay in an outside cafeteria and their real cost.

2. Cost of equipping and operating in-plant medical facilities.

3. Cost of equipping and operating sports teams and leagues, funding and supervising Christmas bonuses, and maintaining incidental employee conveniences.

4. Labor costs for foreign projects may include family living allowances, extra midday rest periods, severance pay, social taxes, and perfect attendance record bonuses.

The effective gross hourly cost for a specific assembler is determined in Table 2.7. A typical job description for an assembler is given by Table 2.6. By knowing the job title and employee it is possible to figure out entitlement for vacation, shift differential, and so on. The expected nonchargeable hours of 5%, line 8, include wash-ups, which are not a part of overhead or not included in production allowances. Line 11, performance subsidy, may be used for award pay rates if provided by management policy. Thus, an effective gross hourly cost of $20.29 is 29% greater than the basic wage. It is also possible to figure gross hourly costs for a class of employees, using averages for entitled vacation, overtime, holidays, and all of the items listed in Table 2.7.

2.4 JOINT LABOR COST

Joint labor costs are those that are shared and act in common for an operation. The engineer *dejoints,* or "unitizes," the cost. An operator simultaneously working on two or more parts, or an operator controlling the output of two or more machines,

TABLE 2.7. CALCULATION OF GROSS HOURLY COST FOR INDUSTRIAL OCCUPATION AND EMPLOYEE

Job title __Assembler__   Department __Mechanical__
Future effective period __Jan - July__   Wage __$14.50/hour__ Shift __1__
Name __Spencer Tucker__   Clock no. __10017__ Subsidy __10%__
Entitled vacation days __10__   Holidays __6__

Item	Annual hours	Annual cost	Annual excess cost
1. Regular paid clock hours	2080	$30,160	
2. Planned overtime hours	192	2784	
3. Overtime cost at 50% of 2	–	1392	$ 1392
4. Subtotal	2272	$34,336	
5. Holidays	48	696	696
6. Entitled vacation	80	1160	1160
7. Paid sick leave		–	
8. Expected nonchargeable hours, 5%	104	1508	1508
9. Subtotal (5 + 6 + 7 + 8)	232	$ 3364	$ 3364
10. Chargeable (4 – 9)	2040	30972	
11. Expected performance subsidy	204		
10% , X 10	–	3097	3097
12. Chargeable standard (10 – 11)	1836	$27875	
13. Non hourly costs			
FICA @ 0.076 X 4			2610
Workmen's compensation, 3% x 15,000			450
Unemployment insurance			300
Supplemental medical insurance			550
Supplemental pension			–
Supplemental life insurance			–
Union Welfare			–
Bonus, gifts			–
Uniforms, tools, etc.			50
Profit sharing			–
Other			–
14. Subtotal of 13			$3960
15. Excess costs (3 + 9 + 11 + 14)			$11813
16. Excess hourly rate (15/10)			5.79
17. Wage			14.50
18. Effective gross hourly cost (16 + 17)			$20.29
19. Increase in hourly base rate (16/18)			29%

are joint-cost examples. Joint costs include not only labor but material and overhead, which will be discussed later.

In a one-operator one-machine industrial operation the worker may load two or more units onto the machine and process them simultaneously. Conversely, from one unit of original material two or more units of output may be possible. Time-study or man-hour data can aid this analysis.

One of the production causes for joint costs is multiple-machine operation. In certain kinds of work it is possible for one operator to tend a number of machines, depending, of course, on the amount of attention that each machine requires. In fact, several machines may need attention at the same time. This waiting is called "machine interference" and is covered by queuing models. The problem arises, therefore, as to how many machines an operator should tend. Although queuing methods are more elegant and exact, simpler methods are preferred when estimating because this information is available.

Machine interference exists whenever too many machines are assigned to one operator, but this may be less costly than operator idleness. Joint labor cost or time exists when an individual operator is involved with the following examples:

1. Similar machines each producing the same output
2. Similar machines each producing different output
3. Dissimilar machines each producing different output

Crew or team work is another example of joint labor cost. Crews are found frequently in construction and chemical manufacturing. In crew work the interest is to balance the work load such that there is no idle time or that all idle time is equal within the crew. Usually, this is unlikely, and the entire crew time is subject to the longest time of one worker.

Dejointing of shared labor divides the time or labor cost. The point of division is called the split point. Labor up to the split point is joint, while afterward it is called unit. Splitting is handled by allocation or units of reference by which the joint costs are divided. Labor joint-cost splitting is usually handled by time or units or a combination of both.

A problem confronting a engineer is given as follows: Suppose that two camera brackets, left and right hand, are machined together. Loading and unloading are simultaneous. The left bracket requires special machining. Gross hourly costs are $20. A time study gives the following data.

Element	Bracket	Standard Minute/Element
1. Loading	Both	0.20
2. Position and drill hole	Both	0.15
3. Rotate and position	Left	0.05
4. Index collet	Left	0.08
5. Counterbore	Left	0.21
6. Eject parts	Both	0.15
7. Clean jig	Both	0.10
Total		0.94

A dejointing solution by time would be

Elements	Left	Right
1,2,6,7	0.30	0.30
3,4,5	0.34	
Total	0.64	0.30

cost for left bracket = $0.64 \times 1/60 \times 20 = \$0.213$/unit

cost for right bracket = $0.30 \times 1/60 \times 20 = \$0.100$/unit

We can also use the two units of output to obtain $0.157 per unit (= $0.94 \times 1/60 \times 20 \times 1/2$).

Costs can be dejointed by market effects that stipulate that cost be a function of price or number of units sold. Marketing strategy can affect allocation accuracy by falsely requiring that some products subsidize others. This policy says that selling price or market value is not proportional to processing costs and thus distorts the estimated cost. In the bracket example, suppose that the left bracket can be sold on a ratio of 4 to 3 units of dollar to the right bracket. We compute the cost as

left bracket = $0.94 \times 1/60 \times 20 \times 4/7 = \$0.179$/unit

right bracket = $0.94 \times 1/60 \times 20 \times 3/7 = \$0.134$/unit

2.5 LEARNING

Reduction in the cycle time is expected as a worker continues to produce similar units of production. This trait is called *learning*. Job performance improvement is measured by the time per unit. The direct labor cycle time declines as the number of units increases, although up to a point. This widely observed phenomena becomes predictable, and companies are able to measure job performance in a training effort.

Look at Fig. 2.5. Assume there are four new employees hired under probationary terms. Employees 1 and 4 are relatively unskilled or slow in job speed. After a period of training, employee 1 is able to produce a unit of product in less than the standard time and is retained. Operator 4, perhaps having less dexterity, skill, or motor ability, is unable to achieve the standard. Operators 2 and 3 have initially greater skill or faster job speed, but only operator 3 achieves the output requirements. Operators 2 and 4 would continue training, or be reassigned or be terminated in some cases.

Though the learning phenomena is widely experienced by anyone who has performed manual and repetitive work, the application of the learning theory to cost analysis is much more broad than this limited application to direct labor and repeti-

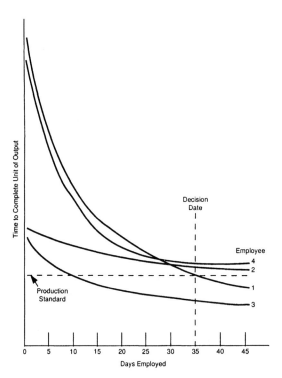

Figure 2.5. Learning application in operator training, performance, and retention.

tive production. The learning improvement is sufficiently predictive and additional models are given in chapters 6, 8, and 11.

SUMMARY

The usual ways to measure labor are time study, predetermined motion-time systems, work sampling, and man-hour reports. Job tickets, especially for smaller organizations, are analyzed and allocated to units of work. For instance, a job ticket may state "136 units turned of part number 8671" and lists "6 man hours." Simple analysis would show 0.044 hour per unit. The engineer would use 0.044 hour the next time this part were run. Although hardly accurate because of the nature of historical work reports, man-hour reports are used because of their simplicity.

Though the engineer may not be directly involved with the measurement of labor, he or she does depend on work measurement. The engineer is satisfied if such labor measurements are objective, as far as that is possible, and is willing to use the information provided that engineering techniques were used in the determination of time. Although those time measurements are of value, it is immensely more important that work measurement data be transformed into information that can be applied prior to the time of the operation design. This transformation of analyzed labor information is discussed in later chapters.

Although past wage and fringe data are useful for cost estimating, there is the caution that future circumstances will be different. Thus, their manipulation for future use is a subject of later chapters.

QUESTIONS

2.1. Define the following terms:

Direct labor	Job ticket
Indirect labor	Personal time
Measured time	Historical time
Time study	Foreman's report
Normal time	Person-hour
Rating	Work sampling
Allowance	Interval
Standard	Job description
Pieces per hour	Gross hourly cost
Man-hour	Dejointing

2.2. Why do we separate labor into direct and indirect categories?

2.3. What is the purpose of rating? A job is time studied and rated greater than 100% and the standard is used for a second worker. What does the fact that the original rating was greater than 100% mean to the second worker?

2.4. What are allowances intended for?

2.5. Why is an operation divided into elements for the time study?

2.6. Why do we use man-hour reports? Name types of work that are appropriate.

2.7. Determine the current FICA rate and the base for the employee and employer. What are the political ramifications?

2.8. List mandatory and voluntary types of fringe costs.

2.9. Specify the nature of work that makes it joint. List allocators that dejoint work.

2.10. Supply a practical illustration of learning for direct labor.

PROBLEMS

2.1. (a) A machine has a cycle time of 1/2 second. What are the production pieces per hour at 100% and 75% efficiency?

(b) The floor-to-floor time (meaning a complete productive cycle) is 31 seconds. Find gross units per hour at 100% and 90% efficiency.

2.2. (a) If the hourly production is 11 units, find the hours per 1000 pieces.

(b) The production is 29.5 units per hour. Find hours per 100, 1000, and 10,000 units.

(c) Find pieces per hour and standard minutes per unit for 15.325 hours per 100 units.

2.3. A time study of an assembly operation is summarized below:

Element	Frequency	Element Minutes	Rating Factor
Part A to 5 in. (125 mm) dowel	1/1	0.037	1.10
Part B to 5 in. (125 mm) subassembly	2/1	0.064	1.15
Short piece to base	1/1	0.089	1.20
Long piece to base	1/1	0.129	1.10
Grab pieces from tote box	1/5	0.185	1.10
Place in conveyor	1/1	0.087	1.10

Allowances for this work are 9.5%. Find (a) the standard minutes per unit, (b) pieces per hour, and (c) hours per 100 units. (d) For a wage rate of $14.75 per hour, what is the labor cost per unit?

2.4. A mail room prepares the company's advertising for mailing. A time study has been done on the job of enclosing material in envelopes. Develop the elemental normal time, cycle standard time, pieces per hour, and hours per 100 units, providing 15% for allowances based on the continuous watch data below. (Readings are in hundredths of minutes.)

Element	Cycle							Rating (%)
	1	2	3	4	5	6	7	
Get envelopes	11		55		105		151	105
Get and fold premium	22	41	65	83	116	134	160	115
Enclose premium in envelope and seal envelope	29	48	73	97	123	141	182	95

2.5. A method has been engineered and an individual has been time studied. The method has produced an average overall time of 2.32 minutes by actual watch timing. The overall pace rating applied to the job is 125%. This company, which uses an incentive system, adds extra time into the production rates to cover personal time, fatigue, and unavoidable delays. The total allowance is 15%.

(a) Determine the rated time or normal minutes, the total allowed time or standard minutes, the standard hour rate for the job per unit and 100 units, and the pieces per hour.

(b) The labor rate per hour is $15.30. What is the standard labor cost per unit?

2.6. What follows are time-study observations. The elemental times are snap back.

Element	Reading	Rating (%)	Allowance (%)
Place bushing in jig	0.03 0.08 0.05 0.04 0.06	117	11
Drill hole	0.31 0.38 0.37 0.39 0.33	100	5
Remove bushing from jig and place part on conveyor belt	0.07 0.08 0.09 0.12 0.11	93	13

    (a) Find the average, normal, and standard times for each element.

    (b) Calculate the production per hour and hour per 100 units.

    (c) By an old method, 0.834 minute of standard time was required to complete this operation. Calculate the increase in output in percent and savings in time in percent.

2.7. A job is time-studied and a summary time sheet is given as follows:

Element Description	Frequency per Unit	Average Time	Rating Factor (%)
Get part off conveyor	1/1	0.040	100
Get incomplete part off	1/1	0.030	105
Connect length and joint	2/1	0.055	110
Connect 3	3/1	0.153	100
Assemble side	1/1	0.234	90
Mate two sides	1/1	0.183	90
Put on conveyor	1/1	0.032	100

    (a) With an allowance factor of 15% to cover P (personal), F (fatigue), and D (delay), what are the normal minutes per piece, the standard minute per unit, and the number of hours per 1000 units?

    (b) If performance against standard has averaged 120%, then what is the incentive hourly rate if day work is paid at $15.75 per hour?

2.8. What follows is a raw continuous-timing two-element time study of a punch press operation where the observations are minutes.

			Cycle		
Element	1	2	3	4	5
Handle part	0.01	0.04	0.08	0.11	0.14
Punch part	0.03	0.06	0.10	0.12	0.16

    (a) What is the average time for each element? The ratings were +8% and −9%. What is the normal time? A man and machine allowance of 20% is used in this plant.

What is the standard minute per piece for this operation? How many pieces per hour? How many hours per 100 units?

(b) If only a man allowance of 20% is used for element 1, then what are the standard minutes per unit? How many hours per 100 units?

2.9. Engineering aides prepare job tickets of their daily work, and after a sufficient period of data collection, classification of the information reveals what follows:

Description of Work	Minutes per Occurrence	Frequency per Print
Post drawing numbers	7.0	1/1
Duplicate drawings	19.0	1/1
Correct computer entries	38.0	2/1
Phone calls	1.5	1/4
Update CAD changes	27.0	1/2

A personal allowance of 15% is used for this work. Ignore holidays and vacation.

(a) Determine the man hour per engineering drawing.

(b) A new product design is anticipated and a separate staff will be collected. This product design will probably result in about 2500 prints over a 1-year period. How many aides will be required, and, at $25 per hour, what amount would you estimate for this activity in a product expense budget?

2.10. A work-sampling survey of an operation, which was designated into 12 categories, has the following observations:

Item	Observations
1	92
2	99
3	37
4	11
5	25
6	14
7	24
8	33
9	3
10	22
11	8
12	32
	400

If this sample covered a span of 25 days for 8 hours per day, what are the percents and expected hours per item of work?

2.11. The pediatrics department in a hospital is work-sampled for 608 hours.

Work Category	Observations
Routine nursing	496
Idle or wait	263
Unit servicing	183
Report	129
Personal time	128
Intervention	102
Unable to sample	91
Other	79
Feeding	52
Bathing	22
Elimination	11
Transporting	8
Housekeeping	7
Ambulation	7

For the 1578 observations find the percent occurrence, percent cumulative occurrence, element hours, and cumulative hours for each work category.

2.12. A work-sampling study is taken of a department with the following information obtained: number of sampling days, 25; number of trips per day, 16; number of people observed per trip, 3; and number of items being sampled, 4. The four sample items are broken down as A, 80; B, 320; C, 1600; and D, 2800.

(a) How many man-days and observations were sampled?

(b) What are the percentage and equivalent hours for the activities?

(c) For a confidence level of 90%, what is the relative accuracy of each of the items?

2.13. Suppose that we want to determine the percentage of idle time of a machine shop by work sampling. Assume that a confidence level of 95% and a relative accuracy of $\pm 5\%$ are desired, where a rough estimate of 25% is suspected for idle time.

(a) How many observations are necessary?

(b) Assume that the relative accuracy is $\pm 2\frac{1}{2}\%$. How many observations are required now?

(c) What happens to the number as relative accuracy becomes less?

2.14. (a) To get a 0.10 interval on work observed by work sampling that is estimated to require 70% of the worker's time, how many random observations will be required at the 95% confidence level? Repeat for 90%.

(b) If the average handling activity during a 20-day study period is 85% and the number of daily observations is 45, then what is the interval allowed on each day's percent activity? Use 90%. Repeat for 99%.

(c) Work sampling is to be used to measure the not-working time of a utility crew. A preliminary study shows that not-working time is likely to be around 35%. For a 90% confidence level and a desired relative accuracy of 5%, what is the number of observations required for this study? Compare to 95% confidence level.

2.15. A shipping department that constructs wooden boxes for large switch gear has five direct-labor workers. A work-sampling study is undertaken, and the following observations of work elements were recorded over a 15-day, 8-hour period:

Item	Count
Set up and dismantle	312
Construct crates	264
Load switch gear in crates	204
Move materials	324
Idle	96

A rating factor of 90% is found. The number of switch gear shipped during this period is 26. This firm uses an allowance value of 10% for work of this kind. Average labor costs $18.75 per hour.

(a) What are the elemental costs?

(b) What is the standard labor cost per box?

(c) What is the actual cost?

2.16. An eight-person CAD department concerned with size A, B, and C drawings is work-sampled by a management consultant over a standard 4-week period. A stick chart is summarized for the categories as follows:

Item	Count
Drafting and tracing	778
Calculating	458
Checking prints	110
Classroom	125
Professional time off	172
Personal time, idle	270

During this period, 55 drawings (A = 20, B = 25, C = 10) were produced with a total payroll of $26,400. Let relative size A = 1, B = 2A, C = 2B.

(a) If the policy is to accept professional and personal time as necessary to the drafting of prints, then what is the man-hour factor?

(b) If personal time and idle time is prejudged at 10% only, then what is the per print size factor?

2.17. Find the effective gross hourly cost for drill press operators paid an average wage of $22 per hour. No overtime is planned. Company policy allows six paid holidays, and the average entitled vacation is 10 days. There is no nonchargeable time or performance subsidy. Sick leave is charged against vacation time. FICA is at current rate, and workmen's compensation is at 2% of the first $20,000. The company pays $400 for unemployment insurance.

2.18. An industrial carpenter is paid $26.50 per hour. The work year consists of 52 40-hour weeks with overtime scheduled for 26 Saturdays. The company allows 8 paid holidays and 10 days of vacation. Four days of sick leave are budgeted and have historically been used. Expected nonchargeable hours are 2%. There is no subsidy for performance. Nonhourly costs include FICA taxes at current governmental rate, workmen's compensation at 2% of regular wages up to $15,000, accident insurance sum of $600, a major supplemental medical plan for $300, and unemployment insurance of $800.

(a) Find the effective gross hourly cost.

(b) What is the job cost if a job will require 10 man hours with a productivity of 90%?

2.19. Determine a company's base annual cost for a top-grade worker. The base hourly rate is $15.75. Use the current FICA costs of the base. State unemployment compensation runs 2%. Health insurance premiums cost $40 per month. The company carries term life insurance that costs $45,000 per year for all employees (there are 60 employees). In addition, the company profit-sharing plan usually pays 5% of the base wage.

2.20. Find a typical hourly cost for direct labor. In this instance the average work week is 48 hours, of which the final 8 hours are premium at one and a half. The wage is $30.30 per hour. Each day a total of 20 minutes is permitted as a coffee break, the final 15 minutes is cleanup, 2 weeks of paid vacation on basis of 40 hours worked and five paid holidays, FICA at the current rate on the basis of $55,500, 1.2% workmen's compensation tax on a basis of $15,000, $25 per month for company-paid medical and life insurance, and a $25 Christmas bonus, all considered pertinent to the calculation.

2.21. A man worked 8 hours on incentive and nonincentive jobs. While on incentive he completed 8 tasks, each with a 1-hour standard time. The nonincentive jobs took 2 actual hours. The incentive plan has a base rate of $13 per hour, though the man's wage rate for nonincentive tasks is $16 per hour. How much did the man earn?

2.22. (a) A worker produces 56 units on an incentive plan during the 40-hour week. Her base hourly rate is $15 per hour, and the standard for one unit of accomplishment is 1.10. What are her weekly earnings?

(b) If FICA is 7.60%, unemployment insurance costs the company 2.1%, the workmen's compensation rate is 2.5%, 10 days of yearly vacation, and $40 per month for insurance, approximate the actual wage and fringe cost per hour? Per unit? Assume a weekly wage for a 52-week year. Yearly vacation is at base rate of $15 per hour.

2.23. A worker is paid the day-work rate if she earns less than 100% incentive premium. A standard is 75 units per hour, and the operator completes 140 good parts in 1.4 hours. The rate is $25 per hour.

(a) Find the piece rate.

(b) What are the earnings and labor efficiency?

2.24. (a) A sheet metal operator is told that her standard is 1.875 hours per 100 units. If her wage is $16 per hour and she makes 800 units in 12 hours, what are her earnings?

(b) If the actual time is 16 hours, what is her guaranteed pay and efficiency for the 100% plan?

2.25. An engineer has efficiency information of 125% for an operation. The standard is 10 hours per unit and the wage is $23.50 per hour. What is the expected cost for 210 units?

2.26. A molding operator tends two machines and plastic products A and B are produced. Machines A and B have a production rate of 400 and 250 strokes per hour. Molds A and B produce three or four buttons per stroke. Operator wage is $11 per hour. Dejoint these labor costs when allocation is based on (a) number of machines, (b) product output of machines A and B, and (c) marketing believes that A to B value is 5:4.

2.27. A time study observed the operation of producing two parts, which are labeled 1 and 2. The elements and their standard minutes are given as follows:

Element	Part	Standard Minutes
Load part in vise	1	0.17
Load second part	2	0.08
Balance and tighten vise	1, 2	0.17
Start machine	1, 2	0.06
Machine	1, 2	1.17
Stop, unload	2	0.13
Start, retighten	1	0.15
Fine machining	1	0.83
Stop, unload	1	0.14
Clean vise	1, 2	0.16

The labor wage is $23 per hour. Dejoint cost using allocation methods of (a) number of parts, (b) time required by part, and (c) potential sales price ratio of 3:2.

2.28. One operator controls four automatic machines. After those machines are set up they produce parts independent of the operator except for occasional inspection. A cam controls the unit time to make one piece and is 4 seconds at 100% efficiency. Ignore the setup time because it is small.

(a) If the actual efficiency is 85%, the operator controls four machines, and the labor wage rate is $15 per hour, then what is the dejointed labor cost per unit?

(b) Repeat for hours per 100 units.

(c) Repeat for dollars per 100 units.

2.29. A technician tests and repairs printed circuit boards. On the average printed circuit boards require 10 minutes for testing and 12 minutes for testing and repairing, if necessary.

(a) If, during a man hour study, 18 units tested OK and 5 failed requiring repair, then what proportion of the labor wage is due to testing and repair?

(b) If the gross labor rate is $21.75 per hour, then what is the labor charge per unit for testing and repair? Use proportional methods to dejoint cost.

MORE DIFFICULT PROBLEMS

2.30. The fire department is concerned about the speed of the crews assigned to the all-purpose trucks. The chief wants to know the time required after receiving the alarm before starting to fight the blaze. To answer the chief's questions, a time study of the activities of one crew on seven different alarms is conducted. On the observation sheet that follows, continuous (no reset) electronic watch readings in hundredths of minutes are recorded, indicating full minutes only when it changed.

Crew Element	Alarm						
	1	2	3	4	5	6	7
Start timing	.00	1.51	2.33	3.24	4.12	5.04	5.91
Get dressed	.12	1.58	2.43	3.36	4.25	5.15	6.01
Board truck	.29	1.82	2.60	3.55	4.41	5.32	6.19
Start engine	.44	1.94	2.76	3.70	4.55	5.48	6.32
Drive to fire*							
Unload hoses	.64	2.00	3.02	3.89	4.78	5.63	6.50
Connect hoses	.89	2.33	3.24	4.12	5.04	5.91	6.75
Unload ladders	1.08						6.95
Position ladders	1.51						7.31
End timing for this alarm	1.51	2.33	3.24	4.12	5.04	5.91	7.31

*Watch stopped because of variable nature of distances.

(a) Determine the average time for the elements.

(b) If the crew is rated at 110% for all elements, find normal time in minute per occurrence.

(c) Determine the cycle-time standard for a 20% allowance.

2.31. A gang nonrepetitive time study is made of a construction crew and what follows is a tabulated summary. It is usually unrealistic to rate this kind of work, and PF&D allowances are seldom a part of the calculation. Crosses in the table indicate that the job element is necessary for the worker. (Pay time and a half for overtime for any assigned work in this job.)

Element	Minutes	Crew and Equipment						
		Foreman	Pump Operator	Hopper Man	4 Workers	2 Vibrator Crew	Truck Driver	Pumps and Fitting Rental
Make ready	60	x			x			
Move to job site	15	x			x			
Pump machine to job	45	x	x	x	x			x
Set up machine	90	x	x	x	x			x
Inspect	30	x	x	x	x		x	x
Adjustment	30	x	x	x	x		x	x
Pump concrete to forms	360	x	x	x	x	x	x	x
Normal delay for set of concrete	80	x	x	x	x	x		x
Dismantle pump	45	x	x	x	x			x
Put tools away	15	x	x		x			x
Return pump	30	x	x		x			x
Clean up	10	x			x			
Cost per hour		14	16	16	12	16	70	50

(a) The work consists of placing 180 yards of concrete by a pipeline. Determine the cost of the job and of placing a yard of concrete.

(b) Initially assume that when not assigned to this job the operator is doing something else profitable. Pay time and a half for overtime for any assigned work in this job. Assume that the union contract requires the entire crew (exclusive of concrete truck and driver) to be at the job site the entire time. What is the actual cost per yard then? What is the cost of this nonproduction? Assume that element 1 starts a new workday.

2.32. We are interested in finding the estimated direct-labor cost for the job description of industrial electrician. The year consists of 52 40-hour weeks, and overtime is seasonal for 12 weeks consisting of Saturday work of 8 hours. The contract allows for nine paid holidays and two weeks of vacation at regular time. Four days of sick leave are paid. Expected nonchargeable hours are 5%. A subsidy for performance will be 15%. Nonhourly costs include FICA taxes at the current federal rate, workmen's compensation at 1% of regular wages up to $ 15,000, accident insurance sum of $200, major medical plans for $500, and Christmas gift of $50. The hourly base is $17.10.

(a) Find the effective gross hourly cost.

(b) What is the job cost for this electrician if a job requires 25 man-hours?

(c) What is the loss or gain if efficiency will be 85% or 115% for the job?

CASE STUDY:
THE ENDICOTT IRON FOUNDRY

"We can't make any profit on that job. There's too much labor cost in it," said Dick Crawford, the foundry superintendent, to George Dobbins, engineer for the Endicott Iron Foundry.

The Endicott Iron Foundry, like its competitors, has always estimated costs on a per pound (kilogram) basis for the delivered casting. Difficult castings are quoted at a higher price per pound (kilogram) than simple castings, but the difference in price (often based on the estimated cost) did not seem to be great enough to warrant the extra labor costs.

Crawford suggests that the company is making little profit or loss on jobs that took considerable labor. What will happen if Endicott starts quoting higher prices for casting requiring extra labor? Should it recover the full cost? What ideas can you suggest at this time to improve the estimates? Should the engineer depend on the knowledge of Dick Crawford as final? What constitutes a loss or profit for the estimate?

3

<hr style="height:4px;border:none;background:black;" />

Material Analysis

Material cost analysis involves complicated bills of material, finishes, standard and nonstandard designs, and extensive inventories of direct and indirect materials. It is not surprising to find industrial and construction companies that have a material list of many millions. A vertically integrated company (i.e., copper ore to finished wire) has the problem of identifying the ore and intermediate cost values for their processed materials. A horizontally integrated company may often transfer materials between plants. Fluctuation of commodity prices adds to the apparent disorganization of material cost. Inflation or deflation of material cost is possible owing to complex interactions. All of those factors produce the bewilderment that faces the engineer in material analysis. Our approach in this chapter identifies the type of material, finds its quantity, and selects a cost policy.

3.1 MATERIAL

Before we consider the topic of *materials,* an understanding of physical materials associated with engineering design is required. The design may call for CA-610, AISI 1020, M type HSS, or A36 materials. Though the terms may be meaningless to some, materials are complicated in other respects. For instance, a casting has a variety of associated costs. There is the cost of melted metal, molding costs, core costs, cleaning costs, heat treatment costs, foundry tooling costs, and so forth. To estimate the cost for a casting, the engineer must be acquainted with the material and heat-treatment specifications, inspection requirements, and the design of the casting. Physical specifications such as tensile strength, yield strength, and elongation and chemical composition must be clarified. Knowledge about engineering materials is so vast that we refer the inquisitive students to other texts. In this book we consider the amount of material, losses, and the cost necessary for the design.

Cost finding includes historical and measured information. In some cases, material costs are uncovered by using company records, though for other materials the engineer must seek out the basic costs.

We define materials as the substance being altered. This may involve iron ore, coke, and limestone to a basic blast furnace industry for producing pig iron, steel ingots to a steel rolling mill for refining and rolling into strip and coil, tin sheet to a can producer manufacturing 6-ounce tins, and cases of 24 cans for a food processor for canning frozen orange juice.

The scope of what constitutes materials depends on the situation. Materials have been purchased, not manufactured, by the plant that uses the materials. Thus sheet steel is a product from a rolling mill but is a material to a sheet-metal forming plant.

Design documents are the *engineering bill of material* and specifications, in addition to the engineer's drawings. Engineers' bill of material or parts list accompany the blueprints and consist of an itemized list of the materials for a design. The list may be prepared on a separate sheet or may be lettered directly on the drawing. For example, a parts list contains the part numbers or symbols, a descriptive title of each part, the quantity, material, and other information such as casting pattern, number, stock size of the materials, weight of parts, or volume of materials. The parts are listed in general order of size or importance or by listing the special-design parts first and the standard parts last. The list proceeds from the bottom upward on the drawings so that new items may be added later. In the case of a product to be constructed for the first time or in the case of a project where the contract must be executed, the materials must be purchased quickly to have them available when needed in the factory or on the job site. In some cases the engineering department determines all the materials and job requirements. Inasmuch as the engineers are conversant with the drawings and specifications and are already familiar with the details of the design and contract, they begin the job of calling out material requirements. The quantity takeoff sheets must be complete and include all descriptions necessary to obtain the materials.

Specifications are considered apart from the bill of material and the engineering drawings. In construction, specifications are specific statements as to construction requirements. In manufacturing, specifications relate to the performances, or materials, or special requirements. Basically, those requirements state the technical details about the item. Specifications are used to discuss the proposed work so that bids may be compiled or quotations may be collected. Specifications are used as a guide or a book of rules during the construction and manufacturing time period. Finally, specifications are legal documents to the construction industry and special requirements to manufacturing.

In addition to the specification of the material, it is important that correct quantities such as units, weight, and volume are known. To have a correct cost such as 86 cents per pound of casting material and then improperly specify the number of pounds required for the casting is a serious flaw. To determine the quantities involved, the engineer examines the plans and specifications and makes the estimate, say, the number of cubic yards of 2800 psi concrete required by type of clean-and-sharp aggregate. In construction, contractors bid against a stated number of cubic yards and their price would be per cubic yard placed. In product-oriented industries, the engineer determines the cost per unit of material after the count of materials has been established. In industry, as in construction, the estimate of materials involves extensive calculation to include allowances for waste, short ends, or losses. After those calculations are completed, often with the aid of data catalogs on weights, allowances, and the like, costs are determined.

Direct materials are subdivided into raw materials, standard commercial items, subcontract items, and interdivisional transfer items. *Raw materials* include fabricated, intermediate, or processed material in a form that will receive direct labor work in conversion to another design. For worked material it is necessary to have designs and planning and then to add direct labor.

Standard commercial materials are a class of materials normally not converted. Rather, they are accepted in a manufactured state (i.e., tires used by an automobile assembly plant). Standard commercial materials may be a significant proportion of total material cost and is considered separate from fabricated raw materials. Standard purchased parts and materials may carry a specific purchasing overhead rather than a larger general overhead rate, although the practice is not uniform. It may not be competitive to add a large markup on the same items that can be purchased by customers. Purchased parts and standard materials may be charged with out-of-pocket expenses for procurement, freight, receiving, handling, inspection, installation, and testing.

Standard commercial parts are costed either by engineering or purchasing. Engineering may provide a bill of material listing of the standard parts to the purchasing department, that will price the parts from catalogs or from quotations and return the information to engineering. Sometimes when a short lead time for sales prevails, engineering may compile the standard part costs. But information from catalogs can be faulty, what with frequent price changes and negotiated price-volume breaks between the company's purchasing agents and seller. For some operations and products the standard purchased parts such as nuts, bolts, and washers are estimated by a multiplying factor correlated to the direct materials. This happens whenever the standard purchased parts are insignificant.

Subcontract items are parts, components, assemblies, intermediate materials, or equipment produced by a supplier or vendor in accordance with designs, specifications, or directions applicable only to the design being estimated. Interdivisional *transfer materials* are materials sold or transferred between divisions, subsidiaries, or affiliates that have common ownership or control. Those sales are ordinarily handled on a cost, no-profit basis. Occasionally, other arrangements may be made for items regularly manufactured and openly and widely sold.

Materials can, alternatively, be classified into commodity, engineering, semi-engineering, and normative. *Commodities* are traded on stock markets or commodities exchanges, and the price is volatile. Examples are foodstuffs, precious metals, timber, and extracted ores such as hematite (Fe_2O_3, iron ore) or chalcopyrite (Cu_2S, copper ore). *Engineering materials* are commodities that have undergone substantial engineering processing (e.g., iron ore to pig iron to hot rolled and pickled American Iron and Steel Institute (AISI) 1020, 40×120 in. (1.2×3 m) steel sheets). *Semi-engineering materials,* insofar as their cost value is concerned, behave as either engineering or commodity. Copper is an example of a semi-engineering material, because occasionally its price is steady like an engineering material and other times the price flip-flops and follows the roller coaster of commodity prices. *Normative materials* have a price control or are price fixed by various governments or cartels, such as oil or certain metals.

In this book we are concerned only with the estimating of engineering materials that have reasonable price stability. Commodity prices are erratic, and methods for accurate forecasting are beyond the scope of this elementary text. The direct materials that we estimate are engineering materials because their cost fluctuation is moderated by the enormous conversion and processing requirements.

Indirect materials are those materials necessary for the conversion of direct materials and are not directly traceable to the design. Lubricating oils and perishable tooling expenses are expenses that are indirect. Some materials can be classified as either indirect or direct. Fuel, such as natural gas, is a raw material (e.g., cracking refinery gases for the production of ethylene or heat-treating furnaces). Gases can be treated as a direct raw material cost or as an indirect utility expense. Convenience of the costing dictates whether it is simpler to classify some material costs as direct or indirect. Operating supplies include such things as brooms, which are too diverse and unimportant to be considered in the cost analysis. Company records are used for indirect materials, and overhead is the usual way to recover their costs.

3.2 SHAPE

The problem of estimating direct materials is broken down into three parts: (1) measurement of the shape, (2) finding the value of cost per unit shape, and (3) the value of any salvage. Qualitatively, the cost is found by using

$$\text{cost of direct material} = \text{shape} \times \text{cost/unit shape} - \text{salvage value} \qquad (3.1)$$

Shape implies mass or area or length or count or one of the many engineering dimensional units. The cost per unit shape is in compatible dimensions. If shape is pounds (kilograms), then \$/lbm (\$/kg) is the appropriate cost rate. *Salvage* is a recovered material having a credit or debit applied against the direct material cost of the design.

During *material takeoff* the engineer determines the theoretical amount or count of the material as finally required by the design. The takeoff could be the bill of materials or could be a separate listing indicating material specification, size, weight, length, shape, and so on. Sometimes this kind of takeoff is termed a *quantity survey,* and the engineer is called a quantity surveyor, a term sometimes found in the United Kingdom. To the exact shape the engineer adds for losses of scrap, waste, and shrinkage.

Scrap is faulty material because of human mistake. Mislocation of drill holes is a shop mistake. The mistake may have been caused by the designer's error, in which case the scrap is caused by engineering.

In manufacturing work the converting or changing of the properties or configuration causes *waste.* This loss is found in all durable-goods manufacturing. The width of the cutoff tool in a lathe cutting-off operation of a round bar becomes chips and is of no value to the design except that it is necessary for part separation. The remaining stock, or the skeleton-like structure that remains after parts have been blanked from sheet metal, or the overburden of castings, which is eventually removed as chips, are other examples of waste. Another interesting term is *offal,* which originally meant the inedible parts of a butchered animal but is now sometimes used to mean waste.

A great deal of engineering effort is concerned with the reduction or elimination of waste. Despite this effort, 1 to 12% is often appropriate as an allowance for waste.

Shrinkage is the loss of materials because of theft or physical laws. Originally, shrinkage dealt with volumetric reduction of lumber owing to drying, but now we mean the economic effects owing to deterioration of materials because of aging, oxidation, chemical reaction, natural spoilage, and so on if reduction in quantity and quality will occur. A polymer raw material may have a limited shelf life, and, if not molded and cured before the onset of deterioration, there is economic, not physical, shrinkage. Rusting of ferrous materials is a common form of shrinkage. Food commodities have serious problems with shrinkage losses, both economic and in the reduction of flavor or moisture.

Those losses are determined by the engineer, usually as a percentage, and added to the theoretical amount. The losses are expressed as

$$S_a = S_t(1 + L_1 + L_2 + L_3) \tag{3.2}$$

where S_a = actual shape in units of area, length, mass, volume, count, etc.

S_t = theoretical shape required for design in units of area, length, etc.

L_1 = loss due to scrap, decimal

L_2 = loss due to waste, decimal

L_3 = loss due to shrinkage, decimal

If a piece is machined, then the amount of stock removed by machining must be added to the final dimensions. The volume is computed from those dimensions.

Consider as a first example a $1\frac{7}{16}$-in. (36.5-mm) OD (outside diameter) carbon steel bar that has a finish length of $1\frac{1}{8}$ in. (28.6 mm). Records indicate that scrap is 1% for this kind of operation. Waste is composed of a $\frac{1}{8}$-in. (3.2-mm) cutoff tool width, and a facing length of $\frac{1}{64}$ in. (0.4 mm) is required for accurate dimension. The machine selected for this turning is limited to a 6-ft. (1.8-m) bar length, but the lathe collet requires a gripping length that reduces the quantity produced from the bar stock by one. Each unit will require 1.266 in. $(= 1\frac{1}{8} + \frac{1}{8} + \frac{1}{64})$ (32.15 mm) of length, and the exact number of pieces will be $72/1.266 = 56.9$. Reducing this to 55 units allows for last-part gripping. The percentage $= 55 \times 1.125/72 \times 100 = 85.9\%$, so the waste $= 14.1\%$. There is no shrinkage for this part, so

$$S_a = (1\tfrac{7}{16})^2 \; \pi/4(1\tfrac{1}{8})\,(1 + 0.01 + 0.141 + 0) = 2.116 \text{ in.}^3 \; (34.7 \text{ cm}^3)$$

Waste in this example includes chips, short ends, cutoff, and facing material. We assumed a finished bar stock diameter equal to the original diameter. If there were chip removal on the circumference, then our volume calculation would find the exact final amount plus those losses.

If the piece is irregular in shape, then it is divided into simple subparts and the minor volumes manipulated to give total volume. Stamping fabrication from coil stock or a blank and trim operation from sheets require slightly different shape calculations. Consider the work piece in Fig. 3.1. In stamping fabrication the distance between blanks is restricted to a minimum $0.75 \times$ thickness, and the margins between the edge of the strip and the blank are restricted to $0.90 \times$ thickness for each side. The width of the strip is calculated as 1.3625 in. $(= 1.250 + 2 \times 0.9 \times 0.0625)$ (34.61 mm). The part long-way dimension or length includes 75% of the stock thick-

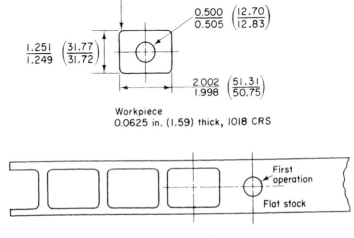

Workpiece
0.0625 in. (1.59) thick, 1018 CRS

Figure 3.1.   Sheet metal component.

ness, and the part dimension and in sheet metal manufacturing is called advance. The advance is 2.0469 (52.0 mm). If $L_1 = 0.25\%$ and $L_3 = 0$,

$$S_a = (1.3625)(2.0469)(0.0625)(1 + 0.0025) = 0.175 \text{ in.}^3 (2.86 \text{ cm}^3)$$

Observe that the corner radius and first operation slug have been ignored.

Other models for sheet metal estimating can be used, depending on the engineering requirements, such as

$$\text{pieces per coil} = \frac{\text{coil length}}{\text{length of design}}$$

$$\text{weight of coil} = \text{gauge} \times \text{width} \times \text{length} \times \text{density} \tag{3.3}$$

$$\text{weight per piece} = \frac{\text{weight of coil}}{\text{pieces per coil}}$$

Equation 3.3 is typical of the many ways to calculate shape. Each design has its own mensuration, which is a field of mathematics that deals with finding length, area, and volume.

The engineer computes the efficiency of conversion of raw materials such as

$$E_s = \frac{S_t}{S_a} \times 100 \tag{3.4}$$

where E_s denotes the shape yield, percent. For the sheet metal component

$$E_s = \frac{2.2914}{1.3625 \times 2.0469} \times 100 = 82\%$$

where the four corner radii and the slug have been included in the calculation for the shape yield.

The final term in Eq. 3.1 deals with salvage. In many cases, scrap and waste are sold to a junk dealer or the original processor, who then credits the firm. In some cases the credit is significant enough to subtract the total unit cost, though in others it is not worth the effort, as in the case of machine chips. Some waste and scrap may even add to the cost because of disposal problems.

Having discussed shape, cost per unit shape, and salvage, the formula is given as

$$C_{dm} = S_a C_{ms} - V_s \tag{3.5}$$

where C_{dm} = cost of direct material, dollars per unit
$\quad\quad C_{ms}$ = cost of material, compatible to S_a units
$\quad\quad V_s$ = salvage cost, dollars per unit

We presume that S_a and C_{ms} are in compatible units. If both sides are multiplied by N_p, the number of pieces produced, then we have total or lot cost, depending on the nature of N_p.

The engineer is able to compute the efficiency of the economic conversion of raw materials, such as

$$E_m = \frac{N_p S_a C_{ms} - N_s V_s}{N_p S_a C_{ms}} \times 100 \tag{3.6}$$

where E_m = material cost yield, %
       N_s = number of salvage units

An example of material estimating is given by the 12-fluid-ounce (0.197 liter) beverage can, which is composed of the body, top, and pull ring. The container body is blanked from 3004-0 aluminum coils (see the layout in Fig. 3.2). An intermediate cup is formed without any significant change in thickness. The cup is drawn in a horizontal drawing machine, and metal is squeezed to a side-wall thickness of 0.0055 in. (0.140 mm). The bottom thickness remains unchanged. The can is trimmed to the

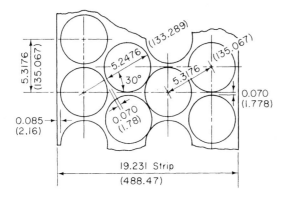

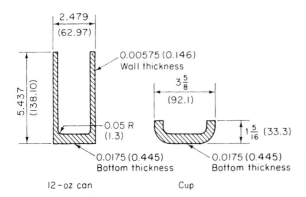

Figure 3.2. Layout of strip for manufacture of the popular 12-oz beverage cans.

final height to give an even edge for later rolling to the lid. The strip stock layout is called a "four-out" advance. Four blanks are punched out simultaneously for the 5.3176-in. (135.067-mm) advance in the die, and the coil is repeatedly advanced and punched in this manner, leaving a skeleton of waste.

A summary of the calculations is given by Table 3.1. Line 1 of Table 3.1 provides the volume in a 12-oz can body, and line 2 indicates the strip volume of advance. Once divided by 4 we have the can body. Line 3 provides the shape yield of the can body to strip. The blank, 5.2476-in. (133.289-mm) OD also has losses in drawing to the final can, and line 5 indicates a yield of 86.3%. Line 6 indicates the cost per pound (kilogram). The density of the material is 0.0982 lb/in.^{3} (2717 kg/m^3). Line 8 indicates the salvage value for the waste, and line 9 shows the recovery on a per can basis. Finally, the metal cost in the can body is calculated on line 11 as $0.0384 per can.

TABLE 3.1.   CALCULATION OF DIRECT MATERIAL COST FOR 12-OZ CAN BODY GIVEN BY FIG. 3.2

---

1. Final metal volume in 12-oz can body = 0.3266 in.^{3} (5355 mm^3)
2. Strip volume per can body = 5.3176 × 19.231 × 0.0175/4 = 0.4474 in.^{3} (7328 mm^3)
3. Metal efficiency = 0.3266/0.4474 = 73.0%
4. Metal volume in 5.2476 (133.975) blank = 0.3785 in.^{3}
5. Can body to blank efficiency = 0.3266/0.3785 = 86.3%
6. Cost of metal = $1.0017/lb ($2.226/kg)
7. Cost of can body = 0.4474 × 0.0982 lb/in.^{3} × 1.0017 = $0.0441
8. Waste salvage value = $0.4830/lb ($1.073/kg)
9. Salvage of waste per can body = (19.231) (5.3176) (0.0175) (1 − 0.73) (0.0982) (0.4830)/ 4 = $0.0057
10. Metal cost in can body = (0.3266) (0.0982) (1.0017) = $0.0321
11. Net cost per can body in strip = 0.0441 − 0.0057 = $0.0384
12. Economic yield = 0.0321/0.0384 × 100 = 83.6%

3.3 MATERIAL COST POLICIES

With the shape determined, material cost analysis proceeds to the second step of finding the cost per unit shape. This is not simple because engineering classification, accounting systems, vendors, and professional, consulting, or association advice adds to the confusion of determining an accurate cost value.

Manufacturing materials often have different technical specifications than those of construction materials (e.g., AISI 1035 versus A36). Further, a manufacturing firm may build a product from a periodically replenished inventory, though a contracting firm or job shop may not order materials until the bid is won. The differences in classification and material ordering policies are superficial for cost analysis purposes. For example, a job shop manufacturing firm or a contractor will not order materials until a particular bid is accepted. A manufacturing firm may order materials in building prototypes even though the equipment may be unsold at the time.

The methods that follow suit the needs of both manufacturing and construction. The word "lot" is more commonplace in manufacturing than in construction and it implies a purchase order of a quantity of discrete items, perhaps as few as one.

Direct material cost analysis within a company is separated into contractual or inventory methods. The contractual method, used for subcontract materials, imply that a buyer-vendor arrangement exists, and the company solicits a material quotation for the design. This excludes those materials informally purchased or items that are widely used and are produced by a number of individual manufacturers because those materials have an ongoing market price. In this case, market competition establishes reasonable prices. Engineers use those market values as they pick a cost, perhaps denominated by pound.

Complex and specialized items that have limited or special application and few suppliers are often subject to wide price variation. Greater care is required in estimating those values.

The *quotation cost method* is the most widely used method. The cost of the material is established by a vendor. The delivery price is considered fixed subject to the guarantees of the mutually agreed-to contract. Quotations can be solicited orally if insufficient time is available or if the purchases are relatively small in value. Written solicitations should generally be tendered where special specifications are involved, where a large number of items is included in a single proposed procurement, or when obtaining oral quotations is not considered economical or legal. Involved in the wording of the tendered quotation are statements about the design and specifications, terms or conditions, delivery date, and price. The obligation to contract at fair and reasonable prices does not diminish as we move down the scale from multimillion-dollar contracts for system acquisition to the dime and quarter prices for nuts and bolts.

The *quote* or *price-in-effect method* is a collaborative legal agreement between the buyer and seller. As usually established the contract allows for adjustment to the original price should the seller incur material costs in excess of those estimated. If the seller's material cost falls below that which was estimated in the contract, then the buyer agrees to the original price and adjustment is unnecessary. If at the time of delivery of the design the seller can prove that the material costs incurred to the seller escalated above the original estimated value, then the buyer will make up the difference by use of a formula or through negotiation. The quote or price-in-effect method may seemingly avoid the troublesome problem of making accurate material analysis and forecasts. However, competition may not allow the luxury of this contract.

Most manufacturers carry inventories, and material costs are affected by the method used to evaluate inventory. Costs vary depending on the method and market conditions. When materials are purchased specifically and are identifiable with a contract, the actual purchase cost is used as the estimate for the design, and the contractual methods as described earlier are appropriate. However, when materials are issued from inventory, it is difficult to find the value because inventories are extensive. Often the raw materials are consolidated for many designs (e.g., a company may standardize on a special grade of steel for all shafts). Furthermore, materials are purchased at different times and in various quantities. Quantity discounts are normally offered as a business practice, and lower prices can be obtained when requirements can be consolidated into a single purchase order.

Here is a specific example: The inventory to be considered for material estimating is given in Fig. 3.3. The horizontal axis is time. The time scale may represent days, weeks, months, or years, and, thus, the model is general. The present is time 0 and is labeled E to designate it as the time of the estimate. Negative integers are past time periods and positive integers are future time periods. Point D is the delivery time for the product. The vertical axis represents units of quantity of the material.

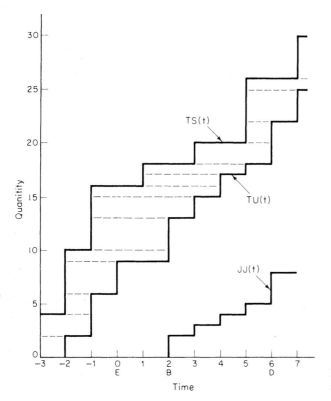

Figure 3.3. Total supply, usage, and material demand.

The three functions designated by TS, TU, and JJ are defined as follows:

$$TS(t) = \text{total supply of the material, time } t$$
$$TU(t) = \text{total usage of the material, time } t$$
$$JJ(t) = \text{job J usage of the material, time } t$$

These step functions will be used to obtain different estimates for the unit material cost for job J.

To the right of point 0 each of the functions represents inventory projections. The step sizes on each of the integers $t = 0, \pm1, \pm2, \ldots$ for the functions TU and JJ represent the material used in production during the period $(t - 1, t)$. The step sizes on each of the integers t for the function TS represent the quantity of material arriving at time t. Hence, the inventory at time $t = 0, \pm1, \pm2, \ldots$ is the difference between the functions TS and TU during the period $(t, t + 1)$.

In this example it is assumed that the total usage curve, TU, is composed of the job J curve, JJ, plus other curves. For example, at $t = 4$, job J uses 50% of the total for that period; at $t = 5$, job J uses 100%. Material estimates are made assuming that job J is won because, as will be expanded on shortly, quantity discounts are assumed to exist for material purchases.

The inventory supply curve TS follows a general review and reorder discipline, such as MRP (material requirements planning) or JIT (just-in-time ordering policy). But this is unimportant to the objective of estimating the cost of the units withdrawn irregularly from inventory.

A fixed lead time of one period is assumed for all orders. The JJ step function totals eight units with initial inventory withdrawal at period 2. *Last inventory withdrawal* occurs at period 6. Units are delivered to the customer during period 6. Total usage was four units when two units were issued for JJ at period 2. One, and three units of the four were taken from inventory purchases made during period -2, and -1. Horizontal dashed lines indicate the units used between TS and TU on Fig. 3.3.

Table 3.2 displays the historical and forecast unit costs the firm faces in purchasing the material. Price breaks are available for various quantity lots. For instance, if one or two units were purchased three periods back from the current time (period $= -3$) of the estimate (at $E = 0$), then each unit costs $10.70. If three or four units were purchased in a lot, then the unit cost is $8.60. If two periods in the future (at period $= 2$), and five or six units are to be purchased, then each unit will cost $9.10. Table 3.2 reflects the period and the quantity purchased.

TABLE 3.2. HISTORICAL AND FORECAST UNIT COSTS

	Quantity			
Period	1–2	3–4	5–6	7–8
Historical				
-4	10.00	8.00	7.00	6.50
-3	10.70	8.60	7.50	7.00
-2	11.10	8.95	7.80	7.20
-1	11.25	9.00	7.90	7.30
0	12.00	9.65	8.40	7.85
Forecast				
1	12.95	10.01	8.73	8.12
2	12.97	10.44	9.10	8.46
3	13.52	10.89	9.49	8.83
4	14.10	11.36	9.89	9.20
5	14.69	11.84	10.32	9.60
6	15.31	12.35	10.75	10.01

Now, with the specific example given as Fig. 3.3 and Table 3.2, consider six methods of evaluating the material unit cost for a firm having inventory, and restocking its inventory from time to time.

1. The *original* cost method assumes that materials are used in the order received and establishes as a cost estimate the unit cost of the oldest material sustained in inventory. In Fig. 3.3 the oldest material in stock at time $E = 0$ came from a lot of six purchased at time -2. From Table 3.2 the unit cost is $7.80. Commonly called FIFO (first in, first out), this method has been popularized by the accounting profession for inventory valuation purposes.

2. The *last-cost method* assumes that the latest materials purchased are the first to be used and establishes as a cost estimate the unit cost of the most recent material in inventory. In Fig. 3.3 the most recent material in stock at time E came from a lot of six purchases at time -1. From Table 3.2 the unit cost is $7.90. This method, also called LIFO (last in, first out), is frequently used by the accounting profession.

3. A unit cost at period $E = 0$ is used in this method as the value of the estimate and thus is time coincident to the preparation of the estimate. The method is known as *current cost*. If price breaks exist, then a lot size must be determined before an estimate is arrived at. Remember, inventory need not be added every period. For example, if the purchase lot is assumed to be 8 at time 0 (even though no material purchases are made in Fig. 3.3), then the estimate from Table 3.2 is $7.85 per unit. Rules may be arbitrarily established, such as the entire requirement is purchased if no other quantity is known for that period.

4. The *lead-time replacement method* adopts as an estimate the replacement cost of the first lot of the material at the time it arrives. This is commonly called the NIFO (next in, first out) method, even though this label is an obvious anomaly. Given a lead-time requirement of one period, the order of two units arrived at time 1, and the unit cost estimate from Table 3.2 is $12.95. This concept is used by engineers where material renewals are significant. For bidding situations there is a delay from submission of the bid to knowing if the estimate wins. Once the estimate wins, material is ordered, and material specifically ordered for job J arrives.

5. The point in time when the production order is delivered to the customer establishes the policy for the *delivery cost method.* As in the other methods, if any quantity is to be purchased at time D, then we use that. We assume a quantity in the absence of a purchase, perhaps equal to the delivery, and use that as the value of the material estimate. In Fig. 3.3 the delivery time is point D, and, if we assume a lot size of eight, then the unit cost is taken as $10.01. There were no scheduled purchases for inventory during period 6.

6. *Money-out-of-pocket methods* refer to a general class of techniques where the overriding philosophy has the estimate reflect actual material expenditure. If the material is purchased in a single lot, then the estimate will simply be the unit cost of the lot. If the material is taken from inventory, then the estimate will be the original purchase cost of the material. If the material is to be purchased in the future, then a forecast of the unit price will be used. Note Fig. 3.4, in which the material is from two purchases and the cost of total usage at period 2 is $31.50 ($= 7.80 + 3 \times 7.90$) and the weighted average is $7.88.

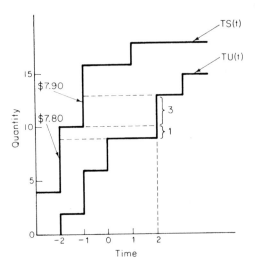

Figure 3.4. Actual material expenditure of usage at period 2 is weighted average.

In the example of Fig. 3.3, the material for job J is actually taken from a number of past and future inventory purchases. The money-out-of-pocket method is a weighted method for forecasted usage of the material in as accurate a manner as possible. Suppose that at time t we have $n(t)$ lots of the material in inventory that were purchased at times $t_j; j = 1, 2, \cdots, n(t)$. At time t define

$f(t_j)$ = fraction of the material inventory coming from the time t_j lot

$p(t_j)$ = unit cost of the t_j lot

$d_j(t)$ = number of units of the material used for job J at time t

The estimate unit material cost is

$$C_{ms} = \frac{\sum_t \sum_{j=1}^{n(t)} p(t_j)f(t_j)d_j(t)}{\sum_t d_j(t)} \tag{3.7}$$

where C_{ms} is the cost of material shape, $/unit shape.

To demonstrate the use of this formula, observe Fig. 3.3 and the data in Table 3.3. From Fig. 3.3 we see that of the four units total usage of period 2, one unit can be considered as purchased at -2 time for $7.80 per unit and three units as purchased at -1 time for $7.90 per unit. The dashed horizontal lines of Fig. 3.3 relate the time of inventory purchase to time of usage. Under the out-of-pocket cost methods, the two units of job J usage for the second period would be costed as

$$(7.80 \times \tfrac{1}{4} + 7.90 \times \tfrac{3}{4}) \times 2 = \$15.75$$

TABLE 3.3. VALUES USED FOR PROBLEM REMOVED FROM FIG. 3.3 AND TABLE 3.2

Time	Units of Inventory Delivered	Unit Price of Inventory Delivered	Usage Units of Inventory of All Jobs	Units of Inventory for Job J
−2	6	$ 7.80	2	—
−1	6	7.90	4	—
0	0	—	3	—
1	2	12.95	0	—
2	0	—	4	2
3	2	13.52	2	1
4	0	—	2	1
5	6	10.32	1	1
6	0	—	4	3
			Total	8

The remaining periods of production for job J ($t = 3, 4, 5, 6$) are similarly costed as

$$(7.90) \times 1 + (7.90 \times \tfrac{1}{2} + 12.95 \times \tfrac{1}{2}) \times 1 + (12.95) \times 1$$
$$+ (13.52 \times \tfrac{1}{2} + 10.32 \times \tfrac{1}{2}) \times 3 = \$67.04$$

Lot cost for job J is $82.79, and for the lot quantity of 8 unit cost is $10.35. Table 3.4 summarizes the estimated unit costs of raw material. The money-out-of-pocket cost method is, we contend, superior to other methods and is recommended for cost analysis.

TABLE 3.4. UNIT COST ESTIMATES FOR INVENTORY MATERIAL DEPENDS ON METHOD OF EVALUATION

Method	Candidate Value for Estimate, C_{ms}, $
Original	7.80
Last	7.90
Current	7.85
Lead-time replacement	12.95
Delivery	10.01
Money out of pocket	10.35

3.4 JOINT MATERIAL COST

Though joint labor cost (discussed in chapter 2) is exclusively labor, *joint material* is usually confronted with elements of material and labor costs. A frequent task is the preparation of a detail cost analysis of a manufacturing process dejointing the cost of common material into finished product. Traditional approaches to joint costs of material involve concepts that serve accounting or marketing needs and do not reflect the true or actual costs of specific interest to the engineer.

Joint materials are those materials that result from the processing of a singular raw material supply. Joint materials are intermingled up to the point at which the materials are divided into separable units. The point of division is called the *split point*. Material and labor costs up to the split point are referred to as joint costs but afterward are called unit cost.

The key element in this definition is the concept of the singular raw material, which, by virtue of a processing step, becomes two or more discrete products. For example, the processing of raw milk into cream and skim milk illustrates the notion of singular raw material into two discrete products that will be individually marketed. Another example is the rough log, which, on sawing and milling, becomes first- and second-grade lumber and sawdust. A molded plastic part where the die has several cavities is a common production joint-cost problem. Similar situations occur in a variety of industries, but especially in process-oriented industries, such as chemical, petroleum, food, metallurgical, and timber. Those industries include processes to create marketable products of what were essentially byproducts of basic processing steps. The problem in processes that result in multiple products is the tracing of the cost contribution of raw material and labor to the individual final products. Often the choice of allocators is not clear or product lines are not direct or other factors appear to prevent cost traceability.

A distinction is necessary between distributing and converting types of joint costs. The distributing type of joint costs is illustrated by the multiple-cavity die problem because plastic pellets are intermingled prior to the molding operation. The characteristic of the plastic is not altered (in a joint-cost sense), and cost traceability is valid albeit complicated. In processing industries a quantity of raw material is transformed into a singular new product or material. A simple example is water + heat process = steam. Splitting of raw material requires that the essential cost nature of the material be changed, resulting in two or more discrete products or materials with differing characteristics and physical measures or values. Figure 3.5 describes this distinction between distributing and converting joint costs.

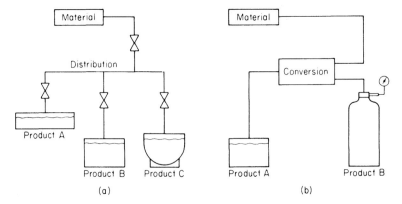

Figure 3.5. Two classes of joint costs: (a) material being "distributed"; (b) material being "converted" into two or more products.

Marketing practice can require that some products subsidize others because the selling price or the market value of a product is not necessarily proportional to the processing costs. An accounting or marketing policy may distort true costs. The allocation of joint cost on the basis of quantity can result in invalid cost data and arises when products are valued for inventory purposes at levels higher or lower than their actual production costs. A similar situation occurs when marketing policy dictates price levels that do not reflect manufacturing costs but, rather, selling.

Often the profitability of one product cannot be gauged by available data, whereas the profitability of a group of products may be properly represented owing to schemes of cost accounting. The reverse may also be true. In the example of a rough log subject to a splitting process, the production of first-grade lumber and other lumber products are not independent actions because the production of first-grade lumber is associated with the production of lesser-grade lumber and sawdust. The processing investment to finish and market secondary lumber products may be subsidized by profits (or operating costs) from first-grade lumber production. Profitability decisions on specific products cannot be made on the basis of accounting data, which may cover a group of products, and may not reflect the true costs of an individual product. Herein arises the need for an exact cost estimate that examines joint product costs. We point out, however, that the traditional approaches are justified for their intended purpose.

Now consider allocators or those units of reference commonly used to prorate or allocate costs. We select a unit of value per unit of reference to separate joint costs into unit cost. Commonly, this is a dollar value per unit of measure: $/ft^2 ($/m^2), $/lb ($/kg), $/kW, $/hr, $/product unit. Fundamental reference units that are readily measured are preferred. The simpler the measured unit, the clearer its use becomes and the more accurate are the conclusions drawn from the analysis. Allocators can be classified as follows:

1. Physical measure (geometry, weight, shape, etc.)
2. Energy (Btu, kW, etc.)
3. Time (second, year, man hour, etc.)
4. Units of finished product, each, 100 units, etc.

The unit of reference needs to be defined for traceability of costs to avoid changing the unit through a process. For example, if pound (kilogram) is used in the material stage, pound (kilogram) is convenient as a finished product rather than foot squared (meter^2). This becomes important with split converting where split products may be in different physical states. Unnecessary unit conversion should be avoided. For example, rather than defining the products of a petroleum cracking process as gallon (liter) and barrel (m^3), a common unit such as pound (kilogram) should be used throughout the process.

Now consider a manufacturing example for a distributing type of joint cost. A plastic material is blended, pigmented, and injection molded. Assume that a part is molded in a multiple-cavity mold and the die has runners that connect the pieces from a sprue. One operator tends two machines that are operating at different rates, and each machine produces a different part, designated A and B. Sets A and B are three- and four-cavity molds, respectively. The essential design features are given by Fig. 3.6.

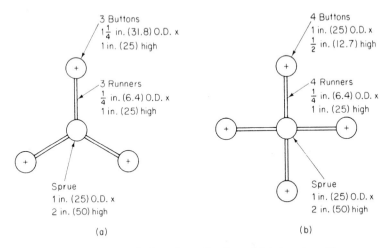

Figure 3.6. Two designs for a distributed joint-cost problem.

Two machines mold round preforms or buttons of a flour-like plastic. The material costs $2.50 per pound ($5.56 per kilogram) and has a density of 0.0275 lb/in.[3] (760 kg/m[3]). Machines A and B have a production rate of 400 and 300 sets per hour. A set is composed of three or four buttons having a sprue and runners. The runners and the sprue are considered waste and are eventually trimmed because only the button is used. The operator is paid $16.50 per hour. Four approaches are given in Table 3.5 and other methods are possible. Assume that the waste is lost and cannot be reground and used again. Usually, the waste degrades to lower quality in terms of specification and thus has a reduction in real economic value. Ponder the joint cost if the sprues and runners are reused at no loss in economic value.

In this example the material cost is uniformly related to production values of output. Labor cost, however, can be apportioned on four basis. Arithmetically, those basis lead to different answers for the contribution of labor cost of joint cost. Four acceptable solutions are given in Table 3.5. Which of the four is best? The answer is a knotty one, and we sidestep the question by noting that details outside the calculations will choose the "best" one.

Traceability is the important element in this example, because the costs of the finished product can be calculated to the original material by using a simple allocator and measure of quantity. This cost traceability is characteristic of simple distributing types of joint cost problems. However, in a converting type of joint cost the converting of a material creates two or more new products in a constant or variable ratio and with relative values and quantities disproportionate to original material values and quantities. The element of direct cost traceability becomes confused if the essential quantity measures vary or become meaningless. This is characteristic

TABLE 3.5.   DISTRIBUTION TYPE OF JOINT COST ANALYSIS

Facts for the Designs

Design A: Volume = 3.6816 in.^{3} (60, 305 mm^3) for 3 buttons
              Weight = 0.1012 lb (0.046 kg) for 3 buttons
              Sprue   = 1.5708 in.^{3} (25,730 mm^3), weight = 0.0432 lb (0.020 kg)
              Runners = 0.1473 in.^{3} (2,413 mm^3), weight = 0.0041 lb (0.002 kg)

Design B: Volume = 1.5708 in.^{3} (25,730 mm^3) for 4 buttons
              Weight = 0.0432 lb (0.020 kg) for 4 buttons
              Sprue   = 1.5708 in.^{3} (25,730 mm^3), weight = 0.0432 lb (0.020 kg)
              Runners = 0.1964 in.^{3} (3,217 mm^3), weight = 0.0054 lb (0.002 kg)

Set A	in.^{3}	mm^3	lb	kg	$
Waste	1.7181	28,140	0.0432	0.020	0.1080
Good product	3.6815	60,310	0.1012	0.047	0.2531
Total shot	5.3996	88,450	0.1485	0.067	0.3712

$$\text{Unit material cost for set A} = \frac{0.3712}{3} = \$0.1237/\text{unit}$$

$$\text{Material in good product for set A} = \frac{0.2531}{3} = \$0.0844/\text{unit}$$

Set B	in.^{3}	mm^3	lb	kg	$
Waste	1.7672	28,950	0.0486	0.022	0.1215
Good product	1.5708	25,725	0.0432	0.020	0.1080
Total shot	3.3380	54,675	0.0918	0.042	0.2295

$$\text{Unit material cost for set B} = \frac{0.2295}{4} = \$0.0574/\text{unit}$$

$$\text{Material in good product for set B} = \frac{0.1080}{4} = \$0.0270/\text{unit}$$

1. Labor apportioned on total lb/hr

   Set A output = 400 (0.1485) = 59.3960 lb/hr (26.942 kg/hr)
   Set B output = 300 (0.0918) = 27.5400 lb/hr (12.492 kg/hr)
   Total for machines A and B = 86.9360 lb/hr (39.434 kg/hr)

   $$\text{Unit labor cost for design A} = \frac{59.396}{86.936}(16.50)\left(\frac{1}{1200}\right) = \$0.0094$$

   $$\text{Unit labor cost for design B} = \frac{27.540}{86.936}(16.50)\left(\frac{1}{1200}\right) = \$0.0044$$

(continued on next page)

TABLE 3.5. DISTRIBUTION TYPE OF JOINT COST ANALYSIS (*cont.*)

2. Labor apportioned on shots/hr and total shots = 700

$$\text{Unit labor cost for design A} = \frac{400}{700}(16.50)\left(\frac{1}{1200}\right) = \$0.0079$$

$$\text{Unit labor cost for design B} = \frac{300}{700}(16.50)\left(\frac{1}{1200}\right) = \$0.0059$$

3. Labor apportioned on units/hr

Machine A units = 1200 units/hr
Machine B units = 1200 units/hr

$$\text{Unit labor cost for design A} = \frac{1200}{1200}(16.50)\left(\frac{1}{1200}\right) = \$0.0069$$

$$\text{Unit labor cost for design B} = \frac{1200}{2400}(16.50)\left(\frac{1}{1200}\right) = \$0.0069$$

4. Labor apportioned on lb/hr of final product produced

$$\text{Total lb/hr for machine A} = \frac{1200}{3}(0.1012) = 40.48 \text{ lb/hr}$$

$$\text{Total lb/hr for machine B} = \frac{1200}{4}(0.0432) = 12.96 \text{ lb/hr}$$

$$\text{Unit labor cost for design A} = \frac{40.48}{53.44}(16.50)\left(\frac{1}{1200}\right) = \$0.0104$$

$$\text{Unit labor cost for design B} = \frac{12.96}{53.44}(16.50)\left(\frac{1}{1200}\right) = \$0.0033$$

Solutions to Joint Cost Problem

Design	Basis for Apportioning of Labor Cost	Material	Labor	Unit Cost of Labor + Material
A	lb/hr input	$0.1237	$0.0094	$0.1331
A	shots/hr	$0.1237	$0.0079	$0.1316
A	units/hr	$0.1237	$0.0169	$0.1306
A	lb/hr output	$0.1237	$0.0104	$0.1341
B	lb/hr input	$0.0574	$0.0044	$0.0618
B	shots/hr	$0.0574	$0.0059	$0.0633
B	units/hr	$0.0574	$0.0069	$0.0643
B	lb/hr output	$0.0574	$0.0033	$0.0607

of converting processes. Those difficulties can be circumvented by following certain rules, and a cost estimate can be calculated.

When the converting type of joint costs occur, the key to unscrambling the cost allocation is the selection of a primary product. With a selection made, unit costs for processed material can be established.

The primary product is the product that forms the financial and physical justification for a company or process to exist. All other products are secondary products regardless of their value and would not exist were it not for the production of the primary product.

An economic profitability analysis should be performed where the primary product cannot be immediately identified or can be changed by minor process changes. For example, in the dairy industry, raw milk can be processed into cream, skim milk, powdered milk, milk, cheese, and butter. All those products can be produced, yet the operation may be set up to optimize the production of only one product, say, cheese. If concentrating on cheese production optimizes the profitability of the product line, then cheese is the primary product.

The identification of a primary product allows the process to be presented schematically as a direct flow from material to finished product, with all secondary products branching off at their split points. The flow of material is an engineering decision, and once a design has been chosen, cost analysis can begin. Converting industries, unlike the manufacturing, fabrication, and the durable-goods industries, deal with this kind of joint-cost problem.

SUMMARY

Direct materials include raw materials, standard commercial items, subcontract items, and interdivisional transfer items. Direct material cost is the cost of material used in the design. The cost should be significant enough to warrant the cost of estimating it. Some direct material, because of the difficulty of estimating, may be analyzed as indirect, although, for accuracy, direct costs are preferred to overhead, which is the method used to include indirect materials.

The engineer begins by calculating the final exact quantity or shape required for a design. To this quantity the engineer adds for losses of scrap, waste, and shrinkage. It is possible to find the direct material cost once the cost of the material is referenced to a shape dimension. Contractual arrangements and inventory schemes affect the method in which the material cost rate is found.

QUESTIONS

3.1. Define the following terms:

Specification	Quantity survey
Bill of material	Offal
Takeoff sheets	Shrinkage
Direct materials	Quote or price in effect
Subcontract materials	Yield
Normative materials	Joint products
Shape	Out-of-pocket material cost
FIFO	LIFO
Current cost	Indirect materials

3.2. Discuss the complications of materials as they relate to engineering design. Does the cost of material affect their engineering selection?

3.3. What kind of records does the engineer use in finding the cost?

3.4. Define material in terms of alteration.

3.5. Divide direct materials into categories. Discuss.

3.6. What makes commodity and semi-engineering materials difficult to estimate? Prepare a list of commodity and semi-engineering materials.

3.7. Give some typical engineering units for shape.

3.8. Indicate the similarities of manufacturing and construction in determining material cost policy. List some of the difference between manufacturing and construction.

3.9. What are the advantages of the money-out-of-pocket method over other inventory methods?

3.10. List allocators for joint material and labor costs.

3.11. Three future trend lines (constant value, deflation, and inflation futures) are assumed for the cost of unit raw material and are shown in Fig. Q3.11. The moment of the engineering estimate is located at time E, exchange of the money between you (the buyer) and the seller at the buy time B, and the delivery of the transformed product to your customer, following its manufacture by a value-added operation, at time D. Write a policy statement defining the point of time for the adoption of a cost value for an engineering product estimate. Does your recommendation of which value is adopted differ for the three sketches?

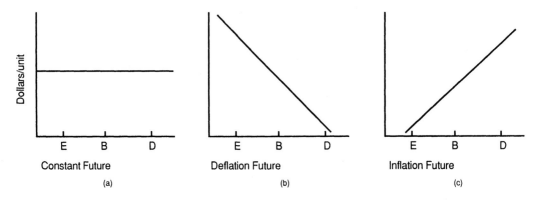

Figure Q3.11

PROBLEMS

3.1. A round shaft 6.5 in. (165 mm) O.D., has a drawing length of 31.675 in. (804.55 mm). The facing dimension necessary for a smooth end finish is $\frac{1}{16}$ in. (1.6 mm). The width of the cutoff tool is $\frac{3}{16}$ in. (4.76 mm). The length of the bar stock supplied to this turning machine is 12 ft (3.6 m). The collet requires 4 in. (100 m) of length for last-part gripping. Material will cost $0.95/lb ($2.11/kg) and density is 0.29 lb/in.^{3} (8024 kg/m^3).

   (a) What is the unit cost of the raw material?

   (b) What cost is lost to waste given that waste is salvaged at 10% of original value?

   (c) Find the shape yield.

   (d) Repeat for bar stock supplied in 16-ft lengths.

3.2. Examine the 2024-T4 aluminum shaft in Fig. P3.2. Raw material is purchased to match the outside dimensions. The bar stock for this part is supplied in 12 ft (3.6 m) lengths. The density is 0.0975 lb/in.^{3} (2700 kg/m^3) and the cost is $1.20/lb ($2.66/kg). Roughly scale for missing dimensions.
  (a) Estimate the cost of the raw material.
  (b) Find the approximate shape yield.

Figure P3.2

3.3. (a) Determine the theoretical and actual material required to produce the bayonet-clip half shown in Fig. P3.3. Raw stock is supplied in 0.875 in. (22.23 mm) diameter. A lathe cutoff tool width is 0.125 in. (3.18 mm). A 0.015-in. (0.38 mm) stock allowance is necessary for facing the spherical end. Allow 4% for shrinkage and bar end losses. The raw material weighs 2.05 lb/ft (3.05 kg/m).
  (b) Find the shape yield.
  (c) What is the unit cost for this hot-rolled steel stock, which costs $0.88/lb ($1.96/kg)?
  (d) Repeat in metric units.

Figure P3.3

3.4. A part is machined as shown by Fig. P3.4. The original stock size of AISI 1045 material is 5 × 2.5 × 2 in. (125 × 63 × 50 mm). Density = 0.29 lb/in.^{3} (8024 kg/m3).
  (a) If this material will cost $1.90/lb ($4.22/kg), then what is the unit cost of the raw material?
  (b) What cost is lost to waste given that the waste is salvaged at 10% of original value?
  (c) Find the yield percentage and suggest ways for improvement.
  (d) Find the approximate values in metric units.

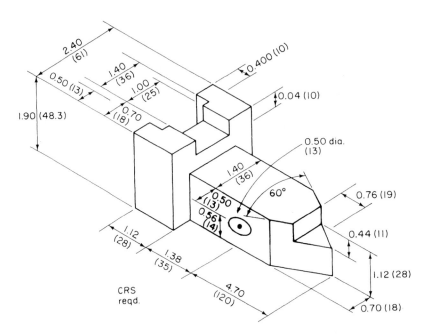

Figure P3.4

3.5. A V-block is manufactured of cast iron. Finish dimensions are shown in Fig. P3.5(a). A wood pattern is made by using a shrink rule for the green sand casting. Hot metal occupies a greater volume than cold, but, upon cooling, the shape is shown by Fig. P3.5(b). Extra material includes $\frac{3}{32}$ in. (2.4-mm) finish stock on all surfaces and a $1\frac{1}{2}\%$ draft on four vertical sides for pattern withdrawal.

(a) Calculate the volume of the raw casting.

(b) Determine the shape yield.

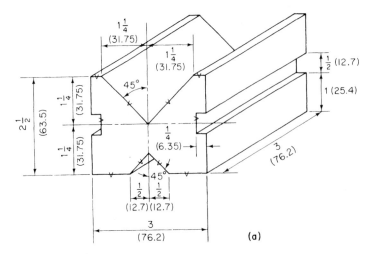

Figure P3.5

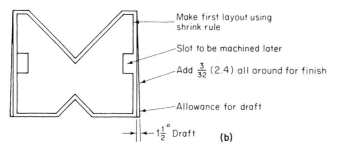

Make first layout using shrink rule

Slot to be machined later

Add $\frac{3}{32}$ (2.4) all around for finish

Allowance for draft

$1\frac{1}{2}°$ Draft   **(b)**   **Figure P3.5** (*cont.*)

3.6. (a) Estimate the unit material cost for the two designs shown in Fig. P3.6. The cold-rolled steel material costs $1.14/lb ($2.54/kg). Density = 0.278 lb/in.^{3} (7692 kg/m^3).
   (b) In addition to the blanking losses, add a 5% loss for overall waste. Find the losses for both designs.
   (c) Salvage is recovered at 10% of original value. Find the economic yield.

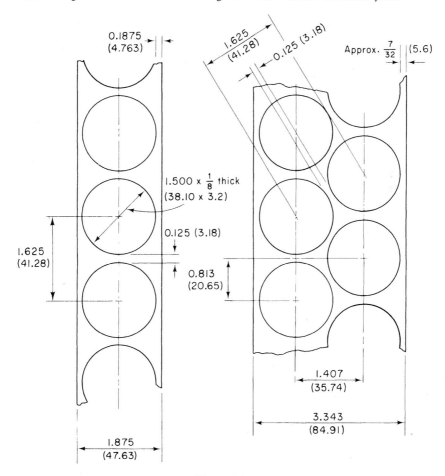

Figure P3.6

3.7. The 12-fluid ounce (0.35 l) beverage can is composed of the body, top, and ring. The container body is blanked from 3004-H19 aluminum coils with the layout given by Fig. P3.7. An intermediate cup is then formed without any significant change in thickness. The cup is then drawn to a side-wall thickness of 0.0055 in. (0.146 mm), and the bottom thickness remains unchanged. The can is trimmed to a final height of 5.437 in. (138.10 mm) to give an even edge for later rolling to the lid. This coil stock costs $1.0728/lb ($2.366/kg), and the density is 0.0981 lb/in.^{3} (2715 kg/m^3). Recovered waste is sold at $0.5300/lb ($1.178/kg).

- **(a)** What is the volume of metal in a trimmed can? (Ignore the 0.05-in. radius.)
- **(b)** Find the strip metal per can body.
- **(c)** Determine the shape yield.
- **(d)** Find the yield of the can body to blank.
- **(e)** What is the cost in trimmed can body?
- **(f)** What is the cost of the can body in the strip?
- **(g)** What is the prorated recovered value per can from waste?
- **(h)** Estimate the net cost per can body.
- **(i)** Repeat with metric units.

0.0135 ± 0.0005 in.(0.343 ± 0.013mm) thickness
coil stock, 3004 – H19 aluminum

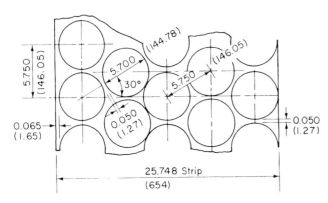

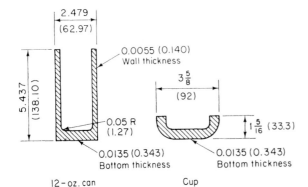

12 – oz. can   Cup   **Figure P3.7**

3.8. Look at Table 3.2 and Fig. 3.3. Find lead-time-replacement, delivery, and money-out-of-pocket unit material cost for the following step-size adjustment to Fig. 3.3.

Period	JJ(t)	TU(t)	TS(t)
2	1	2	2
3	2	3	0
4	0	1	2
5	1	3	4
6	4	4	8

(a) JJ(t) only.
(b) TU(t) only.
(c) TS(t) only.
(d) All parts together.

3.9. The inventory plan for a material is given by Fig. P3.9. Use Table 3.2 and find the following unit costs for six units of stock: (a) original, (b) last, (c) current, (d) lead-time replacement, (e) delivery, and (f) money out of pocket.

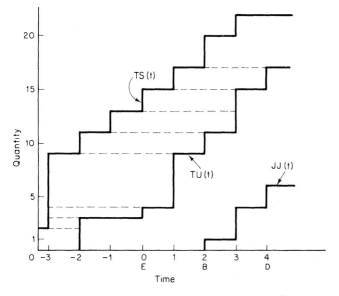

Figure P3.9

3.10. A flange is forged from C-1020 steel $1\frac{1}{2}$ in. (38 mm) O.D. stock (Fig P3.10). Each bar is $7\frac{9}{16}$ in. (190 mm) long, and will produce four flanges. The bar includes a 1-in. (25 mm) tong hold for all four flanges and is later trimmed as waste. The material costs $1.70/lb ($3.78/kg) and has a density of 0.29 lb/in.^{3} (8025 kg/m^3). Waste and scrap is sold at 10% of original value. A labor crew consists of a hammerman (wage = $21.75 per hour) and helper ($19.65 per hour). Each member of the crew performs different elements of the operation, and their joint output is 0.540 hour per 100 units.
(a) Find material yield.
(b) Find the unit material cost adjusted for sold salvage material.

(c) On the basis of output of units, dejoint the cost of labor.
(d) Estimate the total unit cost.
(e) Repeat with SI units.

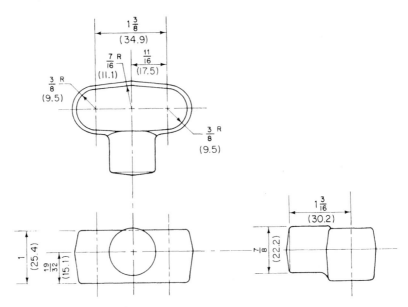

Figure P3.10

3.11. A back plate is used as a "snap-in" cover for a mating cassette case. The back plate is molded of polystyrene resin plastic and weighs 0.02 lb/unit (9 g/unit). There is no waste. Each shot fills eight cavities or units. High-impact polystyrene costs $2.20/lb ($4.88/kg). Density = 0.0365 lb/in.^{3} (1009 kg/m^3). Production time is 8 seconds per shot or 0.028 hour per 100 units. The senior plastic press operator has a direct wage of $28.80. The operator monitors three presses simultaneously. Nothing is known about production on the other two presses during a back plate run.

(a) Expressing the cost on a dollar per 100 units basis, find direct material cost, direct labor cost, and total unit cost.
(b) Repeat in metric units.

3.12. A foundry chooses to cost its castings on the basis of delivered weight per 100 lb. Essentially, each new casting is priced on historical records of a similar design. A description of the procedure is as follows:

1. Poured metal cost per casting = (furnace labor and overhead + cost of metal charged) × casting poured weight.

2. Cost of metal in finished casting = poured metal cost per casting less amount of remelted metal × value of remelted metal.

3. Cost per pound of delivered metal is item 2 divided by finished weight. For our estimate the casting poured weight is (5-lb finished weight) 9 lb, and the cost of charged metal is $1.20 per pound. Furnace labor and overhead is $0.20 per pound. The amount

of remelted metal is expected to be 3.7 lb with a value of $0.60 per pound. Determine the resulting material cost per pound.

MORE DIFFICULT PROBLEMS

3.13. Find the part cost for the following casting. A foundry is to cast the motor cylinder shown in Fig. P3.13. Compute the volume and cost of casting based on volume and the following factors: the shop yield is 54.5% (from experience); the metal loss is 10% or 5.4%, using shop yield; furnace labor and overhead are $0.06 per pound of poured metal; the cost of metal charged is $0.12 per pound; and the amount of remelted metal is 40% and is valued at $0.08 per pound. Density is 0.26/in.³ for cast iron. Allow $\frac{1}{8}$ in. of stock for all machined surfaces. The volume of cylinder is $\pi/4(\text{OD}^2 - \text{ID}^2) \times$ length. The pouring weight is finished weight per shop yield. The remelted metal weight is pouring weight $\times$ remelt factor. Consider what estimating factors are necessary. What items have been omitted in this estimate?

Material: Cast iron, 0.26 lb/in.³ (7194 kg/m³)

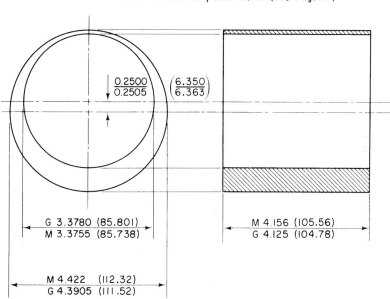

Figure P3.13

3.14. Repeat problem 3.9 for a job L requirement of eight units by using the inventory plan in Fig. P3.14. Use Table 3.2.

3.15. A computer part is molded two at a time of clear polycarbonate plastic. A partially dimensioned sketch (Fig. P3.15) supplies part size, sprue, and two runners.
   (a) Find the weight of one part and shot requirements for sprue, two runners, and two parts. Density = 0.0404 lb/in.³ (1119 kg/m³).
   (b) What is the yield of part material to total material?

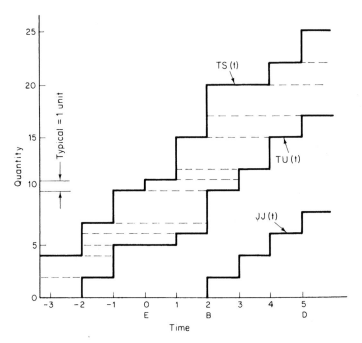

Figure P3.14

(c) The cost of this material is $6.20/lb ($13.78/kg), and waste is recovered at 10% of the original value. Find the net cost per unit, including a fair share of waste.
(d) The cycle time for one operator and one injection molding machine is 45 seconds. Detab time is 10 seconds for both parts, which is performed during molding cycle. For a labor rate of $33.00 per hour find the labor cost per unit.
(e) What is the total cost per unit?

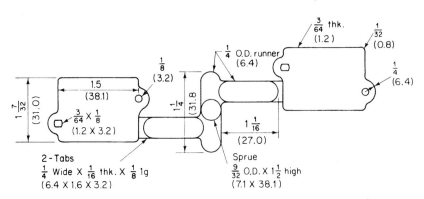

Figure P3.15

3.16. A computer rack handle is shown in Fig. P3.16 and can be constructed from either low carbon steel or a polycarbonate material. Determine costs of particular materials that would be suitable, and find the amount required for annual usages of 50,000 units. Use appropriate assumptions, and write a report indicating your results.

Figure P3.16

3.17. A bar as purchased is 12 ft long and 2 in. O.D. The drawing calls for a design that is 16 in. long in final dimension, but the manufacturing engineer requires a facing dimension of 1/16 in. and a cutoff of 3/8 in. The gripping for the last piece by the collet is 6 in. Scrap historically is 1% and shrinkage is 1/2%. The cost per lb is $0.75 and the density is 0.283 lb/in.^{3} A total of 260 final units are required. How many bars are required? What is the yield? Find the material cost for each piece. The labor performing this job is paid $18.50 per hour, and the production rate is 1.825 hr per 100 units. Analyze the job for direct costs.

3.18. Sheet, cold-rolled carbon steel, 0.1345 × 36 × 96 in., density = 0.29 lb/in.^{3}, is sold for $84.88 per 100 lb in the U.S. Canada, who has adopted SI, desires to buy 5 sheets now. Typical exchange rates in U.S. dollars are spot 0.8378 and one year 0.8506. The SI conversion factor is 0.4535 kg/lb. Each sheet produces 2 units of product with a waste loss of 10%. There is no shrinkage or scrap. Salvage is figured at 5% of the original value.

(a) What is the quoted rate cost to the customer for the material, and total cost for the purchase lot if the transportation costs U.S. $10/shipment from Detroit to Windsor to the customer?

(b) What is the Canadian material cost per unit if the product is to be produced in Windsor?

3.19. A cold rolled sheet is 36 × 96 in. in size. The sheet, commercial grade, oiled, No. 10 thickness (0.1345 in.), costs $85/each free on board from a steel supplier. A manufacturer plans to first shear strips and then shear blanks from the strip. A strip is either 36 or 96 in. long. The shearing process cuts along a line and loses no material, although there may be trim losses (dropoff) after shearing for strips and for blanks. But the placing of strips along the sheet and the choice of the strip dimension followed by the selection of the blanking dimension along the strip can affect the yield of the sheet with respect to the part. A part dimension is 5 by 10 in. What is the maximum shape yield? What is the lowest cost of the material of the part if salvage is sold back at 10% of original material? Assume an infinite quantity run, and develop your answers with respect to one sheet.

CASE STUDY:
DESIGN FOR RUNNER SYSTEM

"What counts in this problem is minimum plastic volume in the runner system." Rich Hall, die designer for General Plastics, mutters to himself. Rich knows that for this plastic mold design it will be impractical to reuse the scrap because the plastic part will be colored and the value of the scrap runners represents a small fraction of virgin material cost. The part to be molded is roughly 25 mm in diameter and 10 mm thick, similar to a preform except for the novelty impressions on the surface.

Rich, recently hired in his job, has learned that full-round runners are preferred. They have a minimum surface-to-volume ratio, thus reducing heat loss and pressure drop. Balance runner systems permit uniformity of mass flow from the sprue to the cavities, because the cavities are at an equal distance from the sprue. Main runners adjacent to the sprue are larger than secondary runners.

Rich has designed three configurations (Fig. C3.1) and will select the one that uses a minimum of runner material. The time factor is not critical, because the three arrangements provide identical number of parts per shot. The sprue volume for the three arrangements is equal. Die data follow. Determine which arrangement has a minimum of material for the runner system. If the plastic cost $1.20/kg and density is 1050 kg/m^3, then what is the prorated loss per unit? Determine the shape efficiency if

$$\text{shape efficiency} = \frac{\text{material in parts}}{\text{material in shot}}$$

Arrangement	Runner Section	Diameter (mm)	Section Length (mm)
A	1	5	25
	2	6	25
B	3	5	12
	4	6	75
	5	8	25
C	6	5	8
	7	6	100
	8	10	175

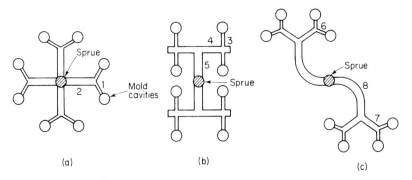

Figure C3.1. Runner system configurations: (a) star pattern; (b) "H" pattern; (c) sweep pattern.

4

Accounting Analysis

Accounting is the means by which the money activities of an organization are recorded. Accountants prepare periodic balance sheets, statements of income, information to aid the control of cost, and essential data useful for the finding of overhead. Although cost accounting data may have been wisely and carefully collected and arranged to suit primary purposes for accounting, raw data are usually incompatible with forecasting, product pricing, and the like.

Accounting specialties are general, public, auditing, tax, government, and cost. Cost and tax accounting are more important to engineering and cost estimating than are the others. Cost accounting emphasizes accounting for costs, particularly the cost of using productive assets. Tax accounting, because of myriad laws and practices, includes the preparation of tax returns and, importantly, the consideration of the tax consequences on business transactions.

The relationship between engineering and accounting is an active one, for both deal with much of the same information. On the one hand, the engineer deals with costs and designs before the spending of money; on the other hand, cost accounting records the cash flow facts. Roughly, engineering looks ahead and accounting looks back, but both are necessary for any successful engineering design. However, before estimating can proceed, accounting analysis must be concluded.

4.1 BUSINESS TRANSACTIONS

A *transaction* is an exchange of wants. An operation, product, project, or system is received or given, and a value, right, or service, collectively referred to as wants, is given or received. The transaction is composed of two elements that are reported in a financial record. This duality leads to double-entry bookkeeping, a practice several centuries old. Tested and found true, double-entry bookkeeping has changed little, even though growth of industry and business has complicated the professional field of accounting. The essential practices show remarkable similarity to the earliest commercial records.

In *double-entry bookkeeping* the results of business transactions are collected in records called accounts. The simplest form of the account is the T account, and the recording of a business transaction is an entry and is shown by Fig. 4.1(a). A *T account* is a graphical representation of an account. An entry on the left side of the account is called a debit (Dr.), and an entry the right side is a credit (Cr.). The terms *debit* and *credit* when used for the bookkeeping of transactions have several meanings. For example, debits increase assets and expense accounts. Credits increase capital, liability, and revenue accounts.

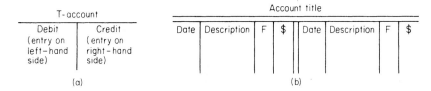

Figure 4.1.   (a) Abbreviated T account for instructional purposes; (b) regular account presentation.

The T account is an abbreviation and is used for textbook illustrations. In practice a more complete account supplies columns for additional data, as is shown in Fig. 4.1(b). The columns provide space for the date of each entry, description, folio (F), or cross reference to indicate the page in another record and amount. If the current status of the account is desired, then it is only necessary to total the debit and credit and show the balance on the larger side. This summing and finding the larger amount is known as *footing*. This columnar arrangement, of course, can be rearranged to suit electronic data processing.

Vouchers, invoices, receipts, bills, sales tickets, checks, and the many documents relating to the transaction support evidence for the entry. With this evidence the original entry is made to a journal that contains the chronological record of the transactions. The information is summarized in Fig. 4.2. In practice, the journals are files, punched cards, reels of computer tape, floppy disks, or other media. A single journal may suffice for entries for a small business. For the typical larger business, however, many types of journals exist, such as cash, sales, purchase, and general journals. This recording in journals is termed *journalizing*.

The transferring of journal entries to appropriate accounts in a *ledger* is the next step. The ledger is a group of accounts. Perhaps one page of the ledger is used for each account. *Posting* is the term applied to the process of transferring the debit and credit items from the journal ledger account. As each item is posted, the number of the ledger account is indicated in the journal folio. Similarly, the journal page number is placed in the folio column of the account of the ledger.

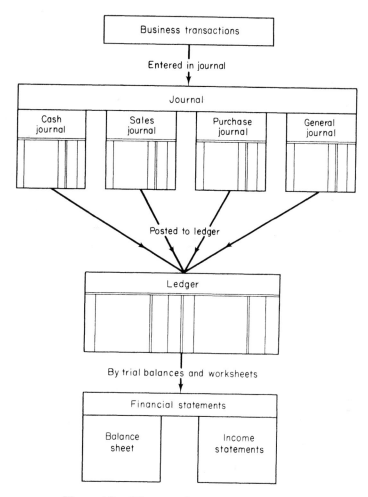

Figure 4.2.   Diagram of accounting procedure.

The original recording of business transactions to the eventual development of the balance sheet and income statement is shown in Fig. 4.2. The accountant is required to adjust the accounting of the business to present the financial situation accurately.

4.2 FUNDAMENTALS

A *money measurement fundamental* requires that business transactions be recorded only in terms of money. This practice expresses many different situations in common units. A fabricating plant can be compared to a construction company. Moreover, designs denominated in number units, such as dollars, deutsche marks, francs, and pesos can be algebraically manipulated to plan the performance of the business. Because of the importance of the money measurement fundamental, it must overlook technology expertise, engineering skill, prominence of design, and brand awareness.

A fundamental insists that business transactions be recorded in *double-entry fashion* as assets and as capital, revenue, or expense where the total debits must equal or balance total credits. This leads to the important accounting equation, where assets equal liabilities plus owner's equity or modeled differently,

$$\text{assets} = \text{liabilities} + \text{owner's equities}$$
$$= \text{liabilities} + \text{net worth} \qquad (4.1)$$

Both creditors and owners have claim to equities. The creditor has first rights, leaving remaining rights to owners. Owner's equities or net worth or proprietorship is the ownership interest in the business assets. Double-entry accounting is basic to balance sheet and income statement presentation.

A *conservatism convention* encourages the recording of financial data as the lower of possible choices. For example, in the evaluation of finished goods inventory (products held by the company for sale), we could value the inventory as either the cost to the business for production and materials or at market value. Conservatism dictates the lower value. In public disclosure the statement is made "the cost or market value, whichever is lower" to reflect this convention.

The *consistency convention* states that business transactions, if accounted for in one way, are recorded and accounted that way in the future. Companies obtain discounts for prompt payments of materials. One firm may use the discounts to reduce the cost of the materials. Another firm may take discounts and record them as income realized on prompt payment. The consistency convention adopts one of those two methods and persists in its use in succeeding accounting. This convention discourages a business from manipulating its figures to reflect favorable condition on one occasion and, as convenient, changing approaches.

The *going concern fundamental* implies that the business is operated in a prudent and rational way. This policy, it is assumed, perpetuates the business over an extended period. This going-concern assumption is the justification for the depreciation principles, discussed later.

A *business entity fundamental* is a simple notion that accounting transactions of a business are for the sake of the business rather than for individuals. If an owner withdraws money from the cash account, then the owner is richer but the business has less. The accountant records the effect of this transaction on the business and ignores the effect on the owner.

A *cost fundamental* recognizes that cost does not necessarily equal value. Stock share prices of a company may have little relationship to the par value per share. Though it is possible to list the assets in preparing a balance sheet, the value is determined by several methods. Consider a car for sale. Early in the day a potential customer asks what we will sell it for. Later, a tax assessor asks the same question. Though the answer might honestly differ, the point is that "value" is subjective. Even so, there are two other ways to evaluate business assets. Market value can be obtained, but this depends on the purchaser's needs. It is possible to find several evaluations of the worth of an asset, however, which returns the selection of value to subjective reasoning. We can value assets on their replacement. A replacement approach leads to a range rather than a single value. Market and replacement value lead to confusing choices, but it is possible to determine the value of the asset as given by original cost. The receipt for the payment of the asset is a document of record and demonstrates cost albeit not value necessarily. The primary appeal of the cost fundamental is objectivity and expediency. When coupled with the going business principle, it is assumed that asset will be used to conduct business.

If a *cash basis* is used for accounting, then income is recorded when the cash is received and expense is recognized when cash is paid out. The cash method is used by small businesses and individuals for personal and family records. Cash accounting is not allowed by the Internal Revenue Service for larger companies.

The *accrual fundamental* is concerned with the majority of business and is generally unfamiliar to us. In accrual, income is recorded when it is earned whether the payment is received during the period or not. Expenses incurred in earning the income are recorded as expense whether or not payment has been made during that period. The profit-and-loss statement includes incomes earned during the period covered by the statement and expenses concurrent to the same period. Many businesses prefer accrual accounting because it matches revenues to the expenses in a specific period. The time of collection of the income or payment of the expense is not a factor. Business transactions that affect income are measured by increases or decreases in net worth or services. Monies received increase net worth and are called revenues. Costs that a business incurs to provide the design, goods, or services decrease net worth and are called expenses.

If a transaction is *irrelevant* to the financial results of the business, then there is discretion as to how and when to record that event. If a piece of paper and pencil were consumed, then, theoretically, the paper and pencil become an expense. However, this penny watching is unwise, because average monthly office supply expenses are a commonsense way of handling this transaction. A convention of relevance permits the accountant to use judgment in handling those expenses.

4.3 CHART OF ACCOUNTS

An account is established for the various business transactions. For instance, a total sales account may lead to a specific customer sales account. Every item of financial information on the balance sheet and income statement has an account.

A chart of accounts is necessary for a business. A chart or listing of accounts intended for a manufacturing company is provided by Table 4.1. Formal charts of accounts are grouped into five principal classes:

1. Assets accounts
2. Liability accounts
3. Net worth accounts
4. Revenue accounts
5. Expenses accounts

Assets. The assets of a manufacturing firm are those things of dollar value that it owns, such as land, buildings, equipment, or inventory, or may be intangible as in the case of trademarks, designs, and patents. Assets may be segregated into current assets, fixed assets, and intangible assets. Current assets have three inventory accounts representing raw materials, in process manufactured goods, and completed products. Fixed assets include office equipment, factory equipment, and buildings, less accumulated depreciation. Land is an asset that does not depreciate. Assets are capable of providing future benefits. Otherwise, they are expense.

As an example of intangible assets, an engineering firm designs new products through its research and development department and obtains patents on the products. The cost of research, engineering, and testing leading to the development of a new product may be significant and in theory could be treated as an asset. However, under present financial tax accounting, those costs are expenses when incurred. Because many research projects may be underfoot at the same time, cost can be incurred over a period of years. Some firms treat engineering costs as a part of current operating expenses. Patents and copyrights are not recorded as assets if developed within the firm. If they are purchased or acquired through merger, then they are shown as assets.

Liabilities. The liabilities of a firm are the debts the firm owes. The category of liabilities is frequently broken down into current and long-term debts. Current liabilities are less than or equal to 1 year. Some of the more common items of business liabilities are (l) accounts payable (debts of the firm to creditors for materials and services received), (2) bank loans (amounts the firm owes to banks for money borrowed), and (3) mortgage payable (debt to investors for money loaned to the business on the security of its real estate or equipment).

Net worth. The net worth of a business is the ownership interest in the firm's net assets. In certain accounting situations the use of proprietorship and capital are synonymous with net worth. In a simple case the net worth of a corporation consists of its capital stock and the retained earnings:

	Net Worth
Capital stock	$45,000
Retained earnings	4,000
Total	$49,000

Broadly speaking, capital stock is the portion of the net worth paid in by the owners. Surplus, another term used within net worth, is that portion that the owners paid for stock, over and above the par value. Retained earnings refer to the accumulated profits and losses of the firm.

A company issues capital stock, which is divided into units of ownership, called *shares,* and the owners of the company are referred to as shareholders. The ownership of a shareholder in the net worth of the company is related to the number of shares he or she owns. The surplus of a company increases as the company earns profit and decreases as the company incurs losses or distributes the profits among the shareholders as dividends. If the losses and dividends of a company since inception exceed its profits, then a negative profit or a deficit within retained earnings results.

Income and expense in business. *Revenues* are generated by sales before the deduction of cost. Expenses represent costs of doing business. Though income is received from the sale of merchandise or products, expenses include salaries, advertising, power and light, telephone, rent, insurance, and interest. The profit-and-loss statement of a firm is a summary of its incomes and expenses for a stated period. If the statement discloses a net profit or loss, then the change represents an increment or decrement in the retained earnings during the period arising from business incomes and expenses and is carried to the net worth section of the balance sheet.

Gross income is the difference between income and expense. Once taxes are removed, we have net profits. Our view on taxes in this text is simplistic because of the many types of taxes, and we refer to taxes as "all kinds."

TABLE 4.1. MANUFACTURING COMPANY CHART OF ACCOUNTS
FOR AN ACCRUAL COMPANY

Asset	*Revenues*
Current Assets	Sales
Cash in bank	Interest earned
Petty cash	Dividends received
Notes receivable, customers	Sale of waste and scrap
Accounts receivable, customers	*Expenses*
Inventories	Manufacturing costs
Prepaid insurance, taxes, interest	Purchases of materials
Supplies	Salaries and wages
Fixed assets	Heat, light, water
Land	Telephone
Building, machinery	Depreciation
Accumulated depreciation	Freight
Furniture and fixtures	Direct labor
Liabilities	Indirect labor
Current liabilities	Factory insurance
Notes payable	Repairs and maintenance
Accrued wages payable	Factory supplies used
Accrued interest payable	Selling Expenses
Accrued taxes	Salaries and commissions
Deferred rent income	Advertising and samples
Dividends payable	General and administrative expenses
Fixed liabilities	Salaries
Mortgage payable	Traveling expenses
Bonds payable	Telephone, postage
Net Worth	Supplies
Capital stock, preferred	*Taxes, All Kinds*
Capital stock, common	
Retained earnings	

4.4 STRUCTURE OF ACCOUNTS

An expanded accounting equation with T accounts is given as

$$\underset{+\ |\ -}{\text{Assets}} = \underset{-\ |\ +}{\text{Liabilities}} + \underset{-\ |\ +}{\text{Net Worth}} + \underset{-\ |\ +}{\text{Revenue}} - \underset{+\ |\ -}{\text{Expenses}} \qquad (4.2)$$

This equation is sometimes referred to as the financial and operating equation, because it has plus and minus signs for each class of T accounts. The plus and minus signs show increases and decreases and are summarized as follows:

DEBIT INDICATES	CREDIT INDICATES
Asset increase	Asset decrease
Liability decrease	Liability increase
Net worth decrease	Net worth increase
Revenue decrease	Revenue increase
Expense increase	Expense decrease

Now, consider an example for the Flying Magnetics Company. The integrated example flows from the transactions to noting the effect on the T accounts, to a trial balance, and finally to the Profit-and-Loss and Balance Sheet statements. Our simplified approach starts with the initial capitialization of the firm, leading to its first reporting of financial documents.

Each transaction is worded to suggest a dual recording of the transaction. The numbered entry connects to the affected accounts. Consider transaction 1 in Table 4.2. A cash account is an asset account, and capital stock is a net worth account. The effect of the cash on the asset is to increase the T account by the recording of a debit to the Cash T account, and net worth is increased with the entry of $50,000 to the credit side of the T account. Each of the transactions follows the financial and operating equation.

By using those rules for applying debit and credit, we record the transaction into a T account. The identity of the account is described further, such as A (= asset). A number of T accounts are presented in Table 4.3. The numbers in parentheses next to the dollars amount in Table 4.3 relate to the transaction numbers shown by Table 4.2. For example, the Cash T account has a debit of $50,000 shown with a preceding 1. This indicates the transaction presented in Table 4.2. The transaction deals with the founding of Flying Magnetics, and $50,000 is paid for capital stock. The dual entry is in the Capital Stock T account, where $50,000 is recorded as a credit, and is noted with a 1 next to the $50,000.

In Table 4.2, one transaction affects two accounts, such as the founding of Flying Magnetics. Both cash and capital stock are influenced, and those accounts are identified into asset and net worth types. The effect of the transaction is to either increase or decrease the account, which is further explained by a recording of either a debit or a credit.

Because of the dual effect of a transaction, the record of the transaction must be equal, and for every debit there must be a credit. It is not necessary that there be

TABLE 4.2. TRANSACTION EFFECTS OF A BUSINESS

Transaction	Accounts Affected	Type of Account	On Account	Is Recorded by a Debit of	Is Recorded by a Credit of
1. Flying Magnetics Co. is founded and $50,000 paid for capital stock	Cash	Asset	Increase	$ 50,000	
	Capital stock	Net worth	Increase		$ 50,000
2. Business buys material from S. W. Specthrie on account, $10,000	Inventory	Asset	Increase	10,000	
	Acct. payable	Liability	Increase		10,000
3. Pay monthly rent on plant, $1500	Rent	Expense	Increase	1,500	
	Cash	Asset	Decrease		1,500
4. Pay S. W. Specthrie on account, $4,000	Acct. payable	Liability	Decrease	4,000	
	Cash	Asset	Decrease		4,000
5. Sells to P. Hall on account $15,000	Acct. receivable	Asset	Increase	15,000	
	Sales	Income	Increase		15,000
6. Pay salaries, $2,850	Salaries	Expense	Increase	2,850	
	Cash	Asset	Decrease		2,850
7. Retires for cash $5,000 of capital stock	Capital stock	Net worth	Decrease	5,000	
	Cash	Asset	Decrease		5,000
8. Collects $2,000 from P. Hall	Cash	Asset	Increase	2,000	
	Acct. receivable	Asset	Decrease		2,000
9. Buy equipment for $3,000 cash	Equipment	Asset	Increase	3,000	
	Cash	Asset	Decrease		3,000
10. Receives $500 rebate on month's rent	Cash	Asset	Increase	500	
	Rent	Expense	Decrease		500
11. P. Hall returns $4,000 of material for credit	Sales	Income	Decrease	4,000	
	Acct. receivable	Asset	Decrease		4,000
12. Pays advertising bill $800	Advertising	Expense	Increase	800	
	Cash	Asset	Decrease		800
13. Buys on credit $12,000 computer from Englewood and Co.	Computer	Asset	Increase	12,000	
	Acct. payable	Liability	Increase		12,000
14. Pays insurance premium $500	Insurance	Expense	Increase	500	
	Cash	Asset	Decrease		500
15. Take depreciation charge $600	Depreciation	Expense	Increase	600	
	Accum. depre.	Asset contra	Increase		600
16. Pay taxes $1,250	Taxes	Expense	Increase	1,250	
	Cash	Asset	Decrease		1,250
Total				$113,000	$113,000

the same number of debit and credit items in any T account. Table 4.3 is a record of transactions that affect the amounts of assets, liabilities, net worth, income, and expenses of a business. Any business transaction may be selected into equal debit and credit elements on the basis of the increase or decrease effect. Purchases are considered an expense because they are an offsetting cost to income sales. That portion of purchases remaining unsold at the close of the period is termed inventory and is classed as an asset.

TABLE 4.3. T ACCOUNTS FOR TABLE 4.2

Cash (A)				Accumulated Depreciation (A)				Insurance (E)		
(1)	$50,000	$1,500	(3)			($600)	(15)	(14)	$ 500	
(8)	2,000	4,000	(4)							
(10)	500	2,850	(6)							
		5,000	(7)							
		3,000	(9)							
		800	(12)							
		500	(14)							
		1,250	(16)							

Inventory (A)			Accounts Payable (L)				Depreciation Expense (E)		
(2)	$10,000		(4)	$4,000	$10,000	(2)	(15)	$ 600	
					12,000	(13)			

Accounts Receivable (A)				Sales (I)				Rent (E)		
(5)	$15,000	$2,000	(8)	(11)	$4,000	$15,000	(5)	(3)	$1,500	$ 500	(10)
		4,000	(11)							

Computer (A)		Salaries (E)		Capital Stock (NW)		
(13)	$12,000	(6)	$2,850	(7)	$5,000	$50,000	(1)

Equipment (A)		Advertising (E)		Taxes	
(9)	$ 3,000	(12)	$ 800	(16)	$1,250

An *open account* has either a debit or a credit balance. A *closed account* has the debit and credit of equal amount and, therefore, has no balance.

F. Baffett		Rent	
$18,000	$12,000	$8,000	$ 2,000
	6,000		6,000

Capital Stock		Englewood & Co.	
$ 5,000	$50,000		$12,000

The F. Baffett and Rent accounts are closed accounts, because the sum of the debits equals the sum of the credits. This is not the case for Captial Stock and Englewood & Co., which are are open.

Each journal entry provides for equal debits and credits, which are posted to the ledger accounts. If the posting is accurate, then the ledger must have equal debits and credits. Additionally, the sum of the debit ledger account balances must equal the sum of the credit ledger account balances. This equality is periodically tested by a trial balance, which is a list of the open ledger accounts as of a stated date. The trial balance shows the debit or credit balance of each account.

The account groups in the trial balance are broadly divided into those used to prepare the balance sheet and the income and expense statement. A few accounts contain both balance sheet and profit statement elements and are called mixed. Those are separated into the two components during worksheet analysis.

Note that for the Cash T account in Table 4.3 that the cash balance is a debit of $33,600, which is entered in the trial balance under Dr. and the cash account. In a similar way all T accounts are footed and their debit or credit balance entered under the trial balance column in Table 4.4

TABLE 4.4.   TRIAL BALANCE

Account	Trial Balance		Income Statement		Balance Sheet	
	Dr	Cr	Dr	Cr·	Dr	Cr
Cash	$33,600				$33,600	
Inventory	10,000				10,000	
Accounts receivable	9,000				9,000	
Computer	12,000				12,000	
Equipment	3,000				3,000	
Accumulated Depreciation	(600)				(600)	
Accounts payable		$18,000				$18,000
Sales		11,000		$11,000		
Salaries	2,850		$2,850			
Advertising	800		800			
Insurance	500		500			
Depreciation expense	600		600			
Rent	1,000		1,000			
Capital stock		45,000				45,000
Taxes	1,250		1,250			
	$74,000	$74,000	$7,000	$11,000	$67,000	$63,000
Profit to retained earnings						4,000
						$67,000

The periodic trial balance of the ledger provides reasonable proof of the arithmetic accuracy of journalizing, posting, and ledger account balancing. The ledger lists account balances from which the balance sheet and income statement are later prepared. In most businesses the trial balance is performed following the end of the month.

An example of a worksheet is given in Table 4.4 and continues the development given in Tables 4.2 and 4.3. Various account titles are listed, and the balances from the ledger are posted in the Trial Balance column as either Debit or Credit. Those entries are carried over to either the Profit-and-Loss or Balance Sheet columns.

Generally, the worksheet adjusts the accounts for accrued expenses, accrued incomes, deferred expenses, depreciation, and bad debts. Once those adjusting entries are disposed, the next step concludes the trial balance by using a worksheet.

Note that in Table 4.4 the first trial balance shows the effect of the ledger accounts. Those in turn are separated and extended horizontally into profit-and-loss and balance sheet entries. The worksheet must balance. If it does not, then an error of some kind is indicated, and it is necessary to find it.

4.5 BALANCE SHEET STATEMENT

The balance sheet is a tabular presentation of the important accounting equation [Eq. (4.1)] and is a summary of the assets, liabilities, and net worth at a point in time. The information for the balance sheet is removed from the worksheet. A balance sheet is shown for XYZ Manufacturing Company and repeats the accounting equation terms. Balance sheets have standard forms, especially giving the title of the firm, title "Balance Sheet," and the date of the evaluation.

XYZ MANUFACTURING COMPANY
BALANCE SHEET
MAY 31, 19xx

Assets		= Liabilities	
Cash	$15,000	Bank Loan	$15,000
Inventory	10,000	Mortgage	15,000
Land	15,000		
Building and equipment	40,000		
		+ Net worth	
		Capital stock	45,000
		Retained earnings	5,000
	80,000		80,000

The Flying Magnetics balance sheet is given in Table 4.5. Important points about the balance sheet are the length of time, handling of the accumulated depreciation, and the asset and liability groups disclosed. Note that in Table 4.5 the closing date is the end of June. The balance sheet does not provide any hint what the assets, liabilities, and net worth were for any date prior to or subsequent to June 30.

Balance sheet assets are not valued on the same basis. Cash, customer receivables, and inventories are valued at cost or net realizable cash value according to the conservatism convention. Land is valued at the amount originally paid for it, and depreciable fixed assets are valued at original cost less the accumulated depreciation. The liabilities are valued at the cash amount required to liquidate at the time of their maturity date. The net worth is a conglomerate value because it represents the difference between assets and total liabilities.

4.6 PROFIT-AND-LOSS STATEMENT

The statement of earnings of the company, known either as the *profit-and-loss* or *income-and-expense* statement, is a summary of its incomes and expenses for a

TABLE 4.5. BALANCE SHEET

Flying Magnetics Company
Balance Sheet
June 30, 19xx

ASSETS

Current assets		
Cash		$33,600
Accounts receivable		9,000
Inventory		10,000
Fixed assets		
Equipment	$ 3,000	
Less accumulated		
depreciation	(600)	
Computer	$12,000	
		14,400
Total assets		$67,000

LIABILITIES AND NET WORTH

Current liabilities		$18,000
Net worth		
Capital stock	$45,000	
Retained earnings	4,000	
		49,000
Total liabilities and net worth		$67,000

stated period. The net profit or loss disclosed represents the net change in net worth during the reporting period arising from business incomes and expenses.

Definition of profit.   *Profit* represents the excess of revenue over cost and is an accounting approximation of the earnings of a manufacturing firm after taxes, cash and accrued expenses (representing costs of doing business), and certain tax deductible noncash expenses such as depreciation are deducted. Loss represents the excess of cost over selling price, such as a product costing $8000 and selling for $6000 has a loss of $2000. The following example describes the effect on business net worth of profit and losses:

INVENTIONS, INC.
BALANCE SHEET
MAY 31, 19XX

Assets		Liabilities	
Customers	$ 4,000	Bank loan	$ 1,000
Gadget A inventory	8,000	Accounts payable	2,000
Gadget B inventory	6,000		
		Net worth	
		Capital stock	15,000
	$18,000		$18,000

If Inventions, Inc., sells the asset gadget A for $10,000 cash, its balance sheet changes to

INVENTIONS, INC.
BALANCE SHEET
JUNE 30, 19XX

Assets		Liabilities	
Cash	$10,000	Bank loan	$ 1,000
Customers	4,000	Accounts payable	2,000
Gadget B inventory	6,000		
		Net worth	
		Capital stock	15,000
		Retained earnings	2,000
	$20,000		$20,000

Net worth increases $2000. If the business sells the asset gadget B inventory for $5000 cash, then its balance sheet looks like this:

INVENTIONS, INC.
BALANCE SHEET
JULY 31, 19XX

Assets		Liabilities	
Cash	$15,000	Bank loan	$ 1,000
Customers	4,000	Accounts payable	2,000
		Net worth	
		Capital stock	15,000
		Retained earnings	1,000
	$19,000		$19,000

The $6000 gadget B inventory is replaced by $5000 cash, and the net assets and the net worth is decreased $1000. From the foregoing illustrations note that profits increase the net worth because profits increase the net assets, and losses decrease the net worth because losses decrease the net assets.

Continuing on with the worksheet and balance sheet, as developed previously for the Flying Magnetics Company, its profit-and-loss (P&L) statement is given in Table 4.6.

These P&L statements should be studied for their heading, income and expense groupings, and length of time and dates covered. Certainly, profits depend on the time of earnings. Net sales measure the net revenue from sales, and allowances for sales returns, freight out, and sales discounts are deducted from gross sales. Cost of goods sold covers the expense of the products sold to the customer. If the cost of goods sold is a gross value, then freight in and purchase discounts may reduce the value.

Operating expenses list recurring usual and necessary costs for conducting the business. Miscellaneous income and expense arise from interest and discounts and other small items of revenue and expense that are unrelated to the major business thrust. Except for depreciation the income statement items result from current period transactions. Depreciation is an allowable noncash tax expense and reduces

TABLE 4.6.   PROFIT-AND-LOSS STATEMENT

Flying Magnetics Company
Profit-and-Loss Statement
June 30, 19xx

Income		
Sales		$11,000
Expenses		
Salaries	$2,850	
Rent	1,000	
Advertising	800	
Insurance	500	
Depreciation	600	
Subtotal		5,750
Gross profits		$5,250
Taxes @ 23.8%		1,250
Profit (to retained earnings)		4,000

total income. Administrative expenses are found in almost all companies and cover the cost of managing the company and, additionally, may include heat, power, rent, insurance, accounting, engineering, legal, and so on. Income taxes, or the provision for income taxes, are an item reducing business income and are identified separately.

4.7 BUDGETING

Budgeting is a frequent engineering task. A *budget* is a written plan covering the activities for a definite future time. Dimensions are in monetary terms for a specific period, such as a quarter or year. Budgets deal with information based on data derived from cost estimating and accounting records and conjectures of future activities. The budgeted cost center should be the smallest unit to which a cost can be clearly traced, provided there is a balance between excessive and too little detail, consistent with the cost of preparing the budget. For example, if all cost centers within the engineering department are physically located together, then heat and light should be charged to the entire department as a practical expedient.

Appropriation, fixed, and variable budgets are common classifications. An appropriation budget may be directed toward proposed expenditures for a machine tool. A fixed budget may be directed toward an operation with only one level of activity for a definite time. This budget may not be adjusted to actual levels and may be satisfactory if the company activities can be predetermined accurately. Budgets may be prepared for one level of activity or for a number of levels of activity. This last is variable or flexible budgeting. A variable budget requires greater knowledge of cost behavior.

Regardless of the level of productive activity, some costs are almost completely fixed per time period, such as depreciation of a building. Others are constant within certain ranges of activity, such as superintendence. Some change as the activity fluctuates, such as consumption of factory supplies.

The cost accounting cycle is framed about the skeleton of the manufacturing process or the physical arrangement or the service for jobs. Because cost accounts are an expansion of general accounts, cost accounts should, as a basic accounting procedure, be related to general accounts. Figure 4.3 presents the relationship between general accounts and cost accounts. The larger an organization is, the greater its span of accounting records. To illustrate: A materials account controls hundreds of different material items; the payroll account controls departmental labor costs and payroll records for each employee; and the factory overhead account controls indirect labor, supplies, rent, insurance, repairs, and many other factory expenses.

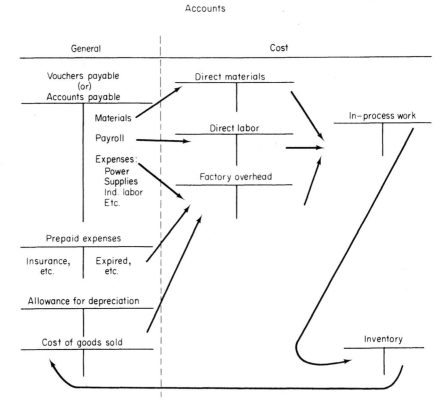

Figure 4.3.   Relationship between general accounts and cost accounts.

As is shown in Fig. 4.3, direct material, payroll (or direct labor), and expenses are transferred to in-process work, to inventory, and eventually to cost of goods sold. In-process work describes material incompletely processed. This material is usually on the factory floor. An inventory account (as shown in Fig. 4.3), if it is finished, has completed processing and is ready for shipment to a customer.

The process of budgeting draws from many sources of information, and as a detailed plan, the budget is the first step in finding overhead. The chart of accounts or the engineering cost codes are used for budget preparation.

Note that in Table 4.7, which is a budget for production centers, the planned number of machines, floor space, and other pertinent data determined for those cost centers are presented. It is necessary to recognize that the January–December period must be further identified as 49 weeks and a two-shift operation. A second closely identified budget is given in Table 4.8. This table connects the direct labor manpower for operation to the identical production centers. The wage and fringe rates are matched to the same annual period. The gross hourly cost, as typically found by Table 2.7, would be entered for the appropriate production center.

TABLE 4.7. PRODUCTION CENTER BUDGET FOR JANUARY–DECEMBER PERIOD

Machine Center	Number of Machines	Floor Space ft^2	Budgeted Hours[a]	Machine Horsepower Hours	Depreciation Charges	Tooling Expenses
Light machining	20	2800	49,300	443,000	$ 52,500	$200,000
Heavy machining	2	3000	6,800	748,000	95,000	80,000
Assembly	15	900	17,000	24,000	—	10,000
Finishing	8	1600	13,600	68,000	22,500	65,000
		8300	86,700	1,283,000	$170,000	$355,000

[a]Budgeted hours are reduced by 15% for nonproductive hours (two-shift operation, 49 weeks).

TABLE 4.8. BUDGETED MANPOWER FOR PRODUCTION CENTERS FOR JANUARY–DECEMBER PERIOD

Machine Center	Number of Direct-Labor Employees	Average Direct Hourly Wage Rate	Average Direct Fringe Hour Rate	Gross Hourly Wage Rate	Budgeted Direct-Labor Hours	Total Direct-Labor Cost
Light machining	29	$16.40	$4.92	$21.32	58,000	$1,236,560
Heavy macnining	4	19.75	5.93	25.68	8,000	205,440
Assembly	10	15.65	4.70	20.35	20,000	407,000
Finishing	8	17.05	5.12	22.17	16,000	354,720
					102,000	$2,203,720

Table 4.8 shows the budgeted direct labor hours and cost. The budgeted direct labor hours are 15% greater than machine hours, as shown by Table 4.7. *Efficiency* and *utilization* relate to two different concepts. Efficiency deals with labor, which can be more or less than 100%. Utilization is found with equipment, which is measured against a budgeted time. In this case, the utilization is 85%, meaning 15% of downtime, for a 49-week and a two-shift operation.

Note in Table 4.9 that factory, engineering, and general management expenses are planned. Each item of the factory overhead is further stipulated as fixed or variable. For instance, the rent of $180,000 is fixed because of contractual requirements. Electrical power is broken into fixed and variable components. Analysis, historical

TABLE 4.9. FACTORY, ENGINEERING, AND GENERAL MANAGEMENT
BUDGET FOR JANUARY–DECEMBER PERIOD

Factory overhead	
Space	
Rent (Fixed)	$180,000
Repairs to factory (Fixed)	60,000
Heat (Fixed)	16,000
	$256,000
Power	
Electricity (Fixed)	$ 50,000
Electricity (Variable)	139,625
Other utilities (Fixed)	15,000
Other utilities (Variable)	35,000
	$239,625
Indirect labor	
Material handling (Variable)	$ 64,000
Inspectors (Variable)	48,000
Supervisors (Fixed)	30,000
Supervisors (Variable)	60,000
Clerical personnel (Variable)	27,000
	$229,000
Equipment, tooling, services	
Depreciation (Fixed)	$170,000
Repairs and maintenance (Variable)	30,000
Perishable supplies (Variable)	38,000
Tooling (nonperishable) (Variable)	355,000
Repairs on tools (Variable)	28,000
Outside services (Variable)	18,600
	$639,600
Engineering and development	
Professional budgeted labor	$105,000
Clerical support labor	67,000
Rent	42,000
Utilities	18,000
Depreciation	15,000
	$247,000
General management	
Professional budgeted labor	$280,000
Clerical support labor	125,000
Rent	67,000
Utilities	18,000
Depreciation	15,000
	$505,000
Total	$2,116,225

information, and judgment will lead to this separation. Identification of fixed costs
is for engineering and general management needs. Variable power is identified with
production requirements, which are affected by output of product. All three tables
are used for various overhead rate determination.

4.8 DEPRECIATION

The purpose of a discussion about depreciation is to present the generalized mechanics of depreciation calculations and some of the thinking in cost analysis. Ambiguity exists about depreciation because forces of politics enter into the picture, such as surtax, credits, accelerated write off, inflation or recession, obsolescence, and new technology. Various accounting terms such as contra-depreciation, reserves for depreciation, allowances for depreciation, and amortization and retirement enlarge the ambiguity.

Depreciation is an accounting charge that provides for recovery of the capital that purchased the physical assets. It is the process of allocating an amount of money over the recovery life of a tangible capital asset in a systematic manner. It is cost, as of a certain time, not value, that changes with time and is allocated and recovered. There are no interest charges or any recognition of a changing dollar value. The depreciation charge is not a cash outlay. The actual cash outlay takes place when the asset is acquired. Depreciation charges are the assignments of that initial cost over the recovery life of the asset and do not involve a periodic disbursement of cash. When the rate at which the asset is depreciated increases, the depreciation charge does not increase the outflow of cash. In fact, the opposite reaction happens because the depreciation charges reduce taxable income and the outflow of cash for taxes. The initial investment is a prepaid operating cost that is expensed or allocated to an operating expense account, typically in overhead. Examples of typical fixed assets are given in Table 4.10.

TABLE 4.10. TYPICAL FIXED ASSET ITEMS

Production and process equipment
    Bins and tanks for inventory and storage of raw materials
    Turret lathes
    Milling machines
    Equipment for manufacturing basic materials into finished products
    Automatic transfer equipment
    Converting-type equipment
    Piping, material-handling trucks, and other moving devices
    Industrial pumps and compressors
    Instruments and controls
    Trucks

Service or nonprocess facilities
    Power generation and distribution
    Furniture
    Business machines
    Motor vehicles
    First aid and safety installations

Buildings and land
    Factory buildings, land
    Warehousing
    Roadways, parking areas
    Shops, maintenance garages

Initial costs are undertaken to acquire assets that contribute to the production of revenue over long periods. The cost of a factory building or a machine, for example, remains a positive factor in generating revenue provided that the building and equipment are used in the manufacture of a sellable product. Because federal laws require income measurement, it is necessary that an appropriate portion of the cost of the building or equipment be charged to or matched with each dollar of revenue resulting from the sale of products produced within its walls and machines. The amount of such cost, matched with revenue, during any one period is the estimated amount of the cost that expires during the period. Thus, a fixed asset, which will not last forever, has its useful value exhausted over a period. In the mineral industries and forestry, depletion, somewhat analogous to depreciation, is a noncash expense representing the portion of a limited natural resource such as oil, shale, and minerals utilized in a product that is sold.

The factors contributing to the decline in utility are considered in the categories of (1) physical wear and tear resulting from ordinary usage or exposure to weather, (2) functional factors such as inadequacy and obsolescence, and (3) governmental actions.

Physical factors are commonplace such as the wear and tear on buildings and corrosion that impairs efficiency and safety. Due to technological progress, obsolescence is a frequent situation where the fixed asset is retired not because the asset is worn out but because the asset is outmoded. The superior efficiency of an asset of a later design is one reason that compels new product designs for the market. Sometimes alterations of the design through research and development techniques make for immediate obsolescence. There is the instance of inadequacy where the asset, although neither worn out nor obsolete, is unable to meet the demands on the asset. An electric power company installs a hydroelectric generator, and, in the course of several years, finds that the demands of an expanding community placed on the generator exceed the ability of the water and hydroelectric power to meet peak loads. Governmental or other forces may prevent operation (e.g., loss of raw material source or a new law prohibiting waste disposal are possibilities).

The government alters the tax laws and frequently affects the depreciation. Changes in the laws governing income taxes and codes are enacted periodically, about every other year. In recent years the federal government coined the term *accelerated cost recovery system* (ACRS). Useful life, recovery period, and salvage value are terms that are carefully defined. Those laws have various motives to encourage economic growth, redistribute income, or provide for social needs.

Engineering estimates are required for the asset, erection, and operating capital and costs for consideration of new projects. In every project there are certain costs that can be tax expensed immediately, such as expenditures for nonphysical assets, some physical assets of extremely short life, and certain installation and startup expenses.

The question of life is an important matter. Some firms are concerned about economic life with little regard for physical *life,* while others, public utilities, for example, are restricted to earning a specified amount on capital invested, and life takes on another ramification. The economic life estimate is affected by tax laws and the actions of Congress. *Book value* is ordinarily taken to mean the original cost of the asset less any amounts that have been charged as depreciation. Book value should not be confused with salvage.

Before calculating depreciation it is necessary to understand property or asset classification. A *depreciable property* is used in the business or held for the obtaining of revenue and has a useful life longer than 1 year. The property will wear out, decay, become obsolete, or lose value from natural causes.

Depreciable property may be tangible or intangible, personal or real. *Tangible property* is seen or touched, such as buildings and equipment. Designs, patents, and copyrights are examples of *intangible property.* Intangible property can be amortized if its useful life can be found. *Real property* is land and generally anything attached to, growing on, or erected on the land. Land, which has an indeterminate life, cannot be depreciated. Personal property, which does not include real estate, is machinery or vehicles, for example. Engineering deals with tangible personal property and tangible real property.

Tax laws affect the practice of depreciation. The IRS provides various definitions. *Recovery property* is subject to the allowance for depreciation. A recovery period, a prescribed length of time, is designated for recovery property.

1. Three-year property has a life of 4 years or less and is used in connection with production and tooling.
2. Five-year property that is not 3-, 10-, or 15-year public utility property. Examples include machinery and equipment not used in research and development, autos, and light trucks.
3. Ten-year property is public utility property with a class life of more than 16 years but less than 20 years. An example could be railroad tank cars.
4. Fifteen-year property includes buildings not otherwise designated as 5- or 10-year property.

A method called accelerated cost recovery is defined as follows:

$$D_j = P(j) \times P \qquad (4.3)$$

where D_j = depreciation in jth year for specified property class
$\quad\;\; P(j)$ = percentage for year j for specified property class
$\quad\;\; P$ = cost of asset, dollars

Typical values of $P(j)$ are given next for years 3, 5, 10, and 15. Note that the sum of each column is 100% and that values are not constant year to year. If percentage values were equal (i.e., $33\frac{1}{3}\%$ for the 3-year property class), then the depreciation is straight line. The method in Eq. (4.3) disregards any expected salvage value.

Many laws are connected with this table. For instance, if an asset is decommissioned any time during the first year of a 3-year recovery period, the entire first year depreciation is not permitted the year the asset is removed from service. If the property is held for a period at least as long as the associated recovery period, then the asset value will be entirely depreciated. If, on the other hand, the asset is disposed of prior to the period, then the asset will not be entirely depreciated. If the asset is held for the recovery period or longer, then it is possible for the book value to be less than the anticipated salvage value.

Recovery Year	Percentage Depreciation			
	3-Year	5-Year	10-Year	15-Year Public Utility
1	33	20	10	7
2	45	32	18	12
3	22	24	16	12
4		16	14	11
5		8	12	10
6			10	9
7			8	8
8			6	7
9			4	6
10			2	5
11				4
12				3
13				3
14				2
15				1

Consider a 3-year recovery property having a cost of $100,000. The property will be sold at the end of the fourth year.

Year	Cost	Percentage Depreciation, $P(j)$	Book Value at Year Beginning	Yearly Depreciation, D_j
0	$100,000		$100,000	
1		33	100,000	$33,000
2		45	67,000	45,000
3		22	22,000	22,000

A firm may choose an alternative method to determine depreciation by using a longer recovery period.

Recovery Period	Optional Recovery Period, $N(k)$
3	3, 5, or 12
5	5, 12, or 25
10	10, 25, or 35
15	15, 35, or 45

The percentage for each and every year is

$$P(j) = \frac{1}{N(k)}$$ (4.4)

where $N(k)$ is the recovery period, years.

Once Eq. (4.4) is used with Eq. (4.3), we have the straight line method, where salvage value is assumed as zero.

Assume a $100,000 asset, which for economic reasons we choose to depreciate over a 5-year life.

Year	Cost	Straight Line Percentage, $P(j)$	Book Value at Year Beginning	Yearly Depreciation, D_j
0	$100,000		$100,000	
1		20	100,000	$20,000
2		20	80,000	20,000
3		20	60,000	20,000
4		20	40,000	20,000
5		20	20,000	20,000

In view of its worldwide popularity and for general understanding, a salvage value is often associated with *straight line depreciation.*

$$D_j = \frac{1}{N(K)}(P - F_s)$$ (4.5)

where F_s denotes the future salvage value of investment, dollars.

Reconsider the investment of $100,000 with a $10,000 salvage at the end of 5 years.

Year	Cost Less Salvage	Straight Line Depreciation (%)	Book Value at Year Beginning	Yearly Depreciation, D_j
0	$90,000		$100,000	
1		20	100,000	$18,000
2		20	82,000	18,000
3		20	64,000	18,000
4		20	46,000	18,000
5		20	28,000	18,000

Another method, sometimes used for depreciation accounting, is the *sum-of-the-years' digits* and provides a declining periodic depreciation charge over an estimated life that is achieved by applying a smaller fraction recursively each year to the cost less its salvage value. In the recursive relationship the numerator of the changing fraction is the number of remaining years of life, and the denominator is the sum of the digits representing the years of life. With an asset having an estimated life of

5 years, the denominator of the fraction is 15 (= 1 + 2 + 3 + 4 + 5). For the first year the numerator is 5, for the second 4, and so forth.

$$D_{sd} = \frac{2}{N}\left(\frac{N + 1 - K}{N + 1}\right)(P - F_s) \tag{4.6}$$

where D_{sd} = sum-of-the-years-digits' depreciation charge, dollars
K = current year
N = life defined for depreciation, years

Reconsider the investment of $100,000 with a $10,000 salvage at the end of 5 years.

Year	Cost Less Salvage	Rate	Book Value at Year Beginning	Yearly Depreciation, D_{sd}
0	$90,000		$100,000	
1		5/15	100,000	$30,000
2		4/15	70,000	24,000
3		3/15	46,000	18,000
4		2/15	28,000	12,000
5		1/15	16,000	6,000

Still another method of depreciation is based on the premise that an asset wears out exclusively as demands are placed on it. Called the *units of production method,* the computation is given as

$$D_{up} = \frac{N(i)}{N_{total}}(P - F_s) \tag{4.7}$$

where D_{up} = unit of production depreciation charge, dollars
$N(i)$ = units for ith year

This method has the advantage that expense varies directly with operation activity. Retirement in those cases tends to be a function of use. An estimate of total lifetime production is necessary. The $100,000 cost and $10,000 salvage value example is estimated to have 200 units of output over the 5-year period.

Year	Cost Less Salvage	Units of Production, $N(i)$	Book Value at Year Beginning	Yearly Depreciation, D_{up}
0	$90,000		$100,000	
1		15	100,000	$ 6,750
2		45	93,250	20,250
3		50	73,000	22,500
4		55	50,500	24,750
5		35	25,750	15,750
		200		

The advantages of accelerated methods of depreciation are compatible with the logic that the earning power of an asset is created during its early service rather than later, where upkeep costs tend to increase progressively with age. *Accelerated methods* offer a measure of protection against unanticipated contingency such as excessive maintenance, and they return the investment more quickly and simultaneously, decreasing the book value at the same rate. A high book value would tend to deter the disposing of unsuitable equipment even when the need for replacement is pressing. Rapid reduction of book values, provided the owner overlooks the tax benefits from capital loss, leaves the owner more free to dispose of inefficient and unsatisfactory equipment.

Depreciation of costs is collected under the control of general ledger accounts, such as accumulated depreciation or allowance for depreciation. Periodic fixed charges resulting from those accounts are analyzed and charged departmentally under the proper cost classification through the application of worksheet analysis. Distribution of depreciation of machinery and equipment is thus made to different departments based on factors such as cost of factory equipment, rates of depreciation applicable to each unit of equipment, and departmental location of each unit. Various property ledgers would show location of cost and accumulated depreciation in the case of plants and buildings. The proration of depreciation on buildings to departments may be based on the cost of the building, the total area of the building, and the area occupied by each department of the building.

An accumulated depreciation account contains the accumulated estimated net decrease in the value of the particular asset account to which it pertains. In most industries the amount shown in the depreciation account does not appear as cash. If a special fund is set aside specifically for this purpose, then it will be called a *sinking fund*. To create a fund of this nature suggests that a fund is actually invested outside the company to earn interest. However, interest rates found on the outside are hopefully less than the earning rate enjoyed by the company. It is wiser to employ the money for some operations. The amount equal to the depreciation will appear as other assets such as working capital, raw materials, or finished products in storage. When it becomes necessary to buy new equipment or replacements, management must convert physical assets into cash (unless sufficient cash is on hand) or to use existing profit to pay for the new equipment.

The appearance of accumulated depreciation on the balance sheet, as with other types of assets, represents capital retained in the business, ostensibly for the ultimate replacement of the capital asset being depreciated. Accumulated depreciation is also known as a "contra" account, whose balance reduces the value of an asset, which is shown at cost. The value of the contra account is enclosed by parentheses, indicating a negative value. For example, note in Table 4.5 where the contra account is ($600).

A number of factors in any depreciation model are subject to estimation: salvage value and, particularly, life. If for some reason the estimates prove faulty, then it is possible to retire an investment before its capital has been recovered. In the circumstance where net income received is less than the amount invested, an unrecovered balance remains. This unrecovered balance is referred to as *sunk cost*, which may be defined as the difference between the amount invested in an asset and the net worth recovered by services and income resulting from the employment of the asset.

As an illustration, consider a case where a capital investment of $5000 is to be recovered in 5 years with a remaining $1000 salvage or $4000 depreciation. On the basis of straight-line depreciation the amount invested and to be recovered per year will be $800. As a result of excessive use the machine was sold after 3 years, for $1400, and had actually consumed $3600 in 3 years or $1200 per year on the average. The sunk cost is equal to the difference in the actual depreciation and the depreciation charge, in this case $3600 − $2400 = $1200. Stated yet another way, sunk cost is determined to be the estimated depreciation value (or book value) minus the realized salvage of the asset. Sunk costs cannot be affected by decisions of the future and must be faced with reality.

The act of exhausting a natural resource and converting it to a saleable product is called *depletion*. The natural resources subject to depletion are oil, natural gases, metal and mineral mines, orchards, fisheries, and forests. Not all natural resources are subject to depletion; for example, soil fertility and urban land. In accounting for depletion, it is the allocation of the value of the quantity of natural resources extracted from a deposit that is considered. Like depreciation, depletion is a noncash expense that is an allowed tax deduction.

4.9 OVERHEAD

By definition, *overhead* is that portion of the cost that cannot be clearly associated with particular operations, products, projects, or systems and must be prorated among all the cost units on some arbitrary basis. Broad details regarding the posting of direct labor cost, time, material, and other indirect costs have been given earlier. What will be discussed here are those overhead aspects that pertain to engineering. The key to this puzzle is the way in which indirect expenses are allocated, unitized, and charged to individual estimates. Direct costs, such as direct labor and direct materials, present little if any allocation problem. Those costs do not exist unless the product is made.

The underestimating or overestimating of overhead rates is serious in view of the proportion of the total estimated cost. As an illustration, consider two different operations where the labor rates are $21.25 per hour. Machine A, a numerical controlled milling machine, is initially worth $150,000. Machine B, a standard general purpose milling machine, is worth approximately $15,000. By using an average burden rate of 200%, it would be indicated that machines A and B would each cost, on a machine hour basis, $63.75 per hour. However, this is false machine hour costing, because the investment in machine A is 10 times that in machine B. The proper cost base and sensible allocation of overheads to handle discrepancies of this sort are necessary. Years ago, machine investment per worker was lower, and it was not uncommon that overhead rates were uniformly distributed over the direct labor base. In recent decades the ratio of fixed cost to variable cost has risen, and the simple expediency of overhead distribution by means of the single rate is misleading.

Traditional cost accounting systems were developed in the early part of the twentieth century for labor-intensive industries. At that time, overhead cost was not a major component of product cost. Direct labor was the major focus of cost containment. It was reasonable to use a single plant-wide rate to allocate overhead charges to product. This rate is based on a single volume related cost driver or on substitutes such as direct labor costs or hours.

If the factory is mechanized, then it would seem reasonable to use machine utilization rather than direct labor. But the emphasis with *productive hour cost,* or *activity based costing,* is to find variables that demonstrate cause and effect for overhead charges. One simple approach identifies two types of cost drivers: volume related and nonvolume related. Volume related cost drivers have a direct positive relationship with production quantity. An example would be direct labor and machine hours. More product quantity means more direct labor and machine hours. An example of nonvolume related cost drivers is the number of products.

Volume Related Variables	Nonvolume Variables
Direct labor hours	Input
Machine hours	Number of vendors
Direct labor costs	Number of engineering drawings
Production volume	Number of orders
Kilowatt hours	Output
Utilities	Number of products
	Inventory levels
	Defect and scrap levels
	Process
	Number of schedule changes
	Amount of rework
	Space utilization
	Downtime
	Process changes
	Number of material moves

A primary distribution of overhead consists of assigning the various overhead costs to several departments or defined divisions within the unit, or factory. Sometimes, cost centers combine or separate departments to form homogeneous groups and are a logical point for the accumulation of costs. In making this distribution no distinction exists between a producing department (milling machine department) and a service department (first aid). This distribution is then followed secondarily by redistribution.

Overhead is allocated to a designated base. A base for distribution of manufacturing overhead may be floor area, kilowatt hours, direct labor dollars or hours, machine hours, or number of employees. The redistribution of costs originally assigned to service cost centers (first aid, for example) to the production cost centers is termed *secondary distribution.* This bookkeeping transfer technique of overhead costs to production departments is for subsequent recovery within operation or product cost. A secondary distribution, in the case of first aid, might be to prorate the total medical center costs to the production department on the basis of the number of employees within that department. If the milling machine department had two times as many employees as the lathe department, then we could reason that the costs that it receives should be borne on a 2:1 ratio.

Building depreciation, building insurance, building maintenance, and building taxes are often distributed on a floor area basis. Electrical power poses a confusing choice, because it may be distributed on the basis of machine weight, machine hours, horsepower hours, or even direct labor hours.

Overhead costs become complicated whenever the costs have a joint or commonness with different levels of variability. Joint costs, or costs incurred jointly, are depreciation, insurance, property taxes, maintenance, and repairs and are dependent on one another.

But joint costs are handled differently for accounting analysis than for the direct labor and direct material discussed in chapters 2 and 3. Accounting joint costs are handled entirely by the overhead process.

The four steps of overhead analysis are graphically given in Fig. 4.4. The horizontal axis is "period," where E is the now time of the estimate. Past periods are −1

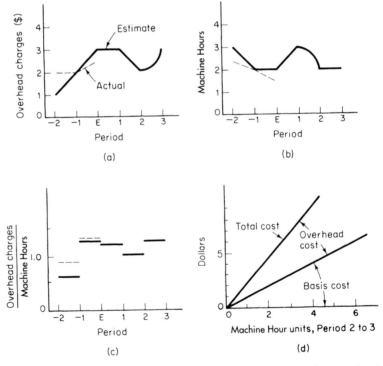

Figure 4.4. Overhead sequence: (a) collecting and plotting overhead charges; (b) collecting and plotting the basis; (c) calculating overhead rate; (d) applying overhead rate.

and −2, and future periods are 1, 2, and 3. Figure 4.4(a) plots overhead charges versus period. Obviously, actual charges can only be collected up to the current time, but charges can be estimated or gathered for future periods. This gathering is sometimes referred to as *pooling.* Note that the behavior of the charges can be linear, flat, or nonlinear. In Fig. 4.4(b) the vertical axis is identified with "machine hours." Machine hours could be replaced by direct labor or direct labor cost. In Fig. 4.4(c) the general definition of overhead is given, or

$$\text{overhead rate} = \frac{\text{overhead charges}}{\text{machine hours}} \tag{4.8}$$

Note that actual charges, machine hours, and rates do not coincide with the estimated value. This is usually what happens, and an effort is made to have coincidence. It is the total area in Fig. 4.4(a) divided by that in Fig. 4.4(b) that gives a constant rate value shown in Fig. 4.4(c). But the overhead rate is applied by engineering. Though an accountant will perform the first three parts of Fig. 4.4, it is usually an engineer who handles the application shown in Fig. 4.4(d).

4.9.1. General Classification of Overhead Methods

The exact nature of the overhead rate differs from one company to the next. It is not uncommon to find various overhead methods used within one company. A classification is given as

1. The *basis* used to determine and apply overhead, such as direct labor dollars, direct labor hours, machine hours, or material costs
2. Whether the rate includes fixed costs, as in *absorption,* or not, as in *direct costing,* which includes variable overhead costs only
3. The *scope* of the rate, such as the plant, a cost center, or to a specific machine
4. Whether the rate applies to all designs (such as product lines) or to one line of the design or to a unit of product

An *actual burden rate* has the merit of distributing the incurred factory overhead among the jobs. The rate is subject to certain defects, because it is unavailable until the close of the accounting period. Historical rates delay cost calculations until the end of the month and often later and fluctuate because of seasonal and cyclical influences acting on the actual overhead costs and on the actual volume of activity for which overhead cost is spread. Our discussion will disregard historical rates in view of the dependency on *predetermined overhead rates.*

Overhead rates are determined in most situations from data developed from operating budgets. We will use Tables 4.7, 4.8, and 4.9 to determine several kinds of rates.

The method of finding overhead as a ratio of the direct labor dollars is one of the oldest and most popular, or

$$R_{dl} = \frac{C_o}{C_{dl}} \tag{4.9}$$

where R_{dl} = overhead rate on basis of direct labor cost
$\quad\quad C_o$ = overhead charges, dollars
$\quad\quad C_{dl}$ = direct labor, dollars

The numerator and denominator may be the factory or department or cost center. Though it would appear that Eq. (4.9) is dimensionless, it is more useful to remember R_{dl} as dollars of overhead charge per dollar of direct labor cost. Now, examine Tables 4.7, 4.8, and 4.9 for the following information:

	Overhead Charges labor fringes	Direct-Labor Costs
Production center		
Light machining	$4.92 × 58,000 = $285,360	$16.40 × 58,000 = $ 951,200
Heavy machining	5.93 × 8,000 = 47,440	19.75 × 8,000 = 158,000
Assembly	4.70 × 20,000 = 94,000	15.65 × 20,000 = 313,000
Finishing	5.12 × 16,000 = 81,920	17.05 × 16,000 = 272,800
		102,000 hr $1,695,000
General Factory		
Space	$ 256,000	
Power	239,625	
Indirect labor	229,000	
Equipment, etc.	639,600	
	$1,881,945	

$$R_{dl} = \frac{1,881,945}{1,695,000} = \$1.11 \text{ overhead charge per dollar of direct labor}$$

Note that engineering and general management overhead costs were not spread against direct labor cost, although it could have been handled that way. We will later spread or distribute engineering and general management against cost of goods sold.

It is sometimes assumed that many factory overhead charges are proportional to labor time. This situation occurs if labor is paid on an hourly basis and the rate per hour is substantially the same for all workers. The rate of overhead per direct labor hours is calculated as

$$R_{dlh} = \frac{C_o}{H_{dl}} \tag{4.10}$$

where R_{dlh} = overhead rate on basis of direct-labor hours
$\quad\quad H_{dl}$ = budgeted direct labor, hours

Scope of overhead includes the factory, department, or cost center or productive machine. The dimensions for Eq. (4.10) would be dollars overhead charge per hour of direct labor.

Now examine Tables 4.7, 4.8, and 4.9 and note that C_o = $1,881,945, which was previously found, and H_{dl} = 102,000 hours. Then

$$R_{dlh} = \frac{1,881,945}{102,000} = \$18.45 \text{ per hour of direct labor}$$

Engineering and general management were not spread against direct labor time. The rate per direct labor hour represents the ideal in methods of overhead proration, if overhead accumulates on a time basis. But many overhead charges do not relate to direct labor time, such as engineering change orders, or engineering design time, for example.

Another method uses prime cost as the denominator, or

$$R_p = \frac{C_o}{C_p} \tag{4.11}$$

where R_p = overhead rate on the basis of prime cost
       C_p = prime cost, dollars

Prime cost is defined as direct materials plus direct labor. Still another case of absorption costing involves dividing the expected volume of product output into the total budgeted overhead to get cost per unit. This method will work for a few industries that have identical product units, such as cement, petroleum, and foodstuffs.

When the machinery used in production is not reasonably equal in value, the *productive hour rate method* is successful. This method has the formula

$$R_{mh} = \frac{C_o}{H_m} \tag{4.12}$$

where R_{mh} = overhead rate on the basis of productive hours
       H_m = machine hours

Equation (4.12) can be computed on a specific machine or machine grouping, as the following example demonstrates. The steps in computing the rate are similar. Budgets shown by Tables 4.7, 4.8, and 4.9 are necessary. But now we consider the physical factors associated with the machine centers, such as number of machines, floor space, machine horsepower hours, depreciation, and tooling expenses. Those data are shown in Table 4.7, which deals with the machine center budget. From Table 4.8 we use the manpower cost requirements to operate those machines, such as number of direct labor employees, wage, fringe rate, budgeted direct labor hours, and the total direct labor cost. From Table 4.9 we use factory overhead, engineering and development, and general management charges.

The machine hour and gross hourly cost spread sheet is shown by Table 4.11. Observe in Table 4.11 the distribution of space costs ($256,000) to each of the four production centers. The floor space cost to each center is on the basis of its area to

TABLE 4.11. CALCULATION OF PRODUCTIVE HOUR COST RATES

Machine Center	Space	Power	Indirect Labor	Depreciation	Nonperishable Tools, Expenses	Services	Engineering and Development	Management	Total
Light Machining	$ 86,361	$ 82,738	$130,216	$ 52,500	$200,000	$ 55,117	$140,451	$287,157	$1,034,540
Heavy Machining	92,530	139,703	17,961	95,000	80,000	38,200	19,373	39,608	522,375
Assembly	27,759	4,482	44,902	—	10,000	2,183	48,431	99,020	236,777
Finish	49,349	12,700	35,722	22,500	65,000	19,100	38,745	79,216	322,533
	$256,000	$239,625	$229,000	$170,000	$355,000	$114,600	$247,000	$505,000	$2,116,225
Allocation Basis	Floor Space	Horsepower Hour	Budgeted Direct-Labor Hours			Depreciation and Tooling	Budgeted Direct-Labor Hours	Budgeted Direct-Labor Hours	
Fraction	256,000 / 8,300	239,625 / 1,283,000	229,000 / 102,000			114,600 / 525,000	247,000 / 102,000	505,000 / 102,000	

Machine Center	Budgeted Assigned Direct-Labor Hours	Machine Hour Cost	Gross Hourly Wage Rate	Productive Hour Cost
Light Machining	58,000	$17.84	$21.32	$39.16
Heavy Machining	8,000	65.30	25.68	90.98
Assembly	20,000	11.84	20.35	32.19
Finish	16,000	20.16	22.17	42.33

the total area. For instance, the allocation fraction 256,000/8300 = $30.84/ft^2 would result in $86,361 (= 2800 × 30.84) for light machining. The $256,000 is total charges for space, and 8300 ft^2 is total area. The allocation factor is described for the several columns, and may be floor space, horsepower hours, direct-labor hours, and depreciation and tooling.

The depreciation and tooling ratio is computed differently. The equipment depreciation (= $170,000), tooling (= $355,000), and services (= $114,600) are removed from that portion of factory overhead. We presume that those services are incurred on the basis of the sum of depreciation and tooling. Light machining services are $55,117 (= 0.22 × 252,500). This ratio is the same as (52,500 + 200,000)/ (170,000 + 355,000) × 114,600 = $55,117.

The results of the *machine hour overhead rate* are $17.84, $65.30, $11.84, and $20.16. For example, $17.84 (= 1,034,540/58,000) is the machine hour cost on the basis of light machining budgeted hours. The *average gross hourly cost* for direct labor is added to each machine center. Comparison of this column in Table 4.11 shows that heavy machining costs about three times as much as finishing.

Productive hour cost rates are the sum of the gross hourly costs and the machine hour costs. Productive hour cost rates afford an accurate method of allocating overhead expenses and, from the engineering point of view, are preferred for estimating manufacturing from specifications and route sheets. The productive hour rate method is best whenever operations are performed by machinery, and is significant in terms of the final cost of the design.

The productive hour rate method is versatile. If, for example, there is no direct labor associated with the operation, then there is no need to add gross hourly costs to the productive hour. Mechanization is an example, because it is possible for equipment to operate unattended. For mechanization, then, the gross hourly costs would be limited, perhaps zero. Consider the case for an operation, such as hand assembly. Assume that the hand assembly did not use any depreciable equipment. The depreciation costs in this case could be zero, indicating that the operation was mostly labor, as one would expect for a labor intensive operation. The versatility of the productive hour cost method is further illustrated by the situation where there is muliple machines per operator. For this case, a fraction of the gross hourly cost would be prorated among the various equipment.

Cost can be identified as either variable or fixed with respect to changes in output. Materials and direct labor are *variable costs*. As a practical expedient it is sometimes also desirable to distinguish overhead costs as variable and fixed. To achieve this distinction, each cost center identifies variable and fixed overhead. The following cost items are a sample of how overhead might be classified into variable or fixed:

Variable Overhead Cost	Fixed Overhead Cost
Indirect materials	Indirect labor
Labor-related costs	Labor-related costs
Power and light	Depreciation
Inspection	Property taxes

Labor-related costs are included as both variable and as fixed costs. It is conceivable, if labor-related costs are related to direct labor, that those costs are variable overhead. Conversely, if labor-related costs are related to indirect labor, a fixed cost, those costs become fixed overhead. Overhead can be established as variable, or

$$R_v = \frac{C_{ov}}{C_{dl}} \qquad (4.13)$$

where R_v = variable overhead rate on basis of direct-labor cost
C_{ov} = variable overhead charges, dollars

By examining Table 4.9, note that the factory overhead items are separated into fixed and variable choices.

$$R_v = \frac{843225 + 508720}{1,695,000} = \$0.798 \text{ per direct labor dollar}$$

The fixed overhead costs are not forgotten but become a consideration of the pricing method, which is covered in chapter 8 as Contribution Pricing. In the accounting procedure of *direct costing*, the overhead charges are separated into fixed and variable choices. Note that direct costing is not the same as direct cost. The distinction to absorption costing is this separation. In absorption, lumped charges are divided by a basis, such as direct labor. In many circumstances we see that the fixed quantity, when divided by a basis more or less variable, causes future difficulties to arise, because overhead may be substantially different than the resulting actual overhead. A variable cost quantity divided by a variable cost basis is more uniform and responsive. A difference between actual and projected overhead is called overabsorption or underabsorption and is a consequence of poor budgeting or operation.

Overhead charges can be separated into those that are factory, management, engineering and development, and sales. Factory overhead charges are spread by using Eqs. (4.10)–(4.13). The total of direct materials and labor and factory overhead charges necessary for the operation of the producing units is sometimes referred to as *cost of goods manufactured*. Management, engineering, and sales expenses are distinct functions in the business enterprise, and we segregate those costs and apply them on the basis of cost of goods manufactured. Those are general expenses for the conduct of many products, and their overhead model is

$$R_{sga} = \frac{C_{sag}}{C_{gm}} \qquad (4.14)$$

where R_{sga} = overhead rate on basis of cost of goods manufactured
C_{sga} = overhead charges for sales, general, and administrative, dollars
C_{gm} = cost of goods manufactured, dollars

The phrase, "sales, general, and administrative," is often abbreviated SG&A, where we understand that engineering, development, and management expenses are also implied.

4.10 JOB ORDER AND PROCESS COST PROCEDURES

Cost accounting procedures are established to provide historical cost information and are known as *job order* or *process cost*. The procedures allocate material, labor, and burden charges to control accounts. If many different types of products are made and there are various customers, then a job order accounting procedure is used. In job or lot production, every "run" of product is assigned a production order number. This number is a convenient way to collect the material requisitions and labor job tickets. A typical job cost sheet describes the costing points to which the run refers. A job order may cover the production of one unit or a number of identical units. Examples might be a large seagoing ship or several similar electric generators when the items are complex or costly. The total quantity is divided in smaller production lots, and the job order for the total contract may be supported by a separate job order for each lot. In job order, the production cycle and the cost cycle are equal in the time allowed for gathering the costs.

Process cost procedures are used whenever a large volume or a repetition of a highly similar process exists. The completed item is a consequence of a series of processes, each of which produces some change in the material. Process cost procedures are found in industries that operate 24 hours per day, such as oil, steel, and chemical. The production cycle continues without interruption, but the cost cycle is terminated for each accounting period, such as a month, to determine the results of operations. Costs are accumulated on process cost sheets that show input and output material and labor.

Upon completion of the job order and the process accounting period, overheads are applied by using actual information if available. Thus, average cost-per-unit information is obtained from both procedures, although both procedures are substantially different approaches. Both procedures fundamentally divide the total manufacturing cost by the number of product units produced and compute the unit costs of the individual processes and total them. In both cases, those historical unit values can be forecast into the future to provide a rough glimpse of estimated cost.

SUMMARY

Cost accounting is important to the performance of diverse estimating functions. Accountants provide overhead rates, some historical costs, and budgeting data. The engineer reciprocates with manpower and material estimates for the several designs. The estimate in many situations will serve as a mini profit-and-loss statement for products or projects. Thus, there is mutual dependence between accountants and engineers.

The engineer is less interested in balance sheets, profit-and-loss statements, and the intimate details of the structure of accounts. However, overhead rates are vital for the estimating functions because the engineer will apply those rates. By definition, overhead is that portion of the cost that cannot be clearly associated with particular operations, products, projects, or systems designs and must be distributed among the cost units in some arbitrary way. The overhead rate is simply

$$\text{overhead rate} = \frac{\text{predicted overhead costs associated with a design}}{\text{predicted direct costs associated with a design}}$$

Accounting costs, if purely historical and derived from job order or process cost procedures, are usually inadequate as estimates of future costs. The design may have changed or costs may have increased or decreased, but one of the essentials for intelligent cost analysis is the continuous flow of reliable information from all activities of the organization. Accounting is an important contributor of past data. The next chapter deals with the future speculation of labor, materials, and accounting data.

QUESTIONS

4.1. Define the following terms:

Asset	Account balance	Charges
Liability	Depreciable value	Income
Net worth	Book value	Expenses
Accounting equation	Service life	Overhead
Balance sheet	Trial balance	P&L
Net assets	Closed account	Machine-hour overhead
Capital stock	Open account	Absorption overhead
Surplus	Declining balance	Direct costing
Debit	Accrual	Activity based costing
Credit	T account	Process costs

4.2. List 10 kinds of liabilities for a business with which you are familiar.

4.3. Why do expense accounts normally have debit balances and income accounts normally have credit balances?

4.4. What is the purpose of a trial balance? Into what main groups are trial balance accounts divided?

4.5. Distinguish between the cash and accrual bases of accounting. Discuss the conditions under which each is acceptable.

4.6. What are the functions of special and general journals?

4.7. What is the purpose of the budget? How would you define a cost center for an engineering budget? How do you prevent the budget from being meaningless?

4.8. Why is equitable distribution of cost essential throughout the organization to cost finding, analysis, and prediction?

4.9. Define overhead. What is the essential and philosophical purpose of overhead?

4.10. What is the difference between direct costing and absorption? How does that affect overhead calculation?

4.11. Prescribe and contrast several methods for the distribution of indirect cost.

4.12. How do job order and process cost procedures differ?

PROBLEMS

4.1. Evaluate the effects of transactions by constructing a daily balance sheet showing an asset side and an equities side. January 1: John Smith starts a sheet metal business producing metal products for the home. The business is called John Smith Sheet Metal. Smith deposits $10,000 of his own money in a bank account that he has opened in the name of the business. January 2: The business borrows $5000 from a bank, giving a note, therefore increasing the assets and cash and the business incurs a liability to the bank. January 3: The business buys inventory in the amount of $10,000, paying cash. January 4: The firm sells material for $300 that cost $200.

4.2. Given the following ledger T accounts from Weichman Mfg., set up a balance sheet for the month of March.

Cash

March	1	Capital	$1000	March	3	Rent	$200
	10	Consulting fee	250		20	Salaries	350
	25	J. A. Wilson on acct.	500				

Customers (Accts. receivable)

March	10	J. A. Wilson	$1200	March	25	Cash on acct.	$500

Supplies on Hand

March	5	Accts. payable	$360	March	31	Supplies used in March	$110

Equipment

March	4	Notes payable	$3200				

Accounts Payable

				March	5	Supplies	$360

Notes Payable

				March	4	Equipment	$3200

Weichman Mfg. Co. Capital

March	3	Cash	$200	March	1	Cash investment	$1000
	20	Salaries	350		10	Consulting	250
	31	Supplies used	110		10	J. A. Wilson	1200

4.3. By using the following ledger, construct a profit-and-loss statement and balance sheet.

Table 4.5, 4.6

Equipment

June	1	Balance	$7210				

Accts. Payable

June	2	Cash Meyers Co.	$350	June	1	Balance	$360
					6	Supplies	600
					28	Misc. exp.	40

Notes Payable

	June	1	Balance	$3200

Capital

	June	1	Balance	$1790
		30	From P&L	1570

Income

June 30	To P&L	$2500	June	4	Accts. pro. A. B. Jones	$2200
				15	Cash I. N. Smith	300

Lease Expenses

June	1	Cash	$200	June 30	To P&L	$200

Misc. Office Expenses

June 10	Telephone, cash	$60	June 30	To P&L	$100
	Elec.	40			

Salaries

June 20	Cash	$350	June 30	To P&L	$350

Supplies Expense

June 30	Supplies used	$280	June 30	To P&L	$280

4.4. Evaluate the effect of each transaction by constructing a balance sheet showing an assets side and a liabilities + net worth side.
  (a) Samuel Specthrie established the SS Company, paying in $250,000 cash for the entire capital stock.
  (b) Paid $50,000 cash for a building site.
  (c) Erected a building costing $200,000, paying $50,000 cash and issuing a $150,000 first mortgage for the balance.
  (d) Borrowed $60,000 cash from First National Bank.
  (e) Bought furniture, costing $15,000, on an open account from Wood Furniture Co.
  (f) Purchased $50,000 of tools from Universal Tool on credit.
  (g) Bought $30,000 worth of computer equipment from Byte Co. for cash.
  (h) Returned $15,000 of faulty tools to Universal Tool.
  (i) Paid $25,000 in reduction of the bank loan.
  (j) Bought $40,000 of U.S. Treasury bonds for cash.

4.5. Evaluate the effects of each transaction by constructing a balance sheet showing an assets side and a liabilities + net worth side.
  (a) The Eastwood Machine Co. is organized with a capital stock of $250,000, which is paid for in cash.
  (b) Bought from Culpepper Co. on credit $100,000 of merchandise.
  (c) Borrowed $80,000 cash from First National City Bank.
  (d) Paid Culpepper $30,000 on account.

(e) Returned $10,000 of defective merchandise to Culpepper.

(f) Loaned $50,000 cash to Robert Gondring.

(g) Paid $20,000 cash for a building site.

(h) Erected a building at a cost of $120,000 cash.

(i) Borrowed $70,000 from Friendly Insurance, giving a mortgage for collateral.

4.6. Short Corporation started the year with the following balances:

Account	Balance as of January 1
Cash	$100,000
Inventory	100,000
New plant and equipment	400,000
Accounts payable	50,000
Owner equity	550,000

Transactions during the year were limited to the following ones: paid $100,000 for labor; purchased $150,000 worth of materials; noted equipment depreciation of $50,000, adding to inventory 300,000 units costing $1 to the manufacturer; sold 300,000 units for $2 each, cash; purchased new equipment costing $200,000. Accounts payable at the end of year were the same as at the beginning of the year. Neglect income taxes. Make an income statement for the year just ended. Make an end-of-year balance sheet.

4.7. Construct a balance sheet for Dynamics Corp. based on the following information:

Retained earnings	$ 610,000
Cash	150,000
Outstanding debt	450,000
Raw materials	100,000
Finished goods	50,000
Current liabilities	40,000
Stock ownership	400,000
Fixed assets	1,100,000
In-process materials	100,000

4.8. Prepare a profit-and-loss statement by using the following account balances of E. Billerbeck for the nine months ended September 30.

Sales	$700,000	Rent	$ 80,000
Sales returns	40,000	Salaries	120,000
Inventory, January 1	120,000	Interest earned	2,000
Purchases	270,000	Sales discounts	10,000
Purchase returns	20,000	Interest expense	5,000
Inventory, September 30	160,000	Taxes, all kinds	47,000

4.9. The following data are the assets, liabilities, incomes, and expenses of Warren Andrews, contractor. (a) Prepare an income and expense statement for the 6-month period ending June 30. (b) Prepare a balance sheet at June 30. (Hint: It is necessary to prepare the income and expense statement first because the profits amount is needed for completing the balance sheet.)

Cash in bank	$ 65,000	Office fixtures owned	$ 50,000
Income from fees	150,000	Bank loan	100,000
Interest income from owned securities	5,000	Accounts payable	70,000
		Interest expense on bank loan	2,500
Rental income on owned properties	30,000	Receivable from clients	90,000
Staff salary expense	120,000	Automobiles owned	40,000
Traveling expense	7,500	Capital stock	415,000
Telephone expense	2,500	Surplus, January 1	85,000
Drafting supplies expense	7,500	Securities owned	200,000
Real estate owned	250,000	Taxes, all kinds	20,000

4.10. What follows are data for assets and liabilities at December 31 and incomes and expenses for the year of Wilmer Hergenrader, industrial consultant. By using this information, prepare (a) an income statement and (b) a balance sheet.

Cash in bank	$ 80,000	Furniture and Fixtures	$ 30,000
Fees earned	420,000	Rental income on building lease	26,000
Interest received from bonds	2,500	Bonds owned	120,000
Land	60,000	Traveling expenses	35,000
Receivable from clients	50,000	Taxes paid	15,000
Staff salaries	270,000	Office expenses	10,000
Building	180,000	Telephone expense	5,000
Bank loan payable	70,000	Accounts payable	25,000
Capital stock	275,000	Automobiles owned	20,000
Surplus, January 1	65,000	Association dues expenses	2,500
Interest expense on bank loan	4,000	Donation to charity	2,000

4.11. The following data, expressed as 10^3, are the closed ledger accounts showing assets, liabilities, income, and expenses of Spotweld, Inc. Prepare a balance sheet and income statement for the year.

Cash in bank	$1,560	Shop fixtures owned	$1,000
Income, parts	3,000	Bank loan	2,000
Interest income, owned securities	100	Accounts payable	1,400
		Interest expense on bank loan	50
Rental income	600	Receivables from buyers	1,800
Salary and wage expense	2,400	Spotwelder owned	800
Office expense	200	Capital stock	8,300
Materials expense	150	Retained earnings January 1	1,700
Building owned	5,000	Securities owned	4,000
Taxes, all kinds	140		

4.12. An engineering firm is chartered to provide consulting services, design, and provide a limited range of products. Its designs have developed into registered patents. Also, some designs are built by subcontractors, because the firm chooses not to produce them itself. But the firm does receive the product income. The firm prepares an income and balance sheet statement quarterly for the owners of the firm. These accounts and their balance have been summarized. These values are $1000, but ignore that. Find the income statement and balance sheet for a firm titled "Acme Engineering Services," Second Quarter.

Cash on hand	$ 3	Receivables	$58
Fees received	73	Royalties on patents	43
Bonds owned	12	Equipment book value	70
Product sale	4	Interest on owned securities	2
Salary expense	24	Office rent	4
Equipment lease	6	Travel and hotel	4
Utilities paid	5	Supplies expense	2
Patent assets	11	Accounts payable	30
Capital stock	25	Retained earning balance	
Taxes, all kinds	5	at start of quarter	27

4.13. An asset has a cost of $43,000, salvage value of $3000, and an asset life of 10 years. Compute the depreciation expense for the first 3 years under (a) accelerated cost recovery, (b) straight line, and (c) sum-of-the-years' digits.

4.14. Company A purchased a machine that costs $250,000. Economic life is estimated as 5 years with a salvage value of $15,000. This company chooses the accelerated cost recovery method for depreciation. Company B, in competition with company A, purchased the same machine at identical cost. The management of company B uses straight-line depreciation. Determine the yearly depreciation charges and the end-of-year book value. Compare the two methods for this competitive situation, and comment on the importance of the depreciation method for cost estimating.

4.15. A construction company will buy a tire-mounted excavating machine for a delivered price of $250,000. Life is 6 years, 12,000 hours, and salvage is expected to be nil. At the end of the life the machine will be retired to secondary and emergency service. Determine the depreciation charge and book value for (a) accelerated cost recovery, (b) straight-line, and (c) sum-of-the-years digit methods.

4.16. A standard-sized sedan purchased for $17,516 will cost its owner $54,212 by the time it has been driven 10 years and 100,000 miles. Besides depreciation to zero, this sum includes $10,396 for repairs and maintenance; $8384 for gas and oil; $7236 for garaging, parking, and tolls; $5400 for insurance; and $5280 for state and federal taxes.
(a) Find the depreciation cost per mile.
(b) What are the yearly and per mile costs?
(c) What are the percentages for the elements?

4.17. Car value and upkeep costs were found to follow this schedule:

Year	Drop in Value (%)	Drop in Value	Upkeep Costs
1	28	$4904	$ 412
2	21	3600	540
3	15	2700	1076
4	11	2000	1372
5	9	1504	1288
6	6	1036	1388
7	4	756	1828
8	3	484	968
9	2	340	1216
10	1	192	308
		$17,516	

(a) How many years does it take for a car to depreciate two-thirds and three-fourths of its value?

(b) If the car is driven 13,500 miles per year, then what are the operating yearly costs in dollars per mile? Plot those operating costs. When is the advantageous time to trade a car assuming that the chief criterion is per mile economy?

(c) Discuss: If the immediate cost of repairing an old car is less than first-year depreciation on a new one, is the best policy to buy a car and drive it until it is ready to be junked?

4.18. Develop direct costing overhead and absorption overhead rates on the basis of direct labor dollars for the following production levels:

Production load	80%	100%	125%
Labor costs	$100,000	$125,000	$156,250
Number of units	20,000	25,000	31,250
Variable overhead cost	$60,000	$75,000	$93,750
Fixed overhead costs	$120,000	$120,000	$120,000

4.19. An assembly area has floor space allocation as follows:

Drop area	300
Conveyor 1	100
Conveyor 2	150
Bench area	800
Total	1350 square feet

Total overhead costs are $8000 for this area. Determine the allocation ratio and overhead costs for each assembly production center.

4.20. A company is composed of five cost centers. Each month a budget is prepared anticipating the primary distribution of certain costs. Let c_w be costs incurred within the cost center such as depreciation and supplies and indirect labor (for producing centers only) and c_m be miscellaneous costs.

Cost Center	c_w	c_m	Area (ft^2)	Direct Labor Dollars	Direct-Labor Hours
Fabrication	$30,000	$1,000	25,000	$20,160	1,600
Assembly	8,000	500	6,000	7,296	640
Finishing	2,000	4,000	1,900	3,744	320
Engineering	4,000	9,000	1,600	—	—
Administration	2,000	1,000	2,100	—	—

Make a secondary distribution for engineering and administration to the producing departments, designated c_o, on the basis of space, dollars, and hours and determine cost center and plant-wide overhead ratios.

4.21. Assume that a job is routed through Machining and Finishing. Machining is heavily mechanized with costly numerical control and other automatic equipment. Finishing has only a few simple tools. Obviously, burden costs are high in Machining and low in Finishing. Job 1 takes 1 labor hour in Machining and 10 hours in Finishing. Job 2 takes 9 labor hours in Machining and only 2 in Finishing. If a single blanket rate based on labor hours is applied to both jobs, then the burden allocation would be the same in both cases (11 hours for each job). What follows illustrates the previous discussion:

	Blanket Rate		Department Rate	
	Machining	Finishing	Machining	Finishing
Budgeted annual burden	$100,000	$ 8,000	$100,000	$ 8,000
Direct labor hours	10,000	10,000	10,000	10,000
Blanket rate per DLH	$5.40			
Department rates per DLH			$10.00	$0.80

Determine the overhead costs for jobs 1 and 2 by using blanket and department rates. Discuss the necessity of selecting the correct base.

4.22. Management is attempting to maintain an overhead rate of 175% for an assembled product. The "problem" area is the machining process. Here, the overhead rate is 225% on the basis of $23,000 direct labor. Total direct dollars excluding machining is $89,000. Overhead charges excluding machining are $150,000. What overhead rate will the company achieve on the basis of this information? Use direct labor dollars as a base.

4.23. Given the following data:

Machine Center	Area	Hours	Horsepower Hours	Depreciation	Tooling	Workers	Direct Labor Budget
Fabrication	25,000	1600	2200	$27,000	$1000	10	$20,165
Assembly	6,000	640	325	7,200	500	4	7,296
Finishing	1,900	320	650	1,800	4000	2	3,744

Item	Amount
Indirect overhead	$16,000
Power	4,200
Indirect labor	4,000
Engineering	13,000
Administration	3,000

(a) Find a plant-wide rate on the basis of direct labor dollars and (b) find the machine center rates on the basis of hours.

4.24. Hercules determines its overhead rate on an annual schedule, and the basis is the direct labor dollar. Some of the facts about Hercules follow:

Cost Center	Shop Area	Hours	Overhead	Direct Labor Dollars
Milling	555 ft^2	520	$7800	$5200
Drilling	145	175	6500	2200

Overhead is exclusive of the depreciation for a brand new milling machine, which cost $10,000 and is depreciated by means of sum of years' digit for 4 years. The material cost for a product is $25. Standards for the shop for the two areas are 2.500 hr/100 units and 1 hr/100 units for milling and drilling for a laser product, respectively. The cost for direct labor in milling is $19.50 per hour and $22.50 per hour for drilling. Find the overhead rate for the two cost centers and the cost for the product for a batch and for one unit for the first year. A batch is 25 units.

MORE DIFFICULT PROBLEMS

4.25. Analyze the following transactions by using the approach given by Tables 4.2, 4.3, 4.5, and 4.6.
 (a) Founded the AJAX supply business, paying $300,000 cash for capital stock.
 (b) Bought inventory from Supplier Caterpillar on credit for the amount of $150,000.
 (c) Paid rent for the month, $4000 using cash.
 (d) Sold four units to M. Meyers for $35,000 cash.
 (e) Paid $50,000 to Caterpillar on account using cash.
 (f) Retired $60,000 of capital stock using cash.

(g) Borrowed $80,000 cash from First National Bank giving note payable.

(h) Sold inventory on account to K. Wilson for $60,000.

(i) Returned $20,000 of inventory to Caterpillar.

(j) Paid salaries and wages, $3000 using cash.

(k) Paid $5000 taxes from cash.

4.26. A manufacturer recently acquired a numerical control long-bed lathe. Owing to the high initial cost and setup charges, the manufacturer is in doubt as to the appropriate depreciation method to apply. The manufacturer considers three methods: straight line, sum-of-the-years' digits, and units of production. For the following data, plot the book value against time and justify your selection as to the manufacturer's best choice. Initial cost, $140,000; installation and programming, $6000; and estimated useful life, 10 years, with 20% salvage value. Production outputs for the 10 years are 600, 800, 1000, 1000, 1200, 1200, 1200, 1100, 1000, and 900, consecutively. Discuss the merits of those allocation schemes.

4.27. (a) Assume the following simple overhead model: $c_b = c_w + c_o + c_m$, where c_b = overhead costs; c_w = costs incurred within department such as depreciation, indirect labor, and supplies; c_o = costs incurred outside the department but allocated to it, such as engineering design, building depreciation, and administration; and c_m = miscellaneous overhead account.

Department	Monthly Costs			Hours	Direct-Labor Dollars
	c_w	c_o	c_m		
Fabrication	$30,000	$7,000	$1,000	1,600	$20,160
Assembly	8,000	5,000	500	640	7,296
Finishing	2,000	4,000	500	320	3,744

Determine departmental overhead rates on the basis of hours and direct labor dollars. Find the overall plant rate on the basis of dollars.

(b) Now a product cost model can be defined as $c_p = c_{dm} + c_e + c_b$, where c_{dm} = direct material cost and c_e = direct labor cost. Let fabrication, assembly, and finish departmental c_e be $6, $2.25, and $0.75, for a unit of product respectively. Also 0.48, 0.20, and 0.064 are standard hour per unit. If c_{dm} = $1 per unit, what is the product unit cost based on departmental overhead rates? If we let c_e = $9 per unit, what is the product cost based on an overall plant rate? Discuss the reasons for the differences in product cost.

4.28. The following problem illustrates variable machine-hour costing. A spread sheet for five production departments is given as follows:

Production Center	Direct Labor	Power	Packaging	Variable Indirect Material	Variable Indirect Overhead	Total
Vertical boring	$ 63,000	$ 6,458	$ 8,253	$ 2,000	$ 4,000	$ 83,711
Light machining	275,500	12,227	36,090	5,000	20,000	348,817
Heavy machining	44,000	20,699	5,764	8,000	3,000	81,463
Bench work	133,900	644	14,920		8,000	137,464
Finishing	66,400	1,877	8,698	3,000	4,000	83,975
Total	$562,800	$41,905	$73,725	$18,000	$39,000	$735,430

In addition to those variable costs there are other variable related labor costs of $20,000 that are unallocated and direct product materials of $462,500. Those product materials are uniform. (a) What is the total out-of-pocket cost or variable cost? (b) Fixed costs are $736,000. What is the full cost of overhead?

CASE STUDY:
MACHINE SHOP

Dennis Schultz is the owner of a modern precision job shop. He is making more estimates in recent years, and the "capture" percentage of estimates won to estimates made is falling. Even worse, he is able to keep his higher priced equipment loaded, but the less expensive equipment is not operating at full capacity. Dennis has said, "I know my direct labor costs are competitive," but he is suspicious of his overhead computation, which is a general plant rate with a basis of direct labor hours. Dennis wants to compute a productive hour rate for his factory. A budget is prepared and shows what follows:

Machine Center	Machine Center Budgeted Hours	Horsepower Hours	Tooling Expenses	Direct-Labor Operators	Gross Hourly Wage	Budgeted Direct-Labor Hours
Lathe	13,328	160,000	$15,000	8	$17.50	15,680
Threading	6,664	66,000	5,000	4	16.00	7,840
Milling	6,664	53,000	12,000	4	21.75	7,840
Drilling	19,992	100,000	8,000	12	16.00	23,520
Bench	26,656	—	—	16	15.75	31,360
Grinding						
Horizontal	3,332	50,000	28,000	2	24.50	3,920
Rotary	1,666	13,000	16,000	1	21.50	1,960
Heat treat	6,664	—	30,000	4	17.50	7,840

Machine Center	Undepreciated Amount/Machine Center	Remaining Depreciable Life, Years	Department Expenses	Utilities and Power	Repairs
Lathe	$ 25,000	5	28,000	$18,000	45,000
Threading	60,000	10	36,000	8,000	21,000
Milling	80,000	8	19,000	6,500	21,000
Drilling	60,000	10	29,000	15,000	15,000
Bench	2,000	5	28,000	1,000	5,000
Grinding					
Horizontal	25,000	10	15,000	6,035	5,000
Rotary	120,000	10	15,000	1,950	15,000
Heat treat	15,000	5	10,000	6,000	8,000

The factory is separate from the engineering and management offices, and Dennis anticipates that those charges will spread at 25% of the productive hour cost per hour. Raw stores, shipping, and inspection are factory overhead charges. Their budget shows what follows:

Indirect Department	Wages	Supplies	Supervision
Raw stores	$120,000	$80,000	$22,000
Shipping	85,000	70,000	26,000
Inspection	210,000	35,000	38,000

The factory layout is shown in Fig. C4.1, and the 200 × 150 ft (60 × 45 m) is equally departmentalized. The undepreciated factory value is $1.6 million, and 20 years of depreciable life remain. Ordinary straight-line methods of depreciation have been used.

(a) Help Dennis by recommending his productive hour cost rate for his machine shop.

(b) Reconsider by applying the accelerated recovery depreciation schedule.

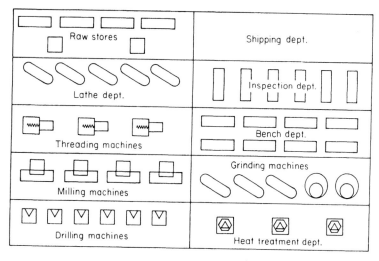

Figure C4.1. Plant layout of Precision Job Shop.

5

Forecasting

Despite all statements to the contrary, "emotional estimating," or the other idiom, "guesstimating," has not disappeared from the engineering scene. Nor has its substitute, estimation by formula and mathematical models, been universally nominated as a replacement. Somewhere between those extremes is a preferred course of action. In this chapter we consider the ways by which cost analysis can be enhanced by graphical and analytical techniques.

Business forecasting comprises the prediction of market demands, prices of material, cost of conversion, availability of labor, and the like. Every major engineering decision is influenced by business forecasts. The usual approach to business forecasting involves extrapolation of past data into the future by using linear or nonlinear curves and mathematical relationships. Future demands are expected to follow some pattern of growth and decay. Most business forecasts are made for the short run period of up to 2 years. Medium-term forecasts cover 2 to 5 years, and long-term forecasts are for more than 5 years.

In this chapter we consider statistical methods, moving averages, and indexing. Our intention is to be basic and to point out special connections to engineering.

Forecasting does not mean estimating. Forecasting is dependence on statistical or mathematical methods and is a small portion of the professional field of cost analysis.

5.1 GRAPHIC ANALYSIS OF DATA

The field of descriptive statistics is concerned with methods for collecting, organizing, analyzing, summarizing, and presenting data and drawing valid conclusions and making reasonable decisions on the basis of analysis of the data. In a more restricted sense, the term *statistics* denotes the data themselves or numbers derived from the data (e.g., averages).

The data gathered for descriptive statistics and graphical presentation may be either discrete or continuous and are usually the result of a series of observations taken over time or another controllable or noncontrollable variable. However, raw data communicate little information. One way to communicate information is to develop a frequency distribution, which compacts the data into manageable proportions.

A *frequency distribution* begins with the collection of observations into a tabular arrangement by intervals. For instance, a market study is made of price for a roll of plastic film, and the number of observations is arranged for the price interval.

Price Interval (dollar/roll)	Number of Observations	Relative Frequency	Cumulative Frequency
12.35–12.75	1	0.003	0.003
12.75–13.15	6	0.019	0.022
13.15–13.55	33	0.102	0.124
13.55–13.95	51	0.157	0.281
13.95–14.35	121	0.373	0.654
14.35–14.75	50	0.154	0.808
14.75–15.15	44	0.136	0.944
15.15–15.65	13	0.040	0.984
15.65–16.05	5	0.016	1.000
Total	324	1.000	

The price range of 12.35–12.75 is called an *interval*, and the end numbers are called *limits*. The size is 0.40 (= 12.75 − 12.35). The first *midpoint* is 12.55 [=(12.35 + 12.75)/2]. *Relative frequency* is the number of observations for each interval divided by all observations. Graphical representations of relative frequency distributions are called *histograms*. Figure. 5.1, which plots percent observations against price (dollar/roll), is a histogram. If the relative frequencies are consecutively summed, then a *cumulative frequency* results. That is shown by the dashed line and right scale of Fig. 5.1. If those data are considered representative of the parent population, then the term *cumulative probability of occurrence* is used. If the midpoints of the histogram cells are joined by a line, and smoothed, the frequency curves will appear as in Fig. 5.2.

Figure 5.2 presents several types of frequency curves obtainable from analysis of data. A common type of title for the *y* axis would be percentage of observations or count. Those graphical plots can be constructed similar to Fig. 5.1.

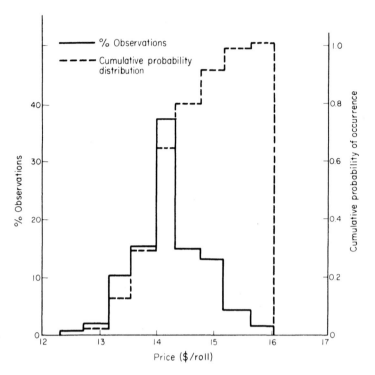

Figure 5.1. Relative frequency curve and cumulative probability of occurrence curve for data.

Curve construction calls for a trained eye. For instance, if a wealth of data exists, then the bimodal or multimodal plot may be evident. But in the absence of an abundance of data, graphical conclusions of this sort are seldom found. Mathematical analyses are preferred for this and for other reasons. After the graphical plot is concluded, we calculate a *measure of central tendency,* such as mean, median, or mode, and a *measure of dispersion,* such as the standard deviation or range.

An *average* is typical or representative of a set of data, and is sometimes called the *arithmetic mean,* and is found as

$$\bar{x} = \frac{x_1 + x_2 + \cdots + x_n}{n} = \frac{\sum\limits_{i=1}^{n} x_i}{n} = \frac{\Sigma x}{n} \tag{5.1}$$

The *median* of data arranged in order of magnitude is the middle value for an odd number of data, or the mean of the two middle values if the set number is even. The *mode* of data is that value that occurs with greatest frequency. The mode may not exist or may not be unique. The set, $-1, 0, 2, 4, 6, 6, 7$, and 8, has the mean 4, median 5, and mode 6. The set, $-1, 0, 2$, and 4, has no mode. The set, $-1, 0, 0, 4, 6, 6, 7$, and 8, is bimodal with the values 0 and 6.

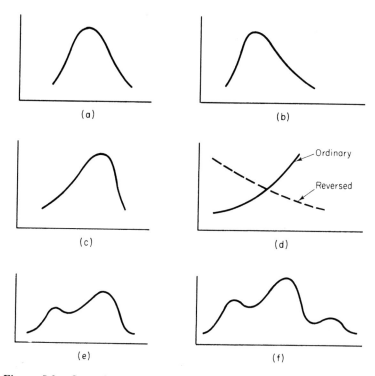

Figure 5.2.   Several types of frequency curves obtainable from analysis of data: (a) symmetrical; (b) skewed to the right; (c) skewed to the left; (d) J shaped; (e) bimodal; (f) multimodal.

The degree to which numerical data tend to spread about a mean value is called the dispersion of the data. Various measures of dispersion are available, the most common being the range and standard deviation. The *range* of a set of data is the difference between the largest and smallest numbers in a set. The *standard deviation* of a set of n numbers, $x_1, \cdots, x_n$, is denoted by s and is defined by

$$s = \sqrt{\frac{\Sigma(x_i - \bar{x})^2}{n - 1}} \tag{5.2}$$

The standard deviation is determined from the mean of the sample. The variance of a set of data is defined as the square of the standard deviation or s^2. The set $(-1, 0, 2, 4)$ has the range of 5, $s = 2.22$, and $s^2 = 4.92$.

Plotting of data is an important step in graphical analysis. Despite the ease with which data can be mathematically and statistically analyzed, plotting of two variable data, y versus x, for example, is useful for the engineer to gain a "feeling" from the actual data. But mathematical relationships void of a visual or real-life experience can lead to misjudgment in estimating. Knowledge, skill, or practice derived from direct observation of or participation in the statistical analysis helps to develop judgment that is so necessary for estimating.

A graph is a pictorial presentation of the relationship between variables. Most graphs are two variables, where we presume that x is the independent or controlled variable and y is the dependent variable. Once raw data are gathered, the next steps select the axes and divisions, locate the points, and draw the straight line. A rule of thumb to follow in plotting by eye is to have half of the points above the line and half of the points below the line, excluding those that lie on the line. With the line drawn we find the graphical straight line equation by using

$$y = a + bx \qquad (5.3)$$

where y = dependent variable

a = intercept value at $x = 0$

b = slope, rise/run

x = independent or control variable

Measurements are taken off the graph because the intercept and slope can be determined from the line drawn through the points. This step, frequently overlooked in cost analysis, improves the gathering of experience that is necessary in engineering.

Both arithmetic and logarithmic axes are popular in cost analysis and are described by Fig. 5.3. Trial plots should attempt different groupings of the axes, such as semilog, or log log in addition to arithmetic axis. That plot that "straightens out" scattered data is considered best. In Fig. 5.4(a), the axes are arithmetic, and slope b is calculated from the right triangle. The line is extended to the y axis at $x = 0$, and at this point the value a is measured from the graph.

The logarithmic copy of Eq. 5.3 is given by Fig. 5.4(b). The slope is found by using the points (x, y) and (x_1, y_1) after the line is plotted by using the rule of thumb stated.

Figure 5.5 is an example of a practical linear plot. Note that the points are left unerased on the graph and indicates the typical variability that may exist with time or cost data. At the nominal 10-in. (250-mm) pipe size, the variability of man hours

Arithmetic scale

| 0 | 1 | 2 | 3 | 4 | 5 | 6 | 7 | 8 | 9 | 10 |

Log scales   Cycle

1	2	3	4	6	8	10	First
10	20	30	40	60	80	100	Second
100	200	...		...	1000	Third
1000	2000	...		...	10,000	Fourth

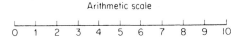

| Cycle 1 | Cycle 2 | Cycle 3 | Cycle 4 |
| 1 | 10 | 100 | 1000 | 10,000 |

Figure 5.3. Arithmetic and logarithmic scales compared.

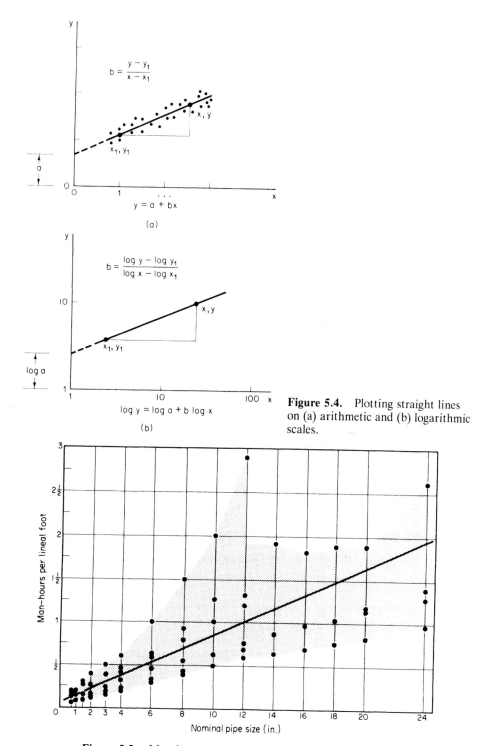

$$b = \frac{y - y_1}{x - x_1}$$

x_1, y_1

x, y

a

$y = a + bx$

(a)

$$b = \frac{\log y - \log y_1}{\log x - \log x_1}$$

x, y

x_1, y_1

$\log a$

$\log y = \log a + b \log x$

(b)

Figure 5.4. Plotting straight lines on (a) arithmetic and (b) logarithmic scales.

Man-hours per lineal foot

Nominal pipe size (in.)

Figure 5.5. Man hours to handle and install standard pipe.

145

is four times from a minimum to maximum, not an unusual event for cost data. Frequently, a graphical line may show a sharp slope change that is not evident in mathematical analysis. Graphical plotting will show this knee jerk in data.

The simple formula $y = a + bx$ is often sufficient for practical day-to-day work, but when the line appears nonlinear or is logarithmic straight, curves other than those by using arithmetic scales are used. Attention turns to mathematical analysis after the plot and the graphical equation are concluded. Graphical plots and equations will vary between two engineers, but their mathematical equations should be identical. The graphical plots select the type of straight line, and the method of least squares analyzes data for their equation.

5.2 LEAST SQUARES AND REGRESSION

The descriptive statistical methods described so far are concerned with a single variable and its frequency fluctuation. However, many problems in the estimating field involve several variables. Simple methods for dealing with data associated with two or three variables will be explained here. Those methods are known as regression models and are important tools.

In *regression*, on the basis of sample data, the engineer wants to know the value of a dependent variable (y) corresponding to a given value of a variable (x). This is determined from a least squares curve that fits the sample data. The resulting curve is called a regression curve of y on x because y is determined from a corresponding value of x. If the variable x is time, then the data show the values of y at various times and are known as a time series. Regression lines or a curve y on x or the response function on time is frequently called a *trend line* and is used for the purposes of prediction and forecasting. Thus, regression refers to the average relationship between variables.

5.2.1. Least Squares

The notion of fitting a curve to a set of sufficient points is essentially the problem of finding the parameters of the curve. The best known method is the method of *least squares*. Since the desired curve or equation is to be used for estimating or predicting purposes, the curve or equation should be so modeled as to make the errors of estimation small. An error of estimation means the difference between an observed value and the corresponding fitted curve value for the specific value of x. It will not do to require the sum of these differences or errors to be as small as possible. It is a requirement that the sum of the absolute values of the errors be as small as possible. However, sums of absolute values are not convenient mathematically. The cause of the difficulty is avoided by requiring that the sum of the square of the errors be minimized. If this procedure is followed, the values of parameters give what is known as the best curve in the sense of least squares difference.

The principle of least squares states that if y is a linear function of an independent variable x, then the most probable position of line $y = a + bx$ will exist whenever the sum of squares of deviations of all points (x_i, y_i) from the line is a minimum. Those deviations are measured in the direction of the y axis. The underlying assumption is that x is either free of error (a controlled assignment) or subject to negligible error. The value of y is the observed or measured quantity, subject to errors that have to be "eliminated" by this method of least squares. The value y is a random variable value from the y population values corresponding to a given x. For each value x_i we are interested in corresponding y_i. Suppose that our observations consist of pairs of values as x_i, y_i, which are assumed to give a linear plot as in Fig. 5.6. The symbol $\bar{y}_i$ ("y-bar") is the average value resulting from the x_i controlled variable. For instance, if the value of $x_i = 5$ and the experiment were repeated 10 times, $\bar{y}_i$ would be the mean value of 10 observations. If a very large number of experiments were made, a histogram and eventually a normal curve, or bell-shaped curve, could be constructed, as is evident in Fig. 5.6. The bell-shaped curve is positioned with respect to the y axis value and is considered a random variable. The area under each normal curve is 1. Our problem is to uncover a and b for a best fit. For a general point i on the line, $y_i - (a + bx_i) = 0$, but if an error ϵ_i exists, then $y_i - (a + bx_i) = \epsilon_i$. For n observations we have n equations of

$$y_i - (a + bx_i) = \epsilon_i \tag{5.4}$$

where ϵ_i is the difference between actual observation and regression value of Eq. (5.4). Summing, we can write the sum of squares of these residuals as

$$\sum_{i=1}^{n} \epsilon^2 = \sum_{i=1}^{n} [y_i - (a + bx_i)]^2 \tag{5.5}$$

For a minimum, we insist on

$$\frac{\partial \sum \epsilon^2}{\partial a} = 0, \qquad \frac{\partial \sum \epsilon^2}{\partial b} = 0 \tag{5.6}$$

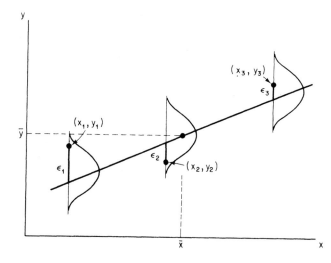

Figure 5.6. Regression line of $y = a + bx$.

or

$$n\Sigma xy - \Sigma x\Sigma y = n\Sigma(x - \bar{x})(y - \bar{y})$$
$$= n(\Sigma xy - \bar{x}\Sigma y + \bar{x}\Sigma y - \bar{y}\Sigma x) \tag{5.7}$$

and solving those two normal equations for a and b we have

$$a = \frac{\Sigma x^2\Sigma y - \Sigma x\Sigma xy}{n\Sigma x^2 - (\Sigma x)^2}$$

and   (5.8)

$$b = \frac{n\Sigma xy - \Sigma x\Sigma y}{n\Sigma x^2 - (\Sigma x)^2}$$

Those values would then be appropriately substituted into $y = a + bx$. This least squares equation passes through $(\bar{x}, \bar{y})$, which is the coordinate mean of all observations.

The calculation of those coefficients may be handled with an electronic calculator or by a computer. Table 5.1 is an example of a tabular form in determining those values. If Y = the index and X = the coded year, the equation of the least squares line is $Y = 84.875 + 2.389 X$. The four left columns of Table 5.1 are necessary for regression. The columns x and y are the data, and x^2 and xy are required computation. Equation (5.8) uses the total of those columns for finding constants a and b.

The limitations of the method must be pointed out. The method of least squares is applicable when the observed values of y_i correspond to assigned (or error free) values of x_i. The error in y_i (expressed as a variance of y) is assumed to be independent of the level of x. If inferences are to be made about regression, it is also necessary that the values of y_i corresponding to a given x_i be distributed normally, as is presented on Fig. 5.6 with the mean of the distribution satisfying the regression equation. The variance of the values of y_i for any given value of x must be independent of the magnitude of x. Though the evidence of this statement can be statistically shown, experience shows that only a comparatively small number of the distributions met within engineering can be described by normal distributions. Cost data are limited at the zero end. Distributions influenced by business, engineering, and human factors are generally skewed. Despite those drawbacks, the least squares method is widely used. Imperfection, apparently, does not reduce popularity.

5.2.2. Confidence Limits for Regression Values and Prediction Limits for Individual Values

If variations around the universe regression are random, then the method of least squares permits the computation of sampling errors and provides for the determination of the reliability of the estimate of the dependent variable from the fitted line. Confidence limits for regression values can be constructed through the extension of simple statistics. The confidence limits for individual regression values and for the

TABLE 5.1. CALCULATION OF REGRESSION COEFFICIENTS

Year, x	Index, y	x^2	xy	$\hat{y}$	$\epsilon = y - \hat{y}$	ϵ^2
0	87	0	0	84.875	2.125	4.516
1	89	1	89	87.264	1.736	3.014
2	90	4	180	89.653	0.347	0.120
3	92	9	276	92.042	−0.042	0.002
4	93	16	372	94.431	−1.431	2.047
5	99	25	495	96.820	2.180	4.752
6	97	36	582	99.209	−2.209	4.879
7	100	49	700	101.598	−1.598	2.554
8	101	64	808	103.987	−2.987	8.922
9	106	81	954	106.376	−0.376	0.141
10	106	100	1,060	108.765	−2.765	7.645
11	109	121	1,199	111.154	−2.154	4.640
12	115	144	1,380	113.543	1.457	2.123
13	118	169	1,534	115.932	2.068	4.277
14	122	196	1,708	118.321	3.679	13.535
105	1524	1015	11,337			63.167

For $Y = a + bx$, the constants

$$a = \frac{\Sigma y \Sigma x^2 - \Sigma x \Sigma xy}{n \Sigma x^2 - (\Sigma x)^2} = \frac{(1524)(1015) - (105)(11{,}337)}{15(1015) - (105)^2} = 84.875$$

$$b = \frac{n \Sigma xy - \Sigma x \Sigma y}{n \Sigma x^2 - (\Sigma x)^2} = \frac{15(11{,}337) - (105)(1524)}{15(1015) - (105)^2} = 2.389$$

$\hat{Y} = 84.875 + 2.389 X$ or if X is year, then index $= 84.875 + 2.389 X$. The forecast mean value for year $x_i = 15$ is $Y = 84.875 + 2.389(15) = 120.71$.

$$S_y = \left(\frac{\Sigma \epsilon^2}{n - 2}\right)^{1/2} = \left(\frac{63.167}{13}\right)^{1/2} = 2.204$$

$$S_{y_i} = S_y \left[\frac{1}{n} + \frac{(x_i - \bar{x})^2}{\Sigma(x - \bar{x})^2}\right]^{1/2} = 2.204 \left[\frac{1}{15} + \frac{(15 - 7)^2}{280}\right]^{1/2} = 1.198$$

For degrees of freedom $= 13$ and a 5% level of significance, $t = 2.160$. The confidence interval is $120.71 \pm 2.160(1.198) = (118.122, 123.298)$. For a single estimated value y_i for $x_i = 15$,

$$S_{y_i} = S_y \left[1 + \frac{1}{n} + \frac{(x_i - \bar{x})^2}{\Sigma(x - \bar{x})^2}\right]^{1/2} = 2.204 \left(1 + \frac{1}{15} + \frac{64}{280}\right)^{1/2} = 2.508$$

The prediction interval is $120.71 \pm 2.160(2.508) = (115.293, 126.127)$. The confidence interval for the slope and intercept is

$$S_b = \frac{S_y}{[\Sigma(x - \bar{x})^2]^{1/2}} = \frac{2.204}{(280)^{1/2}} = 0.132$$

$$\text{slope interval} = 2.389 \pm 2.160(0.132) = (2.103, 2.674)$$

$$S_a = S_y \left[\frac{1}{n} + \frac{x^2}{\Sigma(x - \bar{x})^2}\right]^{1/2} = 2.204 \left(\frac{1}{15} + \frac{49}{280}\right)^{1/2} = 1.083$$

$$\text{intercept interval} = 84.875 \pm 2.160(1.083) = (82.536, 87.214)$$

straight line are quadratic in form around the sample line of regression. The confidence band for the slope is fan shaped with the apex at the mean and the confidence limits for the intercept are parallel lines.

The variance of an estimate permits the forming of confidence limits of the estimate. The approach, similar to the variance of a sample, in this case reckons the deviations from a line instead of a mean. The variance of y, estimated by the regression line, is the sum of squares of deviations divided by the number of degrees of freedom available for calculating the regression line, or

$$s_y^2 = \frac{\Sigma \epsilon_i^2}{\nu} \tag{5.9}$$

where s^2 is the variance around the regression line and ϵ_i is as defined previously by Eq. (5.5). Only two bits of information are required to determine the regression line: means $(\bar{x}, \bar{y})$ and either slope b or intercept a. With n as the number of paired observations, ν is defined as

$$\nu = n - 2 \tag{5.10}$$

where ν represents the degrees of freedom. Also,

$$s_y^2 = \frac{\Sigma \epsilon_i^2}{n - 2} \tag{5.11}$$

Defining

$$s_{\bar{y}}^2 = \frac{s_y^2}{n} \tag{5.12}$$

where $s_{\bar{y}}^2$ is the variance of the mean value of y or $\bar{y}$.

It is now possible to write the confidence limits for $\bar{y}$. A table of the Student t distribution and of the values of t corresponding to various values of the probability (level of significance) and a given number of degrees of freedom ν is found in appendix 2. With this t value we state that the true value of $\bar{y}$ lies within the interval

$$\bar{y} \pm t s_{\bar{y}} \tag{5.13}$$

The probability of being wrong is equal to the level of significance of the value of t. Because the regression line must pass through the mean, an error in the value of $\bar{y}$ leads to a constant error in y for all points on the line. The line is then moved up or down without change in slope.

If limits for an individual value are desired, then a different approach must be asserted. The statement that usually describes the limits for individual values goes like this: If we use the sample line of regression to estimate a particular value for y, then we add to the error of the sample line of the regression some measure of the possible deviation of the individual value from the regression value. For individual values a new set of parabolic loci may be viewed as prediction limits. Figure 5.7 presents the prediction loci for individual values as well as the confidence loci for aver-

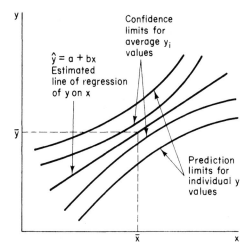

Figure 5.7. Confidence bands for average values and prediction limits for individual values.

age values. Note that the prediction limits for *y* get wider as *x* deviates from its mean, both positively and negatively. This means that predictions of the dependent variable are subject to the least error when the independent variable is near its mean and are subject to the greatest error when the independent variable is distant from its mean.

If we require an estimate of the confidence limits corresponding to any x_i, we calculate the limits for $\hat{y}_i$ (strictly speaking, an estimate). The variance of the estimate of this mean value is

$$s_{\bar{y}_i}^2 = s_y^2 \left[\frac{1}{n} + \frac{(x_i - \bar{x})^2}{\Sigma(x - \bar{x})^2} \right] \tag{5.14}$$

where $s_{\bar{y}_i}^2$ denotes the variance of each mean value of *y*. A new $s_{\bar{y}_i}^2$ is computed for each *x* value. The confidence interval for the mean estimated value of y_i corresponding to specific x_i is

$$y_i \pm t s_{\bar{y}_i} \tag{5.15}$$

The interior confidence limits of Fig. 5.7 correspond to Eq. (5.15).

For a predetermined level of significance we are able to predict the limits within which a future *mean estimated value* of y_i will lie with an appropriate chance of error.

We can find the prediction interval of a *single* estimated value of $\hat{y}_i$ using the variance of a single value, which has as its variance

$$s_{y_i}^2 = s_y^2 \left[1 + \frac{1}{n} + \frac{(x_1 - \bar{x})^2}{\Sigma(x - \bar{x})^2} \right] \tag{5.16}$$

where $s_{y_i}^2$ is the variance of individual value of *y*. This variance is larger than s_y^2 because the variance of the single value is equal to the variance of the mean plus the variance of $\hat{y}$ estimated by the line or

$$s_{y_i}^2 = s_{\bar{y}_i}^2 + s_y^2 \tag{5.17}$$

Each x requires a separate value of $s_{y_i}^2$. The prediction interval for a single value is greater, too, or

$$y_i \pm t s_{y_i} \tag{5.18}$$

In Fig. 5.7 the external lines are computed by using Eq. (5.18). The terms *confidence* and *prediction interval* have different meanings. A confidence interval deals with an expected *average* Y value. A prediction interval deals with a *single* Y value. The prediction interval is greater in magnitude.

The variance of the intercept a is a particular case of the variance of any mean estimated y_{y_i}. If we substitute $x_i = 0$ in Eq. (5.14) then the variance of intercept a is

$$s_a^2 = s_y^2 \left[\frac{1}{n} + \frac{\bar{x}^2}{\Sigma(x - \bar{x})^2} \right] \tag{5.19}$$

Its confidence band is given by

$$a \pm t s_a \tag{5.20}$$

Note Fig. 5.8(a), which corresponds to Eq. (5.19).

The variance of slope b is given as

$$s_b^2 = \frac{s_y^2}{\Sigma(x - \bar{x})^2} \tag{5.21}$$

Its confidence band is given by

$$b \pm t s_b \tag{5.22}$$

Note Fig. 5.8(b), which corresponds to Eq. (5.21).

The confidence band for the slope is represented by a double fan-shaped area with the apex at the mean.

The number of degrees of freedom is $\nu = n - 2$ for all confidence intervals in this section.

Table 5.1 demonstrates the manual calculation of regression values, namely, the parameters a and b. The $\hat{y}$ column is calculated by using the regression equation

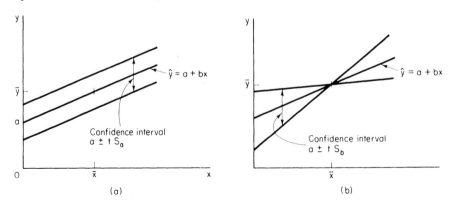

Figure 5.8. (a) Confidence interval for intercept a; (b) confidence interval for slope b.

$\hat{y} = 84.875 + 2.389x$. For example, if $x = 1$, then $\hat{y} = 87.264$. This calculation is repeated for each observation of x. The $\epsilon = y - \hat{y}$ column is the error difference between the observed and regression value. The right column is the square of the error. The sum 63.167 is used to calculate s_y. Confidence and prediction intervals are found for $x = 15$ and are given in Table 5.1.

5.2.3. Curvilinear Regression and Transformation

The world of linearity, largely an imaginary one, is a tidy and manageable world about which generalizations can be asserted with confidence. Many prefabricated tools are known that can be used with those assumptions. The function $y = a + bx$ is a characterization frequently employed for known nonlinear situations. Computers and a greater awareness are forcing a reevaluation of linearities. A listing for engineering purposes would show the following nonlinear relationships:

$$y = ae^{bx} \qquad \text{semilog} \qquad (5.23)$$

$$y = ab^x \qquad \text{exponential} \qquad (5.24)$$

$$y = ax^b \qquad \text{power} \qquad (5.25)$$

$$y = a + \frac{b}{x} \qquad \text{reciprocal } x \qquad (5.26)$$

$$y = \frac{1}{(a + bx)} \qquad \text{reciprocal } y \qquad (5.27)$$

$$y = \frac{x}{(a + bx)} \qquad \text{hyperbolic} \qquad (5.28)$$

$$y = a + b_1x + b_2x^2 + \cdots \qquad \text{polynomial} \qquad (5.29)$$

The exponential and power fit, Eqs. (5.24) and (5.25), are used frequently. The regression formulas of section 5.1 can be made to work with those curvilinear models. First, those equations appear like $y = a + bx$ if transformed. The exponential function $y = ab^x$ transforms to

$$\log y = \log a + x \log b$$

The three exponential curves $y = ae^{bx}$, $y = 10^{a+bx}$, and $y = a(10)^{bx}$ are equivalent. If any two points are chosen and the constants for each curve determined so it passes through these points, then the curves are equal.

Let $y = \log y$, $a = \log a$, $b = \log b$, and $x = x$. This plots as a straight line when plotted on arithmetic-logarithmic scales. Similar to Eq. (5.8), the values of the parameters of Eq. (5.24) can be solved by using

$$\log a = \frac{\Sigma x^2 \Sigma \log y - \Sigma x \Sigma x \log y}{n\Sigma x^2 - (\Sigma x)^2}$$

$$\log b = \frac{n\Sigma x \log y - \Sigma x \Sigma \log y}{n\Sigma x^2 - (\Sigma x)^2}$$

$$(5.30)$$

The power function $y = ax^b$ is transformed from a curved line on arithmetic scales to a straight line on log-log scales if we let $y = \log y$, $a = \log a$, and $x = \log x$. Intercept and slope equations can be found by making those substitutions into Eq. (5.8).

$$\log a = \frac{\Sigma(\log x)^2 \Sigma \log y - \Sigma \log x \Sigma(\log x \log y)}{n\Sigma(\log x)^2 - (\Sigma \log x)^2}$$

$$b = \frac{n\Sigma(\log x \log y) - \Sigma \log x \Sigma \log y}{n\Sigma(\log x)^2 - (\Sigma \log x)^2}$$

(5.31)

The power equation is used to model the learning concept. If we repeat the production of a product, then it takes less time to manufacture subsequent products, which is known as learning.

Assume that we have five sets of experiences, or the unit number x and the man hours y, to make that unit. Data are (10, 510), (30, 210), (100, 190), (150, 125), and (300, 71). Thus, the 10th and 300th units require 510 and 71 hours, respectively. Table 5.2 shows the calculation. The first two columns are the original units, and the four right columns are the transformed calculations. The transformed and original unit equations also give identical results. Let $x = 500$th unit; then,

$$\log y = 3.1921 - 0.515 \log 500 = 1.8021$$

$$\text{antilog } y = 63.4 \text{ hours}$$

Also

$$y = 1556(500)^{-0.515} = 63.4 \text{ hours in original units}$$

which shows the agreement.

Each of those functions can be statistically evaluated for the fitted data, such as

Coefficients of the equations

Listing of calculated estimates

Standard error of y, $\Sigma \varepsilon_i^2 = (y_i - \hat{y})^2$

Squared correlation coefficient

95% confidence limits values for y

Computer programs do this calculation where the work is large.

Normally, the best fit line is judged by the smallest value of the standard error. Other computed values provide an intuitive feel for the accuracy of the correlation. Polynomial regression, that is, where for any x the mean of the distribution of y is given by $a + b_1x^2 + b_2x^2 + b_3x^3 + \cdots + b_px^p$ is used to obtain approximations whenever the functional form of the regression curve is a mystery. Engineers plot data a variety of ways hoping to "straighten out" an arithmetic curve by means of semilog or log-log plots. If sets of paired data straighten out on semilog paper, the engineer would conclude that the form is exponential or $y = ae^{bx}$ and would be tempted to apply least squares methods that used this form to the data.

TABLE 5.2. LEAST SQUARES ANALYSIS OF ACTUAL COST DATA TO FIND INITIAL VALUE AND LEARNING SLOPE

Given: $y = ax^b$

or

$$\log y = \log a + b \log x$$

which is of the form

$$y = a + bx$$

Let $y = \log y$, $a = \log a$ intercept, $b = b$ slope, $x = \log x$, and $n =$ sample size.

Unit, x	Man Hours, y	$x = \log x$	$y = \log y$	$(\log x)^2$	$\log x \log y$
10	510	1.0000	2.7076	1.0000	2.7076
30	210	1.4771	2.3222	2.1818	3.4301
100	190	2.0000	2.2788	4.0000	4.5576
150	125	2.1761	2.0969	4.7354	4.5631
300	71	2.4771	1.8513	6.1360	4.5859
		9.1303	11.2568	18.0532	19.8443

$$\log a = \frac{\Sigma(\log x)^2 \Sigma \log y - \Sigma \log x \Sigma(\log x \log y)}{n\Sigma(\log x)^2 - (\Sigma \log x)^2}$$

$$\log a = \frac{(18.0532)(11.2568) - (9.1303)(19.8443)}{5(18.0532) - (9.1303)^2} = 3.1921$$

$$b = \frac{n\Sigma(\log x \log y) - \Sigma \log x \Sigma \log y}{n\Sigma(\log x)^2 - (\Sigma \log x)^2}$$

$$= \frac{5(19.8443) - (9.1303)(11.2568)}{5(18.0532) - (9.1303)^2} = -0.515$$

Then

$$\log y = 3.1921 - 0.515 \log x \text{ (in logarithm units)}$$

antilog $a = 1556$ (initial value at $x = 1$)

$$y = 1556x^{-0.515} \text{ in original units}$$

What is the estimate for $x = 350$ units?
In logarithm units:
$$\log y = 3.1921 - 0.515 \log 350 = 1.8819$$
antilog $y = 76.19$ hours

In original units:

$$y = 1556(350)^{-0.515} = 76.19 \text{ hours}$$

Suppose that coded data were obtained as follows, where y, or dollars, was assumed to be related to small gas engine horsepower as $y = a + bx^2$.

x	Observed Value, y_i	Error-Free Value, y	Deviation $y_i - y$
0	1	a	$1 - a$
1	1.4	$a + b$	$1.4 - (a + b)$
2	1.8	$a + 4b$	$1.8 - (a + 4b)$
3	2.2	$a + 9b$	$2.2 - (a + 9b)$

The sum of the squares of the deviations is

$$\Sigma\varepsilon^2 = (1 - a)^2 + (1.4 - a - b)^2 + (1.8 - a - 4b)^2 + (2.2 - a - 9b^2)$$

For a minimum we are required to satisfy

$$\frac{\partial\Sigma\varepsilon^2}{\partial a} = 0 \quad \text{and} \quad \frac{\partial\Sigma\varepsilon^2}{\partial b} = 0 \tag{5.32}$$

or $(1 - a) + (1.4 - a - b) + (1.8 - a - 4b) + (2.2 - a - 9b) = 0$ and $(1.4 - a - b) + 4(1.8 - a - 4b) + 9(2.2 - a - 9b) = 0$. This reduces to the normal equations

$$4a + 14b = 6.4$$

$$14a + 98b = 28.4$$

Hence, $a = 1.175$ and $b = 0.122$. When substituted back into the general form, the fitted equation becomes $y = 1.175 + 0.122x^2$. A better fit is obtained from an equation of a different form or $y = a + bx + cx^2$.

The previous problem is straightforward, but the application of the method of least squares to nonlinear relations usually requires a good deal of computational effort, computers notwithstanding. In most cases we can transform or rectify a nonlinear relation to a straight line relation. This manipulation simplifies handling of the data and permits a graphical presentation that may be revealing for certain facts. With a rectified straight line, extrapolation is simpler, and the computation of certain other supportive statistics, such as the standard deviation or confidence limits, is simpler.

Now, those previous functions indicated a regression of y on x that was linear in some fashion. Sometimes a clear relationship is not evident, and a general polynomial is selected. A predicting equation of the polynomial form [see Eq. (5.29)] requires a set of data consisting of n points (x_i, y_i). We estimate the coefficients $a, b_1, b_2, \cdots, b_p$ of the pth degree polynomial by minimizing

$$\sum_{i=1}^{n} [y_i - (a + b_1x + b_2x^2 + \cdots + b_px^p)]^2 \tag{5.33}$$

according to

$$\frac{\partial \Sigma \varepsilon^2}{\partial a}, \frac{\partial \Sigma \varepsilon^2}{\partial b_1}, \frac{\partial \Sigma \varepsilon^2}{\partial b_2}, \cdots, \frac{\partial \Sigma \varepsilon^2}{\partial b_p} = 0 \qquad (5.34)$$

which is the least squares criterion by minimizing the sum of squares of the vertical distances from the points to the curve. This results in $p + 1$ normal equations of the shape.

$$\Sigma y = na + b_1 \Sigma x + \cdots + b_p \Sigma x^p$$

$$\Sigma xy = a\Sigma x + b_1 \Sigma x^2 + \cdots + b_p \Sigma x^{p+1} \qquad (5.35)$$

$$\vdots$$

$$\Sigma x^p y = a\Sigma x^p + b_1 \Sigma x^{p+1} + \cdots + b_p \Sigma x^{2p}$$

where summation notation has been eliminated. This is now $p + 1$ linear equations in $p + 1$ unknowns $a, b_1, \cdots, b_p$.

Consider the following example: An aircraft manufacturer has accumulated data on the chem-milling operation of steel panels, and paired sets of data are dollars per square foot, depth of cut, inches—$(0.01, 7.0)$, $(0.02, 8.4)$, $(0.03, 9.2)$, $(0.04, 10.1)$, $(0.05, 10.3)$ $(0.20, 26.2)$—and resultant tabulations, very similar to previous least squares calculations in Table 5.1, reveal $\Sigma x = 0.35$, $\Sigma x^2 = 0.0455$, $\Sigma x^3 = 0.008225$, $\Sigma x^4 = 0.00016979$, $\Sigma y = 71.2$, $\Sigma xy = 6.673$, and $\Sigma x^2 y = 11.0225$. With those values we are ready to solve the following system of three linear equations,

$$6a + 0.35b_1 + 0.0455b_2 = 71.2$$

$$0.35a + 0.0455b_1 + 0.008225b_2 = 6.673$$

$$0.0455a + 0.008225b_1 + 0.0016979b_2 = 11.0225$$

for which $a = -31.73$, $b_1 = 1660$, and $b_2 = -7025$ and the predicting equation becomes

$$y = -31.73 + 1660x - 7025x^2$$

In practice it may be difficult to determine the degree of the polynomial to fit data, but it is always possible to find a polynomial of degree at most $n - 1$ that will pass through each of n points, although we want the lowest degree that describes our problem.

5.2.4. Correlation

The method of regression shows the relationship to one independent variable that can be considered linearly related. There is a closely related measure, called correlation, that tells how well the variables are satisfied by a linear relationship. If the values of the variables satisfy an equation exactly, then the variables are perfectly correlated. When two variables are involved, the statistician refers to simple correlation and simple regression. When more than two variables are involved, it is referred to as multiple regression and multiple correlation. In this section, only simple correlation is considered.

Figure 5.9 indicates the location of points on an arithmetic coordinate system. If all the points in a scatter diagram appear to lie near a line as in Fig. 5.9(b) or 5.9(d), the correlation is presumed to be linear, and a linear equation is appropriate for regression or estimation. If there is no relationship indicated between the variables as in Fig. 5.9(c), then there is no correlation (i.e., the data are uncorrelated). In Fig. 5.9(b), the correlation coefficient is negative linear. For Fig. 5.9(d) a positive linear correlation coefficient is found.

With a fitted curve from data it is possible to distinguish between the deviations of the y observations from the regression line and the total variation of the y observations about their mean. A calculated difference between the two variations gives the amount of variation accounted for by regression, and, the higher this value, the better the fit or correlation. For $y = a + bx$, no correlation exists if $b = 0$ as in Fig. 5.9(c) and the line plots as a horizontal line. Thus, x and y are independent.

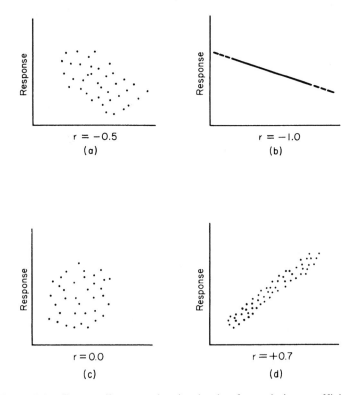

Figure 5.9. Scatter diagrams showing levels of correlation coefficients: (a) negative linear correlation; (b) negative linear correlation; (c) no correlation; (d) positive linear correlation.

There is no correlation of x on y if x is independent of y. Those two statements are mathematically

$$\frac{n\Sigma xy - \Sigma x \Sigma y}{n\Sigma x^2 - (\Sigma x)^2} = 0 \tag{5.36}$$

and

$$\frac{n\Sigma xy - \Sigma x \Sigma y}{n\Sigma y^2 - (\Sigma y)^2} = 0 \tag{5.37}$$

For no correlation the product of Eqs. (5.36) and (5.37), or the product of the slopes, is zero. Conversely, for perfect correlation all the points lie exactly on each of the two regression lines and their lines coincide, or

$$\frac{n\Sigma xy - \Sigma x \Sigma y}{n\Sigma x^2 - (\Sigma x)^2} \times \frac{n\Sigma xy - \Sigma x \Sigma y}{n\Sigma y^2 - (\Sigma y)^2} = 1 \tag{5.38}$$

The square root of this product is called the *correlation coefficient,* and it is denoted by r:

$$r = \frac{n\Sigma xy - \Sigma x \Sigma y}{\{[n\Sigma x^2 - (\Sigma x)^2][n\Sigma y^2 - (\Sigma y)^2]\}^{1/2}} \tag{5.39}$$

which is a computation equation form. Correlation is concerned only with the association between variables, and r must lie in the range $0 \le |r| \le 1$. By using the data from Table 5.1 the correlation coefficient may be found as an example. Recollecting that $n = 15$, $\Sigma xy = 11{,}337$, $\Sigma x = 105$, $\Sigma y = 1524$, $\Sigma x^2 = 1015$, and $\Sigma y^2 = 156{,}500$, then

$$r = \frac{(15)(11{,}337) - (105)(1524)}{\{[(15)(1015) - (105)^2][15(156{,}500) - (1524)^2]\}^{1/2}} = +0.98$$

which is a high score for correlation. The magnitude of the correlation coefficient r determines the strength of the relationship, and the sign of r tells us whether the dependent variable tends to increase or decrease with the independent variable. The engineer may realize that r is a useful measure of the strength of the relationship between two variables, only if, however, the variables are linearly related. The value of r will be equal to $+1$ or -1 if and only if the points of the scatter lie perfectly on the straight line, which is unlikely in cost analysis.

The interpretation of the correlation coefficient as a measure of the strength of the linear relationship between two variables is a purely mathematical interpretation and is without any cause or effect implications.

5.2.5. Multiple Linear Regression

It may happen that the method of least squares for estimating one variable by a related variable yields inadequate success. Although the relationship may be linear, frequently there is no single variable sufficiently related to the variable being estimated to yield good results. The extension is natural to two or more independent variables.

Because linear functions are simple to work with and estimating experience shows that many sets of variables are approximately linear related, or assumed so for a short period, it is reasonable to estimate the desired variable by means of a linear function of the remaining variables. Problems of *multiple regression* involve more than two variables but are still treated in a manner analogous to that for two variables. For example, there may be a cost relationship between differential profit (DP), investment (I), and market saturation (MS), which can be described by the equation DP $= a + b_1I + b_2MS$, which is called a linear equation in the variables DP, I, and MS. The constants are noted by a, b_1, and b_2. This kind of analysis can explain variations in one dependent variable by adding the effects of two or more independent variables.

It is not limited to problems involving a time trend. Time is a catch all and takes into account gradual changes that may be due to different factors both known and suspected. For three or more variables, a *regression plane* is a generalization of the regression line for two variables, as was previously considered. We are concerned with the linear regression function of the form

$$y = a + b_1x_1 + b_2x_2 + \cdots + b_kx_k \tag{5.40}$$

where $x_0 = 1$, a is a constant, and b_1, b_2, $\cdots$, b_k are partial regression coefficients. This is a plane in $k + 1$ dimension. We do not say that the result so obtained is the best functional relationship. We simply state that, given this assumed function and criterion, we have chosen the best estimate of the parameters.

Regression analysis requires the following assumptions:

1. The x_j values are controlled and/or observed without error. Perfection remains a difficult requirement within cost estimating practices, but it is nominally met.
2. The regression of y on x_j is linear.
3. The deviations $y - [j|x_j]$ are mutually independent.
4. Those deviations have the same variance whatever the value of x_j.
5. Those deviations are normally distributed.
6. The data are taken from a population about which inferences are to be drawn.
7. There are no extraneous variables that make the relationship of little intrinsic value.

The plane in the $k + 1$ dimension passes through the mean of all observed values, similar to the two-variable case, and

$$\bar{y} = a + b_1\bar{x}_1 + b_2\bar{x}_2 + \cdots + b_k\bar{x}_k \tag{5.41}$$

Reworking Eqs. (5.40) and (5.41), we have

$$a = \bar{y} - b_1\bar{x}_1 - b_2\bar{x}_2 - \cdots - b_k\bar{x}_k \tag{5.42}$$

$$y - \bar{y} = b_1(x_1 - \bar{x}_1) + b_2(x_2 - \bar{x}_2) + \cdots + b_k(x_k - \bar{x}_k) \tag{5.43}$$

As before, the coefficients are determined by using the method of least squares. To illustrate, consider the case of two independent variables, and we have

$$y = a + b_1x_1 + b_2x_2 \tag{5.44}$$

with n sets (y, x_1, x_2) of points at this point. In each set the error is given as

$$\varepsilon = y - (a + b_1x_1 + b_2x_2) \tag{5.45}$$

and the sum of squares of errors in the n sets is

$$\Sigma\varepsilon^2 = \Sigma[y - (a + b_1x_1 + b_2x_2)]^2 \tag{5.46}$$

We minimize $\Sigma\varepsilon^2$ as before, which requires that the partial derivatives of $\Sigma\varepsilon^2$ with respect to a, b_1, and b_2 be zero:

$$\frac{\partial(\Sigma\varepsilon^2)}{\partial a} = \Sigma[y - (a + b_1x_1 + b_2x_2)] = 0$$

$$\frac{\partial(\Sigma\varepsilon^2)}{\partial b_1} = \Sigma x_1[y - (a + b_1x_1 + b_2x_2)] = 0 \tag{5.47}$$

$$\frac{\partial(\Sigma\varepsilon^2)}{\partial b_2} = \Sigma x_2[y - (a + b_1x_1 + b_2x_2)] = 0$$

Subscripts for summation were dropped for convenience. If we keep x_2 constant, then the graph of y versus x_1 is a straight line with slope b_1. If we keep x_1 constant, then the graph y versus x_2 is linear with slope b_2. Because y varies partially because of variation in x_1 and partially because of variation in x_2, we call b_1 and b_2 the partial regression coefficients of y on x_1 keeping x_2 constant and of y on x_2 keeping x_1 constant. The normal equations corresponding to the least squares plane for the y, x_1, and x_2 coordinate systems are

$$\Sigma y = na + b_1\Sigma x_1 + b_2\Sigma x_2$$

$$\Sigma x_1y = a\Sigma x_1 + b_1\Sigma x_1^2 + b_2\Sigma x_1x_2 \tag{5.48}$$

$$\Sigma x_2y = a\Sigma x_2 + b_1\Sigma x_1x_2 + b_2\Sigma x_2^2$$

The solution of this system of three simultaneous equations gives the values of a, b_1, and b_2 for Eq. (5.44) and is referred to as y on x_1 and x_2. This is a regression plane, but more complicated regression surfaces can be imagined with four-dimensional, five-dimensional, etc., space. The problem given earlier as a polynomial has been solved again by multiple linear regression. Table 5.3 presents the procedure to distinguish the gross product in manufacturing, $\$10^9$, the index of output per man hour, and worker productivity. Those were assumed to be the influencing variables.

5.2.6. Computer Statements

The arithmetic in the simple linear or nonlinear regression equations is digestible; but when an equation is to consider many variables, the situation seems impossible.

TABLE 5.3. MULTIPLE LINEAR REGRESSION

Gross Product in Manufacturing, y (10^9)	Index of Output per Man Hour, x_1	Productivity, x_2	Computations Required in Solution for y on x_1 and x_2
92.6	81.5	1.48	$\Sigma y = 1756.30$
102.0	83.7	1.64	
105.0	86.1	1.74	$\Sigma x_1 = 1520.1000$
111.9	87.6	1.84	$\Sigma x_1^2 = 156,872.75$
103.8	91.2	1.89	$\Sigma x_2 = 32.37$
116.7	96.3	1.96	$\Sigma x_2^2 = 72.01092$
116.4	95.0	2.07	
117.8	100.0	2.20	$\Sigma x_1 y = 180,565.04$
109.7	103.9	2.28	$\Sigma x_2 y = 3860.8860$
121.8	107.2	2.34	$\Sigma x_1 x_2 = 3357.3750$
122.0	108.8	2.44	
122.0	113.1	2.49	
134.1	118.4	2.57	
138.5	121.6	2.67	
142.0	125.7	2.76	

The normal equations are

$$17,563.30 = 15a + 1520.10b_1 + 32.37b_2$$

$$180,565.04 = 1520.10 + 156,872.75b_1 + 3357.3750b_2$$

$$3860.886 = 32.37a + 3357.3750b_1 + 72.0109b_2$$

for which

$$a = 29.1181$$

$$b_1 = 0.7052$$

$$b_2 = 7.658$$

and the multiple linear equation becomes

$$y = 29.1181 + 0.7052x_1 + 7.6458x_2$$

Computers, capable of making short work out of massive computations, make it possible to find linear and nonlinear regression equations of 150 variables or more with all the accompanying statistical measures of reliability and to select the best equation meeting the statistical attributes. This has without a doubt created many benefits to engineering but, unfortunately, also many pitfalls. Perhaps the greatest danger is encountered at the outset when the source and type of data are being selected. Despite the excellence of computation, final results are entirely dependent on the reliability of data used, the interpretation of the computations, and judgment as to reasonableness of the conclusion. Thus, an idea of *causation* must precede statistics.

5.3 MOVING AVERAGES AND SMOOTHING

The engineer may be concerned with periodic observations of labor, material cost, overhead, product demand, and other prices. The characteristics of those observations are described as constant, variable, trend cycle, seasonal, or regular. A *trend cycle* or seasonal term suggests a *time series* to a set of observations taken at specific times. Examples of time series are the monthly demand for motorcycles and the total of monthly costs of expenses in a manufacturing department.

The simplest of the time series cases is the algebraic model using the mean. For the trivial case of constant mean, or nearly so over the time interval for which the forecast is required, the dependent variable is nonsensitive. Samples of pictorial graphs where time is the abscissa and a response is the ordinate are given in Fig. 5.10. The simple case, constant with no trend, has already been discussed. The linear model with a trend is found more widely than the quadratic and exponential models.

Movements are generally considered to be cyclical only if they recur after constant time intervals. An important example of cyclical movements are the so-called business cycles representing intervals of boom, recession, depression, and business recovery. Seasonal movements are well known and refer to identical or nearly identical patterns that a time series appears to follow during corresponding months of successive years. Those events may be illustrated by peak summer production preceding the Christmas demand. Certain of those effects are sometimes superimposed on other effects (e.g., Fig. 5.10(f) where a linear and a cyclic pattern are superimposed).

Consider the following data:

PLASTIC SHEET COST DATA

Years Ago	Price ($/100 lb)
6	$60.20
5	60.50
4	68.70
3	60.24
2	60.55
1	62.32
Now	75.71

The engineer may assume that the computation of moving average at any single point in time should ideally place no more weight on current observations than those achieved some time previously. This is the major logic for the moving average. A reasonable estimate, given by the average price per plastic sheet, is $64; and the forecast for any future observation could be that same value. The actual average of N most recent observation computed at time t is given by

$$M_a = \frac{x_t + x_{t-1} + \cdots + x_{t-N+1}}{N} \tag{5.49}$$

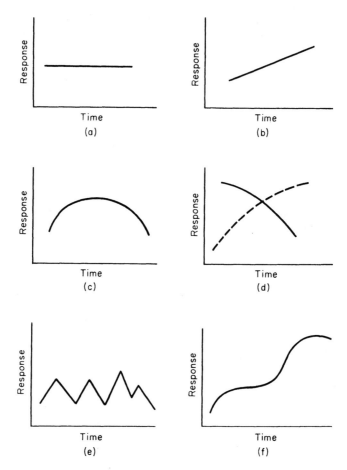

Figure 5.10. Typical time-series models: (a) constant (no trend); (b) linear; (c) quadratic; (d) exponential; (e) cyclic or seasonal; (f) linear and cyclic.

where M_a = moving average value

x = observation

t = time series period number

N = number of observations

with the restriction that the number of terms in the numerator be equal to N and x_t be the latest term added. Another arrangement, of course, is to use the most recent three, four, or five observations and divide the sum by 3, 4, or 5. For example, the model can be arranged to another form

$$M_a = M_{t-1} + \frac{x_t - x_{t-N}}{N} \qquad (5.50)$$

for $t, N = 1, 2, 3, \cdots$, and $t > N$. If a 3-year moving average were required, then the data would be arranged as follows:

Date t	Data, x_1	3-Year Moving, Total	3-Year Moving, Average
1	60.20	$\cdots$	$\cdots$
2	60.50	$\cdots$	$\cdots$
3	68.70	189.40	63.13
4	60.24	189.44	63.15
5	60.55	189.49	63.16
6	62.32	187.11	62.37
7	75.71	198.58	66.19
8	$\cdots$	$\cdots$	$\cdots$
$\cdots$			

The matter of computing a *moving average* is simple and sufficiently straightforward for computer processing or for manual computation.

One of the difficulties with moving averages is that the rate of response is sometimes difficult to change. The rate of response is controlled by the choice of N of the observations to be averaged. If N is arbitrarily chosen large, then the estimate is stable. If N is selected small, then fluctuations due to random error or some other cause can be expected. The engineer is able to take advantage of those properties, for if the process is considered constant, then the engineer may wish to choose a large value of N to have accurate estimates of the mean. However, if the process is fluctuating, small values of N provide faster indications of response.

For most estimating problems some type of moving average is desired that reflects historical and current trends. A smoothing function may be defined as

$$S_t(x) = \alpha x_t + (1 - \alpha)S_{t-1}(x) \tag{5.51}$$

where $S_t(x)$ = smoothed value
α = smoothing constant, $0 \le \alpha \le 1$.

The α is similar but not exactly equal to the fraction $1/N$ in the moving average method. Whenever this operation is performed on a sequence of observations it is called exponential smoothing. The new smoothed value is a linear combination of all past observations. Statistically speaking, the expectation of this function is equal to the expectation of the data, which is its average.

When the *smoothing constant* α is small, the function $S(x)$ behaves as if the function provides the average of past data. When the smoothing constant is large, $S(x)$ responds rapidly to changes in trend. Though no precise statements can be made regarding this smoothing, what follows generally describes the effect of smoothing constant on time series data:

	Variation in α Values		
Drift in Actual Date	Small $\alpha = 0$	Little $\alpha = 0.5$	Large $\alpha = 1$
None	none	none	none
Moderate	very small	small	moderate
Large	small	moderate	large

What follows is an example of the exponential smoothing for $\alpha = 0.2$.

EXPONENTIAL SMOOTHING

Date, t	Data, x_i	Smoothed Data, $S_i(x) = 0.2x_i + 0.8S_{t-1(x)}$
...	...	...
10	...	67.38
11	63.2	66.54
12	68.3	66.89
13	65.7	66.66
14	78.4	69.00
...	...	...

There are initial conditions that must be established for exponential smoothing because a previous value of the *smoothing function* $S_{t-1} = M_{t-1}$. If there are no past data to average, then smoothing starts with the first observation, and a prediction of the average is required. The prediction may be what the process intended to do. Those predictions can also be based on similarity with other processes that have been observed for some time. If there is a great deal of confidence in the prediction of initial conditions, then a small value of the smoothing constant, $\alpha \rightarrow 0$, would be satisfactory. On the other hand, if there is very little confidence in the initial prediction, then it is appropriate to have α as a larger value, $\alpha \rightarrow 1$, so that the initial conditions are quickly discounted. This argument is the counter to the argument about flexibility of response to a change of the process. If the engineer believes that the real process is like the prediction, then there is little reason to have a change. On the other hand, the contrary viewpoint would have a quick response between the prediction and the real process.

Cycles may be interpreted either of two ways: as deviations from established long-term or as significant fluctuation due to some time series effect. In the case where the cycles are interpreted as deviations from a trend, the peaks and troughs are normally referred to as errors from the estimate and are caused by a collection of unknown factors. However, dependence on long-term data, such as industry-wide sales for a particular product, can also be faulty.

TABLE 5.4. THOUSAND MOTORCYCLE SALES IN THE UNITED STATES

	3 Years Ago	2 Years Ago	Last Year	This Year	
January	46	45	60	54	
February	62	78	91	89	
March	78	111	121	132	
April	99	125	145	154	
May	124	154	172	180	
June	118	132	161	155	
July	102	122	142	135	
August	96	119	139		
September	75	93	111		
October	51	82	96		
November	39	51	62		
December	68	96	133		
Total	950	1208	1433	1584	(projected)

Table 5.4 and Fig. 5.11 picture the monthly rise and fall of the cycles about an increasing linear demand. The high point for annual sales occurs during March (purchased for the summer months ahead) and to a lesser magnitude during December. The least squares line for those data is total monthly sales $= 105,070 + 1392X$, where X is the base month 0 at October 2 years ago.

Months following this time reference would be $1, 2, 3, \cdots$, and months previous would be $-1, -2, -3, \cdots$. The historical data conclude with July of the present

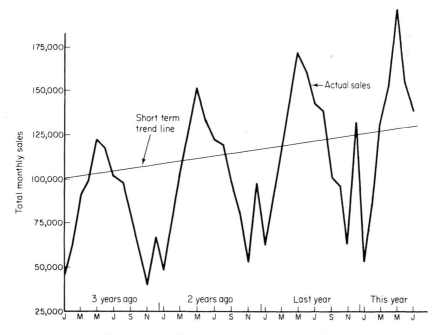

Figure 5.11.   Actual and short-term trend line.

year, and the time index for the next month of August would be 22. The demand would be August sales = 105,070 + 1392(22) = 135,694 motorcycles. Noticing the deviation of the cycles from the trend line, little faith can be given to this estimate for August. Adjustments are called for in view of the monthly variation. What follows discusses one method where arithmetic adjustments can be made to account for the seasonal or monthly pattern. The quarterly sales are converted to percentage of the yearly sales.

PERCENTAGE OF TOTAL YEARLY SALES BY QUARTERS

| | Quarter | | | | |
	First	Second	Third	Fourth	Total
3 Years ago	19.4	35.6	28.5	16.5	100
2 Years ago	19.4	34.0	27.6	19.0	100
Last year	19.0	33.6	27.4	20.2	100
Total	57.8	103.2	83.3	55.7	300
Average, %	19.3	34.3	27.8	18.6	100
Range, %	−0.3 +0.1	−0.9, +1.3	−0.4, +0.7	−2.1, +1.6	

The first quarter averages 19.3% of the year's sales with a range of −0.3 and +0.1%. If the range is assumed to represent random fluctuation, then the engineer may wonder if this randomness can account for the difference between the first quarter and the fourth. If the engineer compares the first quarter against the second quarter, then there is greater justification to suppose a seasonal pattern of sales. The logic of a March quarter sales peak for a product essentially a summer vehicle could overcome timidity in view of a lack of data. However, if the range of the data were, say, to illustrate a point, 10% either way for the four quarters, the engineer would be in a weaker position to put much credence in seasonal factors.

Nonetheless, the differences are assumed to be relevant, and the next step is to translate the percentages into factors. For example, the first quarter averaged 19.3% of the year, and an average quarter would be 25% of the year. The percentages are translated to a base of 1.0 by multiplying by the number of periods in a year. (If the comparison is on the basis of 3 months, then multiply by 4; if on months of the year, then by 12, and so on.) The first quarter with an average of 19.3% of the year would have a seasonal factor of 4 × 0.193 = 0.772:

| | Quarter | | | |
	First	Second	Third	Fourth
Seasonal factor	0.772	1.372	1.112	0.744

The quarterly sales figures would be converted to the average quarter by dividing by 0.772, 1.372, 1.112, and 0.744, respectively. The next step is the incorporation of exponential smoothing, which uses formula (5.51):

$$\alpha(\text{adjusted sales}) + (1 - \alpha)(\text{previous smoothed value}) \tag{5.52}$$

If the adjusted sales were 249,000 units for the second quarter and the previous smoothed value was actually 216,000 units, then the forecast (with an α of 0.6) for the second quarter is $0.6 \times 249,000 + 0.4 \times 216,000 = 236,000$ units. Although the selection of α has been discussed, it is worthwhile for any real-world problem to forecast sales for a sample of items by using different α values and to compare the results. The value that results in the lowest absolute forecast error is superior.

The forecast can be made after the parameters have been chosen (α value and seasonal factors). The actual sales are backcast several periods in the past to provide a basis for smoothing. The sales are seasonally adjusted by dividing by the seasonal factor, and the smoothing technique is applied to give the next forecast value. The forecast is then readjusted to put the seasonal factor back in by multiplying by the factor. The process is continued for each period as sales are reported. Table 5.5 uses $\alpha = 0.6$. The figure of 216,000 units for the first quarter for "3 Years Ago" was provided by the use of the least squares equation for the previous 3 months. In starting up the system, the initial figure can be based on the average of the previous 3 or 6 months or another value acceptable to the engineer.

TABLE 5.5. FORECAST USING EXPONENTIAL SMOOTHING AND SEASONAL ADJUSTMENT (100 UNITS OF MOTORCYCLES)

	3 Years Ago				2 Years Ago			
	First	Second	Third	Fourth	First	Second	Third	Fourth
Seasonal factors	0.772	1.372	1.112	0.744	0.772	1.372	1.112	0.744
Actual sales	186	341	273	158	234	411	334	229
Adjusted sales	241	249	246	212	303	300	300	308
Smoothed value[a]	216	236	242	224	271	289	295	303
Forecast sales	167	324	270	167	209	397	330	226
Error	−19	−17	−3	+9	−25	−17	−4	−3

	1 Year Ago				This Year			
	First	Second	Third	Fourth	First	Second	Third	Fourth
Seasonal factors	0.772	1.372	1.112	0.744	0.772	1.372	1.112	0.744
Actual sales	272	478	392	291	275	489		
Adjusted sales	352	348	353	391	356	356		
Smoothed value[a]	332	342	349	374	363	359	360	
Forecast sales	256	469	389	278	280	493	401	
Error	−16	−9	−3	−13	+5	4		

[a] $\alpha = 0.6$.

A typical calculation for the adjusted sales of the second quarter for "This Year" is given as

$$\text{actual sales for second quarter} = 489{,}000 \text{ units}$$

$$\text{adjusted sales} = \text{actual sales} \div \text{seasonal}$$
$$\text{factor for the current period}$$

$$\text{adjusted sales} = 489{,}000 \div 1.372 \doteq 356{,}000 \text{ units}$$

$$S_t(x) = 0.6(356{,}000) + 0.4(363{,}000)$$

$$\doteq 359{,}000$$

$$\text{forecast sales for second quarter} = \text{smoothed value} \times \text{seasonal factor}$$

$$= 359{,}000 \times 1.372 = 493{,}000 \text{ units}$$

The comparison to the actual sales is fortunate at this point because the deviation approximates 400 for this period. Actually, up to this point, the calculations did not provide forecasts, because there were real-life data for a comparison. The adjusted seasonal forecast can now attend to a legitimate forecast for the forthcoming third period of "This Year":

$$\text{smoothed value} = \alpha(\text{second period smoothed value})$$
$$+ (1 - \alpha)(\text{first period smoothed value})$$
$$= 0.6(359{,}000) + 0.4(363{,}000) = 360{,}000 \text{ units}$$

$$\text{forecast sales} = 360{,}000 \times 1.112 \doteq 401{,}000 \text{ units}$$

In any real problem it is advisable to keep a running count of the magnitude and sign of the error to alert engineering management to deficiencies in the planning.

5.4 COST INDEXES

A *cost index* compares cost or price changes between periods for a fixed quantity of goods or services. The engineer forecasts the cost of a similar design from the past to the present or future period without going through detail costing. If discretion is followed in choosing the proper index, then a reasonable approximation of cost will result. Extrapolation through time series analysis of cost indexes is possible for future periods.

Index numbers have been used for those purposes for a long time. An Italian, G. R. Carli, devised the index numbers about 1750. He investigated with indexes the effects of the discovery of America on the purchasing power of money in Europe.

Index numbers are useful in other respects. With time for estimating usually scarce, the engineer is forced to make immediate use of previous designs and costs, which are based on outdated conditions. Because costs vary with time due to changes in demand, economic conditions, and prices, indexes convert costs applicable at a past date to equivalent costs now or in the future. A cost index is merely a dimensionless number for a given year showing the cost at that time relative to a certain base year. If a design cost at a previous period is known, then present cost is determined by multiplying the original cost by the ratio of the present index value to the index value applicable when the original cost was obtained. This may be stated formally as

$$C_c = C_r\left(\frac{I_c}{I_r}\right) \tag{5.53}$$

where C_c = present or future or past cost, dollars

C_r = original reference cost, dollars

I_c = index number at present or future or past time

I_r = index number at time reference cost was obtained

The engineer considers the design, region, elements of the index, and original estimate in selecting an index to upgrade an estimate. If major items are ignored in the index as compared to the estimate, then adjustments in the composition of the index become necessary. An index seldom considers all factors such as technology progress or local and special conditions.

Consider the following example. Construction of a 70,000 square foot warehouse is planned for a future period. Several years ago a similar warehouse was constructed for a unit estimate of $162.50 when the index was 118. The index for the construction period is forecast as 143, and construction costs per square foot will be

$$C_c = 162.50\left(\frac{143}{118}\right) = \$196.93/\text{ft}^2$$

Though general purpose indexes are published and have become accepted, their construction, alteration, and application are worthy subjects because it may be better for the firm to develop its own index. Arithmetic development of indexes falls into several types: (1) adding costs and dividing by their number, (2) adding the cost reciprocals and dividing by their number, (3) multiplying the costs and extracting the root indicated by their number, (4) ranking the costs and selecting the median value, (5) selecting the mode cost, and (6) adding actual costs of each year and taking the ratio of those sums. The weighted arithmetic method is the most popular. But, in most cost-estimating situations, the best method develops the index by using a tabular approach. Though formulas are straightforward, determination of engineering indexes bears little resemblance to formulas because of the variety, number, and complications involved.

A cost index is a dimensionless number representing the change in cost of material or labor or both over a period of time. Prices, which are the input of an index, must relate to specific material or labor. An index for lumber is based on the price of a specific quantity and type of lumber, such as board foot of 2 × 8 in. (50 × 200 mm) utility grade pine. Quantity and quality must remain constant over the periods so that price movements represent a true price change rather than a change in quality or quantity. This is difficult for indexes that are charted over many periods.

A *cost index* expresses a change in price level between two points in time. A cost index for lumber in year 2000 is meaningless alone. An index for material A has no relationship to the index for material B. Similarly, the cost indexes for material A in two geographical areas are not be directly comparable.

To compute a price index for a single material, a series of prices are gathered for a period of time, a specific quantity and quality of the material. Index numbers are usually computed on a periodic basis. The prices gathered for the material may be average for the period (month, quarter, half-year, or year) or they may be a single observed value, as found from invoice records for one purchase.

Assume that the following prices have been collected for a standardized unit of a laser glass material. Let the origin of the data be labeled period 1.

	Period					
	1	2	3	4*	5	6
Price	$43.75	$44.25	$45.00	$ 46.10	$ 47.15	$ 49.25
Index	94.9	96.0	97.6	100.0	102.3	106.8

*Bench mark period.

Index numbers are computed by relating each period to one of the prices selected as the denominator. If period 4 is the bench mark period, then period 3 price divided by period 4 price = 45.00/46.10 = 0.976. When period 4 price is expressed as 100.0, period 3 price is 97.6. The index can be expressed on the basis of 1 or 100 without any loss of generality. The bench mark period is defined as that period that serves as the denominator in the index calculation.

Movements of indexes from one period to another are expressed as percent changes rather than changes in index points:

Current index	106.8
Less previous index	102.3
Index point change	+4.5
Divided by previous index	102.3
Equals	+0.044 = +4.4%

The average periodic change resulting from these indexes can be found by using

$$r = \left[\left(\frac{I_e}{I_b} \right)^{1/n} - 1 \right] \times 100 \qquad (5.54)$$

where r = average percentage rate per period

I_e = index value at end of period, number

I_b = index value at beginning of period, number

n = number of periods

For an index beginning with 94.9 and ending with 106.8 over a 5-year period, the average rate is 2.4%. If the average index rate is expected to persist, then Eq. (5.54) is re-formed to give

$$I_e = I_b\left(1 + \frac{r}{100}\right)^n \qquad (5.55)$$

This will give an approximate future value. Suppose that we want an index for period 8, or

$$I_8 = 94.9(1 + 0.0239)^7 = 112.0$$

If cost $C_3 = \$3700$ is known, use of Eq. (5.53) will give for $n = 8$,

$$C_8 = 3700\left(\frac{112.0}{97.6}\right) = \$4246$$

A composite index is often required, say, for adjustment of "quote-or-price-in-effect" types of purchase contracts. Equally important is the updating of estimates of complicated assemblies, buildings, and plants.

Assume that a product called "10-cm disk aperture laser amplifier" is selected for a composite index. The 10-cm disk amplifier was produced only during period 1 and cost tracking of selected items has continued. To worry about all amplifier components is too involved, so major items are picked for individual tracking and prices have been gathered for 4 years. Table 5.6 is an example of a tabular construction of an index. The "material" column identifies those items that are significant cost contributors to the laser. Usually, a representative set is selected if the product is complex. However, all the materials may be chosen for a simple product. The "quantity" column is proportional to the requirements of the product. The "quality specification"

TABLE 5.6. CALCULATION OF INDEX FOR MATERIAL, QUANTITY, AND QUALITY SPECIFICATIONS

Material	Quantity	Quality Specification	Period 1[a]	2	3	4
1. Laser glass	3–10 cm disk	Silicate	$26,117	$24,027[b]	$22,345	$21,228
2. Stainless steel turnings	18 kg	AISI 304	1,913	2,008	2,129[c]	2,278
3. Aluminum extrusion	4 kg	3004	418	426	439	456
4. Fittings	3 kg	MIL STD 713	637	643	656	657
5. Harness cable	4 braid, 4 m	MIL STD 503	2,103	2,124	2,134	2,305[d]
6. Annular glass tube	12 m	Tempered $\frac{3}{16}$-in. wall PPG-27	4,317	4,187	4,103	4,185
Total			$35,505	$33,415	$31,806	$31,109
Index (%)			100.0	94.1	89.6	87.6

[a]Bench mark year
[b]Change in laser glass subsequently observed in year 2. Note Table 5.7
[c]Change in quantity of stainless steel observed in year 3. Note Table 5.8
[d]Change in specification of material observed in year 4. Note Table 5.9

column identifies the technical nature of the material. Cost finding begins once those three columns are determined. Year 1^a is the first determination of the cost facts, and the index 100.0 is determined once we divide the total by itself. Cost facts are collected for each subsequent period, and each total is divided by the bench mark total to obtain the index. Thus, the indexes for Table 5.6 are 100.0, 94.1, 89.6, and 87.6. A general decline in prices is suggested by those indexes. Apparently, there is improvement in technology reducing prices.

One may argue that materials, quantities, and qualities are not consistent. Indeed, if technology is active, then a decline in the cost and index is possible. Indexes should reflect basic price movements alone. Index creep results from changes in quality, quantity, and the mix of materials or labor. Assume now that three changes occur and that those are denoted by superscripts b, c, and d in Table 5.6. Effects of material mix, quantity, and quality changes need to be handled. This requires recomputing the index to keep it current. The effects of those three changes are made in the year noted.

In period 2 of Table 5.7 there is a quality change in glass, improving from silicate to fluorophosphate. Instead of $24,027 for silicate glass, a value of $37,621 is quoted for a comparable quantity. To substitute $37,621 in place of the old silicate value would cause distortion unless the bench mark year was appropriately repriced for the new glass. If a new glass price is unavailable for period 1, then the bench mark $37,621 can be adjusted by the overall index as uncovered, such as $37,621/0.941 = \$39,980$. But it is more correct to adjust the new glass by the history between periods 1 to 2 for silicate glass, or $26,117/24,027 \times 37,621 = \$40,893$. This value is entered in Table 5.7 for period 1, a new bench mark total is computed, and new indexes are computed. This going backward is termed backcasting.

TABLE 5.7. RECALCULATING INDEXES FOR QUALITY CHANGE IN PERIOD 2

Material	Quantity	Quality Specification	Period	
			1	2
1. Laser glass	3–10 cm	Fluorophosphate	$40,893	$37,621
2. No change			1,913	2,008
3. No change			418	426
4. No change			637	643
5. No change			2,103	2,124
6. No change			4,317	4,187
Total			$50,281	$47,009
Index (%)			100.0	93.5

There is a change in the weight of stainless steel in period 3 of Table 5.6. Instead of 18 g, a new design has increased to 23 kg. The backcasting adjustment from 18 to 23 kg is calculated for the bench mark year and for intervening years. At period 3 of Table 5.8, the new price entry would be $23/18 \times 2129 = \$2720$. This value

TABLE 5.8. RECALCULATING INDEXES FOR QUANTITY CHANGE IN PERIOD THREE

Material	Quantity	Quality Specification	Period 1	Period 2	Period 3
1. No change			$40,893	$37,621	$39,423
2. Stainless steel	23 kg	AISI 304	2,444	2,566	2,720
3. No change			418	426	439
4. No change			637	643	656
5. No change			2,103	2,124	2,134
6. No change			4,317	4,187	4,103
Total			$50,812	$47,567	$49,475
Index (%)			100.0	93.6	97.4

is deflated to periods 1 and 2 by using the index values for the material. Those entries are shown in Table 5.8. Totals and new indexes are shown again.

In period 4 of material 5 (Table 5.6) a substitution is made as harness cables are changed to flexible printed circuit wires. The value is substituted into period 4 of material 5 by using a vendor's quotation. New quantity and quality specifications are stated, and bench mark and intervening values are backcast. The last schedule of index changes is shown in Table 5.9. All indexes were subsequently reviewed and revised again. Those changes for indexes of designs that are technologically active are necessary to keep the index values accurate.

TABLE 5.9. RECALCULATING INDEXES FOR MATERIAL-MIX CHANGE

Material	Quantity	Quality Specification	Period 1	Period 2	Period 3	Period 4	Period 5
1. No change			$40,893	$37,621	$39,423	$42,617	$48,507
2. No change			2,444	2,566	2,720	2,910	3,085
3. No change			418	426	439	456	479
4. No change			637	643	656	657	689
5. Flexible printed circuit cable	4 m	MIL STD 711	3,743	3,733	3,762	3,861	3,900
6. No change			4,317	4,187	4,103	4,185	4,311
Total			$52,452	$49,176	$51,103	$54,686	$60,971
Index (%)			100.0	93.8	97.4	104.3	116.2

The effects of changes in mix, quantity, and quality on the index scheme are called "technology creep." Unless indexes are evaluated at time of occurrence, they are unsuitable for high technology products. Product indexes can be maintained by noting the changes when they occur, inputting changes for previous data, and backcasting the previous year's indexes. Every so often it may be necessary to reset the bench mark year whenever delicate effects are influencing the index and are not being removed.

The several kinds of indexes are

1. Material
2. Labor
3. Material and labor
4. Regional effects for material, labor, or composite mixes
5. Design
6. Quality

Virtually any combination of materials, labor, services, products, and projects can be evaluated for an index because the intent is to show relative price changes. An interesting contrast is a quality index. Instead of noting price changes, the purpose is to remove price effects and show quality changes between the periods.

Several characteristics distinguish indexes. In the construction of the index there is a choice in the selection of the information. Wholesale prices or retail prices, wages or volume of production, proportion of labor to materials, and the number of separate statistics used are typical alternative choices. Indexes apply to a place and time, that is, period covered or region considered, base year, and the interval between successive indexes, yearly or monthly. Additionally, indexes are varied as to the compiler and sources used for data. A variety of objectives create diversity for the many indexes.

A variety of cost indexes is available to the engineer. Virtually every industrialized nation regularly collects, analyzes, and divulges indexes. A government index listing is given by the *Statistical Abstract of the United States,* a yearly publication that includes material, labor, and construction. A yearly publication of the U.S. Department of Labor is the *Indexes of Output Per Man Hour for Selected Industries.* This volume contains updated indexes such as output per man hour, output per employee, and unit labor requirements for the industries included in the U.S. government's productivity measurement program. Each index represents only the change in output per man hour for the designated industry or combination of industries. The indexes of output per man hour are computed by dividing an output index by an index of aggregate man hours. For an industry the index measures changes in the relationship among output, employment, and man hours. The Bureau of Labor Statistics publishes monthly producer prices and price indexes and covers some 3000 product groupings.

SUMMARY

As is now evident, forecasting is analysis by using imperfect information. Estimating, when coupled with the forecasting process, takes the analysis of imperfect information and adds the ingredients of judgment and knowledge about engineering designs to provide the setting for decisions. In forecasting processes it should be remembered that data provide the information, although imperfect and incomplete, that pretend to be the situation in the future. Thus the forecasting methods of statistics and analysis are used to uncover the relationships that exist for products, production costs, prices, sales, and technology. The application of time series, single and multiple linear regression, correlation, and graphical analysis are the principal statistical techniques used in forecasting for the future. Moving averages involve a consideration of time and fluctuations that occur during those periods. Cost indexes are useful adjustments that permit estimating over lengthy periods.

QUESTIONS

5.1. Define the following terms:

Relative frequency	Best curve	Moving average
Histogram	Confidence limits	Time series
Mean	Prediction limits	Smoothing constant
Standard deviation	t table	Cycles
Regression	Transformed units	Trend line
Least squares criterion	Power function	Indexes
Intercept	Correlation	Backcasting
Slope	Multiple linear regression	Technology creep

5.2. "Statistics never lie, yet liars use statistics" is a common statement. Discuss.

5.3. Why are graphical plots preferred initially over mathematical analysis of data?

5.4. Is cost estimating more concerned with empirical evidence or theoretical data? Illustrate both.

5.5. Cite instances when a cumulative curve would be necessary.

5.6. Discuss what regression analysis is. What are its underlying assumptions?

5.7. What is minimized in a least squares approach?

5.8. What is meant by correlation analysis? What does $r = 0$ or 1 imply?

5.9. Distinguish between correlation and causation. Could you have causation without correlation?

5.10. What is the purpose of a moving average? How does smoothing relate to a moving average?

5.11. If the engineer is confident of past data, then would the smoothing constant be large or small?

5.12. What are the differences between cycles and trend cycles?

5.13. Define a cost index.

5.14. Why would an engineer use a cost index? What qualifications are necessary before a manufacturing index is chosen?

5.15. Three moving-average trend lines are assumed for a standard cold-rolled-sheet steel material to be used for your manufactured product and are shown as Fig. Q5.15. Write a paragraph that evaluates the pros and cons of the proposed moving average for the three cases.

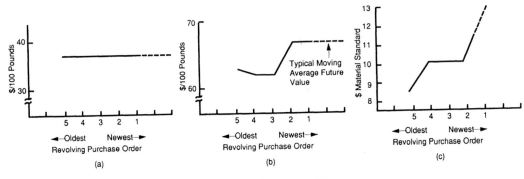

Figure Q5.15

PROBLEMS

5.1. A labor survey is conducted for the area firms, and the following classification is given:

Wage	Number	Wage	Number
8–10	5	17–19	36
11–13	11	20–22	52
14–16	28	23–25	15

Construct the frequency distribution table, relative frequency curve, and cumulative frequency curve. (Note that wage interval end numbers are open.)

5.2. Time studies were examined for a specific handling element, and their observations are given as follows:

Time	Observations	Time	Observations
0.04	2	0.16–0.20	14
0.04–0.08	12	0.20–0.24	13
0.08–0.12	14	0.24–0.28	12
0.12–0.16	15	0.28–	6

Construct the frequency distribution table, relative frequency curve, and cumulative frequency curve.

5.3. Find the mean, median, mode, range, standard deviation, and variance of the following sets of data:

(a) 3, 5, 2, 6, 5, 9, 5, 1, 7, 6   (b) 41.6, 38.7, 40.3, 39.5, 38.9

(c) 2, −1, 0, 4, 6, 6, 8, 3, 2   (d) 16, 23, 82, 41, 16, 0, −3

(e) $\frac{1}{8}, \frac{1}{16}, \frac{1}{4}, -\frac{1}{2}, 0.125, 0$ (f) 11, 13, 13, 16, 16, 16, 20, 21, 23

5.4. Market surveys have determined the U.S. value of the ship-building industry over a 10-year period.

	Year									
	1	2	3	4	5	6	7	8	9	10
10^6 value	1467	1216	1360	1400	1518	1678	1818	2160	2200	2460

(a) Graph the data on arithmetic coordinates by using the "one-half" rule. Find the graphical equation $y = a + bx$ of the line. Forecast the value in year 11 by using the curve and equation.

(b) Repeat part (a) using the least squares equation.

5.5. The index for the union wage rate for carpenters has followed this pattern:

	Year									
	1	2	3	4	5	6	7	8	9	10
Index	100.0	104.4	107.8	112.3	117.2	122.5	132.8	145.6	159.3	171.6

(a) Plot the straight line of the data on arithmetic coordinates by using the "one-half" rule. Find the graphical equation $y = a + bx$ of the line. Forecast the value in year 11 by using the curve and equation.

(b) Visually determine the year for which there is a change in slope, and construct two new straight lines for the data, finding their graphical equations. Also forecast the value in year 11.

(c) Find the regression equation for the data.

5.6. The life of a cutting tool is determined by standard tests, and the two principal variables are cutting velocity in feet per minute (V) and tool life in minutes (T). The tool life equation is $VT^n = K$ where n is the slope and K the intercept and are statistically determined using regression methods. Assume that the $[(V, T) = (y, x)]$ data are $(1000, 1)$, $(100, 10)$, and $(10, 100)$. Find the regression equation. What velocity is expected for $T = 5$ min?

5.7. Aerospace companies control the production and costing of their major products, such as airplanes, major assemblies, and vendor materials, by a technique known as "learning." This function, which was discovered as an empirical truth in the 1930s, states that as production continues, the time, or cost to produce one more unit declines at a constant rate between doubled units. The function that models this relationship is $T = KN^s$ where T = time or cost of a specific unit, and N is the unit number. K and s are empirical quantities determined by regression calculations. The firm determines the following data ($y = T, x = N$) such as $(450, 15)$, $(325, 30)$, and $(200, 45)$. Determine the regression equation in original and transformed units. Find the estimate for the 60th unit by both methods.

5.8. (a) A model of the form $y = ax^b$ is to be fitted to (x, y) data $(1, 100)$, $(10, 10)$, and $(100, 0.1)$. Find the equations in transformed and original units. What are the values for $x = 150$?

(b) Calculations for data, which are not given, are as follows:

$$n = 4$$

$\Sigma y = 14$	$\Sigma x = 10$
$\Sigma y^2 = 54$	$\Sigma x^2 = 30$
$\Sigma lny = 4.787$	$\Sigma lnx = 3.178$
$\Sigma(lny)^2 = 6.200$	$\Sigma(lnx)^2 = 3.129$
$\Sigma xy = 40$	$\Sigma lnxlny = 4.516$
$\Sigma xlny = 13.487$	$\Sigma ylnx = 13.405$

The regression model is believed to be $y = ab^x$. Find the regression parameters in transformed and original units. What is the estimated quantity for $x = 10$ in both original and transformed units? (Hint: Note that quantities are given in arithmetic and natural logarithms. Be selective in your choice of which quantities to substitute into the standard equations. Also, there is no distinction for the regression equations for either base 10 or natural-base logarithm quantities.)

5.9. (a) Find t_α if $\alpha = 0.10$ and $v = 10$ degrees of freedom.
(b) Find t_α if $\alpha = 0.05$ and $n = 20$.
(c) Compare t_α and Z (from appendix 1) for $\alpha = 0.05$ and $n = 120$.
(d) If $t_\alpha = 2.8$, then find α for $v = 16$.

5.10. (a) Find the mean and individual dependent value for Table 5.1 for year 16.
(b) Determine the variance of the mean value y and the individual y for $x = 16$ from Table 5.1.
(c) Determine the confidence and prediction limit for parts (a) and (b) for 95%.

5.11. A price index for labor is given for 7 years:

				Year			
	1	2	3	4	5	6	7
Index	100.0	106.0	111.1	117.2	121.3	125.2	128.0

(a) Graph the data by using the one-half rule.
(b) Compute trend values and find a least squares line fitting the data and construct its graph.
(c) Predict the price index for year 8 and compare with the true value 132.6. What is the range for the individual year 8 by using a 95% prediction interval?

5.12. From the following data determine the values of parameters a and b by using the method of least squares. Assume that the data can be plotted as $y = a + bx$.

Year	Retail Price Index, y
0	95.1
1	97.7
2	98.4
3	100.0
4	101.1
5	102.2
6	103.5
7	104.9
8	106.6
9	109.7

Find the 90% prediction interval for the year 10 value. Estimate the 95% confidence interval for the slope.

5.13. A time study is conducted on a spot-welding operation. Three elements are summarized:

			Load		Unload	
Time Study	Number Spots	Spot Time	L + W + H	Time	L + W + H	Time
1	8	0.28	29	0.08	37	0.10
2	3	0.13	46	0.10	46	0.17
3	9	0.34	101	0.27	106	0.28
4	14	0.87	60	0.21	60	0.32
5	36	2.06	53	0.19	53	0.51

(a) Find regression equation for number of spots (x) versus decimal minute spot time (y). Determine the 95% confidence limits for 15 spots.

(b) Determine regression equation for load L + W + H (x) versus time (y). Find the 95% confidence limits for L + W + H = 50.

(c) Determine the regression equation for unload L + W + H (x) versus time (y). Find the 95% confidence limits for $x = 100$.

5.14. Product cost learning has been found to follow the function $T_u = KN^s$, where $T_u =$ unit time for Nth unit, $K =$ man hours estimate for unit 1, and $s =$ the slope of the improvement rate. Transform this into a log relationship, and determine the log regression line. Also determine the equation in original units. Estimate the man hours at $N = 50$ in both log and original units.

(a)	N	Man hours, T		(b)	N	Man hours, T
	5	155			7	210
	8	143			13	160
	13	137			21	142
	17	97			26	128
	25	75			31	121

5.15. Direct labor for a 12-kW four-cylinder diesel generator set is plotted on arithmetic graph paper and shown as Figure P5.15.

(a) Replot these data on semilog paper.

(b) Find the regression equation.

(c) What is the expected value for year 5 from your log plot and equation? Use the theoretical equation $T = Ks^x$, where T is the man hours, x the year, K the intercept, and s the slope.

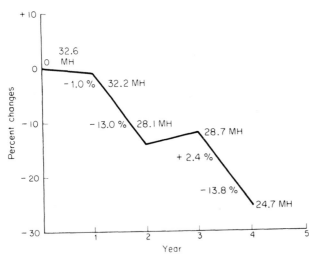

Figure P5.15. Evidence of direct labor learning for 12-kW four-cylinder diesel generator set.

5.16. A firm that manufactures totally enclosed capacitor motors, fan-cooled, 1725-rpm, $\frac{5}{8}$-in. (15.9-mm) keyway shaft has known costs on $\frac{1}{4}$, $\frac{1}{3}$, $\frac{3}{4}$, and 1 hp (0.20, 0.25, 0.55, and 0.75 kW) of $67.50, $75.00, $88.50, and $105.00. Determine the cost equation by using (a) the semilog equation, and (b) the power equation. (c) Which equation using the test of $\Sigma(y - \hat{y})^2$ is best?

5.17. Find the correlation for Problem 5.5.

5.18. A company that has been converting manual production methods to automatic methods wants to evaluate its efforts. Is there correlation for this number of employees and shipment value? What assessment do you make?

Shipments ($\times 10^6$)	Number of Employees
$573	1573
606	1550
648	1530
720	1550
765	1540
798	1550
848	1560

5.19. Analyze the following data by using multiple linear regression:

	Cost/1000 Units for Length		
Weight	1.50	1.75	2.00
60.5	5070	4770	4540
52.6	4540	4215	3930
42.3	3660	3660	3120
33.3	2390	2390	2100

5.20. In a study of scrap and production rate, the following information is collected:

	Production, x (units/period)						
	1000	2000	3000	3500	4000	4500	5000
Scrap, y (% of production)	5.2	6.5	6.8	8.1	10.2	10.3	13.0

(a) Determine an exponential model and a second degree polynomial equation.
(b) Which model gives the smallest error sum of squares?

5.21. Material costs were collected from historical purchase orders, and the information is categorized by period and quantity. The quantity represents the volume of units purchased, but the data are listed as dollars per unit. (See Table 3.2 for a practical example.)

	Quantity		
Period	1	2	3
-3	10.00	8.00	7.00
-2	10.70	8.60	7.50
-1	11.10	8.95	7.85
0	11.25	9.00	7.90

(a) Use multiple linear regression to find the unit cost equation y where $x_1 =$ period and $x_2 =$ quantity.
(b) Find the unit cost for period 1 and quantity 2.

5.22. A company uses extensive amounts of plastic film for one of its automatic processes. This purchased film constitutes the major cost for productive-hour cost.

Budget Period	Unit Price	Budget Period	Unit Price
1	$60.20	11	$63.20
2	60.50	12	68.30
3	68.70	13	65.70
4	60.20	14	78.40
5	60.50	15	81.20
6	62.30	16	82.10
7	75.70	17	82.10
8	73.10	18	84.60
9	80.10	19	83.10
10	77.40	20	82.10

(a) Plot a four-period moving average and actual unit price.
(b) By using an exponential smoothing function determine the smoothed data for $\alpha = 0.25$.

5.23. (a) Plot the following data:

Period	Value	Period	Value
1	7.5	11	4
2	8	12	2.5
3	9.5	13	1.8
4	9.7	14	0.8
5	10	15	0.2
6	9.9	16	0.1
7	9.8	17	0.3
8	9	18	0.9
9	7.6	19	2
10	6	20	4

(b) By using a smoothing function with $\alpha = 0.1$ determine smoothed data and overplot the raw information. What familiar function does the information resemble?

5.24. The cost of a purchased component used in the assembly of a product is analyzed on a time series. Make a table for a 3-year moving total and average price. Plot the average price versus time and describe the movement.

					Years Ago					
	10	9	8	7	6	5	4	3	2	1	Now
Cost (cents/each)	23.2	24.1	26.3	25.7	26.8	27.2	28.0	27.8	28.0	28.5	28.3

5.25. A warehouse construction job is planned in 2 years. A similar layup steel-walled warehouse was constructed for a unit price of $105 per square foot of wall when the index was 107. The index now is 143, but in 2 years it is expected to be 147. What will be the unit estimate for the construction? What is the bid cost of a warehouse with 700,000 square feet of wall?

5.26. A production man-hour index based on 1977 = 100 is given as 1959, 41.9; 1969, 48.0; 1979, 107.6; and 1989, 205.8. Convert the indexes to a 1989 basis = 100.

5.27. A construction index for Los Angeles is

Year	Index
5	1200
15	3108

and a structure built in year 10 cost $1,000,000. How much will the same structure cost in year 20?

5.28. Indexes for buildings in Denver and New York are as follows:

Year	Denver	New York
1	400	500
5	600	750

(a) If the building cost $400,000 in year 1, then what is its cost in year 6?
(b) If the building costs $400,000 in year 2, then find its cost in year 6.

5.29. A company receives an order for a large quantity of drill steel to be used in mining. The delivery point is equally spaced between its plants in the United States and France. Material and transportation costs are assumed to be equivalent for both countries. The efficiency index is 1.3 French man hours = 1.0 U.S. man hours for the same job. Indirect cost percentages are 120% France = 75% United States. The direct unit cost (labor only) is $99.65 France = $153.53 United States. Which of the two plants should this company build and ship from?

5.30. Determine the fabrication base cost for a pressure vessel. A pressure vessel is a cylindrical shell capped by two elliptical heads. The base cost estimates the vessel fabricated in carbon steel to resist internal pressure of 50 psi with average nozzles, manways, supports, and design size. Design: diameter, 8 ft; height, 15 ft; shell material, stainless 316 solid; and operating pressure, 100 psi. Estimating data: Construction 3 years hence with a 5% material increase per year, carbon steel material costs = $24,000, factor for non-carbon steel material = 3.67, factor for nonstandard pressure = 1.05. What is the expected cost?

5.31. A cast-steel foundry uses indexes to price its raw materials. The materials are purchased over time and stored in open inventory, but they are priced out on a current index basis to remain competitive. Find the indexes for periods 2 and 3 with the reference period as 1. Speculate on the next period index.

Item	1-Ton Finished Casting, Proportion (%)	Period		
		1	2	3
1. Pittsburgh scrap steel, No. 1 heavy	80	$37.00	$37.50	$38.00
2. Metal alloy No. 1	15	48.00	48.50	50.00
3. Metal alloy No. 2	5	57.00	56.25	55.00

5.32. A construction firm uses indexes to price wall-footing forms. Material quantities and labor man hour data and/or material unit costs have been gathered for three periods.

Item	Quantity	Rate for Period		
		1	2	3
1. Materials				
Sides: 2 x 12 in.	200 BF	0.06	0.057	0.053
Stakes and braces	75 BF	0.055	0.053	0.052
2. Labor				
Carpenter	3.5 hr	24.60	27.52	31.02
Laborer	1.75 hr	21.85	23.66	26.12
3. Stripping forms, laborer	1.5 hr	21.85	23.66	26.12

(a) Find the indexes and determine the value for year 5.
(b) A cost is $17,500 at period 2. Calculate the cost for period 5.

MORE DIFFICULT PROBLEMS

5.33. Raw data of unemployment are gathered for two separate years in community A and are shown next.

RAW DATA OF UNEMPLOYMENT FOR COMMUNITY A

Age Range	March First Year	March Second Year
14–15	20	26
16–17	87	93
18–19	636	709
20–21	206	191
22–23	202	50
24–25	81	229
26–27	15	37
28–29	13	29
30–31	19	73
32–33	25	83
34–35	38	47

Age Range	March First Year	March Second Year
36–37	53	42
38–39	36	85
40–44	89	30
45–49	101	97
50–54	86	107
55–59	111	67
60–64	117	173
65–69	144	180
70+	101	102
Total	2180	2450

(a) Construct the frequency distribution table, relative frequency curve, and cumulative frequency curve for both years. Describe their appearance.

(b) Repeat part (a) by using intervals 14–19, 20–24, 25–34, 35–44, 45–54, 55–64, and 65+. Describe the appearance of the curve. Comment about nonuniform-sized classes and deliberate or unintentional statistical distortion.

5.34. The life of a cutting tool is determined by testing under standard conditions and observing flank wear. Taylor's tool-life equation is $VT^n = K$, where V is the velocity in ft/min (m/s), T the time in minutes, n the slope, and K the intercept. The test log is AISI 4140 steel; depth of cut, 0.050 in. (1.27 mm) feed; 0.010 in./revolution (0.25 mm/rev); and tool wear limit, 0.005 in. (0.13 mm) of flank wear. Once the flank wear length is reached, the time is recorded against a controlled cutting surface in feet per minute (m/s). The data are as follows:

Cutting Speed, (surface feet/min)	Tool Life, T (min)
400	7, 9, 8
450	6, 8.5
500	5.5, 7.5, 6
550	4, 7, 6
600	5
650	3
700	2.5, 3, 3.4
750	3

(a) Plot the equation as $\log V = \log K - n \log T$ and find the graphical equation. Find the equation using regression methods.

(b) Forecast the value for 300 fpm (1.52 m/s) and 800 fpm (4.06 m/s).

(c) Repeat in metric units.

5.35. Solve Problem 5.34 by using regression methods. Find the equation in transformed and original units. What is the value at $V = 600$ fpm (3 m/s)?

5.36. A study of past records of installed insulation cost (including labor, material, and equipment to install) for central steam-electric plants revealed the following data:

Equipment Cost ($\times 10^5$)	Insulation Cost ($\times 10^4$)	Equipment Cost ($\times 10^5$)	Insulation Cost ($\times 10^4$)
$ 5.5	$ 3.5	$14.1	$ 9.2
10.7	5.5	14.8	9.3
34	28	15.5	14.1
2	1.4	15.3	13.8
6	6.4	21.3	15.0
1.5	2.1	34.0	15.8
8.1	7.2	24.1	9.8
10.1	6.4	26.0	15.8

(a) Plot a chart of this information on arithmetic coordinates; on log-log coordinates. Which do you think is more suitable for cost analysis purposes? Major equipment costs have been estimated as $300,000 for a new project. What is the estimate for the installed insulation cost?

(b) Find the arithmetic regression equation and correlation r. What is the insulation cost value for $300,000 of equipment?

(c) Find the logarithmic equation and the value for $300,000.

5.37. Requirements for identical materials are consolidated, and purchase requests are issued for a quantity each period. The following table is a summary of the purchased cost history of a single material; y is identified as cost per unit. There is no information for values not shown.

Period	Quantity		
	1	2	3
-3	10.00	—	—
-2	—	—	7.50
-1	—	8.95	—
0	—	—	7.90

(a) Use multiple linear regression and find the equation where $x_1 =$ period, $x_2 =$ quantity, and $y =$ material cost per unit.

(b) Find unit cost for period 1 and a quantity of 2.

5.38 Hot-rolled $\frac{7}{32}$-in. (5.6-mm) coils of low-carbon steel have the following cost history:

Year	Quarter	Cost
1	1	$ 7,761
	2	7,844
	3	7,844
	4	7,844
2	1	9,577
	2	9,688
	3	9,668
	4	9,668

Year	Quarter	Cost
3	1	9,668
	2	13,391
	3	13,059
	4	12,786
4	1	12,786
	2	12,952
	3	14,195
	4	14,195
5	1	14,195
	2	14,195
	3	14,958
	4	14,958
6	1	15,854
	2	16,127
	3	15,953
	4	15,953
7	1	15,622
	2	17,031
	3	17,247
	4	17,247
8	1	18,043
	2	18,673

Plot these time series data point to point, and eyeball a straight line through the data. Identify the cycles. Provide a smoothing equation.

5.39. Basic union wage rates and indexes for major construction trades are given.

Trade	Weight	Month 1	Month 2	Month 3
Carpenter	31.4	100.0	104.4	107.8
		20.40	21.30	22.00
Electrician	13.6	100.0	102.2	104.3
		23.00	23.50	24.00
Laborer	10.1	100.0	106.1	110.9
		14.70	15.60	16.30
Plumber	15.3	100.0	107.6	110.3
		22.30	24.00	24.60
Painter	11.5	100.0	105.7	108.5
		17.60	18.60	19.10
Others	18.1	100.0	105.0	107.9
		19.80	20.80	21.30
Weighted average index		100.0	105.0	107.9

(a) What is the next period index for carpenters? For the total trade?

(b) A new job is to be estimated, but it will have no painting component. Construct a revised index free of the painting trade and determine next period index.

(c) Discuss: While the rates relate to the amount paid the tradesman for 1 hour, the indexes do not encompass the efficiency with which that hour is utilized in successive years.

CASE STUDY:
MARKET-BASKET INDEX

Construct a "market-basket" index of these items. The base year is 1. The prices of these items were collected under similar circumstances over a 5-year period.

Item	Price, Yearly					Total
	1	2	3	4	5	
1. Milk, homogenized, $\frac{1}{2}$ gal	$ 1.13	$ 1.23	$ 1.62	$ 1.63	$ 1.71	$ 7.32
2. Ice cream, regular vanilla, $\frac{1}{2}$ gal	1.67	1.92	1.87	1.96	2.16	9.58
3. Eggs, grade A large, 1 dozen	0.95	1.16	1.17	1.21	1.43	5.92
4. Margarine, 1 lb, regular Blue Bonnet, Parkay	0.85	0.99	1.11	1.36	1.47	5.78
5. White bread 1 lb loaf, sliced	0.75	0.83	1.02	1.10	1.14	4.54
6. Instant coffee, 10 oz jar	2.83	3.13	3.47	4.61	5.83	19.87
7. Flour, 5 lb, all-purpose Pillsbury	1.67	1.69	1.73	2.01	2.27	9.37
8. Ground beef, less than 25% fat, 3 lb package	7.15	8.13	8.47	9.16	10.11	43.02
9. Potatoes, U.S., one 5 lb bag	1.15	1.18	1.39	1.43	1.89	7.04
Total	$18.15	$20.26	$21.85	$24.47	$28.01	$112.74

(a) Determine yearly indexes.
(b) What is the average percentage rise over the 5-year period?
(c) What are the major and minor items contributing to this increase?
(d) Construct a time-series plot of the index.
(e) Determine a least squares equation fit for this time series.
(f) If those items constitute 3% of the grocery bill for a hypothetical time period and family, what are the gross dollars lost to inflation?
(g) By using inquiry methods at your local grocery store determine current prices for those items and compare to year 1.

6

Preliminary and Detail Methods

Estimating methods are remarkably similar even though designs may differ. That is seen in this chapter, which introduces general methods for all designs. A preliminary method is used in the formative stages of design. Attention turns to detail methods as designs and information become complete. Estimating methods are discussed and advantages and shortcomings are noted. Estimating methods range from experience and judgment as the dominant requirement to ones with mathematics.

6.1 DESIGN AND EVALUATION

A design is without specific form and shape in the early stages of its evolution. For instance, engineers may have progressed through problem definition, concepts, engineering models, and evaluation with only the final design step remaining. The *preliminary estimate* is requested during the initial evaluation. With a lack of facts and specific information the engineer is asked to provide this first estimate. By using various methods, rules of thumb, and simple calculations, a quick and relatively inexpensive estimate is provided. Obviously, the accuracy of the estimate depends on the amount and quality of information and the time available to prepare the estimate.

The preliminary estimate may cause the firm to take some sort of action. An estimate, such as an operation cost, product price, project bid, or system effectiveness, can have serious financial implications. More frequent, however, is the case where the preliminary estimate screens designs and aids in the formulation of a budget. The preliminary estimate is used by engineering and management to commit or stall additional design effort, to appropriate requests for capital equipment, or to cull out uneconomic designs at an early point. Although decisions based on the preliminary estimate may not lead to legal obligations or authorization for capital spending, mistakes can be costly by eliminating potentially profitable designs.

A preliminary estimate is an estimate made in the formative stages of design. Overlooked in this definition is the accuracy, type of design, nature of the organization, dollar amount, and the purpose for the estimate. A precondition of accuracy for preliminary estimates cannot be imposed because special designs or objectives create a unique set of requirements. An estimate involving, say, pennies is no less of a challenge than an estimate involving millions of dollars because the same methods are used at both ends of the dollar scale. Other terms used in practice include conceptual, battery limit, schematic, order of magnitude, and imply preliminary estimate. Their purpose is to screen and eliminate unsound proposals without extensive engineering cost. If those estimates lead to a continuation rather than a dismissal decision, additional methods are required.

Now we confine our attention to methods more thorough in preparation and accurate as well as costly in design and estimating. At this time the designer would have extended preparations, and the engineer constructs an estimate on enlarged quantities of verified information. In some cases the *detail estimate* is a re-estimate, because only limited updating needs to be done.

The man hours devoted to estimating the operation, product, project, or system design naturally varies with circumstances. The case for increased accuracy from a detail estimate is often made. Some practitioners claim that detail estimates are within ±5% about a future actual value. Whether the particular value is ±5% or ±50% is not significant now. However, we assert that methods generally more accurate have an increase in cost of preparation. Thus, data are purified, design has increased detail, and in actual estimating situations management stipulates that the estimate be within an interval about the future actual or standard value.

Detail estimating methods are more quantitative. Arbitrary and excessive judgmental factors are suppressed, although the factors are never eliminated, and emphasis shifts to comprehensiveness. Whether the model is a computer program, recapitulation columnar sheet, a set of computational rules, or a functional model, the intent is the same. We want to find the value of the detail estimate through formal and rigid rules.

6.2 OPINION

Personal *opinion* is inescapable in estimating. It is easy to be critical of opinion, but in the absence of data and shortage of time, there may be no other way but to use opinion in the evaluation of designs. The engineer is selected for the job because of his or her observational experience, common sense, and knowledge about the designs. The mettle of the engineer is tested in judging the economic want of a design. Engineers respected as "truth tellers" are sustained in this activity. Others are not. Time, cost, or quantities about minor or major line items are estimated by using this inner experience. That the engineer be objective in attempting to measure all future factors that affect the out of pocket cost is understood. Opinion estimating is also done collectively.

6.3 CONFERENCE

The *conference method* is a nonquantitative method of estimation and provides a single value or estimate made through experience. The procedure, although having many forms, involves representatives from various departments conferring with engineering in a round-table fashion and jointly estimating cost as a lump sum. Sometimes, labor and material are isolated and estimated with overhead and profit added later through various formula methods by the engineer.

The conference method may also be used within the engineering department. Engineers having specialized knowledge confer on a design and determine a cost figure without counsel from other departments.

The way the method is managed depends on the available information. Various gimmicks can be used to sharpen judgment. A hidden card technique has each of the committee experts reveal a personal value. This could provide a consensus. If agreement is not initially reached, discussion and persuasion are permitted as influencing factors. Sometimes this is called estimate-talk-estimate (or E-T-E). The hidden card idea prevents a brain storming session, which generally tends to provide optimistic estimates. The engineer will serve as a moderator during a conference on estimating and will provide questions such as, "What is the labor and material cost for this part?"

A ranking scheme along the "good-better-best" lines of Sears, Roebuck & Co. can be applied in a cost sense. The engineers rank two or more designs and then give their cost value. The ranking method seems to work provided that the set is not large. Major drawbacks to the conference method are the lack of analysis and a trail of verifiable facts leading from the estimate to the governing situation. Although little faith and accuracy is assigned to the method, the lack of procedural rigor seldom deters usage.

6.4 COMPARISON

The *comparison method* is similar to the conference method except that the comparison method attaches a formal logic. If we are confronted with an unsolvable or excessively difficult design and estimating problem, we designate it problem *a* and construct a simpler design problem for which an estimate can be found. The simpler problem is called problem *b*. This simpler problem might arise from a clever manipulation of the original design or a relaxation of the technical constraints on the original problem. Thus, we attempt to gain information by branching to *b* as various facts may already exist about *b*. Indeed, the estimate may be in final form, or portions may exist and there need only be a minor restructuring of data to allow comparison. The alternative design problem b must be selected to bound the original problem a in the following way:

$$C_a(D_a) \leq C_b(D_b) \tag{6.1}$$

where $C_{a,b}$ = value of the estimate for designs a and b
$\quad\quad D_{a,b}$ = design a or design b

Also, D_b must approach D_a as nearly as possible. We adopt the value of our estimate as something under C_b. The sense of the inequality in Eq. (6.1) is for a conservative stance. It may be management policy to estimate cost slightly higher at first, and, once the detail estimate is completed with D_a thoroughly explored, we comfortably find that $C_a(D_a)$ is less than the original comparison estimate.

An additional lower bound is possible. Assume a similar circumstance for a known or nearly known design c, and a logic can be expanded to have

$$C_c(D_c) \leq C_a(D_a) \leq C_b(D_b) \tag{6.2}$$

Assume that designs b and c satisfy the technical requirements (but not the economic estimate) as nearly as possible.

Consider an example of comparison estimating. It is desired to decrease the rolling speed of a ball bearing relative to the rotational speed of the shaft. Among other factors the maximum operating speed of a ball bearing depends on its size and the rolling speed of the balls within the raceways. A preliminary design solution is to add an intermediate ring with raceways on its inner and outer peripheries for light radial loads. This would cut the relative rolling speed to approximately half that of the balls in an unmodified bearing. Figure 6.l(a) indicates a possible arrangement of design A. The adaptation of an additional outer roll of balls achieves many of the similar features required of design A and is known as design B [Fig. 6.l(b)]. Velocities are reduced, and other technical advantages are achieved. The design requirements for B satisfy most of A. The cost of B can be determined as the outer roll of

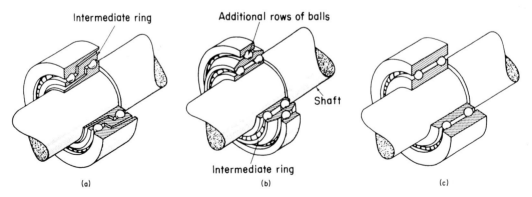

Figure 6.1. Comparison of (a) unknown design to (b) higher-cost design and (c) lower-cost design.

balls and is in many ways similar to single-roll ball-bearing technology. The technology for conventional double row ball bearings [Fig. 6.1(c)] has known costs and is our design C. With costs determined for designs B and C, comparison becomes possible. Precautions that consider up-to-date costs, processing variations, similar production quantities, and spoilage rates are factors that call for insight in picking a specific value for design A between the B and C range. The comparison method is sometimes called similarity or analogy.

6.5 UNIT

The *unit method* is the most popular of the preliminary estimating methods. Many other titles describe the same thing—average, order of magnitude, lump sum, module estimating—and involve various refinements. Extensions of this method lead to the factor estimating method, discussed later. Examples of unit estimates are found in all activities:

For instance

Cost of house construction per square foot of livable space
Cost of metal casting per pound
Cost of electrical transmission per mile
National-norm cost of university education per student year
Chemical plant cost per barrel of oil capacity
Factory cost per machine shop man hour

Though typically vague in those contexts, the strongest assumption necessary is that the design to be estimated is like the composition of the parameter used to determine the estimate. Notice that the estimate is "per" something. Thus, if the unit value of a casting is $1.22 per pound ($2.72 per kilogram), a 5-lb (2.3 kg) casting will cost $6.10 each.

Unit estimates are easily figured. Consider the manufacturing operation of metal turning. By using job tickets the total time for several jobs and many parts for a lathe are known. Divide this total time by the number of inches (centimeters) turned. Thus we have a unit estimate, or hours/inches (cm) of length. Even though this is simplistic when expressed in this context, unit estimating methods are widely used.

A unit estimate is defined as the mean, where the divisor is the principal cost driver, or

$$C_a = \frac{\Sigma C_i}{\Sigma n_i} \tag{6.3}$$

where C_a = average cost per unit of design
C_i = value of design i, dollars
n_i = design i unit (lb, in., kg, mm, count, etc.)

Consider the design of cast-iron sphere where there are three observations:

Design Weight, lb (kg)	Cost
2 (0.9)	$ 2
3 (1.4)	3
4 (1.8)	6
9 (4.10)	$11

The cost per pound (kilogram) is $1.22 (= 11/9)($2.71 kg). A new casting design is estimated by finding sphere weight and by multiplying this value by $1.22/lb.

Though the estimating value is obviously improved with more observations, the value suffers from other more fundamental faults. Note Fig. 6.2(a), where the data are plotted and a regression equation would give $C = -2.33 + 2.00(lb)$. Effectively, the unit value is nothing more than the general slope b where $a = 0$, or $y = 0 + bx$. For in the estimating of $1.22/lb to a design, the cost is zero for no weight because 1.22 is a linear multiplier. An improvement is made when the data are fitted to $y = a + bx$. When using $C_a = 1.22$, the unit method either overestimates or underestimates the design when compared to $y = a + b_1 x$. The only location where there

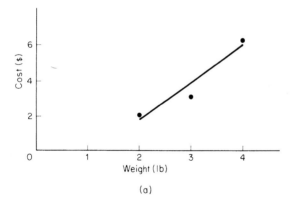

(a)

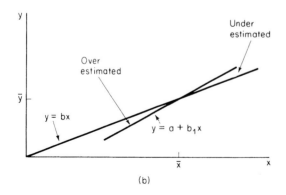

(b)

Figure 6.2. (a) Plot of three observations; (b) unit model compared with $y = a + b_1 x$ results in an over and under estimate error except at $\bar{x}, \bar{y}$ of the data.

is no policy error of the choice of the estimating model occurs at $\bar{x}$ and $\bar{y}$ of the data, or weight = 3 lb and C_a = \$3.67, which for $y = a + b_1x$ and C_a are equal. Note that the regression equation has a negative intercept, which as a practical matter suggests negative fixed costs, an unlikely situation. When $a < 0$, the engineer is signaled for faulty data or the equation model is suspect.

The unit method fails to use the principle of economy of scale. For example, a 2-lb sphere costs \$2.44, and a 4-lb sphere costs \$4.88. At least a linear regression improves on the value as \$1.67 and \$5.67, which is not simple doubling.

6.6 COST- AND TIME-ESTIMATING RELATIONSHIPS

Cost-estimating relationships (CERs) and time-estimating relationships (TERs) are mathematical models or graphs that estimate cost or time. CERs and TERs are popular jargon within the U.S. Department of Defense and other national organizations.

The word parameter is often used in the CER context. A parameter usually refers to empirical coefficients in an equation [e.g., $C_a = -2.33 + 2.00(\text{lb})$]. But after the development of a CER, and in the application phase, we substitute the particular parameter values for the design and calculate a cost, such as the parameter of a casting weighing 5 lb (11.1 kg) and costing \$7.67. Simply, CERs and TERs are statistical regression models that mathematically describe the cost of an item or activity as a function of one or more independent variables. Previous discussion of those principles were covered in chapter 5, but forecasting deals with time serial and minor analytical problems. However, CERs and TERs are formulated to present estimates for end items. Rules of thumb, such as the unit method, are not recognized as CERs.

Those approaches are not new, but for product, project, and system estimates the ideas use output design variables (i.e., physical or performance parameters such as weight, speed, power thrust, etc.) to predict cost since the design parameters are available early. Estimates are preliminary, and their analysis adopts historical information. Those equations are statistical relationships between cost or time and physical or performance characteristics of the past designs. Sometimes, those characteristics or parameters are termed *cost drivers.*

As in all functional estimating models there must be a logical relationship of the variable to cost, a statistical significance of the variables' contribution, and independence of the variables to the explanation of cost. This is sometimes referred to as *causality.*

CERs are developed by using a variety of steps. We suggest this approach: obtain actual costs, interview experts who have knowledge of cost and design, find cost-time drivers, plot data roughly and understand anomalies, replot and conclude regression analysis, review for accuracy and communication, publish, and distribute the CERs to engineers and cost analysts. Those steps constitute the basis for much of cost analysis.

6.6.1. Learning

It is recognized that repetition with the same operation results in less time or effort expended on that operation. This improvement can be modeled with ordinary estimating techniques. The improved performance is called learning. The first applications of learning were in airframe manufacture, which found that the average number of man hours spent in building an airplane declined at a constant rate over a wide range of production. Direct labor learning is discussed in chapter 2.

Other names abound for learning model, including manufacturing progress function and the experience or dynamic curve. They suggest that cost can be lowered with increasing quantity of production or experience. Knowing how much product cost can be lowered and at what point learning is applied in estimating procedures are reasons for studying learning prior to product estimating methods.

The learning model rests on the following observations:

1. The amount of time or cost required to complete a unit of product is less each time the task is undertaken.
2. Unit time will decrease at a decreasing rate.
3. Reduction in unit time follows a specific estimating model such as $y = ax^b$, an equation studied in chapter 5.

To state the underlying hypothesis, the direct labor man hours necessary to complete a unit of product will decrease by a constant percentage each time the production quantity is doubled. Though the hypothesis stresses only time, practice has extended the concept to other types of measures. A frequent stated rate of improvement is 20% between doubled quantities. This establishes an 80% learning and means that the man hours to build the second unit will be the product of 0.80 times that required for the first. The fourth unit (doubling 2) will require 0.80 times the man hours for the second, the eighth unit (doubling 4) will require 0.80 times the fourth, and so forth. The rate of improvement (20% in this case) is constant with regard to doubled production quantities, but the absolute reduction between amounts is less.

The notion of constant reduction of time or effort between doubled quantities can be defined by a unit formula:

$$T_u = KN^s \tag{6.4}$$

where T_u = effort per unit of production, such as man hours or dollars required to produce the Nth unit

N = unit number

K = constant, or estimate, for unit 1 in dimensions compatible to T

s = slope or a function of the improvement rate

The slope is negative because the effort decreases with increasing production. Equation (6.4) plots as a curved line on arithmetic coordinates but as a straight line on logarithmic coordinates. Taking logarithms of both sides we get

$$\log T_u = \log K + s \log N \qquad (6.5)$$

which, in terms of $y = \log T_u$, $x = \log N$, and $a = \log K$, has the form $y = a + bx$, the equation of a straight line. Figures 6.3 and 6.4 show the arithmetic and logarithmic plot.

To understand the graphic presentation of the learning curve on logarithmic graph paper, first compare the characteristics of arithmetic and logarithmic graph paper. On arithmetic graph paper equal numerical differences are represented by equal distances. For example, the linear distance between 1 and 3 will be the same as from 8 to 10. On logarithmic graph paper the linear distance between any two quantities is dependent on the ratio of those two quantities. Two pairs of quantities having the same ratio will be equally spaced along the same axis. For example, the distance from 2 to 4 will be the same as from 30 to 60 or from 1000 to 2000.

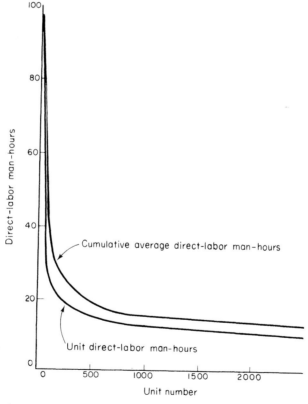

Figure 6.3.   Eighty percent learning curve with unit 1 at 100 direct-labor man hours, arithmetic coordinates.

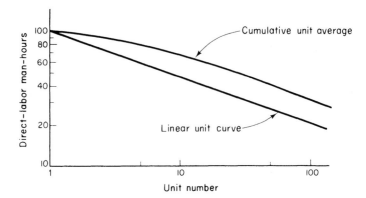

Figure 6.4. Eighty percent learning curve with unit 1 at 100 direct-labor man hours, logarithmic coordinates, and unit line assumed linear.

The learning curve is usually plotted on double logarithmic paper, meaning that both the abscissa and the ordinate will be a logarithmic scale. For an exponential function $T_u = KN^s$ the plot will result in a straight line on log log paper. Because of this, the function can be plotted from either two points or one point and the slope (e.g., unit number 1 and the percentage improvement). Also, by using log log paper the values for a large quantity of units can be presented on one graph, and these can be read relatively easily from the graph. Arithmetic graph paper, on the other hand, requires many values to sketch in the function.

Equation (6.4) implies a constant reduction for doubled production. For any fixed value of s,

$$T_1 = K1^s, \qquad T_2 = K2^s$$

and

$$\frac{T_2}{T_1} = \frac{K(2^s)}{K(1^s)} = 2^s$$

Also, $T_2 = K(2^s)$, $T_4 = K(4^s)$, and $T_4/T_2 = 2^s$, and, similarly, $T_8/T_4 = 2^s$. Every time quantity is doubled the time per unit is a constant 2^s of what it was. Because s is negative, the time per unit decreases with quantity. It is common practice to express learning in terms of the gain for double production. An 80% learning requires only 80% of the time per unit every time production is doubled. Note Fig. 6.5 for this relationship. Define

$$\Phi = \frac{T_{2N}}{T_N} = \frac{K(2N)^s}{K(N)} = 2^s \tag{6.6}$$

where Φ is the decimal ratio of time per unit required for doubled production. Taking logarithms, we have

$$\log \Phi = s \log 2$$

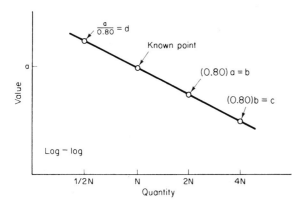

Figure 6.5. Constructing learning line on basis of known data for 80% curve. Note: 50 to 100 is the same distance as 500 to 1000 or 5 to 10 on log-log paper.

or

$$s = \frac{\log \Phi}{\log 2} \qquad (6.7)$$

A table of decimal learning ratios to Φ is given as follows:

Φ	Exponent, s
1.0 (no learning)	0
0.95	−0.074
0.90	−0.152
0.85	−0.234
0.80	−0.322
0.75	−0.415
0.70	−0.515
0.65	−0.621
0.60	−0.737
0.55	−0.861
0.50	−1.000

Application of Eq. (6.4) can be illustrated by an 80% learning curve with unit 1 at 1800 direct labor man hours. Solving for T_8 the number of direct labor man hours required to build the eighth unit gives

$$T_8 = 1800(8)^{\log 0.8/\log 2}$$

$$= 1800(8)^{-0.322} = \frac{1800}{1.9535} = 922 \text{ hours}$$

For unit 1 requiring 1500 hours and a projected learning of 75%, the time for unit 90 is

$$T_{90} = 1500(90)^{-0.415} = \frac{1500}{6.471} = 232 \text{ hours}$$

It may be desired to find whether learning has materialized and, if it has, to determine the function for which other unit estimates can be found. Where only two points are specified it may be desirable to find the learning curve that extends through them. Let the two points be specified as (N_i, T_i) and (N_j, T_j). At each point

$$T_i = KN_i^s \qquad T_j = KN_j^s$$

Dividing the second into the first gives

$$\frac{T_i}{T_j} = \left(\frac{N_i}{N_j}\right)^s \tag{6.8}$$

and taking the log of both sides we have

$$\log \frac{T_i}{T_j} = s \log \frac{N_i}{N_j} \tag{6.9}$$

$$s = \frac{\log T_i - \log T_j}{\log N_i - \log N_j}$$

K may be found by substituting s into $T_i = KN_i^s$ and solving for K:

$$\log T_i = \log K - s \log N_i \tag{6.10}$$

This has a linear form like $y = a + bx$, one of the linear models described in chapter 5. On log log paper the intercept is K and the slope of the line is equal to $-s$.

An example will now illustrate this concept. A company audited two production units, the 20th and 40th, and found that about 700 hours and 635 hours were used, respectively. Now, at the 79th unit, they want to estimate the time for the 80th unit:

$$s = \frac{\log 700 - \log 635}{\log 20 - \log 40}$$

$$= \frac{2.8451 - 2.8028}{1.3010 - 1.6021} = \frac{0.0423}{-0.3011} = -0.1406$$

and

$$\log \Phi = s \log 2$$

$$= (-0.1406)(0.3010) = 9.9577 - 10$$

Taking the antilog of both sides gives $\Phi = 0.907$. The percentage learning ratio is 90.7%. By using the data for the 20th unit, we obtain

$$700 = K(20)^{-0.1406}$$

$$\log 700 = \log K - 0.1406 \log 20$$

$$\log K = 2.8451 + 0.1406(1.3010) = 3.0279$$

Taking the antilog, $K = 1066$ hours. The learning curve function is

$$T_u = 1066N^{-0.1406}$$

The unit time for any unit can now be calculated directly. For the 80th unit

$$T_{80} = 1066(80)^{-0.1406} = 576 \text{ hours}$$

Usually, learning is calculated by using regression models since historical data of many points are required. For instance, see Problems 5.7 and 5.14.

The unit formulation can be extended to other types of functional models. It may be necessary to determine a cumulative average number of direct labor man hours. This is found by the cumulative total T_c from unit 1 to N and is

$$T_c = T_1 + T_2 + \cdots + T_N = \sum_{u=1}^{N} T_u \tag{6.11}$$

let T_a denotes the average total effort from unit 1 to N.

$$T_a = \frac{\sum_{u=1}^{N} T_u}{N} = \frac{T_c}{N} \tag{6.12}$$

A good approximation of the cumulative average number of direct labor man hours for 20 or more units is given by

$$T_a \doteq \frac{1}{(1 + s)} KN^s \tag{6.13}$$

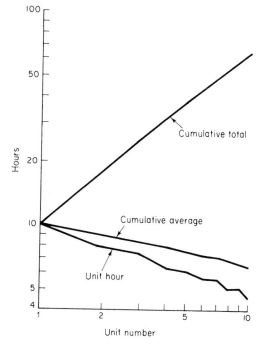

Figure 6.6.   Example of raw data where cumulative average appears to be a straight line.

Unit 1 is estimated to require 10,000 hours, and production is assumed to have an 80% learning curve. What will be the unit direct labor man hours, the cumulative direct labor man hours, and the cumulative average direct labor man hours for unit 4?

$$T_1 = 10,000$$

$$T_2 = 10\ 000(2)^{-0.322} = 8000$$

$$T_3 = 10\ 000(3)^{-0.322} = 7021$$

$$T_4 = 10\ 000(4)^{-0.322} = 6400$$

The cumulative direct labor man hours using Eq. (6.11) are

$$\sum_{u=1}^{N=4} T_u = 31,421$$

and the average by using Eq. (6.12) is

$$T_a = \frac{1}{4}(31,421) = 7855$$

Also using Eq. (6.13)

$$T_a \doteq \frac{6400}{1 - 0.322} = 9440$$

Equation (6.13) is not accurate in this quantity range because it is on the top left hump of the curve. This can be seen by examining Fig. 6.4.

What has been shown so far is only one half of the learning. The previous development assumed that the unit learning line is linear.

Two other curves, T_c and T_a, were derived from this basic one. Sometimes those formulas are called the Boeing concept. Another frequently encountered learning curve model is the cumulative average, which is the original model first described by T. P. Wright in 1936; that is, as the total quantity of units doubles, the average cost per unit decreases by a constant percentage. In this case the average learning line is linear and the unit line curves under the average until 20 to 30 units, at which point they become parallel.

Figure 6.7 is a graph of the two sets of curves. If the two learning theories start at the same K intercept and have the same learning rate, different values will result. Figure 6.7 shows that both theories have a common line, T_u and T_a', but their derivative lines depart. It is the common line that is regressed against actual data. It is necessary to state which basic theory is chosen for any application. Various proponents adopt one practice or the other. Either will work, but values may not be directly comparable.

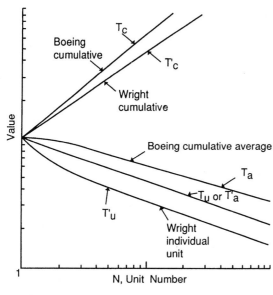

Figure 6.7. Comparison of two theories: (a) unit line T_u is assumed linear and (b) cumulative average line T_a' is assumed linear.

Define

$$T_a' = \frac{KN^{s+1}}{N} = KN^s \tag{6.14}$$

where T_a' is the average effort per unit given T_a' is a straight line.

$$T_c' = KN^{s+1} \tag{6.15}$$

where T_c' is the cumulative total effort from unit 1 to N.

$$T_u' = KN^{s+1} - K(N - 1)^{s+1} \tag{6.16}$$

where T_u' is the unit effort given that T_a' is the linear line.

Unit 1 is estimated to require 10,000 hours, and production is assumed to have an 85% learning curve. What will be the cumulative average hours, unit hours, and cumulative hours at 45 units?

$$T_a' = 10,000(45)^{-0.234} = 4103 \text{ average hours}$$

$$T_u' = 10,000(45)^{-0.234+1} - 10,000(44)^{-0.234+1}$$

$$= 186,654 - 181,503 = 3151 \text{ hours}$$

$$T_c' = 10,000(45)^{-0.234+1} = 184,654 \text{ hours}$$

Because of their frequent use, the formulas are converted into tables provided in appendix 3. Columns 1–4 are developed by using Eqs. (6.4), (6.11), and (6.12) and by having the unit line as the straight line.

What follows is typical for 93%, where $s = $ log 0.93/log 2 $= -0.105$.

UNIT ASSUMPTION IS STRAIGHT LINE

(1) N	(2) T_u	(3) T_c	(4) T_a
1	$K = 1.0000$	1.0000	1.0000
2	0.9300	1.9300	0.9650
3	0.8913	2.8213	0.9404
4	0.8649	3.6862	0.9216
5	0.8449	4.5312	0.9062
10	0.7858	8.5604	0.8560
15	0.7531	12.3856	0.8257
Equation	KN^s	$\sum\limits_{1}^{N} T_u$	T_c/N

Appendix 3 also provides values for the learning curve where the cumulative average is the straight line on logarithm scales. Tabulation for the 93% slope is as follows:

CUMULATIVE AVERAGE ASSUMPTION IS STRAIGHT LINE

(1) N	(2) T'_a	(5) T'_c	(6) T'_u
1	$K = 1.0000$	1.0000	1.0000
2	0.9300	1.8600	0.8600
3	0.8913	2.6740	0.8140
4	0.8649	3.4596	0.7856
5	0.8449	4.2246	0.7650
10	0.7858	7.8579	0.7073
15	0.7531	11.2969	0.6767
Equation	KN^s	KN^{s+1}	$KN^{s+1} - K(N-1)^{s+1}$

The learning curve has been successfully applied to engineering change-order calculation, follow-on estimating, break-even analysis, and spare-parts production. Those practical applications are discussed in chapter 8. Learning deals with the economy of quantity principle, which differs from economy of size or scale.

6.6.2. Power Law and Sizing Model

The power law and sizing model is frequently used for estimating equipment. This model is concerned with designs that vary in size but are similar in type. The un-known costs of a 200-gallon kettle can be estimated from data for a 100-gallon kettle provided both are of similar design. No one would expect that the 200-gallon kettle would be twice as costly as the smaller one. The law of economy of scale assures that. The power law and sizing model is given as

$$C = C_r \left(\frac{Q_c}{Q_r}\right)^m \tag{6.17}$$

where C = cost value sought for design size Q_c
$\quad\quad C_r$ = known cost for a reference size Q_r
$\quad\quad Q_c$ = design size expressed in engineering units
$\quad\quad Q_r$ = reference design size expressed in engineering units
$\quad\quad m$ = correlating exponent, $0 < m \leq 1$

If we let $m = 1$, then we have a strictly linear relationship and deny the *law of economy of scale*. For chemical processing equipment m is frequently near 0.6, and for this reason the model is sometimes called the sixth-tenth model. The units on Q are required to be consistent because it enters only as a ratio.

Assume that 6 years ago an 80-kW diesel electric set, naturally aspirated, cost \$160,000. The plant engineering staff is considering a 120-kW unit of the same general design to power a small isolated plant. If the value of $m = 0.6$, then

$$C = 160,000 \left(\frac{120}{80}\right)^{0.6} = \$204,000$$

The model can be altered to consider changes in price due to inflation or deflation and effects independent of size, or

$$C = C_r \left(\frac{Q_c}{Q_r}\right)^m \frac{I_c}{I_r} + C_1 \tag{6.18}$$

where C_1 is the constant unassociated cost. The price index for this class of equipment 6 years ago was 187 and now is 194. Assume we want to add a precompressor, which, when isolated and estimated separately, costs \$18,000 or

$$C = 160,000 \left(\frac{120}{80}\right)^{0.6} \left(\frac{194}{187}\right) + 18,000 \doteq 230,000$$

The determination of m is important to the success of this model, and methods of curvilinear regression and rectification given in chapter 5 are cogent. If the statistical analysis assumes constant dollars, then the index ratio I_c/I_r is used for increases or decreases for inflation or deflation effects.

The model usually does not cover those situations where the estimated design Q_c is greater or less than Q_r by a factor of 10. Some typical exponents for equipment cost versus capacity are given next.

Equipment	Size Range	Exponent, m
Blower, centrifugal (with motor)	1–3 hp	0.16
Blower, centrifugal (with motor)	$7\frac{1}{2}$–350 hp	0.96
Compressor, centrifugal (motor drive, air service)	20–70 hp drive	1.22
Compressor, reciprocating (motor drive, air service)	5–300 hp drive	0.90
Dryer, drum (including auxiliaries, atmospheric)	20–60 ft^2	0.36

The value of m is important in several ways. If $m > 1$, then we deny the economy of scale law, as shown with the centrifugal compressor. In fact, when $m > 1$, we have *diseconomy of scale*. If $m = 0$, then we can double the size without affecting cost, an unlikely event.

An equation expressing unit cost C/Q_c can be found, or

$$C\frac{Q_r}{Q_c} = C_r\frac{Q_r}{Q_c}\left(\frac{Q_c}{Q_r}\right)^m = C_r\left(\frac{Q_c}{Q_r}\right)^{-1}\left(\frac{Q_c}{Q_r}\right)^m$$

$$\frac{C'}{Q_c} = \frac{C_r}{Q_r}\left(\frac{Q_c}{Q_r}\right)^{m-1}$$

(6.19)

Because total cost varies as the mth power of capacity in Eq. (6.17), C/Q_c will vary as the $(m - 1)$st power of the capacity ratio.

6.6.3. Other CERs

Other relationships exist, such as

$$C = KQ^m$$

(6.20)

where K = constant for a plant, equipment, part
$\quad\quad Q$ = capacity expressed as a design dimension

Difficulties exist in using Eq. (6.20) because it is necessary to know the capital cost of an identical plant. The scale factor m is not constant for all sizes of the design. Generally, scale up or scale down by more than a factor of 5 should be avoided.

The problems of Eq. (6.20) can be overcome by differentiating between the direct capital costs, which are subject to the economy of scale rule, and the indirect or fixed element, which is not. Another relationship can be expressed as

$$C = C_v\left(\frac{Q_c}{Q_r}\right)^m + C_f$$

(6.21)

where C_v = variable element of capital cost, dollars
$\quad\quad C_f$ = fixed element of capital cost, dollars

A multivariable CER is also possible. For instance,

$$C = KQ^m N^s \qquad (6.22)$$

where the symbols are as previously shown. In those cases the coefficients are determined by regression methods. Equation (6.22) is a relationship dealing with the cost analysis principles of quantity and scale.

CERs can be plotted together with all points, giving the engineer a feeling for the data and the line. Removing the points destroys the sense of variation. Are the data a scattergram or are the data lying neatly on a line? Figure 6.8 is an example of the plot of pressure vessels, where the x axis is "gallons per minute" multiplied by "feet head."

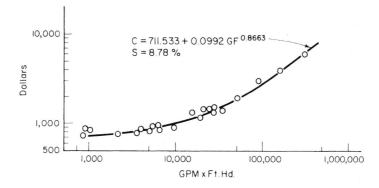

Figure 6.8.   Example of CER plot and equation by using physical variables.

6.7 PROBABILITY APPROACHES

As usually prepared, the estimate represents an "average" concept. However, the estimate does not reveal anything about the probability of the *expected values* and uses information called certain, or deterministic.

Much has been written on the topic of statistical decision theory, also called Bayesian analysis. Despite the subject's importance, delving into details here is beyond the scope of this text.

6.7.1. Expected Value

For this simpler discussion we assume that the engineer can give a probability point estimate to each element of uncertainty as represented by the economics of the design. This assignment has nonnegative numerical weights associated with possible events, such that if an event is certain its associated probability equals 1. If two events A and B are mutually exclusive, the probability of the event "either A or B" equals the sum of the probability for each of the events. Those probabilities are really a numerical judgment of future events, and the techniques for deriving probabilities follow: (1) analysis of historical data to give a relative frequency interpretation, (2) convenient approximations like the normal, and (3) introspection, or what is called opinion probability.

Opinion probabilities call for judgmental expertise and a pinch of luck. Better success is assured when past data are analyzed. On the other hand, data may be unavailable, and it should be remembered that while data are past, those probabilities should be indicators of the future. Sometimes both past data and a reshuffling of probabilities are jointly undertaken. This type of discrimination is not new to professional cost analysis practice.

For the most part, engineers have preferred to deal with the simplest case of certainty. When "material cost is $1," we imply, but leave unstated, that the probability is 1, a certain event. Any other material cost has the probability of zero. Despite this practice it seldom exists. The category involving risk is appropriate whenever it is possible to estimate the likelihood of occurrence for each condition of the design. Those probabilities describe the true likelihood that the predicted event will occur. Formally, the method incorporates the effect of risk on potential outcomes by a weighted average. Each outcome of an alternative is multiplied by the probability that the outcome will occur. This sum of products for each alternative is entered in an expected value column, or mathematically for the discrete case,

$$C(i) = \sum_{j}^{n} p_j x_{ij} \qquad (6.23)$$

where C = expected value of the estimate for alternative i
p_j = probability that x takes on value x_j
x_{ij} = design event

The p_j represents the independent probabilities that their associative x_{ij} will occur with $\sum_{j=1}^{n} p_j = 1$. The expected value method exposes the degree of risk when reporting information in the estimating process.

Consider the following example: An electronics manufacturing firm is evaluating a portable television that has special design features. Market research indicated a substantial market available for a small lightweight set if priced at $850 retail. This implies that the set will have to be sold for approximately $650 to wholesalers. Several questions need to be answered before the decision can be made to enter the market. Three important ones are: What will be the first year's sales volume in units? How much will the sets cost to produce? What will be the profit? To answer those questions, marketing furnishes an estimate of the first year's sales in units. Subsequently, the engineers provide a total cost per unit. Marketing presented the following forecast:

Annual Sales Volume	Probability of Event Occurring
15,000	0.2
20,000	0.2
25,000	0.6*

*Most frequent case.

After inspecting the sales forecast, engineering provides the following estimates:

Cost per Unit ($)	Probability of Event Occurring
450	0.7*
500	0.2
550	0.1

*Most frequent case.

The illustration can now be divided into (1) risk not apparent and (2) risk apparent. Assume that marketing and estimating use the most probable figure from their studies and do not report any uncertainty. The profit is calculated as

$$profit = (650 - cost)volume$$

where cost $\quad = \$450$
volume $= 25{,}000$ units
profit $\;= (650 - 450)25{,}000 = \$5{,}000{,}000$

Conversely, assume that the organization encourages a policy of reporting risk in estimates. The probability numbers are used without the effects of editing. Because three possibilities exist for cost and three for volume, we calculate nine profit possibilities.

The reader may argue that simple mathematics need not be complicated. Both engineering and marketing could have computed an expected value as 470 ($= 450 \times 0.7 + 500 \times 0.2 + 550 \times 0.1$)   and   22,000   ($= 15{,}000 \times 0.2 + 20{,}000 \times 0.2 + 25{,}000 \times 0.6$). The expected profit becomes 22,000 (650 − 470) = \$3.96 million. In the case of risk not apparent the total profit is overstated. The calculations for this example require that the cost per unit be independent of volume and that cost be inversely related to volume, which is generally not the case.

(650 − Cost)Volume	Joint Probability of Occurrence	Expected Value Profit
(650 − 450)15,000 = 3,000,000	0.14	420,000
(650 − 450)20,000 = 4,000,000	0.14	560,000
(650 − 450)25,000 = 5,000,000	0.42	2,100,000
(650 − 500)15,000 = 2,250,000	0.04	90,000
(650 − 500)20,000 = 3,000,000	0.04	120,000
(650 − 500)25,000 = 3,750,000	0.12	450,000
(650 − 550)15,000 = 1,500,000	0.02	30,000
(650 − 550)20,000 = 2,000,000	0.02	40,000
(650 − 550)25,000 = 2,500,000	0.06	150,000
	Total = 1.00	Profit = 3,960,000

One of the difficulties in applying expected value is the inability of engineers to guess opinion probabilities. Seldom are there sufficient experiences that allow for a gradual development of this ability.

Suppose that two designs are to be judged and that cost is the criterion. Figure 6.9 illustrates four cases where the cost estimates are shown as probability distributions. For case a all probable costs of A are lower than B and there is no difficulty in choosing A. The situation for case b has a possibility that the actual cost of
A will be higher than B. Given that this is a small chance, the engineer selects A.
As the amount of overlap increases, the expected cost estimate C_a may not be given
a clear choice. For case c the average cost estimates are equal, although their distributions are not. Certainly, the cost distribution of B is greater and there is an associated risk of having a lower cost than A. On the other hand, A is less variable. The
engineer's assessment of cost return and risk will serve as the guide here. In case d
the expected cost estimate of B is lower but less certain than A. If only the expected
value estimate is used in this case, then the engineer likely chooses the more desirable alternative B.

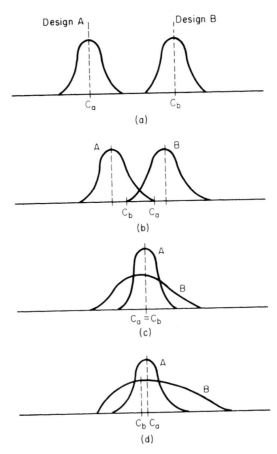

Figure 6.9. Variations of cost distributions.

6.7.3. Range

Engineers knowing the weaknesses of information and techniques recognize that there are probable errors. The indication that cost is a random variable opens up the topic of range estimating. A random variable in statistical parlance is a numerical valued function of the outcomes of a sample of data. Another approach to single valued estimating involves making a single estimate and bracketing this estimate for each cost element. This forms the basis for range estimating.

The following procedure is based on a method developed for PERT (program evaluation and review technique) and involves making a most likely cost estimate, an optimistic estimate (lowest cost), and a pessimistic estimate (highest cost). Those estimates are assumed to correspond to the beta distribution in Fig. 6.10. This particular figure is skewed left. Symmetric and skewed-right distributions are also possible. The total area under the curve is 1. The mathematics of the beta probability density are complicated, but their expected values lead to simplified equations.

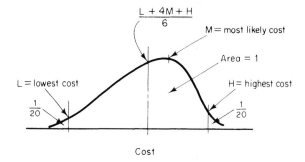

Figure 6.10.   Location of estimates for PERT-based beta distribution.

With the three estimates made, a mean and variance for the cost element can be calculated as

$$E(C_i) = \frac{L + 4M + H}{6} \tag{6.24}$$

$$\text{var}(C_i) = \left(\frac{H - L}{6}\right)^2 \tag{6.25}$$

where $E(C_i)$ = expected cost for element i, dollars
     L = lowest cost, or best case, dollars
     M = modal value of cost distribution, dollars
     H = highest cost, or worst case, dollars
     $\text{var}(C_i)$ = variance of element i

If a dozen or more elements are estimated this way and are assumed to be independent of each other and are added, then the distribution of the total cost is approximately normal. This follows from the central limit theorem. The mean of the sum is the sum of the means, and the variance of the sum is the sum of the

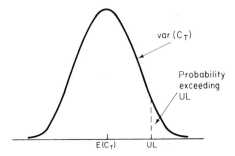

Figure 6.11. Normal curve assumed to be a result of sum of many individual beta probabilities.

variances. The distribution of the sum of costs will be normal despite the individual shape of cost elements. This is shown by Fig. 6.11.

$$E(C_T) = E(C_1) + E(C_2) + \cdots + E(C_n) \tag{6.26}$$

$$\text{var } (C_T) = \text{var } (C_1) + \text{var } (C_2) + \cdots + \text{var } (C_n) \tag{6.27}$$

where $E(C_T)$ = expected total cost, dollars
  $\text{var } (C_T)$ = variance of total cost, dollars

Table 6.1 shows an application of this approach. The expected cost is \$10,600, and the variance representing probable error is 22,300. Now, various statements regarding the interval for the future value can be made, such as some cost will not be exceeded and probability boundaries can be established.

With the normal curve assumed it is possible to examine this distribution and ask the question, What is the probability that a certain cost will be exceeded, or $P[\text{cost} > E(C_T)]$? This is found by using

$$Z = \frac{UL - E(C_T)}{[\text{var } (C_T)]^{1/2}} \tag{6.28}$$

where Z = value of the standard normal distribution, appendix 1
  UL = upper limit cost arbitrarily selected, dollars

TABLE 6.1. CALCULATION OF EXPECTED COST AND VARIANCE FOR THE RANGE-ESTIMATING METHOD

	Lowest Cost, L	Most Likely Cost, M	Highest Cost, H	Expected Cost, E(C_i)	Variance, var (C_i)
1. Flashlamp	\$ 370	\$ 390	\$ 430	\$ 393.33	100.00
2. Data grid	910	940	1,030	950.00	400.00
3. Computer	200	210	270	218.33	136.11
4. Optical isolator	170	180	190	180.00	11.11
5. Power supply	260	290	350	295.00	225.00
6. Switch	171	172	176	172.50	0.69
7. Capacitor	875	925	975	925.00	277.78
...	...	...	...	...	...
n. Frame	2,000	2,100	2,600	2,166.67	10,000.00
			$E(C_T) =$	\$10,610.66	var $(C_T) =$ 22,301.39

The location of *UL* is shown by Fig. 6.11 in relation to $E(C_T)$. The square root of var (C_T) is the standard deviation. The value of Z is entered in appendix 1, and an upper tail probability can be determined. Consider Table 6.1 and find the probability that the cost will exceed $10,850, which is

$$Z = \frac{10,850 - 10,611}{(22,301)^{1/2}} = 1.66$$

$$P(\text{cost} > 10,850) = 0.05$$

According to PERT practices, the optimistic and pessimistic costs would be wrong only once in 20 times if the activity were to be performed repeatedly under the same conditions. Usually, the job is estimated only once, so those requirements are never met. We suggest that those costs be viewed as best, most likely, or worst cost. Further, it is unlikely that a distribution of major actual cost elements can ever be constructed. Costs, if they are internal to a firm, are seldom independent of each other, although for a project estimate of various contractors' costs, independence is more valid. The expected value method discussed earlier required selection of opinion probabilities, which we believe is unsound. However, the range method requires three values of cost; and if rational methods are employed, then the range method is acceptable as a preliminary technique. This method is discussed in chapter 9, where it is useful for contingency analysis.

6.8 STANDARD TIME DATA

Time study, man hour reports, and work sampling are methods of work measurement, which were introduced in chapter 2. But time data in this raw form are not usable by the engineer for design estimating. Frequently, those data include bad methods, inefficient conditions, or nonaverage operators. Sometimes those measurements are restricted to a narrow range of observed work. Worst of all, the time studies are after the design is complete and are not available during the design period. Standard time data are available during the design period and thus can be used in a variety of ways.

Standard time data are a catalog of standard tasks used for estimating a variety of work. Parenthetically, standard time data are arranged in a systematic order and are used over and over. The advantages over direct observation methods such as time study, predetermined motion time data, work sampling, and man-hour reports are lower cost and greater consistency. Description is provided in advance of the need for the estimating data, and more engineers can use standard time data.

Standard time data are ordinarily determined from any of the various methods of observing work. In manufacturing, time study and predetermined motion data are the major sources. Whatever the source of work data, the development of standard data is similar.

Starting with the time studies, the engineer creates new information that is widely useful for estimating, both in manufacturing and construction.

The several steps in the finding of standard time data are as follows:

1. Collect several or more time studies for an operation.
2. Arrange the elements of the time studies into common groups.
3. Regress the element time against one or more independent and causal variables, and find an equation of the element, such as a time estimating equation (TER).
4. Determine if each element TER is constant or variable according to conditional rules.
5. If the element TER is constant, then find a typical value that represents the element.
6. If the element TER is variable, then convert the TER into a table.
7. Collect all constant and variable elements, and publish tentative standard time data.
8. Check the accuracy of the standard time data against old and new time studies.
9. If the accuracy is acceptable, then publish and distribute the information to other engineers for their use in estimating designs.
10. The engineer uses the standard time data for estimating new designs and products requiring manufacturing. The standard data are applied by using *lookup rules of thumb*.

A dozen or so time studies of a single manufacturing process are necessary before standard time data can be constructed. Although the number of basic studies used for standard time data is a statistical question, accuracy is a function of the number of observations.

Engineers use regression analysis to extend those raw time study data into a more digestible form. However, it is not the original time measurements that the engineer ultimately desires. Rather, it is a set of engineering performance data, or standard time data, or more briefly standard data, that the engineer uses for estimating.

Standard time data are now defined as time values arranged in a concise form from which a standard time can be found that estimates the cost of an operation performed under usual conditions. The key word in this definition is "usual." It is unwise to devise standard time data without first knowing conditions, specifications, methods, and procedures. If those precautions are heeded, then the time values are based on the local conditions in which the standard is to apply.

Assuming that the basic time studies were correctly determined (see chapter 2), one of the first steps facing the engineer is to determine which of the several elements of the time studies are constant or variable. Estimating data for a process are composed of constant and variable elements for several reasons. Because accuracy is necessary for estimating, variable elements provide sensitivity to the major drivers of an operation. Constant elements, while necessary for the absolute quantity of the estimate, are nonresponsive to operation drivers but are required each and every time the operation is run.

A constant is accepted by convenience and assumed to have less than an arbitrary percentage P_1 slope between the limits of work scope. Very little work is ever constant, but to get products estimated we overlook some variability. If the time slope, say, a positive 20%, from the minimum to the maximum is found, then the analyst arbitrarily chooses to classify this element as a constant if the P_1 limit value were 50%. In this case a constant value is found by averaging the values of the process element.

If the slope of the element is greater than 50%, then the element is classified as a variable. The dependent variable time is related to an independent variable by a graph, a table, or regression formulas. A particular element may be one of many elements that define a set of standard time data for some production operation. Moreover, if the element is P_2 or less of the standard time of the total expected mean operation time, it could be classified as a constant irrespective of its elemental variability. If the element is greater than P_2, then it is classified as a variable element.

A TER element is judged "constant or variable" on the outcome of two rules. The technician will measure time study conditions that include extremes, minimum to maximum, of the design. The two tests are as follows:

Test 1. An element is conditionally variable if

$$\frac{\hat{y}_{max} - \hat{y}_{min}}{\hat{y}_{min}} \times 100 \geq P_1\% \qquad (6.29)$$

where $\hat{y}_{max}$ = maximum dependent value related to time driver x_{max}
  $\hat{y}_{min}$ = minimum dependent value related to time driver x_{min}
  P_1 = percentage, arbitrarily fixed

The $\hat{y}$ values are preferred over actual raw y data because the least squares time value minimizes observational errors. If test $1 < P_1$, then the element is constant, otherwise the second rule comes into play.

Test 2. An element is variable if

$$\frac{\hat{y}_{ave}}{\hat{y}_t} \times 100 \geq P_2\% \qquad (6.30)$$

where $\hat{y}_{ave}$ = average dependent value for x_{ave}
  $\hat{y}_t$ = total dependent value for $\bar{x}$ of operation elements
  P_2 = percentage, arbitrarily fixed

Test 1 is concerned with inner element variability, and test 2 evaluates the effect of the element average to the total average for the operation. An element could be variable on the basis of rule 1, but it might have so little effect on the average outcome that it would be more practical to consider it constant and overlook small variability. Both of those rules are subject to discretion by selection of P_1 and P_2 variability.

After the TER standard time data elements are judged constant or variable, the analyst organizes the information into tables.

Standard time data tables are developed in the following way. Constant elements are combined, and an average time value is calculated. One or several elements will be found constant for the operation. An abbreviated description of the elements is written, together with the best value, and is prepared as printed or typed material. Obviously, constant values are easy to use, but they are insensitive in estimating cost. A trade off between ease of estimating and sensitivity is necessary.

Each variable element is converted to a table. What is adequate and accurate for one variable element could be too detailed for the next. The simplest method of tabulating the variable element is to divide the time driver into classes (i.e., small, medium, large, etc.) and then select a time value for each class. This can be improved by subdividing the y into several or many unequal cells, or equal percentage jumps, and finding the x value corresponding to each specific y. One efficient way of doing this is given by

$$\hat{y}_{i+1} = \hat{y}_i(1 + P_3)$$

$$x_{i+1} = \frac{\hat{y}_{i+1} - a}{b} \tag{6.31}$$

where x_{i+1} = independent variable value, recursively determined
$\hat{y}_i$ = minimum starting dependent value
P_3 = percentage for step, arbitrarily fixed

Assume a TER $\hat{y} = 0.0642 + 0.0133x$, where $y_i = 0.131$ for $x_{min} = 5$. For $P_3 = 20\%$ and $y = 0.131 \times 1.2 = 0.157$, for which $x = 7.0$; $0.157 \times 1.2 = 0.189$, for which $x = 9.4$; and $0.189 \times 1.2 = 0.227$, for which $x = 12.2$. Summarizing, we have

x	5.0	7.0	9.4	12.2
$\hat{y}$	0.131	0.157	0.189	0.227

The $\hat{y}$ increases in absolute time from 0.026 (= 0.157 − 0.131) to 0.038 (= 0.227 − 0.189), and the neighboring steps jump at 20% increments.

Construction of a standard time table is by trial and error. Ease of understanding, speed in labor estimating, compression of tables into the fewest number, and accuracy are desirable features. After the tables are tentatively constructed, the tables are compared back to the original time measurements by using the three rules of table lookup. Three measures to evaluate accuracy are

$$E_g = \frac{\Sigma H_s - \Sigma H_a}{\Sigma H_a} \times 100 \tag{6.32}$$

where E_g = gross overall percentage error
H_s = time for design from tables by using lookup rules of thumb
H_a = time from original measurements

$$E_{ad} = \Sigma\left(\frac{H_s - H_a}{H_a} \times 100\right)\Big/N \qquad (6.33)$$

where E_{ad} = average deviation percentage error
$\quad\quad N$ = number of measurement studies

and

$$E_{at} = \Sigma\frac{(H_s - H_a)}{N} \qquad (6.34)$$

where E_{at} is the average time error per estimate.

Accuracy in those formulas relates the original measurement to the table lookup value. If those three measures of error are found unacceptable, the process is repeated with different P_1, P_2, and P_3 until satisfactory tables are constructed. Now, consider the construction of standard time data for a manufacturing operation.

Here is an example: Spot welding is a form of resistance welding in which two sheets of metal are held between copper electrodes. A welding cycle is started when electrodes contact the metal under pressure before current is applied, then a squeeze time under pressure with current, and finally a pressure dwell without current. Five time studies have been tabulated. Among the many choices of an independent variable, girth, or length plus width plus height (= L + W + H), and the number of spots are selected for handling and welding. Five elements are identified. The number of spots is specified as x_1. In spot welding, a second and third part, if necessary, are assembled and welded to the first or major part. Girth for handling the part is shown as x_2. Spot-welding time includes position part for spot weld, spot weld, and occasional cleaning of the electrodes. The load-major-part time includes moving part from skid to the electrodes. The secondary part and third part time includes moving the part from skid and positioning on major part with clamps and fixtures, and the variables are identified as x_3 and x_4. The unload time starts after the last spot and includes removing fixtures and clamps and moving the welded part onto the skid. Time is normal (i.e., observed time study elemental time that has been rated). The time in units of minutes is associated with their causal variables.

Table 6.2 is a listing of time studies with time drivers, $x_1, \cdots, x_5$, identified for each of the five elements. For element 1 the time driver x_1 is the number of spots, and $x_2, \cdots, x_5$ are the values of the girth L + W + H, which are identified as the causal variables for elements 2–5.

There are five dependent values, normal time and without allowances, at this stage. For example, y_1 is the spot time, and $y_2, \cdots, y_5$ are the times for handling of the major part, second and third part, and unloading of the spot-welded assembly. Those values are given in Table 6.3.

Thus, 0.28 minutes identified under spot-weld minutes corresponds to time study 1 where the number of spots is 8. The 0.28 was measured by a time study.

TABLE 6.2. INDEPENDENT VARIABLES AND VALUES FOR TIME STUDIES

Time Study	Number Spots, x_1	L + W + H Major Part, x_2	L + W + H Secondary Part, x_3	L + W + H Third Part, x_4	L + W + H Final Part, x_5
1	8	29	16	—	37
2	3	46	8	—	46
3	9	101	17	5	106
4	14	60	28	33	60
5	36	53	23	44	53

TABLE 6.3. DEPENDENT VALUES FOR TIME STUDY ELEMENTS

	y_1	y_2	y_3 Load Additional Parts Min	y_4	y_5	Total Time Study Min
Time Study	Spot-Weld Min	Load Major Part Min	Second	Third	Unload Min	
1	0.28	0.08	0.12	—	0.10	0.58
2	0.13	0.10	0.14	—	0.17	0.54
3	0.34	0.27	0.21	0.10	0.28	1.20
4	0.87	0.21	0.59	0.61	0.32	2.60
5	2.06	0.19	0.29	0.57	0.51	3.62

The linear regression equations are given in Table 6.4. For example, $\hat{y}_1 = -0.1156 + 0.0608x_1$, which relates the normal time to the number of spot welds, x_1. The regression equation and independent variable are shown for each element.

Assume that the engineer arbitrarily selects $P_1 = 50\%$, $P_2 = 15\%$, and $P_3 = 20\%$. Table 6.4 shows the steps to select constant or variable elements. The values x_{min} and x_{max} are determined from the time studies, and corresponding $\hat{y}_{min}$ and $\hat{y}_{max}$ values are found by using the equations. Data points x_{ave} and $\hat{y}_{ave}$ are calculated. Test 1 shows that only element 5 is constant. Test 2 confirms that elements 1 and 4 are finally variable, because elements 2 and 3 are classified as constant by this test. Table 6.4 shows the regression equations and tests 1 and 2 for finding constant and variable elements.

With elements 2, 3, and 5 as constant, their total regression mean time is 0.741 normal minute. Now we use Eq. (6.31). Element 1 is variable, and though requiring increasing steps of 20%, the number of spot welds is also required to be integer. Because the independent variable "number of spots" is an integer, it is necessary to show natural numbers for spot welds. Element 4 is variable and increases by 20% from minimum $\hat{y}$ value. Trial values for the variable elements are given by Table 6.5.

For ease and consistency in using standard time data tables, lookup rules of thumb are followed. Standard time tables are used in a special way for estimating. The following practices are rules of thumb that guide the selection of times for estimates.

1. If the value of the independent variable, or time driver, of the design is less than shown in a standard time data table, then the lookup will choose the

TABLE 6.4. REGRESSION EQUATIONS AND CONSTANT AND VARIABLE ELEMENTS FOR SPOT-WELD MEASUREMENTS

Element	Independent Variable	Regression Equation
1. Spot-weld	Number of spots, x_1	$-0.1156 + 0.0608x_1$
2. Load major part	Major part girth, x_2	$0.0139 + 0.0027x_2$
3. Load second part	Second part girth, x_3	$-0.1282 + 0.0216x_3$
4. Load third part	Third part girth, x_4	$0.0642 + 0.0133x_4$
5. Unload	Final part girth, x_5	$0.1907 + 0.0014x_5$

Element	x_{min}	x_{ave}	x_{max}	$\hat{y}_{min}$	$\hat{y}_{ave}$	$\hat{y}_{max}$
1	3	19.5	36	0.067	1.070	2.073
2	29	65	101	0.092	0.189	0.287
3	8	18	28	0.045	0.261	0.477
4	5	24.5	44	0.131	0.390	0.649
5	37	71.5	106	0.243	0.291	0.339
				$\hat{y}_t =$	2.201	

	Test 1		Test 2	
Element	$\dfrac{\hat{y}_{max} - \hat{y}_{min}}{\hat{y}_{min}} \times 100\ (\%)$	Test 1 Type Element	$\dfrac{\hat{y}_{ave}}{\hat{y}_t} \times 100\ (\%)$	Test 2 Type Element
1	2995	Variable	49	Variable
2	212	Variable	9	Constant
3	960	Variable	12	Constant
4	396	Variable	18	Variable
5	40	Constant	—	—

TABLE 6.5. TRIAL VALUES FOR VARIABLE ELEMENTS

Number of Spot Welds	Normal Time	Number of Spot Welds	Normal Time	Number of Spot Welds	Normal Time
3	0.067	9	0.432	21	1.162
4	0.128	10	0.493	24	1.344
5	0.188	12	0.614	28	1.588
6	0.249	14	0.736	33	1.892
7	0.310	16	0.858	39	2.257
8	0.371	18	0.979		

Load Third Part, L + W + H	Normal Time	Load Third Part, L + W + H	Normal Time
5.0	0.131	24.6	0.391
7.0	0.157	30.5	0.469
9.4	0.189	37.5	0.563
12.2	0.226	46.0	0.676
15.6	0.272	56.2	0.811
19.7	0.326		

minimum value as listed. This minimum is equal to or greater than the positive fixed intercept a of the TER.

2. If the time driver falls between listed table values, then the engineer will choose the next higher value. This rule discourages tabular interpolation, a time-consuming step often ineffective in improving accuracy.

3. If the time driver exceeds the tabular listed value, then the maximum of the values is used. This presumes a natural process barrier that may have been uncovered during measurement.

Those rules, while admittedly inflexible, encourage later verification of an estimate by someone other than the original engineer. Consistency is improved.

The trial values for the constant and variable elements are compared with measured values by using the table lookup rules of thumb and are shown in Table 6.6. Each time study will have a constant, and the time for the number of spots are removed from the trial standard time table, along with the time for the third part handling, which are summed to give H_s. This is then compared to the time study total time, and various errors are determined.

TABLE 6.6.   COMPARISON OF TRIAL VALUES TO TIME STUDY TIME

Time Study	Constant	+	No. Spots	+	Third Part	=	H_s	Time Study Total Time, H_a
1	0.741	+	0.371	+	0	=	1.112	0.58
2	0.741	+	0.067	+	0	=	0.808	0.54
3	0.741	+	0.432	+	0.131	=	1.304	1.20
4	0.741	+	0.736	+	0.554	=	2.031	2.60
5	0.741	+	2.257	+	0.664	=	3.662	3.62
							8.917	8.54

Errors can be assessed, or

$$E_g = \frac{8.917 - 8.54}{8.54} \times 100 = 4.4\%$$

$$E_{ad} = \frac{91.7 + 49.6 + 8.7 - 21.8 + 1}{5} = 25.9\%$$

$$E_{at} = \frac{0.532 + 0.268 + 0.104 - 0.569 + 0.042}{5} = 0.075 \text{ min}$$

Usually, standard time data are tested against new measurements in addition to original information. Though gross errors are small, average deviations are larger, and the average magnitude of estimating exceeding measurement is 0.075 min. Improvements can be expected with more studies and other trial values of P_1, P_2, and P_3. For this example we assume that the data meet evaluation criteria.

At this point the normal time data are converted to standard minutes ready for widespread use. If the allowances are 14%, then the allowance multiplier F_a is 1.163

[= 100/(100 − 14)]. Finally, the standard time data are distributed to engineers. Table 6.7 is a small example of detail standard time data. The engineer visualizes the elements knowing the design and operation and selects time values for the labor estimate. Additional applications of standard time data are demonstrated in chapter 7.

TABLE 6.7. EXAMPLE OF DETAIL STANDARD TIME DATA

Setup hours   1.2 hr
Operation elements in standard minutes
1. Load major part, load secondary part and clamp if necessary,
   and unload spot-welded part   0.86
2. Load third part, if required, and clamp

L + W + H in.	min	L + W + H in.	min
5.0	0.16	24.6	0.45
7.0	0.18	30.5	0.55
9.4	0.22	37.5	0.65
12.2	0.26	46.0	0.79
15.6	0.32	56.2	0.94
19.7	0.38		

3. Position parts, spot-weld, and occasional electrode cleaning

Spots	min	Spots	min	Spots	min
3	0.08	9	0.50	21	1.35
4	0.15	10	0.57	24	1.56
5	0.22	12	0.71	28	1.85
6	0.29	14	0.86	33	2.20
7	0.36	16	1.00	39	2.62
8	0.43	18	1.14		

Once the standard time data are prepared and distributed, it is used in a variety of ways. For example, consider the situation where three parts are to be spot welded. From the design, the third part, considered the minor part, is found to have a girth L + W + H = 27.2 in. There are 26 spots. A lot of 75 units is to be estimated for operation time. The lookup rules are adopted to increase speed, consistency, and accuracy in providing the estimate. By using Table 6.7, we estimate total lot hours as

Table Number	Description	Table Value
	Setup	1.2 hr
1	Constant elements	0.86 min
2	Load third part, 27.2 = L + W + H	0.55
3	26 spots	1.85
Total cycle		3.26 min

The time for the handling and clamping of the third part, having 27.2 L + W + H in. of girth, falls between two values in Table 6.7 (24.6 and 30.2). The higher value is selected by using the second lookup rule of thumb. Similarly, the number of spots required for this design falls between two entries, and the higher value is used for element 3. For a lot of 75 units it is necessary to have a setup of 1.2 hours and then to begin the cycle production. Total lot hours = 1.2 + 75(3.26/60) = 5.28 hours.

Standard time data are in advance of the product being produced in a manufacturing operation. Thus, the design can be estimated before a time study is available.

6.9 FACTOR

The *factor method* is a basic and important method for project estimates. Other terms such as ratio and percentage methods are about the same thing. Essentially, the factor method determines the estimate by summing the product of several quantities, or

$$C = (C_e + \sum_i f_i C_e)(f_I + 1) \qquad (6.35)$$

where C = value (cost, price of design)
$\quad C_e$ = cost of selected major equipment
$\quad f_i$ = factor for the estimating of buildings, instrumentation, etc.
$\quad f_I$ = factor for the estimating of indirect expenses such as engineering contractor's profit, and contingency
$\quad i$ = 1, $\cdots$, n factor index

The factors f_i are uncovered by historical, measured, or policy methods. Data from internal reports, industry at large, or from the government are the principal source.

A natural simplification leads to the preliminary unit estimating model $C = fC_e$, where one factor is used to find the composite cost. This factor-estimating formula has variants such as $C = \sum_i f_i D$, where D is the particular design parameter (e.g., material machining time, combat force, and capital investment) and f_i is the factor in units compatible to the design.

Though the unit-cost estimating method is limited to a single factor for calculating overall costs, the factor method achieves improved accuracy by adopting separate factors for different cost items. For example, the approximate cost of an office building can be estimated by multiplying the area by an appropriate unit estimate such as the dollars per square foot factor. As an improvement, individual cost per unit area figures can be used for heating, lighting, painting, and the like, and their value C can be summed for the separate factors and designs.

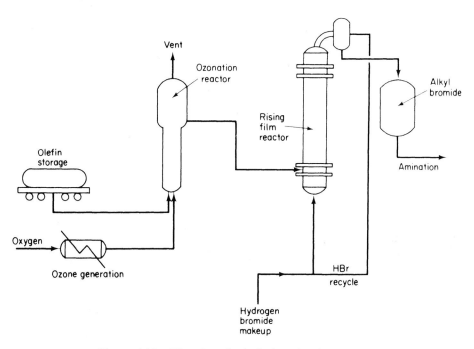

Figure 6.12.   Flowchart for hydrobromination process.

Plant cost can be estimated by the factor method. The hydrobromination process is one step in the sequence of making an intermediate product for a liquid soap formulation. The hydrobromination flow chart, given by Fig. 6.12, is a microprocess plant.

The flowchart and the specification sheet are input data to the engineer. The basic item (or items) of the process is identified with the flowchart. This basic item should be a major item for a building such as the structural shell or tons of concrete for a highway or process equipment in a chemical plant. After the cost of the basic item is determined, the next step is to find the cost relationship of other components as a percentage, ratio, total cost, or factor of the basic item.

For the building structural shell, the factors would be brick, concrete, masonry, architectural and reinforced concrete, carpentry, finish trim, electrical work, plumbing, and so forth. For the chemical process project, equipment erection, piping and direct materials, insulation, instrumentation, and engineering is correlated to process cost. These correlations are statistically found by using the methods in chapter 5. In some instances the variations in the cost of components being analyzed are independent of the variation in the cost of the basic item. Those independent components (e.g., roads, railroad siding, and site development for a new plant) must be estimated separately by other methods.

There are practical considerations in applying the factor method. For the chemical plant problem, variation between estimates are due to size of the basic equipment selected, materials of construction, operating pressures, temperatures, technology such as fluids processing, fluids-solids processing, or solids processing, location of plant site, and timing of construction.

Practically speaking, if the physical size of the basic item becomes larger and therefore more costly, the factor relating, say, the engineering design cost, is marginally smaller. If the basic item is constructed with more expensive materials such as stainless steel or glass-lined materials, then the factors become nonproportional to the item being estimated.

This factor behavior is described by Fig. 6.13. The basic item cost is the entry. A vertical intersection with the line leads to a factor, as read from the y axis. If the x entry value is doubled, then we assume that the factor is not doubled, thus assuring economy of scale. This is also apparent because the factor lines are nonlinear in Fig. 6.14. Frequently, a popular industrywide or government bench mark year is used and all data are referenced to this base year. Indexes are important in the factor method.

The first step estimates the basic item costs. The rising film and ozonation reactors are chosen in the flowchart. Because this process is high pressure and high temperature, we would expect the piping, supports, instrumentation, foundation, and so on to correlate to the reactors. If the basic item costs, the reactor material being carbon steel instead of stainless steel, then the factor reflects the distinction.

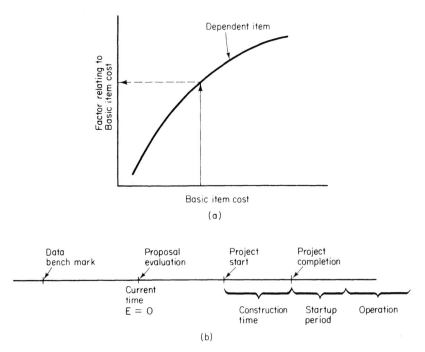

Figure 6.13. (a) Economy of scale for factor method; (b) time line for project evaluation.

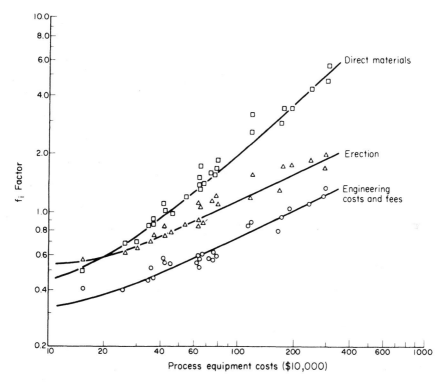

Figure 6.14. Typical factor chart based on process equipment costs for bench mark year.

The two reactors are estimated for current time costs, which correspond to the time of the estimate at $E = 0$. The costs could be estimated internally, or vendors can be asked for bids. In either case, once the basic items are estimated, we have

Major Process Item	Cost
Rising film reactor	$2,200,000
Ozonation reactor	700,000
	$C_e = \$2,900,000$

The estimate of the basic items is in current time terms. In doing this estimating a company maintains indexes to backcast the C_e value to data bench mark year. Once the basic item cost is backcast to the period of the factor curve, the f_i factors are read off the curve. At this point the indexed base item cost value is multiplied by f_i. Next, the bench mark cost is indexed forward to that point in time when the cost is expected out of pocket, project start, completion, or when progress payment periods are stipulated by the contract.

Factor data for plants are often regionally located in a popular area, where significant activity is expected. As an example, chemical plants are charted for the Gulf Coast of the United States. Construction at other sites requires regional indexing to adjust for differences in labor and material cost.

By using Eq. (5.53), we revise it to show

$$C_r = C_c\left(\frac{I_r}{I_c}\right) \tag{6.36}$$

where C_r = bench mark cost, dollars
       C_c = current cost, $2,900,000
       I_r = 100 for bench mark year
       I_c = current index as 114.1 for equipment
       C_r = $2,900,000(100/114.1) = $2,541,630

With this adjusted major component cost as the entry, Fig. 6.14 yields engineering costs and fees, 1.1; erection, 1.7; and direct materials, 4.1.

Note Table 6.8, where the factors become a multiplier for the bench mark cost, or $4,320,770 = 1.7 × 2,541,630 for erection. Next, this cost is inflated to the project start point or the progress payment time stipulated by the contract clauses. This would, for instance, be

$$4,320,770 \times \left(\frac{189}{100}\right) = \$8,166,250$$

The data in Fig. 6.14 are Gulf Coast information, though our hydrobromination plant is constructed in the Midwest where direct materials are 8% more expensive, erection costs are 13.2% more costly, and engineering costs and fees are identical. Further, those regional indexes account for the mix of the components. The regional factors are applied to current values. The Gulf Coast cost becomes 8,166,250 × 1.132 = $9,244,200, adjusting for geographical distinctions of cost. Finally, the capital investment, $34,630,210, is provided as the estimate. A factor for indirect costs, contingency, and profit can be applied to this raw capital investment.

The example deals with plant equipment estimating, but the factor method is used more broadly to estimate labor, materials, utilities, and indirect costs as a multiple of some other estimated or known quantity. Operating costs as a percentage of plant investment or as a percentage of the product selling price are popular.

6.10 POST ESTIMATE ANALYSIS

Engineering provides estimates about designs to make future decisions. There is the premise that the action recommended by the estimate will add to the benefits of the enterprise to make it worth the trouble. Typical situations in which engineering weighs the proposition "will it add to the benefits" include the following ones: A

TABLE 6.8. FACTOR ESTIMATE FOR A CHEMICAL PLANT PROCESS

Estimated Item	Current Time Cost	Current Index	Bench mark Cost	f_i	Bench mark Cost	Project Start Index	Gulf Coast Project Start Cost	Midwest Regional Index	Project Start Plant Site Cost
Equipment									
Major process items	$2,900,000	114.1	$2,541,630	1.	$ 2,541,630	118.0	$ 2,999,120	—	$ 2,999,120
Erection				1.7	4,320,770	189.0	8,166,250	1.132	9,244,200
Direct materials				4.1	10,420,680	127.0	13,234,260	1.08	14,293,000
Engineering costs and fees				1.1	2,790,000	145.0	4,053,890	1.0	4,053,890
Building site development									
Site development									1,250,000
Process building									2,100,000
Railroad spur									40,000
Utilities									650,000
Total									$34,630,210

firm considers a new plant producing an intermediate soap chemical or a special product has been designed and will be marketed shortly. Not all situations assure that the originator will be better off. Consider this case.

An electric utility serving several separated geographical regions employs transmission repair crews for high voltage line failures. The manager has permission to add another maintenance operator to a crew. In district A the average of the cost of repair per operator is estimated as $780, and in district B the unit estimate is $470. Our manager chooses to assign the operator to district B. But it is possible that the difference in the estimates occurred because the size of the repair crew in B was better adapted. If so, the new line mechanic may add little to cost reduction, and in A the mechanic may lead to a greater decrease in cost. The reasoning that leads to the right choice is based on marginal cost and is important to engineering. Too often decisions are made on the basis of average cost or average return rather than on marginal cost or marginal return.

Marginal costs are the added costs because of adopting a change in operations or making an engineering change order for a product. The change in operations usually implies increasing or decreasing production. The term "marginal cost" is interchangable with differential cost or incremental cost. In a broad sense all estimates are marginal estimates, because they create changes from a current course of action. Only after a detail estimate is made can the engineer exploit the advantages of marginal analysis.

We explain the arithmetic of marginal analysis with a simple illustration.

Quantity	Estimated Total Cost	Analyzed Marginal Cost	Analyzed Average Cost
0	0	0	—
1	$1000	$1000	$1000
2	1900	900	950
3	2700	800	900

The hypothetical cost is zero before any production. Marginal cost, by convention, is also zero, and average cost is undefined at this point. At two units, marginal cost is the amount added by the production from one to two units, or $1900 - 1000 = \$900$. Average cost (arithmetic mean) is as usually determined.

The principles of marginal cost are better explained by calculus, which is the more precise way to cope with those matters. The brute force way to explain marginal analysis calculates extensive tables, where total cost is enumerated for successive units (i.e., 61, 62, 63, . . .) in the region of interest. However, the approach recommended is to have a scattering of cost-estimated points and then proceed to fit linear and nonlinear regression lines. To illustrate, we fit a first-, second-, and third-order polynomial regression line through the estimating data for the product. Total product cost data are given in Table 6.9. We analyze the marginal properties of those data from fitted regression models.

TABLE 6.9.   TOTAL COST ESTIMATE FOR PRODUCT

Six-Month Production	Total Cost
5,000	$ 6,875
10,000	10,460
15,000	13,260
20,000	16,400
25,000	21,125
30,000	29,250

Higher-order polynomials sometimes create problems. Usually, it is acceptable to fit a polynomial regression equation of low order. The number of data points for the equation may be three to eight times the order of the equation. A simple model relating variable and fixed costs to production rate n is given as

$$C_T = nC_v + C_f \tag{6.37}$$

where C_T = total cost dollars per period
n = number of production units per period
C_v = lumped variable cost per unit
C_f = lumped fixed cost per period

If at various values for n, C_v is the same, then we have the linear cost model. If C_v is permitted to vary, then we are concerned with nonlinear models. For the constant C_v, nC_v is a straight line increasing at a constant rate per unit, and C_v is the slope of the variable cost line. The linear model statistically fitted for the product data is $C_T = 0.84n + 1527$, where $C_v = 0.84$. The fixed cost is $1527 and is the vertical-axis intercept.

The average cost model is given as

$$C_a = \frac{nC_v + C_f}{n} = \frac{C_T}{n} \tag{6.38}$$

where C_a is the average cost, dollars per unit. If C_v is constant, we have the linear case again. Equation (6.38) is still valid if C_v is nonconstant and we have the nonlinear case, which is the usual fare.

For linear cost situations marginal cost equals C_v and is a constant. For the general case we define marginal cost as

$$C_m = \frac{dC_T}{dn} \tag{6.39}$$

where C_m = marginal cost, dollars per unit
dC_T/dn = derivative of total cost function with respect to quantity n

By similar reasoning, if the output n is increased by an amount Δn from an established level n and if the matching increase in cost is ΔC_T, then the increase in cost per unit increase in output is $\Delta C_T/\Delta n$. *Marginal cost* is the limiting value of this ratio as Δn gets smaller (i.e., marginal cost as the derivative of the total cost function). It measures the rate of increase of total cost and is an approximation of the cost of a small additional unit of output from the given level. Sometimes marginal cost is called the slope or tangent of the total cost curve at the point of interest.

For the data of Table 6.9 fitted to a second-order polynomial or $C_T = 6594 + 0.08n - 0.0000217n^2$, marginal cost dC_T/dn becomes

$$C_m = 0.08 - 0.0000434n$$

The marginal cost is a linear line inclining downward with a slope 0.0000434. A third-order polynomial fit of the same data is

$$C_T = 840 + 1.527n - 7.418 \times 10^{-5}n^2 + 1.827 \times 10^{-9}n^3$$

and the marginal cost function becomes

$$C_m = 1.527 - 14.836 \times 10^{-5}n + 5.481 \times 10^{-9}n^2$$

from which a marginal curve is plotted directly. The marginal cost curve is a parabola and is shown in Fig. 6.15.

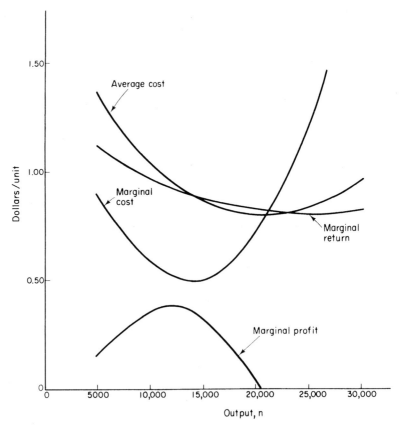

Figure 6.15.   Average, marginal cost, and marginal return relationships.

Plotting the average cost and marginal cost curve, the marginal cost line intersects the lowest point of the average cost curve. At this particular value of n, average cost is minimum and equal to marginal cost. This is shown in Fig. 6.15.

The decision for the maintenance operator was made on the average cost estimate. It should have been made on the marginal yield, which it promises. If district A has a greater marginal cost reduction by the addition of the maintenance operator, it is the preferred location. The choice between the two districts should be based on the greatest negative C_m.

The parallel to marginal cost is marginal return. It associates many of the same concepts. In an operation design, *marginal return* is marginal savings; in a product design, marginal revenue; in a project design, marginal rate of return; and finally, in system design, marginal effectiveness.

We assume that a specific economic measure can be chosen that will reflect important differences among different values of the design variables. In a market situation, *total return* is the total money revenue of the producers' supplying the demand and the total money outlay of the consumer's providing the demand. For our purpose a total return function is estimated with knowledge about input price and demand and that it increases or decreases with increasing output according to whether the demand is elastic or inelastic. Elastic demand implies a demand that increases in greater proportion than the corresponding decrease in price, and vice-versa.

The total return on revenue is given in Table 6.10 to continue our product problem. A zero sale estimate is a trivial requirement, but it aids the construction in a least squares sense. The total return column contains no reference to the price and is computed by dividing total return per unit of output. We tacitly require that production equal sales volume. In a similar fashion we fit polynomial regression curves to those data and analyze those determined models on the basis of marginal principles.

TABLE 6.10. ESTIMATED TOTAL REVENUE FOR PRODUCT

Estimated Sales Quantity	Total Revenue
0	0
5,000	$ 6,000
10,000	11,125
15,000	15,200
20,000	19,300
25,000	23,500
30,000	27,000

For a product design we construct a model of the form

$$R_T = nR_v + R_f \tag{6.40}$$

where R_T = total return dollars per period
    n = number of units per period
    R_v = lumped variable return per unit
    R_f = lumped fixed return per period

This is a linear or nonlinear model because either R_v is a constant multiplier or a nonlinear term. If we consider the sale of a product, then no receipts are obtained until the sale of at least one product or $nR_v > 0$. For a no-sale condition, $R_f = 0$. If the total return curve is linear, then the marginal return is a constant and equals R_v. For a product sold commercially this is the sales price. For an average return the model becomes

$$R_a = \frac{nR_v + R_f}{n} = \frac{R_T}{n} \tag{6.41}$$

where R_a is the average return dollars per unit.

Marginal return is defined in a way similar to marginal cost, or

$$R_m = \frac{dR_T}{dn} \tag{6.42}$$

where R_m = marginal return, dollars per unit
    dR_T/dn = derivative of total return function with respect to quantity n

The data in Table 6.10 fitted to a second-order polynomial become $R_T = 90 + 1.138n - 8.13 \times 10^{-6}n^2$. This second-order polynomial does not pass through the origin. It might be assumed that at $n = 0$ there should be no revenue, which is correct. However, the data in Table 6.10 are least squares approximations, and a displaced origin is not uncommon with estimated data. Further,

$$R_T = \begin{cases} 0 & \text{if } n \leq 0 \\ R_v n + R_f & \text{if } n > 0 \end{cases}$$

The marginal return of the second-order polynomial becomes

$$R_m = 1.138 - 16.26 \times 10^{-6}n$$

and thus marginal return is a linear line inclining downward with a slope -16.26×10^{-6}.

For a third-order polynomial fit,

$$R_T = -48 - 1.23n - 1.64 \times 10^{-5}n^2 + 1.833 \times 10^{-10}n^3$$

the marginal return function would be $R_m = 1.23 - 3.28 \times 10^{-5}n + 5.499 \times 10^{-10}n^2$ from which a marginal return parabolic curve is plotted in Fig. 6.15. In the usual circumstance the second-order marginal model is considered more accurate than a first-order model.

Marginal estimating depends on calculus. The marginal models are for the most part derived functions from fitted data. Occasionally, natural physical functions are employed. For either case the functions C_T or R_T are assumed continuous and differentiable.

A good deal of information is uncovered from those basic models. Returning to the maintenance operator problem, the decision should be made on the marginal yield that it promises. Indeed, the addition of maintenance operators can proceed until the net yield is zero. The greatest negative marginal cost indicates the design with greatest cost reduction for least input resources (a negative cost change is a gain). Given adequate resources, both crews should be expanded until the yields are zero. If resources for expanding crews are not present, then the next strategy adopts the same marginal cost for both districts. This may be done by reassigning operators from one district to another district.

With curves and mathematical models, such as those described for the product, it is possible to optimize sales revenue and cost. The question is asked, What shall we optimize? Maximizing sales revenue may not guarantee maximum profit, nor does minimizing cost balance other factors for maximum profit. In some cases those actions may reduce profits. One of the traditional ways to study this interaction is by *break-even analysis*. A break-even point is defined as the point of production or operation at which there is neither a loss nor profit nor savings. Several points of neutrality are possible, and the location of those points is a straightforward exercise after the curves and models have been formulated from the data.

By using Eqs. (6.37) and (6.40) the linear approximation of a point of indifference occurs whenever $C_T = R_T$. Ignoring profits tax and solving for n, we have

$$n = \frac{C_f - R_f}{R_v - C_v} \tag{6.43}$$

Model (6.43) is for the class of problems where R_f is not zero. For a product that is commercially sold, no income is received until products are sold and R_f as a fixed residual income does not exist, so $R_f = 0$. In the case of fitted models to estimated data, R_f may exist to satisfy the best fit in a least squares sense. Moreover, when using linear models $C_T = 0.84n + 1527$ and $R_T = 0.887n + 1284$ the break-even point is

$$n = \frac{1527 - 1284}{0.887 - 0.84} = 5170 \text{ units}$$

This is pictured in Fig. 6.16. The denominator is unit profit without fixed cost or return. But, because fixed cost is absorbed in the production and sales volume, the net unit profit when divided into fixed dollars yields the break-even point in terms of dollars and units sold.

Several comments about linear break-even analysis are worth noting. Those relationships are valid only for the short-term. Short term for one activity may be only weeks, and another firm may view short term as being an extended period of months and even years. To establish arbitrary rules about what is short or long term is dangerous unless specific cases are analyzed. As was demonstrated by the example, the mathematics and graphics involving linear (or nonlinear) costs and income provide information for break-even production.

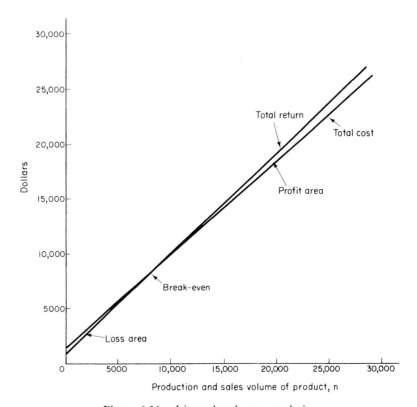

Figure 6.16.   Linear break-even analysis.

In the linear example a narrow wedge of profits begins at 5170 units and continues indefinitely. The defect of pure linear models is obvious: Reduced revenue or unit price as an additional quantity is produced can occur from a variety of economic happenings. Increased unit costs of production beyond that which is normal are typical as production increases indefinitely.

Although linear methods can be used for production above or below normal capacity and for cost-cutting tactics, the simplest way is to examine the plots of the original data. This can be done by straightforward plotting of Tables 6.5 and 6.6. This is shown in Fig. 6.17 where two break-even points, a lower and upper, are indicated. C_v is an increasing function, and R_v is a decreasing function. With this dual intersection of the return and cost lines, additional analysis of the meaning of optimum profit is called for.

The intersection of the slopes of the total cost and total return function is given as

$$\frac{dR_T}{dn} = \frac{dC_T}{dn} \tag{6.44}$$

The finding of this intersection is the point at which marginal profit equals zero or $dZ/dn = dR_T/dn - dC_T/dn = 0$. If n increases beyond this point, the cost of

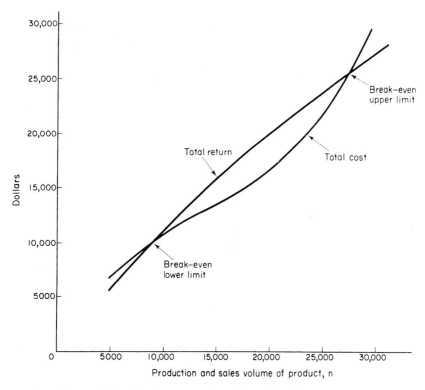

Figure 6.17.   Total return and total cost functions.

each unit exceeds the revenue from each unit and total profits begin to decline as they do beyond the upper break-even point in Fig. 6.17. The profits do not become zero until the profit limit point, which is the beginning of a loss area. When the *marginal profit* is zero we have the *critical production rate* and maximum profit. This critical production does not necessarily occur at the rate corresponding to the minimum average unit cost, nor does it necessarily occur at the point of maximum profit per unit.

A plot of $dR_T/dn - dC_T/dn$ is given in Fig. 6.15. The point at which $dR_T/dn = dC_T/dn$ is also zero for marginal profit or dZ/dn. The critical production rate may be determined directly from differentiation of the profit function Z. This critical production rate is zero at $n = 20,800$ units, which is the point of maximum gross profit. But it is not the rate corresponding to minimum average unit cost or the maximum profit per unit. The maximum profit per unit may be estimated from the level dome of the marginal profit curve, or approximately 12,200 units. By using Fig. 6.15 minimum marginal cost is estimated as 14,500 units, and maximum marginal return is at the minimum sales projection of 5000 units.

Although several points of operation can be obtained from minimizing total or unit cost, maximizing total or unit return, or maximizing gross or unit profit, we should not conclude that all is lost if those exact objectives are never precisely realized. There is a redeeming feature for the engineer if the actual and the ideal do not coincide. Fortunately, most optimums are relatively flat near the optimum. For those domes, operation on either side is insensitive. Naturally, a converse example can be found that demonstrates a sharp peak as optimum. In this case, operation at this point becomes important.

SUMMARY

Estimates are constructed on available information. If there is no information, then there can be no estimate. Conversely, an estimate is unnecessary if actual costs are available. The engineer operates within those limits. In this in-between region we separate estimating methods into preliminary and detail.

Preliminary methods are less numeric than detail methods. Accuracy of the estimate is improved by attention to detail, but balancing this is the speed and cost of preparation that favor preliminary methods. The methods are preliminary because they correspond to a design not well formulated. Frequently, a preliminary estimate leads to the management decision of further consideration for the design. Before detail methods of estimating are attempted, information that has been deliberately collected, structured, and verified to suit the method of estimating must be available. Most estimates are made by using detail methods.

In the next four chapters the methods of detail estimating are examined further. The concern becomes one of using various cost-analysis relationships, ranging from simple cost factors to functional models, to estimate operations, products, projects, and systems.

Here we return to the question, How do we select the right method for a particular application? Regrettably, no textbook can supply an infallible set of rules: You will have to rely on experience and continuing analysis.

QUESTIONS

6.1. Define the following terms:

Preliminary estimating	Straight-line average
Detail estimating	Power law
Conference method	Economy of scale
Hidden card	Standard time data
Unit method	Opinion probability
Cost driver	Lookup rules
CERs and TERs	Errors of standard data
Parameter	Point of indifference
Learning	Economy of quantity
Cumulative average	Marginal cost

6.2. Discuss the timing of preliminary estimates. Is the timing a precise point in a well-ordered organization?

6.3. Use the conference method to estimate
  (a) The price for a clean air car using a turbine drive.
  (b) The ticket price for a 5000 mile one-way air trip.
  (c) The cost of a year's college education in the year 2000.
  (d) The time to dig a trench $2 \times 4 \times 10$ ft in soft clay.
  (e) The improvement in the efficiency of handling mail of a central post office near you after converting to automated methods.

6.4. What advantages can you cite for the conference method? Disadvantages?

6.5. Assume that design A can be redesigned into designs B and C with known costs. What safeguards can you suggest to assure that A is properly estimated?

6.6. Discuss some substitute unit measures for "cost per square foot of residential construction." What are your local values?

6.7. Give the steps for the construction of standard time data.

6.8. Point up the human frailties in determining opinion probabilities. Would you think that one would underestimate or overestimate those point probabilities?

6.9. Describe the kinds of applications suitable for the factor method. What kinds are suitable for standard time data?

6.10. What is meant by the law of economy of scale? Apply it to the factor method. Does the unit method follow this law?

6.11. Outline the method to be followed for post-estimate analysis. What distinguishes linear from nonlinear margins?

6.12. Define mathematically the point of indifference, the upper and lower break-even points, the maximum profit point, and the profit limiting point.

6.13. Three marginal cost per unit versus quantity curves are assumed in Fig. Q6.13. They are L-shaped, U-shaped, and two quadratic curves. The two quadratic marginal cost curves are redefined into an envelope curve, which is the consolidated cost line. These marginal cost curves were determined for a manufacturing operation. Write a paragraph that describes the behavior of cost elements that will supply these marginal cost curves.

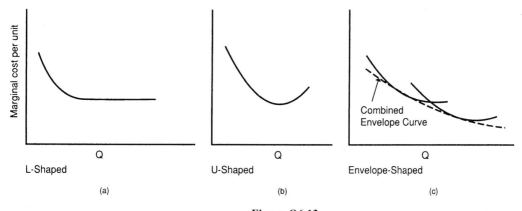

Figure Q6.13

PROBLEMS

6.1. Use the conference method to estimate some product or sample statistic familiar to the entire team but whose exact value is unknown; for instance, cost of interstate highway construction per mile or number of automobiles in your town. Then check its value.

6.2. A manufacturer of factory-built houses has collected raw data for flooring and has converted them to a common square foot basis:

Material	Size	Material Cost	Labor Cost
Laminated oak blocks	$\frac{1}{2} \times 9 \times 9$ in.	$3.30	$1.85
Parquet	$\frac{5}{16} \times 12 \times 12$ in.	3.10	2.65
Strip flooring, oak	$\frac{25}{32} \times 2\frac{1}{4}$ in.	3.60	2.25
Resilient tile			
A grade	$9 \times 9 \times \frac{1}{8}$ in.	1.60	1.70
B grade		1.85	1.70
C grade		1.90	1.70
Vinyl asbestos	$9 \times 9 \times \frac{1}{16}$	2.05	1.70
Softwood, C grade	S4S, 1×6 in.	2.65	1.80
Linoleum	$\frac{1}{8}$ in., plain	2.50	1.75
Slate, irregular flags	Irregular flags	7.00	5.00
Terrazo	$\frac{1}{4}$ in. thick	5.75	4.95

Using only your opinion, where would you cost cork tile? Where would you cost carpet? What are the nonfeasible alternatives? Those houses are delivered to the site by company trucks. Conduct an analysis after your opinion to help sharpen your value.

6.3. An inventor has fashioned a new mailbox design with an interesting feature. When the mailman closes the hinged cover, a spring-loaded flag pops up at the back of the box telling the owner that he has received mail. A survey of popular catalogs has revealed the following data:

Size	Weight	Material	Cost	Features
$14 \times 7 \times 4$	4 lb, 4 oz	Steel	$ 23.60	Holds magazine, wall mount
$13 \times 7 \times 3$	3 lb, 6 oz	Steel	15.60	Holds magazine, wall mount
$14 \times 6 \times 4$	4 lb, 15 oz	Aluminum	35.96	Holds magazine, wall mount
$14 \times 7 \times 4$	4 lb, 12 oz	Steel, aluminum	27.60	Holds magazine, wall mount
$6 \times 2 \times 10$	4 lb, 8 oz	Forged iron	32.40	Letter-sized
$6 \times 2 \times 10$	2 lb, 11 oz	Steel	17.60	Letter-sized
$5 \times 2 \times 11$	2 lb, 11 oz	Steel	10.20	Liberty bell emblem
10×3	1 lb, 8 oz	Brass	23.60	Mail slot
10×3	1 lb	Steel	7.08	Mail slot

Size	Weight	Material	Cost	Features
10 × 3	1 lb, 5 oz	Aluminum	23.60	Mail slot
18 × 7 × 6	8 lb, 8 oz	Steel	26.76	Red signal flag, mount post
19 × 7 × 6	10 lb	Steel	39.16	Black wrinkle finish
18 × 6 × 9	10 lb	Steel	37.16	Name holder
18 × 7 × 6	8 lb, 1 oz	Steel	9.48	Size no. 1
18 × 7 × 6	8 lb, 1 oz	Steel	19.60	Rural mailbox
18 × 7 × 6	4 lb	Aluminum	15.40	Rural mailbox
17 × 9 × 10 unit		Steel	200.00	Apartment type

(a) In using a comparison method, where would you place the inventor's design?

(b) Construct a rough unit estimate from the data to help in the evaluation.

(c) Discuss various reasons that might lead to a different analysis.

6.4. The cost of a residential home with 2000 square foot livable space and a basement and garage is $146,000. The house dimensions are such that 1200 square feet are on the first floor. The volume including house, garage, and basement is 30,200 cubic feet. Determine unit estimates.

6.5. For the following problems, assume that the unit line is linear.

(a) Find the first unit value when the 100th unit is 60 hours with 81% learning.

(b) Find the value for unit 6 when the value for unit 3 is 1000 hours at 74% learning.

(c) If the unit value at number 1 is $2000, then find the unit dollars for units 20 and 40 with learning rates of 93% and 100%.

(d) Units 1–100 have accumulated 10,000 hours. Calculate the unit time for number 50 with a learning ratio of 71%.

(e) If the cumulative average time at unit 100 is 100, then find the unit, cumulative, and average time at unit 101 for a learning rate of 92%.

6.6. For the following problems, assume that the average line is linear.

(a) Find the cumulative average value for the 20th unit when the 10th unit has an average of 100 hours for 68% learning.

(b) The cumulative average cost for the 100th unit is $10,000. Find the average cost for the 50th and 200th unit for a learning rate of 87%.

(c) If 400 cumulative average hours are required for the third unit and 350 hours for the sixth unit, find the slope, T_a', T_u', and T_c' for the tenth unit. What can you say about those calculations?

(d) If 800 average man hours are required for the tenth unit and 750 hours for the 20th unit, find the percentage learning ratio, T_a', T_c', and T_u' for the 5th and 40th units.

(e) If the unit time at unit 100 is 100, find the unit and cumulative and average times at unit 101 for a learning rate of 92%.

6.7. Unit man hours are reduced by 20% every time production is tripled. Find the slope.

6.8. Find the slope for the following conditions:

(a) A company assumes that the unit line is linear. Two points uncovered by a shop study show a pair $(N_1 = 15, T_1 = 80)$, $(N_2 = 25, T_2 = 70)$ of unit values. Find the slope, T_a and T_c for N_1 and N_2.

(b) A cumulative average linear line study has revealed that at 10 units, $T_a' = 1550$ hours, while at 25 units, $T_a' = 1010$ hours. Find s, T_u' and T_c' for 50 and 100 units.

6.9. A company assumes that the unit line is linear. Two points uncovered by a shop study show a pair $(N_1 = 1, T_1 = 80)$, and $(N_2 = 2, T_2 = 70)$ of unit values. Find the slope and T_a and T_c for N_3.

6.10. Calculate learning tables for 76% learning where only the unit line is considered. Consider units 1, 2, and 3.

6.11. The engineering department has determined an estimate of 10,000 hours at unit 100. They believe an improvement of 10% can be expected from unit 1 onward. Find the unit time to build units 1 and 2. Assume that the unit line is linear.

6.12. **(a)** For management proposal A, cost at unit 1 is $8000, and the learning rate is 80%. For management proposal B, unit 1 cost is $6000 and the learning rate is 90%. Which is the best proposal at unit 10? At unit 30? Assume that proposal B was adopted. Engineering discovers that at unit 15 the actual cost was really $3975, and for unit 1 it was as initially estimated. What is learning performance? Assume that the unit line is linear.

(b) An audit of learning curve performance revealed that at the 18th product unit the direct product cost was $181,000, and at the 24th product unit it was $169,000. What is the estimated direct product cost for the 27th product unit?

6.13. Use appendix 3 and plot T_u, T_a, and T_c for 1, 5, 10, 20, and 100 units for 80% learning on arithmetic and logarithmic graph paper. Find the graphical values for unit 500.

6.14. Construct a learning theory table similar to the appendix for the unit and average straight lines for $N = 1, 2,$ and 3 for 88%.

6.15. Use appendix 3 and plot T_a', T_c', and T_u' for 1, 5, 10, and 20, and 100 units for 80% learning on arithmetic and logarithmic graph paper. Find graphical values for unit 500.

6.16. Calculate new tables such as those in appendix 3 for units 1 to 4 (a) for 83% learning and (b) for 63%.

6.17. Calculate new tables such as those in appendix 3 for units 1 to 4 (a) for 73% learning and (b) for 77%.

6.18. An 80-kW diesel electric set, naturally aspirated, cost $160,000 8 years ago. A similar design, but 140 kW, is planned for an isolated installation. The exponent $m = 0.6$ and index $I = 187$. Now the index is 207. A precompressor is estimated separately at $19,000.
(a) By using the power law and sizing model, find the estimated equipment cost.
(b) Repeat for $m = 0.7$.

6.19. A cost estimate is desired for a 34-cm disk laser amplifier. A previous reference design of 22 cm has been made for a cost of $16,459. Inflation indexes for this design are unavailable, but the annual Consumer Price Index of 7.5% will be used for a 3-year escalation. A calculation of $m = 0.7$ was found.
(a) Find the cost of the new design.
(b) Repeat for $m = 0.8$.

6.20. A product is known to follow the power law, which can be used to estimate the cost of new design. Four sizes of the turbine have been analyzed as to cost:

kW	Cost ($\times 10^6$)
3500	$1.8
5000	2.5
7000	3.2
8000	3.5

A similar design, except that it is larger at 8750 kW, is to be estimated. Determine the slope m and estimate the cost of the new unit. (Hint: Use log C = log C_r + m log Q/Q_r and let Q_r = 3500. Plot the relationship on logarithmic graph paper.)

6.21. A firm that manufactures totally enclosed capacitor motors, fan cooled, 1725 rpm, $\frac{5}{8}$ in. O.D. keyway shaft has known costs on $\frac{1}{4}, \frac{1}{3}, \frac{3}{4}$, and 1 hp as $90, $100, $118, and $140. A 3450-rpm, 1-hp motor costs $134. Determine m for the 1725-rpm motor series, and estimate the cost for a $1\frac{1}{2}$-hp motor. Should the 3450-rpm motor be included in the sample to estimate the 1725 rpm motor?

6.22. (a) A company sells two different designs of one item. A study discloses that 65% of its customers buy the cheapest design for $75. The remaining 35% pay $110 for the expensive model. What is the expected purchase price?

(b) A salesperson makes 15 calls without a sale and 5 calls with an average sale of $200. What is the expected sale per call?

(c) A machine tool builder takes old lathes as tradeins for new models and sells the returned lathes through a secondary party outlet. Analysis shows that the markup is $2500 on 70% of the lathes and $4000 on the rest. What is the expected markup?

(d) An insurance company charges $20 for an additional $50 increment of insurance (from $100 to $50 deductible). What is their assessment of the risk for the increment of insurance?

6.23. A student is interested in selling his car instead of trading it in. His estimating model, he reasons, is the sale price of a new car: (depreciation + major maintenance cost). Other costs for driving are the same regardless of whether he drives a new car or not. The original sticker price is $13,137 and a major maintenance cost is $2000. His opinion probability for a major maintenance cost is given as follows:

Life	Cumulative Probability of Major Maintenance	Cumulative Decline in Depreciation
2	0.2	0.49
3	0.4	0.64
4	0.7	0.75
5	1.0	0.83
6	0.1	0.89
7	0.2	0.93
8	0.4	0.96
9	0.7	0.98
10	1.0	0.99

When should he sell his car? Initially assume that the next car, whenever he buys it, will be equal to his first car's price. Next, assume that the new car's price increases by 3% compounded per year.

6.24. A company is considering redesign of one of its basic products. The engineering cost and the manufacture of new tooling will cost $25,000. Three alternatives are to be evaluated, and the engineers determine the profit and opinion probability of success. Production costs for designs A and B are the same:

Profit	Probability	Profit	Probability	Profit	Probability
$200,000	0.4	$400,000	0.3	$300,000	0.2
250,000	0.3	450,000	0.5	450,000	0.3
300,000	0.3	500,000	0.2	600,000	0.5

Determine the expected value of the profit — new tooling for each of the three alternatives. Which one is preferred?

6.25. (a) Use the technique of range cost estimating to find the expected total mean cost and variance.

Cost Element	Optimistic Cost	Most Likely Cost	Pessimistic Cost
Direct labor	$79	$95	$95
Direct material	60	66	67
Indirect expenses	93	93	96
Fixed expenses	69	76	82

(b) What is the probability that cost will exceed $325?

6.26. A five-element cost program has been summarized as follows:

Cost Item	Optimistic Cost	Most Likely Cost	Pessimistic Cost
1	$ 4	$ 4.5	$ 6
2	10	12	16
3	1	1	1.5
4	4	8	12
5	2	2.5	4

(a) Determine the elemental mean costs, the total cost, and the elemental and total variances.

(b) Find the probability that cost will exceed $26.

6.27. A TER is given as $\hat{y} = 0.050 + 0.015 x$ where $\hat{y}$ = minutes and x = weight or lb. Let $P_3 = 20\%$ and the minimum x, y time study point is $(2.0, 0.080)$. (a) Find the standard data table value for 3 steps, and (b) what is the time for 3.5 lb?

6.28. Five time studies have been conducted of a "handling" element. Weight is considered to be the primary driver. The weight and average elemental time are given as follows (wt., min): $(3, 0.20)$, $(1, 0.15)$, $(5, 0.40)$, $(4, 0.30)$ and $(7, 0.50)$. The average total operation uses 1 min., or $\hat{y}_t = 1$ min. Is "handling" a variable or constant element? If it is a variable element, then construct a table. Let P_1, P_2, and $P_3 = 100\%$, 10%, and 50%.

6.29. Construct standard data such as in Table 6.7 for the spot-welding regression equations. Use the three rules of thumb for applying standard time data tables, and find the errors. Let allowances be 15%. (Hint: Start with Table 6.4.)

(a) $P_1 = 50\%$, $P_2 = 15\%$, and $P_3 = 10\%$
(b) $P_1 = 50\%$, $P_2 = 10\%$, and $P_3 = 25\%$
(c) $P_1 = 50\%$, $P_2 = 20\%$, and $P_3 = 25\%$

6.30. (a) A spot-weld design has a primary, secondary, and third part, where the third part girth is 20 in. A total of 34 spot welds is necessary. The gross hourly direct labor wage is $25. Find the estimated cost per unit, cost per 100 units, and pieces per hour. Follow the three rules of thumb in applying Table 6.7. What purposes do those rules serve?

(b) Repeat for another design where third-part girth is 10 in. and the number of spots is 19.

(c) Repeat for another design where third-part girth is 24 and the number of spots is 33.

6.31. (a) A spot-weld design has a primary, secondary, and third part, where the third part girth is 14 in. A total of 22 spot welds are necessary. The gross hourly direct labor wage is $35. Find the estimated cost per unit, cost per 100 units, and pieces per hour. Follow the three rules of thumb in applying Table 6.7. What purposes do those rules serve?

(b) Repeat for another design where third-part girth is 20 in. and the number of spots is 29.

(c) Repeat for another design where third-part girth is 12 and the number of spots is 37. Let lot quantity be 175 and find lot cost.

6.32. An element, "move casting, and aside part to tote pan," has been time studied and information recorded. Three possible independent variables are weight, girth, and fixture locating points.

(a) Determine three linear regression equations. Indicate criteria to allow selections of the best time driver for the preparation of standard time data. Which of the three equations is the "best line"? For your best line, provide a table of 25% increasing steps in time.

(b) Repeat for metric dimensions of weight and girth.

Time Study	Time, min.	Weight		Girth	Fixture Locating Points
		lb	kg	L + W + H (in. (mm))	
1	3.4	15	(6.8)	7 (178)	3 (3 buttons)
2	1.4	5	(2.3)	6 (152)	4 (nest)
3	2.8	12	(5.4)	7 (178)	3 (2 edges, 1 pin)
4	2.2	10	(4.5)	3 (76)	4 (nest)
5	4.8	20	(9.1)	11 (280)	1 (pin only)
6	4.2	17	(7.7)	9 (230)	2 (2 edges)

6.33. Develop standard data for an operation of "sawing and slitting" $\frac{1}{4}$ in. O.D. maple dowels in a furniture factory. Several time studies are available from which normal times are removed for a recap sheet. It is assumed that methods engineering has standardized workplace layout. Time is in decimal minutes. Elements 1 and 4 are variable. Use steps of 25% for variable time.

Element	1	2	3	4	5
1. Pickup dowel	0.023	0.023	0.035	0.041	0.051
Reach distance	8 in.	8 in.	13 in.	16 in.	17 in.
2. Position for slotting	0.031	0.035	0.029	0.030	0.032
3. Move 14 in. and piece					
aside in tote bos	0.024	0.021	0.031	0.030	0.030
4. Saw slot to length	0.093	0.126	0.103	0.210	0.193
Length	$\frac{1}{4}$ in.	$\frac{1}{4}$ in.	$\frac{1}{4}$ in.	1 in.	$\frac{7}{8}$ in.

Allowances are 15%. Develop the standard data in a tabular format.

6.34. A residential builder constructed a 2000 ft^2, two story home on sandy loam soil for $124,000. This cost was exclusive of lot, taxes, and utility hookup costs. What is the unit estimate for the next home? An additional breakout of costs revealed the following percentages for this same job:

Item	Percent	Item	Percent
Rough lumber	13.0	Earthwork	2.6
Rough carpentry	9.2	Flooring	6.4
Plumbing	13.6	Hardware	2.1
Finish lumber	1.2	Heating	3.0
Finish carpentry	2.8	Insulation	1.5
Cabinets	6.4	Lighting	0.5
Concrete	5.8	Painting	7.1
Wallboard	6.0	Roofing	2.2
Electric wiring	5.5	All other	11.1

A 3200-ft^2, two-level home is to be built. Estimate total and item costs. What elements would you adjust if this were a single-story home? Up or down?

6.35. The hydrobromination bid received for another rising film and ozonation reactor is $1,750,000, and the current index is 121.5. Use the method discussed in section 6.9.
(a) What is the bench mark cost?
(b) What are the factors for engineering, erection, and direct materials?
(c) An overall index of 123 is assumed for costs during construction. The regional factor is 7% more than the Gulf Coast for all factors. Items independent of the estimating process are found to cost $2,000,000. What is the total factor estimate for a chemical plant?

6.36. (a) An opportunity sale of an additional 200 units is received above an existing production level. The buyer is prepared to pay $350 for those 200 units. Cost for the current 500 units is $950. Total cost for producing 700 units is $1277. What is the marginal total cost? What is the unit marginal cost? What is the profit or loss? What is the minimum opportunity sale price to break even?

(b) The total cost of producing a given output varies according to the following estimated data:

Output	Total Cost	Output	Total Cost
0	$ 0	5	$192
1	100	6	205
2	150	7	230
3	175	8	280
4	187	9	380

Plot the estimated total cost curve, average cost curve, and marginal cost curve.

(c) A cast iron sphere has the following design weight (x) and cost (y): (2, 2), (3, 3), (4, 5), and (5, 8). Plot the cost curve, average cost curve, and marginal cost curve. Do those data comply with the economy of scale law?

6.37. Given the marginal return function $= 500 + 0.10n$, marginal cost $= \$1000$, point of maximum profit $= \$5000$, and fixed cost $= \$10,000$, find the demand function for price, quantity per period, and total cost function for the same period.

6.38. A motorcycle manufacturer is considering producing a new model of motorcycle designed for a specific market. Management has determined the total fixed cost for the plant and design to be $\$10^5$. The cost of producing the cycle after the initial investment has been made is estimated as a function $V(n)$ of the total number n of cycles produced.

(a) If $V(n) = 200n - 3 \times 10^{-3}n^2 = 10^{-7}n^3$, then determine the marginal cost.

(b) What is the minimum marginal cost? How many cycles does this correspond to, and what is the average cost per cycle at this level of production?

(c) If the total return from sales $S(n)$ is given by $S(n) = 250n$, find the marginal return, the total profit, and the marginal profit.

(d) How many cycles should be produced to maximize profit?

MORE DIFFICULT PROBLEMS

6.39. A food factory is planned near a large area of truck farms. Those farms sell to stores, but a surplus is generally available. A small general-purpose production plant is planned, and the following equipment is engineered and priced out by bidding:

Process	Cost	Process	Cost
Roller grader	$10,000	Extractor	$ 36,000
Peeling and coring	18,000	Two cookers	7,000
Pulping	30,000	Can seaming	54,000
Instrumentation	57,000	Packaging	22,000
Conveying	37,000	Six motors	35,000
Filler press	14,000	Heat exchanger	44,000
Tanks	10,000	Total	$374,000

Factors to obtain installed cost, including cost of site development, buildings, electrical installation, carpentry, painting, contractor's fee, foundation, structures, piping, engineering overhead, and supervision, are listed for various equipment as

Process centrifuges	2.0	Can machines	0.8
Compressors	2.0	Cutting machines	7.0
Heat exchangers	4.8	Conveying equipment	2.0
Motors	8.5	Ejectors	2.5
Graders	1.6	Blenders	2.0
Tanks	2.1	Instruments	4.1
Fillers	0.8	Packaging equipment	1.5

Estimate a plant cost based on this information. What other way would there be to improve on the quality of this estimate?

6.40. Market studies of three locations, X, Y, and Z, show a demand of 20,000, 7000, and 6000 units of the same product per day in each location for 6 days per week. At the time of this estimate, two alternatives are being planned to meet this demand. In plan A, one plant with a daily capacity of 40,000 units could be located equidistant between the locations and produce with a fixed cost of $2000 per day and a variable cost of $0.15 per unit. Alternatively, plan B has mini operations in each location with capacity of 24,000, 8000, and 8000 units, respectively. Fixed costs could be $1500, $700, and $700 per day, respectively. Because of reduction in transportation costs and differences in labor costs, variable costs per unit are $0.10, $0.11, and $0.12 per unit.
(a) Which alternative is more desirable?
(b) If sales were to increase to production capacity, then does the best choice change?
(c) What variable cost per unit may plan B increase to for a break even to A at demand capacity? At full capacity?

6.41. Develop standard time data by using the following information. Maximum and minimum part sizes are 32 and 288 in.^{2} Elements have been summarized into four linear equations and are

Element	Normal-Minute Equation
1	89.2 + 0.069 (square inch)
2	11.0 + 0.590 (square inch)
3	0.312 (square inch)
4	49.4 + 0.764 (square inch)

Use these rules:

1. A constant element is defined as one with less than 20% increase in time from minimum to maximum size of the independent variable.

2. A variable element is defined as one with more than a 20% increase in time from minimum to maximum size and a time greater than 15% of the total time for average size work; otherwise, the element is constant.

3. Tabular time values for variable time increase 25% for each step, starting with minimum size and concluding with maximum size.

4. Tabulated values are expressed in standard hours. Allowances for personal, fatigue, and delay are 10%.

CASE STUDY:
MARGINAL LABOR AND TOOL COST

The Don Boyle Co. makes zinc die castings. It has one 300-ton machine with a trim press located at the end of the quench conveyor. The machine is a hot chamber type and is capable of running automatically.

Andy James, Inc., manufactures leather goods. The purchasing agent for Andersons, Inc., calls Andy James (and several other manufacturers) for a quote of a 3-in-wide belt. The belt is to be made with an adjustable buckle. The James engineers design a zinc die-cast chrome-plated buckle, and call Don Boyle (and several other die casters) to quote on the buckle. Andersons has estimated a volume of 75,000 belts annually for the next 3 years, when the item will be phased out.

Don Boyle must quote Andy James a unit price and a tooling price. He has a 300-ton machine capable of casting up to 16 buckles at one shot. His trim press is also capable of handling a shot this size. How does he quote this part to stay competitive? A 16-cavity die can minimize the unit price, but the tooling cost could be enormous. A single cavity die maximizes the unit price but minimizes the tooling price. Which alternative should Don choose?

A single-cavity casting die, Don estimates, would cost about $4500. A trim die would cost an additional $900. Direct labor and overhead to cast one shot, no matter how many parts it has, is 6 cents. In other words, to cast the required 225,000 buckles in the next 3 years, the lowest cost for tooling is $5400. However, the unit cost (excluding tooling) is 6 cents per part, or $13,500 for the total of 225,000 units (no allowance is made for scrap).

Now, the marginal costs and marginal revenues must be examined. Marginal revenue is defined as the increment of total revenue (plus or minus) that results when the number of cavities is increased by one unit (or fraction thereof). Marginal cost likewise is defined as the increment of tooling cost that results from a similar increase in cavities. (Total revenue does not refer to the amount of money to be paid by Andy to Don. That would be a pricing problem, whereas this discussion concerns the problem of an efficient design.) Total revenue here is the total dollars saved by using a two-cavity die compared with a one-cavity die, or three cavities against two, and so forth. This total revenue may be called unit savings.

To cast the buckle for Andersons, the minimum starting point is single-cavity tooling. Total revenue at this point is zero. Don has not yet experienced a saving as a result of his tooling choice.

A two-cavity design, Don estimates, would cost an additional $600 for the die casting die and $200 for the trimming die. The marginal cost is $800, and this does not include unit cost. Unit costs are considered only in respect to changes that create a savings or loss. A two-cavity die will produce two parts per shot, reducing the unit cost to 3 cents per part, or $6750 for the 3-year requirement. This is a reduction from $13,500 to $6750, or a unit saving of $6750.

For each additional cavity added to the tooling Don estimates that the tool cost will increase by an additional $800. With 16 buckles per shot, the maximum capacity of the machine has been reached. Anything over 16 buckles per shot would require a larger die casting machine.

Where does Don maximize the profits from a tooling standpoint? How many cavities does this decision call for? What is the full cost for this decision? How does raw material affect your decision?

7

Operation Estimating

The heart of operation estimating is, in a substantive measure, a talent for breaking a task into essential elements. A design formulated sufficiently for this analysis is available, and the engineer uses cost equations and standard time data that model the details of the design.

In manufacturing, the design is a part print and a processing plan; in chemical industries, a flowchart and layout with engineering calculations; in engineering construction, a set of drawings and specifications. Those designs are used for purposes such as estimating, planning, scheduling, methods improvement, and production or construction. The designs communicate the description and sequence of operations and are the hub on which individually and collectively many decisions, both great and small, are made.

Outmoded practices of operation estimating take the total cost of operating the plant for a time and divide by a total production quantity. Other practices determine a labor and material cost for an operation from a test run or prototype. Historical records of like operations are used. Those methods are unsuitable for predictive cost estimates.

The engineer begins by subdividing the design task into large portions of labor and material. Progressively finer detail is determined until a description of labor and materials is very broad. At this point, dollar extensions of labor and material are made to reflect the cost of the design.

Thousands of various labor and material operations exist. Regrettably, our choice of explanation extends only to a few. Trade books, handbooks, company sources of data, and so on must be consulted for data for any real-life estimating. Some practical sources of information are listed in the references.

7.1 OPERATION COST

Before an *operation estimate* is started, some authority must initiate this activity; that is, a request for quote, work order, production planner, foreman's request, work-simplification savings, employee suggestion, or marketing request. Those requests are written on a request-for-estimate form, or RFE, which is the commissioning order to begin work. The RFE will describe or include the design, period of activity, specification, quality, and give other assumptions.

Economic evaluation follows many forms. In all these approaches, what is necessary is *microanalysis,* which is a facility for subdividing an operation design into both physical and economic elements. Because detail estimates are relatively costly to prepare, the risk of achieving success must be offset by the cost of preparation. If risk were not minimal, then the estimates for the operations would not be started. This risk is measured by a capture percentage, which we define as (estimates won)/(estimates made) × 100. For job shops the capture percentage may be low, and quotations are prepared by using preliminary procedures. If the capture percentage is high, or the cost of estimating cost is not proportionately high to the benefits from success, then detail estimating is mandatory.

Though the labor and material may differ from design to design, techniques for evaluation are based on the same principles and practices. Operations are necessary to produce a change in value. The basic combination of a person and tool is the primary ingredient. The economic value of material is altered through people and tool activities. The measure of the change in economic value is cost.

Cost in this instance implies a consumption of labor, materials, and tools to increase the value of some object. The use of expensive fixed assets involving capital cost in operations, while a consumption activity of wealth, is deferred until chapter 9, where methods are introduced. In some cases, automatic equipment produces units of output and may not require labor. Those operations, however, consume materials and utilities and constitute the operation cost estimate. But operation designs require tools, fixtures, or test equipment. Without those indirect materials and equipment the operation is not possible. Those tools, which we refer to as nonrecurring initial fixed costs, are estimated along with direct labor and direct material.

Operation evaluations are limited as to the time horizon. The immediate future period rather than an extended time period is intended. The nature of labor is for a period of time such as "units of labor time per piece" or "units of labor per month or year."

The purpose of operation estimates is to establish the cost for components and assemblies of a product, to initiate the means of cost reduction, to provide a standard for production and control, and to compare different design ideas. Estimates may be used to verify operation quotations submitted by vendors and to help determine the economic method, process, or material for manufacturing a product. Whenever an operation is material or labor intensive, rather than capital intensive, the methods described in this chapter are primary. Figure 7.1 is a description of the manufacturing items estimated for an operation. The term *prime cost* is an accounting

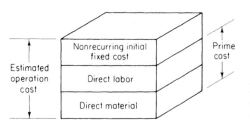

Figure 7.1. Descriptive layer chart of items included in an operation estimate.

term meaning direct labor and material cost. *Cost* is the measure of economic want for the operation.

7.2 MANUFACTURING WORK

Manufacturing of durable goods, such as toys to turbines, can be broadly classified as mass or moderate or job lot production. In mass production, sales volume is established and production rates are independent of single orders. In moderate production, parts are produced in large quantities and, perhaps, irregularly over the year. Output is more dependent on single sales orders. Job lot industries are more flexible, and their production is closely connected to individual sales orders. Lot quantities may range from 1 to 500, for example. For cost estimating, those distinctions are superficial because a similar logic is necessary for mass, moderate, or lot quantity operations despite the sales volume.

In manufacturing, operations are conducted at a machine, process, or bench. Those workstations involve direct labor. A machine is capable of metalworking: examples would be a computer-controlled lathe, automatic milling machine, or drill press. A process tends to be chemical or fusion in character, such as spray painting, silver plating, welding, or casting. Bench work is an assembler performing joining, fastening, and assembly at a bench or conveyor. Figure 7.2 is a simple sketch of turning, milling, and drilling metal working operations.

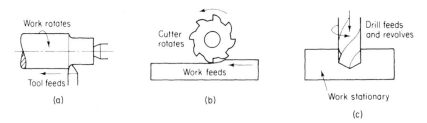

Figure 7.2. Various metal-working operations: (a) turning; (b) milling; (c) drilling.

Engineering drawings, marketing quantity, specifications of the workpiece material, machines, processes and benches, and standard time data are required before detail estimating can begin. Preparation of the operations sheet occurs simultaneously with direct labor estimating. Once the operations are listed, the engineer refers to standard time data, which coincides with the machine, process, or bench identified by the operations sheet for doing the work. Each operation is detailed into elements that correspond to the standard time data. (This is discussed in chapter 6.) Those elements may be presented on a computer monitor, and the form is standardized by the company, or marginal jottings or scratch pad calculations may be followed. Computer spreadsheets are used also.

For the operation and workstation, and after visualizing the elements of an operation, the engineer selects times from the standard time data for setup and cycle.

Setup includes work to prepare the machine, process, or bench to produce parts or run pieces. Starting with the workstation in a neutral condition, setup includes operator punching in or out, paperwork, obtaining tools, positioning unworked materials nearby, adjusting, and inspecting. After the parts are cycled, setup includes tearing down, returning tooling, and cleaning up of the workstation to a neutral condition ready for the next job. The setup does not include time to make parts because that is included in the cycle time. Estimating setup time is done for job shops and moderate quantity production. In mass production, setup costs are of less unit importance, although its absolute value remains unchanged for large quantities. Setup is handled as an overhead charge for mass production but is estimated for each lot of production quantity.

Cycle time or *run time* is the work to complete one unit after the setup is complete. Unlike setup, cycle is the repetitive portion of the production quantity.

7.3 METAL CUTTING OPERATIONS

Involved in a metal cutting operation cycle is load work (LW), advance tool (AT), machining, retract tool (RT), and unload work (UW). These elements are shown in Fig. 7.3(a) and are labeled "one work cycle." Tool maintenance is required after a number of parts are machined, and includes removing the tool and replacing or regrinding the tool point and reinserting the tool ready for metal cutting. This is shown in Fig. 7.3(b). The mathematical model most frequently used in the study of machining estimating describes the cost of a single-point-tool rough-turning operation.

Setup standard time data for metal-cutting operations are listed in hours because this is customary in the United States. For instance, Table 7.1 shows greatly abbreviated and typical values for computer-controlled lathes, mills, and drill presses. The time driver is the number of different tools and holding fixtures. Setup times, which include allowances, are posted once for each occurrence on the operations sheet. Setup element times are additive; that is, time for a work holder and for a tool are added, those values being read from standard time data and posted on the operations sheet.

Operation unit cost is a function of handling, machining, tool changing, and the tool cost.

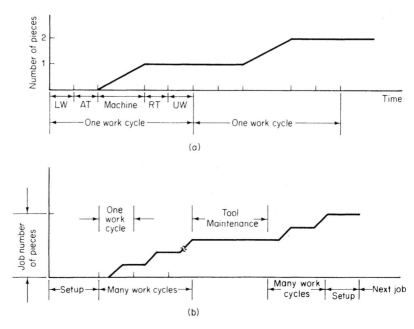

Figure 7.3. Operation items including setup, cycle, and tool maintenance.

TABLE 7.1. SETUP VALUES, EXPRESSED IN HOURS

Punch in and out, study drawing	0.2
Turret lathe	
First tool	1.3
Each additional tool	0.3
Collet fixture	0.2
Chuck fixture	0.1
Milling machine	
Vise	1.1
Angle plate	1.4
Shoulder-cut milling cutter	1.5
Slot-cut milling cutter	1.6
Tight tolerance	0.5
Drill press	
Jig or fixture	0.1
Vise	0.05
Number of numerically controlled turrets	
First turret	0.25
Additional turrets	0.07/each

Handling time is the minutes to load and unload the workpiece from the machine and can also include the time to advance and retract the tool from the cut and the occasional dimensional inspection of the part. Handling time is unrelated to cutting speed and is a constant for a specified design and machine. Table 7.2 is an example of cycle standard time data shown in decimal minutes, which includes the

TABLE 7.2.   HANDLING AND OTHER CYCLE-TIME ELEMENTS, EXPRESSED IN MINUTES

All machines	
Start and stop machine	0.08
Change speed of spindle	0.04
Engage spindle or feed	0.05
Air-clean part and fixture	0.06
Inspect dimension with micrometer	0.30
Brush chips	0.14
Turret lathe	
Advance turret and feed stock	0.18
Turret advance and return	0.04
Turret return, index, and advance	0.07
Cross slide advance and engage feed	0.08
Index square turret	0.04
Cross slide advance, engage feed, and return	0.14
Place and remove oil guard	0.09
Milling machine	
Pick up part, move, and place; remove and lay aside	
5 lb	0.13
10 lb	0.16
15 lb	0.19
Open and close vise	0.14
Seat with mallet	0.11
Wipe off parallels	0.26
Pry part out	0.06
Clamp, unclamp vise, $\frac{1}{4}$ turn	0.05
Quick clamp collet	0.06
Clamp and unclamp hex nut	0.21 each
Numerically control turret drill press	
Pick up part, move, and place; pick up and lay aside	
To chuck	
2.5 lb	0.10
5.0 lb	0.13
10.0 lb	0.16
Clamp and unclamp	
Vise, $\frac{1}{4}$ turn	0.05
Air cylinder	0.05
C-clamp	0.26
Thumb screw	0.06
Machine operation	
Change tool	0.06
Start control tape	0.02
Raise tool, position to new x-y location,	
advance tool to work	0.06/hole
Index turret	0.03/tool

AM Cost Estimator, 1988 Edition, Phillip F. Ostwald, Penton Publishing Co., Cleveland, OH.

personal, fatigue, and delay allowances. Decimal minutes are adopted for cycle work rather than seconds or hours because it is more widely understood. We define

$$\text{handling cost} = C_o t_h \tag{7.1}$$

where C_o = direct-labor wage, dollars per minute

t_h = time in minutes for handling

Figure 7.4(a) is an example of this element plotted against cutting speed. C_o does not include overhead.

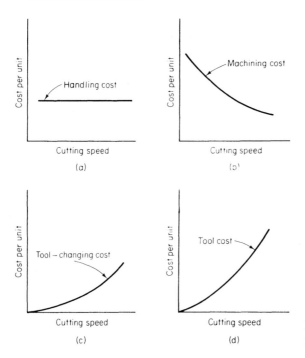

Figure 7.4. Graphic costs for four parts of machine turning economics.

Machining time is the minutes that the tool is actually in the feed mode or cutting and removing chips. We define

$$t_m = \frac{L}{fN} = \frac{L\pi D}{12Vf} \qquad (7.2)$$

$$\text{machining cost} = C_o t_m \qquad (7.3)$$

where t_m = machining time, min

L = length of cut for metal cutting, in.

D = diameter, in.

V = cutting speed, ft/min

f = feed rate, in./rev

$$N = \text{rotary cutting speed} = \frac{12V}{\pi D}, \text{ rpm} \qquad (7.4)$$

Each work material will have unique turning and milling cutting speeds and feeds as determined by testing or experience. Those values will differ for rough or finish machining and for tool materials. A roughing pass will remove more material when compared with a finish pass but does not satisfy dimensional and surface finish requirements. The material removal rate is min/in. for drilling and tapping. Table 7.3 is a small example of standard time data for machining. Two tool materials are high speed steel (HSS) and tungsten carbide. Stainless steel, medium carbon steel, and cast iron are work materials. The HSS value for rough turning of stainless steel is 150 fpm and 0.015 in. per revolution (ipr).

TABLE 7.3. MACHINING SPEEDS AND FEEDS

	Turning and Facing (fpm, ipr)			
	High-Speed Steel		Tungsten Carbide	
Material	Rough	Finish	Rough	Finish
Stainless steel	150,0.015	160,0.007	350,0.015	350,0.007
Medium carbon steel	190,0.015	125,0.007	325,0.020	400,0.007
Cast iron, gray	145,0.015	185,0.007	500,0.020	675,0.010

	HSS		HSS	
	Plain Milling		End Mill Slot 1 in.	
Material	Rough	Finish	Rough	Finish
Stainless steel	140,0.006	210,0.005	85,0.002	95,0.0015
Medium carbon steel	170,0.008	225,0.006	85,0.0025	95,0.002
Gray cast iron	200,0.012	250,0.010	85,0.004	95,0.003

	Drilling		
Drill Diameter*	Stainless Steel	Medium-Carbon Steel	Cast Iron, Gray
$\frac{1}{4}$	0.55	0.20	0.20
$\frac{5}{16}$	0.61	0.23	0.23
$\frac{3}{8}$	0.65	0.25	0.25

	Tapping	
Threads/in.*	Steel	Stainless Steel
32	0.18	0.33
20	0.15	0.30
16	0.19	0.33
10	0.32	0.48

*Times are minutes per inch for power-feed drilling; threading times are minutes per inch.
AM Cost Estimator, 1988 Edition, Phillip F. Ostwald, Penton Publishing Co., Cleveland, OH.

As the circular cutting velocity, fpm, increases, the unit machining time and cost decreases, and this is shown in Figure 7.4(b).

The length of cut depends on the geometry to be machined. In lathe turning, the length, which is being machined, is at least equal to the length dimension and is usually longer because of additional stock for roughing or finishing. In a lathe-facing element, the length of cut depends on the diameter, and the length is equal to the distance from the center of the barstock circle to the outside diameter (or the stock O.D.). Thus, the facing length is at least equal to one half of the diameter.

The diameter, D, may be either the workpiece diameter, as found in lathe turning or facing, or the tool diameter, as found in milling work. When a lathe-turning operation is visualized, the largest unmachined barstock dimension is the diameter for formula (7.2). For milling and drilling, the diameter is the cutter dimension. If a 6-in. milling cutter is used, the diameter is 6 in., and it is this dimension that determines the time for milling.

The cutting speed velocity, V, is surface feet per minute (meters per second). Its value depends on many factors, and standard time data will consider those effects. Table 7.3 gives a very small sampling used for the examples and problems.

TABLE 7.4. TAYLOR'S TOOL LIFE PARAMETERS, AND TOOL CHANGING, INDEXING, AND COSTS

	High-Speed Steel		Tungsten carbide	
Material	K	n	K	n
Stainless steel	170	0.08	400	0.16
Medium carbon steel	190	0.11	150	0.20
Cast iron, gray	75	0.14	130	0.25

Time to index a turning type of carbide tool, 2 min
Time to set a high-speed tool, 4 min
Large milling tool replacement, 10 min
Remove drill, regrind and replace, 3 min
Cost per tool cutting corner for turning, carbide, $3
Cost for high-speed steel tool point, $5
Cost per milling cutter, 6-in. carbide, $250
Drill, $3

AM Cost Estimator, 1988 Edition, Phillip F. Ostwald, Penton Publishing Co., Cleveland, OH.

Cutting tools become dull as they continue to machine. Once dull, they are replaced by new tools or are removed, reground, and reinserted in the tool holder. Empirical studies relate tool life to cutting velocity for a specified tool and workpiece material. Two popular tool materials are HSS and tungsten carbide. Most studies of tool life are based on the famous Taylor's tool life cutting speed equation.

$$VT^n = K \tag{7.5}$$

where T = average tool life, minutes per cutting edge

n, K = empirical constants resulting from regression analysis and field studies, $0 < n \le 1, K > 0$

The average tool life can be found as

$$T = \left(\frac{K}{V}\right)^{1/n} \tag{7.6}$$

For an example of finding the empirical coefficients n and K, refer to problem 5.6, which deals with a tool wear out study. Table 7.4 provides Taylor tool life data, tool changing, and indexing times and costs.

If a tool life equation is $VT^{0.16} = 400$, then we can find either V or T given the other variable. If $V = 200$ fpm, we expect 76 min of average machining life.

The third cost is the tool changing cost per operation. Define it as

3) $$\text{tool changing cost} = C_o t_c \frac{t_m}{T} \tag{7.7}$$

where t_c is the tool-changing time, minutes.

The tool-changing time, t_c, is the time to remove a worn-out tool, replace or index the tool, reset it for dimension and tolerance, and adjust for cutting. Time depends on whether the tool to be changed is a disposable insert or a regrindable tool for which the whole tool must be removed and a new one reset. In lathe turning and milling, there is the option of an indexable or regrindable tool. The drill is only reground. Standard time data are shown for the tool changing and tool costs and are given in Table 7.4. In Figure 7.4(c), we see the relationship of tool changing cost to cutting speed.

Define the following as

$$\text{tool cost per operation} = C_t \frac{t_m}{T} \tag{7.8}$$

where C_t denotes the tool cost, dollars.

Tool cost, C_t, depends on the tool being a disposable tungsten carbide insert or a regrindable tool for turning. For insert tooling, tool cost is a function of the insert price and the number of cutting edges per insert. For regrindable tooling, the tool cost is a function of original price, total number of cutting edges in the life of the tool, and the cost to grind per edge. As the cutting speed increases, the cost for the tool increases, as is shown in Figure 7.4(d). Table 7.4 provides tool costs and changing times.

The total cost per operation is composed of those four items. Machining cost is observed to decrease with increasing cutting speed and tool and tool changing costs increase. Handling costs are independent of cutting speed. Thus, we can say that unit cost C_u is given as

$$C_u = \Sigma\left[C_o t_h + \frac{t_m}{T}(C_t + C_o t_c) + C_o t_m\right] \tag{7.9}$$

On substitution of t_m and T by Eqs. (7.2) and (7.6), and after taking the derivative of this equation with respect to velocity and equating the derivative to zero, the minimum cost is found once we have V_{min}.

$$V_{min} = \frac{K}{\left[\left(\frac{1}{n} - 1 \right) \left(\frac{C_o t_c + C_t}{C_o} \right) \right]^n} \tag{7.10}$$

which gives the velocity for the unit cost of a rough turning operation. We do not recognize revenues that are produced by the machine in this development. Had we found the marginal cost and marginal revenue, similar to section 6.10, the intersection of marginal revenues and marginal costs provides a higher value of velocity than that demonstrated by Eq. (7.10). This intersection is the profit maximum point. Consequently, Eq. (7.10) identifies the minimum acceptable velocity.

Occasionally, to avoid bottleneck situations, we need to accelerate production at cutting speeds greater than that recommended for minimum cost. In those expedited operations, we assume the tool cost to be negligible, or $C_t = 0$. If the costs in the basic model are not considered, then the model gives the time to produce a workpiece, and we develop

$$T_u = t_h + t_m + t_c \frac{t_m}{T} \tag{7.11}$$

where T_u is minutes per unit. The production rate (units per minute) is the reciprocal of T_u. The equation gives the cutting speed that corresponds to maximum production rate and is

$$V_{max} = \frac{K}{\left[\left(\frac{1}{n} - 1 \right) t_c \right]^n} \tag{7.12}$$

The tool life that corresponds to maximum production rate is given by

$$T_{max} = \left[\left(\frac{1}{n} - 1 \right) t_c \right] \tag{7.13}$$

Consider an operation optimization problem of machining 430F stainless steel, 1.750 in. (44.45 mm) O.D. barstock. The cutting length is 16.50 in. plus a $\frac{1}{32}$ in. for approach giving 16.53 in. (419.8 mm). The turning operation will use a tungsten carbide, insertable and indexable eight corner tool (about the size of a dime) that costs $3 per corner. Time to reset the tool is 2 min. Handling of the part is 0.16 min, and the operator wage is $15.20 per hour. The Taylor tool life equation is $VT^{0.16} = 400$ (in U.S. customary units) for the tool and workpiece material. Feed of the rough turning element is 0.015 ipr (0.38 mm/rev) for a depth of cut of 0.015 in. (3.8 mm) per side. (Those data are included in Tables 7.1, 7.2, 7.3, and 7.4.) If the four items of the cost equation are plotted with several velocity values as the x variable, we have Fig. 7.5, which shows the optimum at 200 fpm (0.2 m/s).

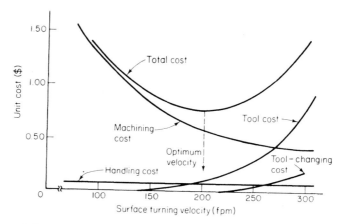

Figure 7.5. Study of the various effects of cutting speed on handling, machining, tool, and tool changing cost to give optimum unit cost.

Similarly, by using Eq. (7.10) we have for the information

$$V_{min} = \frac{400}{\left[\left(\frac{1}{0.16} - 1\right)\left(\frac{0.25 \times 2 + 3}{0.25}\right)\right]^{0.16}} = 201 \text{ fpm}$$

and $T_u = 2.74$ at V_{min}, $V_{max} = 275$, and $T_{max} = 10.5$ min.

7.4 METAL PART CASE STUDY

A metal pinion, shown in Fig. 7.6, and a long-term quantity of 1000 units is planned with five lot requirements of 200 units. Raw material is 1.750 in. (44.45 mm) O.D. 430F stainless steel. Raw stock is purchased in 12-ft (3.7 m) lengths. Note that in Figure 7.6 the finished dimension is 18.750 in. (476.25 mm), with a 0.015 in. (0.38 mm) facing length plus a 0.125 in. (3.18 mm) cutoff distance, resulting in a minimum length of 18.89 in. (479.81 mm).

The important procedure that identifies the sequence of the metal pinion throughout the factory is known as an *operations sheet.*

Table 7.5 identifies workstation, operation number, description, setup time, cycle time in hours per 100 units, and lot hours together with other technical and cost estimating information. The engineer determines the routing or sequencing of the workstations to produce the pinion.

With the design, Figure 7.6, the engineer studies the material requirements. Raw material amounts depend on scrap, waste, and shrinkage losses, which in turn are related to manufacturing operations. Each 12-ft bar will produce 7.6 pieces (= 144/18.89). Allowing for last piece gripping we expect 7 units per bar. Shape yield from Eq. (3.4) gives about 91% after considering waste losses for facing, cutoff width, and end gripping for the last pinion. This material costs $1.24/lb ($2.76/kg).

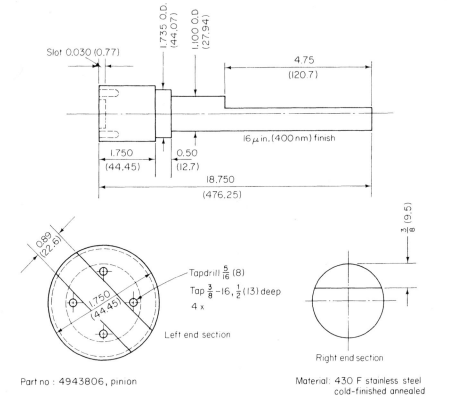

Figure 7.6. Typical part studied for cost estimating.

Each 200 quantity lot will require 29 bars, and each bar foot weighs 8.178 lb (3.71 kg). The unit cost of material is $17.38 = (8.178 \times 12 \div 7 \times 1.24)$, which is posted in Table 7.5.

The engineer selects the elements for each operation that correspond to standard time data elements. Knowing the manufacturing required to machine the metal pinion, the engineer posts machine specifications, machining elements, descriptions, workstation, and operations. Handling is added after the machining elements are calculated for stock removal, dimensions, feeds, and speeds. Information is selected from standard time data for the machine. Other constant elements are added to cover start and stop, inspection, and machine adjustment. The evaluation of setup concludes the operation estimate. This information is collected for operation 10 in Table 7.6.

Each lathe-machining element, for example, will use a single tool on a square or hexagon turret. Handling of the turrets to advance, engage, and retract from the workpiece depends solely on machining elements. Similarly, setup will depend on the number of tools, tolerance, and so on, which are dependent on machining elements.

TABLE 7.5. OPERATION SHEET FOR TYPICAL PART, GIVING ESTIMATE

Part no. 4943806
Part name Pinion

Ordering quantity 1000
Lot requirement 200

Material 430F Stainless Steel 1.750 ± 0.003 in. cold finished 65-12 ft. bars = 1000 pieces
Unit material cost $17.38

Workstation	Operation no.	Description of operations (list tools and gauges)	Setup hour	Cycle hour/ 100 units	Lot Hours	Productive cost	Lot operation cost	Unit operation cost
Turret lathe	10	Position Face 0.015 Turn rough 1.45 Turn rough 1.15 Turn finished 1.110 Turn 1.735 Cut off to length 18.730 (Carbide tools)	3.2	10.067	23.33	39.16	913.76	4.57
Vertical mill	20	End mill 0.89 slot with 3/4 H.S.S. end mill (Collet fixture)	1.8	7.850	17.50	90.98	1592.15	7.96
Horizontal mill	30	Slab mill 4.75 x 3/8 (Nesting vise, H.S.S. tool)	1.3	1.500	4.30	90.98	391.21	1.96
N.C. turret drill press	40	Drill 5/8 holes-4x 0.66 Tap 3/8-16 (Collet fixture)		5.245	11.15	39.16	436.63	2.18

Total productive hour unit cost $16.67
Material and labor cost $34.05

TABLE 7.6. DETERMINING CYCLE AND SETUP ESTIMATES FOR OPERATION 10, TABLE 7.5

A. Calculation of machining elements

Element	Dim. (in.)	Depth of Cut, (in.)	Length of cut L_d, (in.)	Safety Stock (in.)	Length (in.)	Diameter (in.)	Velocity (fpm)	Feed (ipr)	Time t_m
Facing		0.015	0.875	1/32	0.906	1.750	350	0.015	0.08 min
Rough turn	1.45	0.15	16.5	1/32	16.53	1.750	350	0.015	1.44
Rough turn	1.15	0.15	16.5	1/32	16.53	1.45	350	0.015	1.20
Fine finish	1.10	0.025	16.5	1/32	16.53	1.15	350	0.007	2.03
Turn	1.735	0.0075	0.5	1/32	.53	1.750	350	0.007	0.10
Cutoff		0.125	0.875	1/32	0.906	1.750	350	0.015	0.08

t_m = sub total machining time =   4.93 min

B. Selection of handling and other machine standard time data

Start and stop machine	0.08 min
Advance turret and position raw stock	0.18
Place and remove oil guard	0.09
Speed change, assume 4 times, 4 × 0.04	0.16
Advance, index, and return turret, 6 times, 6 × 0.08	0.48
Inspect part with micrometer, irregular element, $0.30 \times \frac{1}{5}$	0.06
Air-clean part	0.06

t_h = subtotal handling and other machine data = 1.11 min
total cycle   = 6.04 min

C. Setup development

Punch in and out, study drawing	0.2 hour
First tool	1.3
Five additional tools	1.5
Collet fixture	0.2

Total setup = 3.2 hour

D. Entry values for operations sheet

Setup = 3.2 hours
hr/100 = 10.067

Consider the elements of operation 10, which uses a lathe machine. A turret lathe is a turning metal-cutting machine. Tools can be positioned in readiness on the turrets for consecutive machining cutting elements. Usually there are two turrets, a six- or more stations-type turret, mounted on the ways and a four-position square-type mounted on a carriage. Ten or more different cutting tools can be preset and mounted on those turrets.

Figure 7.7 is a description of the consecutive machining elements. A facing cut is made in Fig. 7.7(a). Note Table 7.6, where the design factors are given to calcu-

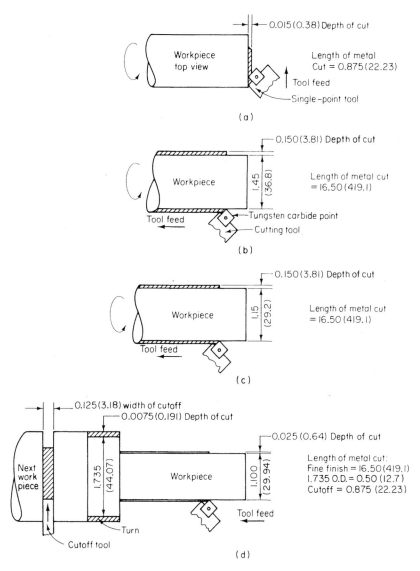

Figure 7.7. Machining cuts for operation 10 by using turret lathe: (a) facing element; (b) first pass, rough-turning element; (c) second pass, rough-turning element; (d) final turret lathe elements.

late t_m using tungsten carbide. Values of t_m are calculated by using Table 7.3 circular velocity V and feed f and Eq. (7.2). Because those computations are frequent, special slide rules are available or computers or electronic calculators are programmed to do this work quickly. Observe that values of t_m are listed in Table 7.6(a). Note that the diameter is successively reduced as a result of two rough and one finish turning passes. Those differing values of diameter theoretically require different values of the rpm or N, according to Eq. (7.4), or

$$N = \frac{12 \times 350}{\pi \times 1.75} = 764 \text{ rpm}$$

versus

$$N = \frac{12 \times 350}{\pi \times 1.45} = 922 \text{ rpm}$$

Each of those rpms calls for a speed gear adjustment, which requires time. If the time saving is small, an average rpm will be used and no speed adjustment made. But for long cuts the savings in machining time is worth the time to change the rpm. That is the plan we use in our examples.

Knowing the number of different cuts, each cut will require a separate tool and an "advance, engage, and retract" turret handling time for either the hex or square turret. Handling and other manual elements are now selected from standard time data in Table 7.2 and listed in Table 7.6(b).

Setup is estimated after the cycle time is determined. Note that setup values are shown in Table 7.6(c). The entry for the cycle is listed in the hours per 100 units column. For example, $(6.04/60) \times 100 = 10.067$ hours per 100 units. The setup of 3.2 hours and 10.067 hours per 100 units are transferred to Table 7.5, operation 10.

In operation 20, a $\frac{3}{4}$-in. (19-mm) end mill will machine the 0.89-in. (22.6-mm) wide by 0.030-in. (0.77-mm) deep slot. A collet fixture grips the pinion diameter. The width of the slot is slightly wider than the end mill diameter. An end mill cutter is similar to a drill except that the bottom is flat and has four cutting flutes on the bottom and side. A vertical milling machine grips the end mill. Two passes are necessary for the width, and a rough cut will remove 0.025 in. (0.64 mm) depth and the finishing cut will remove 0.005 in. (0.13 mm) depth with a high-speed steel tool. Figure 7.8 describes the machining.

In operation 30, a 4-in. (100-mm) plain milling cutter, high-speed steel, with a 6-in. (150-mm) face and eight teeth will slab mill the 4.75-in. (120.7-mm) dimension given on the metal pinion. This flat on the metal pinion is relative to the slot cut in operation 20, and a special nesting vise is designed and constructed to ensure dimensional compliance. Because of the greater rigidity of this tool, along with the horizontal arbor milling machine, only one pass is necessary to reach finish dimension. Figure 7.9 shows an end view of the basic features of the machining operation.

In operation 40, four holes are drilled and tapped. A collet fixture with bushing guides the drilling of the four holes. The fixture design satisfies the hole locational requirements. Figure 7.10 describes the drilling and tapping element.

The end milling, slab milling, and drilling-tapping elements have different length of cut requirements. Figure 7.11 is an illustration. The length of cut L is the distance that the cutting tool is penetrating at feed f velocity. This is much less than

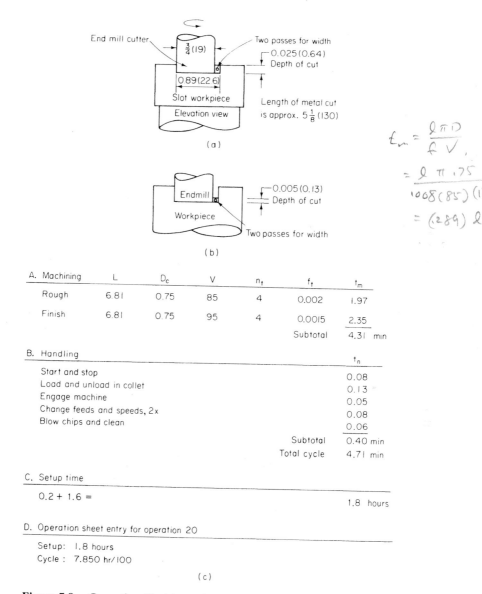

$$t_m = \frac{\ell \pi D}{f V},$$

$$= \frac{\ell \pi .75}{.008(85)(12)}$$

$$= (.289)\ell$$

Figure 7.8. Operation 20: (a) rough end-mill slotting element; (b) finish end-mill slotting element; (c) calculation of estimate.

the "rapid traverse" velocity, which may be as high as 200 to 500 in./min. The general relationship for the length of cut is

$$L = L_s + L_a + L_d + L_{ot} \qquad (7.14)$$

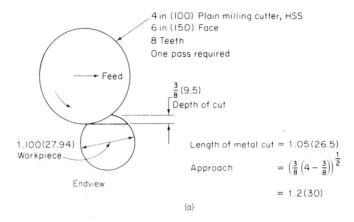

$t_m = \dfrac{L\pi D}{12\,Vf}$

$= \dfrac{3.45\,\pi\,4}{(12)(140)(.008)}$

$= .54$

Figure 7.9. Operation 30: (a) slab milling operation, end view section; (b) calculation of estimate.

where L = length of cut at velocity f, in.
       L_s = safety length, in.
       L_a = approach length owing to cutter geometry, in.
       L_d = design length of workpiece, in.
       L_{ot} = overtravel length, in.

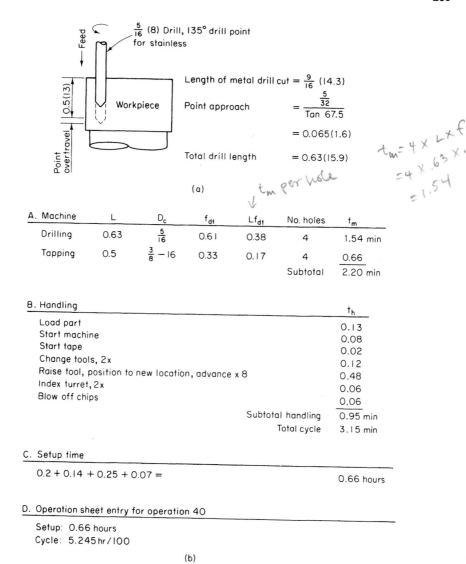

$$t_m = 4 \times L \times f$$
$$= 4 \times .63 \times .61$$
$$= 1.54$$

A. Machine	L	D_c	f_{dt}	Lf_{dt}	No. holes	t_m
Drilling	0.63	$\frac{5}{16}$	0.61	0.38	4	1.54 min
Tapping	0.5	$\frac{3}{8} - 16$	0.33	0.17	4	0.66
					Subtotal	2.20 min

B. Handling	t_h
Load part	0.13
Start machine	0.08
Start tape	0.02
Change tools, 2x	0.12
Raise tool, position to new location, advance x 8	0.48
Index turret, 2x	0.06
Blow off chips	0.06
Subtotal handling	0.95 min
Total cycle	3.15 min

C. Setup time

0.2 + 0.14 + 0.25 + 0.07 =	0.66 hours

D. Operation sheet entry for operation 40

Setup: 0.66 hours
Cycle: 5.245 hr/100

(b)

Figure 7.10. Operation 40: (a) drilling element, end view section; (b) calculation of estimate.

L is the value used in finding t_m for machining calculations. Safety length, L_s, is necessary for any possible stock variations. If the stock is oversize and the cutter is at rapid traverse velocity, then damage could result with the cutter's striking the workpiece. Safety stock values are $0 < L_s \leq \frac{1}{2}$ in.

Approach length depends on cutter workpiece geometry. Note Figure 7.11, which shows the three cases of operation 20, 30, and 40.

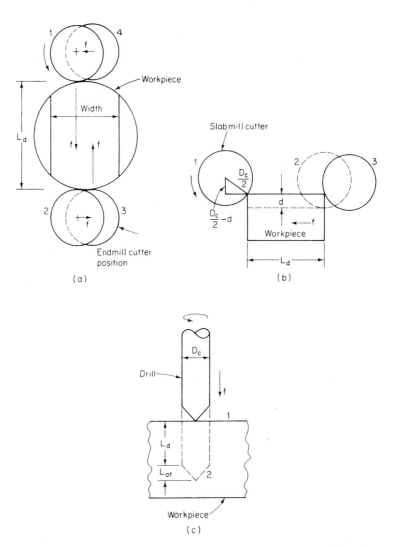

Figure 7.11. Approach and overtravel length for (a) end milling (top view), (b) slab milling (elevation view), and (c) drilling (elevation view).

In operation 20, a $\frac{3}{4}$-in. (19.1-mm) end mill is vertically milling the 0.89-in. (22.6-mm) slot. Note that the cutter will start in position 1 and mill to 2, 3, and 4 and finally back to position 1, ready to begin end milling the next workpiece. L_a is approximately two diameters plus the distance from 2 to 3 and 4 to 1 or $2(0.89 - 0.75) = 0.28$ in. Let $L_s = \frac{1}{64}$ in. (0.4 mm) $= L_{ot}$, and $L_d = 1.75$ (44.5 mm). Then cutting length, noting Fig. 7.11(a), $L = \frac{1}{64} + 4(0.75) + 0.28 + 2(1.75) + \frac{1}{64} = 6.81$ in. (173 mm). In this calculation the design length is about 50% of total length. Operation 30 can be simplified and shown as Fig. 7.11(b). The chip removal of the cutter is on the periphery of the 4-in. (100-mm) O.D. It has an approach given by

$$L_a = \sqrt{\left(\frac{D_c}{2}\right)^2 - \left(\frac{D_c}{2} - d\right)^2} = \sqrt{d(D_c - d)} \tag{7.15}$$

where D_c = cutter diameter, in.
   d = depth of cut, in.

For this operation we assume that $L_a = L_{ot}$. If the operation were only roughing, then we could let $L_{ot} = 0$, since the full depth would have been removed at position 2. In a roughing pass, the cutter is not required to leave the stock, only that the full diameter has completed the cut length. If the stock surface is considered a finished surface, then the cutter is required to move entirely off the pinion so as to avoid a potential dragging tooth, which could cause a scratch and lower the quality of the surface. But the operation design calls for only one pass, so the cutter will feed from position 1 to 3. At position 3 it will reverse and return to position 1, at rapid traverse velocity. For operation 30, we let

$$L_s = \tfrac{1}{16} \text{ in. (0.4 mm)}$$

$$L_d = 1.05 \text{ in. (26.7 mm)}$$

$$L = \tfrac{1}{16} + 2[0.375(4 - 0.375)]^{1/2} + 1.05 = 3.45 \text{ in. (87.5 mm)}$$

Operation 40 requires four $\tfrac{5}{16}$-in. (8-mm) holes. Drilling will have an overtravel distance required for the length of the 135° conical drill point. Stainless steel uses a different conical point than other materials; the usual included point angle is 118°.

$$L_{ot} = \frac{D_c}{2 \tan 67.5} = 0.2D_c \tag{7.16}$$

where D_c is the drill diameter, in.
   For drilling we use $L_s = \tfrac{1}{6}$ in. (1.6 mm) and $L_a = 0$. For operation 40 the length for machine feed is

$$L = \tfrac{1}{16} + 0 + \tfrac{1}{2} + 0.2(\tfrac{5}{16}) = 0.63 \text{ in. (15.9 mm)}$$

Tapping usually does not require additional distances and only the tap length L_d is necessary.
   With the cutting lengths determined, we turn next to finding cycle and setup times by using standard time data. As before, the steps proceed with the various machining elements, handling, setup, and converting estimated values to setup and cycle hours per 100 units. For a multitooth milling cutter, Eq. (7.2) is converted to

$$t_m = \frac{L\pi D_c}{12Vn_t f_t} \tag{7.17}$$

where n_t = number of teeth on cutter
   f_t = feed in. per tooth

For the rough milling element of operation 20, $L = 6.81$ in., $D_c = \frac{3}{4}$ in., $V = 85$ fpm, $n_t = 4$, and $f_t = 0.002$ in./rev.-tooth; then $t_m = 1.97$ min. For the finish milling element the same cutter and setup are used, except that $V = 95$ fpm, $f_t = 0.0015$ iprt, and $t_m = 2.35$ min. Those are marked in Fig. 7.8, where other standard time data entries are made. Finally, for operation 20 the setup and cycle time values are posted on the operations form.

Operation 30 is horizontal milling and varies somewhat from operation 20. The metal cutting values are $L = 3.45$ in., $D_c = 4$ in., $V = 140$ fpm, $n_t = 8$, $f = 0.006$ iprt, and $t_m = 0.54$, which is entered in Fig. 7.10. Other values for handling and setup are selected from Tables 7.1 and 7.2, and finally setup and cycle time are posted to the operation sheet for operation 30.

The drilling and tapping element of operation 40 uses a different formula for machining, because those kinds of metal cutting are expressed as rates, or

$$t_m = Lf_{dt} \tag{7.18}$$

where f_{dt} is the drilling or tapping feed rates, min/in.

The metal cutting rates are classified by standard time data, and typical values are given in Table 7.3. Entry information is drill diameter and workpiece material. Other handling time is collected for Fig. 7.10, and final cycle information is listed in Table 7.5.

7.5 NONRECURRING INITIAL FIXED COSTS

Tools and test equipment are necessary to adapt general-purpose machinery to a specific operation or product design. The engineer evaluates tool designs for their cost at the moment of the operation sheet preparation. Officially, we term those devices as nonrecurring initial fixed costs; that is, one-time-only costs despite the number of lots or continuous production quantity requirements. Those costs are unlike setup costs, which occur for each batch. The pinion or product cannot be manufactured without those tools, and thus they are front end or initial costs. Those designed devices are classified as fixed costs rather than variable and are a depreciable asset subject to taxation laws. Additionally, the tools are classed as permanent tools, as contrasted to perishable tools, such as lathe-turning tools, milling cutters, and drill bits. Perishable tooling incurs an indirect operational expense and is an overhead charge. Examples of nonrecurring initial fixed costs are jigs, fixtures, molds, dies, electronic testing and quality control gauges, and special-handling devices; that is, items that require engineering design and are costly.

Engineers are confronted with dollar signs at every stage in manufacturing. The designer should know enough of cost analysis to determine whether temporary tooling would suffice even though funds are provided for more expensive permanent tooling. The designer should be able to defend decisions when amortizing the tooling on a single run as opposed to amortizing tooling costs against probable future reruns.

Many types of jigs, fixtures, tools, gauges, and dies have common details that permit simplification in methods of estimating. The factor method is usually employed; that is, the factors are larger than are used for labor estimating from standard time data, which is too time consuming. In some cases the engineer may resort to the comparison method, which is the quickest but not the most accurate way.

In terms of test equipment and inspection devices, it too is weighed on equal footing with ordinary tooling. The engineer needs to know what is available for production and what is special that may require design or purchase. Special processing tools may be required for general-purpose or special-purpose equipment.

Tooling before assembly in operation and postoperation inspection could include gauges, holding devices, cutting tools, universal jigs, and specially designed and constructed fixtures. Inspection requirements include tools for the receiving department, in process inspection, postfabrication inspection requirements, and postassembly requirements. Environmental needs where engineering specifications call for peculiar ambient, weather, or shock tests cannot be overlooked.

Those initial costs will be estimated either before design or following design but before manufacture. If estimating precedes design, then an experienced engineer will compare the new design with other designs for which historical information exists. This comparison method is quick but not necessarily accurate.

An alternate and accurate approach is to use standard time data. Tool-estimating information can be developed for a variety of designs. A typical and much reduced sample of estimating data for manufacturing drill jigs is given in Table 7.7. The data are dimensioned in hours and specifically are for the manufacturing of fixtures and jigs using hardened steel components precision machined and assembled by a journeyman tool and die maker.

The data in cycle standard time data have the dimension of decimal minutes. But tool estimating data are different, because, though only one tool is to be built, the data need not have a too fine level of resolution. For example, a block to be machined, ground, drilled, reamed, and pinned in place on a larger block is too much detail to be estimated for one tool. Instead, the engineer visualizes the overall character of the tool and selects a time from the provided numbers. This simple practice is preferred because several engineers can use the same data, and accuracy and consistency are improved. Figure 7.12 is a sample drill jig, and Table 7.7 gives the estimating data for milling fixtures, drill jigs, and work holders.

Table 7.8 is a sample estimate for a drill jig, where the part is to have one accurately located hole. The simple analysis shows that various elements are picked from the standard time data in Table 7.7, and eventually the total hours are multiplied by the productive hour rate of $65 per hour. Material cost is found by finding the weight of the tool and then by multiplying by the $6 per pound.

Cost of tooling is significant. Tooling cost is apportioned to the product by amortization or the tool builder may make a direct sale to customer or the cost of the tool may be applied to general overhead application.

Amortization is an estimating calculation, meaning the division of the cost by the quantity of the operational lot or market total.

TABLE 7.7.   STANDARD TIME DATA[a] FOR WORK HOLDERS AND DRILL JIGS

1. Flat Baseplates

| | | Dimension | | |
Type	Thickness	Width	Length	Hour
Drilling Fixtures				
Small	$\frac{3}{4}$–1	3	3	5
Medium	$\frac{3}{4}$–1	7	7	8
Large	$\frac{3}{4}$–1	12	12	10
Milling Fixtures				
Small	$\frac{5}{8}$–$1\frac{1}{2}$	3	6	10
Medium	$\frac{5}{8}$–$1\frac{1}{8}$	6	9	12

2. Angular Baseplates

Dimension	Base	In Combination	C-Angle
$3 \times 3 \times 3$	10	5	7
$6 \times 6 \times 6$	15	9	12

3. Two Baseplates for Box Jig

Dimension	Hr
$\frac{3}{4} - 1 \times 3 \times 3$	11
$\frac{3}{4} - 1 \times 7 \times 7$	14

4. Drill Plates and Support Blocks

Type	Dimensions	Hr
Drill plates	$\frac{3}{4} \times 3 \times 3$	3
Box jig	$1 \times 3 \times 3$	5
Support blocks	$\frac{5}{8} \times 1\frac{1}{2} \times 3$	2
Collet, round	to 4 in. OD	7

5. Miscellaneous

	Hr
Hand knobs, each	1.5
Guide bushings, each	2.0
Hinge plate, each	12.0
Clamp with screw	7.5
Feet	2.5
Pin locators	1.75

[a]Times are in hours.

A.M. Cost Estimator, Philip Ostwald, Penton Publishing, Cleveland, OH, 1988.

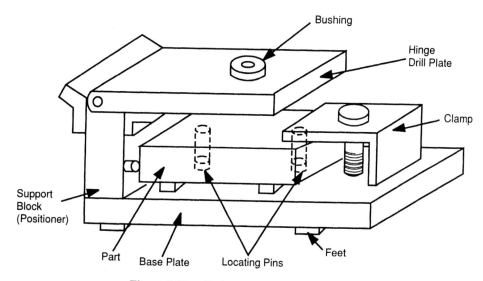

Figure 7.12.   Basic components of a fixture.

TABLE 7.8.   TOOL ESTIMATE FOR FIGURE 7.12, DRILL FIXTURE

Tool type:	Drill fixture, with hinge top, one bushing
Tool material:	Hardened high carbon steel, 8 lb
Description of part:	Flat block with one accurately located hole

Description of Tool Elements	Hours
Angular baseplate, C-angle	7
Guide bushing	2
Hinge plate	12
Clamp with screw	7.5
Feet, 2	5
Pin locators, 5 × 1.75	8.8
Total hours	42.3

Productive hour cost for tool making	$65/hour
Cost for tool, 42.3 × $65	$2750
Material cost @ $6.00/lb	48
Total cost	$2798

Selecting this quantity presents a trade-off problem. If this quantity is too small, then amortization will be overstated, causing greater cost and increased price, perhaps resulting in fewer sales of the product. If the quantity is too large, then other errors may result or the tooling could be made obsolete owing to unfortunate design changes required by the part and then sunk costs are created. The cost of the tooling may not be completely amortized, owing to actual production falling short of expectations. Problems involved in evaluating trade-off policies are left for the student. But the simplest expedient is to add the prorated cost of the tooling to operation cost.

A model for amortization is given as

$$C_{ot} = \frac{C_{nif}}{N} \tag{7.19}$$

where C_{ot} = operation cost for tooling, dollars per unit

C_{nif} = nonrecurring initial fixed-cost dollars

N = quantity used for amortization, which may be lot, model, or year

In some business situations a manufacturer may sell the tools directly to the customer. Vendors will design and build the tool and separately quote the cost of the tooling independent of the unit price to a customer. A separate tooling charge sidesteps the trade-off problem involved with amortization, because the customer receives the tooling on delivery of the product. In other situations the engineer will add tooling costs to overhead, which spreads the cost to all products. This is not a recommended practice.

The engineer is able to evaluate several alternative designs. A part can be manufactured with or without tooling. Tooling will allow less direct labor time in the operation but will increase initial costs. This break even can be evaluated by using

$$C_{nif} = \frac{Na(1 + t) - SU}{I + T + D + M} \tag{7.20}$$

where N = number of units manufactured per year

a = savings in labor cost per unit compared to another operation

t = percentage of overhead applied on labor saved

SU = yearly cost of setup, dollars

I = annual allowance for interest on investment, decimal

T = annual allowance for taxes, decimal

D = annual allowance for depreciation, decimal

M = annual allowance for maintenance, decimal

This formula is used in interesting ways. A design can be manufactured with a number of methods. Is it cheaper to use a tool or employ direct labor that performs the same operation without a tool? This question, one that deals with comparative economics between competing methods, has many variations.

Consider the tool that was previously estimated for the block with a single hole. There are 50,000 units to be produced, and a previous nontooled method indicates that $0.15 will be saved per unit if a tool were available that guides the drill. A previous nontooled method is estimated, and estimates show that the tooled method will save $0.15 = a$. Savings in direct labor translate to 20% comparable savings in overhead, or $t = 20\%$. A setup of the tooled method is expected to cost $215, or $SU = \$215$. This company requires an annual interest of 12.5%, taxes are 9%, depreciation is 25%, and annual maintenance of 18% is anticipated. The cost of a tool is then estimated to be

$$C_{nif} = \frac{50,000 \times 0.15(1.20) - 215}{0.125 + 0.09 + 0.25 + 0.18} = \$12,167$$

This equation suggests that the cost of the tool of $12,167 will pay back the potential savings in direct labor. The tool that is estimated to cost $2798, as shown in Table 7.8, is less than the tool cost necessary for the break even. The new tool is recommended for the operation. The new tool will reduce the cost of producing the part.

7.6 OPERATION COST

Operation cost is found after the operations sheet is completed. Those important steps start with the posting of the setup and cycle times that are expressed in hours and hours per 100 units. Lot hours are found by using the relationship

$$\text{lot hours} = SU + N \times H_s \tag{7.21}$$

where lot hours = total time for the operation
SU = setup hour for operation
N = lot quantity
H_s = standard hours per unit

For example, operation 10 in Table 7.5 indicates 3.2 setup hours, hours per 100 units = 10.067, and $N = 200$ lot requirement. Thus, lot hours = 23.33 ($= 3.2 + 200 \times 0.10067$).

The next step is to post the productive hour cost (PHC), which is sum of the machine hour rate and the gross hourly cost for labor to operate the equipment. The value of $39.16 is selected from chapter 4, Table 4.11, which gives the methods for determining the productive hour cost. Then,

$$C_{dlo} = \text{lot hours} \times PHC \tag{7.22}$$

where C_{dlo} = operational labor and equipment cost
PHC = productive hour cost for equipment, dollars per hour

Productive hour cost depends on a number of factors that influence the cost of labor and equipment. The development of methods that form the PHC were provided in chapter 4, but the essential thought is that the PHC fairly allocates overhead costs to the equipment. Of the several methods, the PHC is considered the preferred method.

In operation 10 in Table 7.5 this is really one calculation, or $913.76 (= 23.33 $\times$ 39.16) is the lot cost for the operation. Then

$$C_u = \Sigma C_{dlo}/N + C_{dm} + \Sigma C_{ot} \qquad (7.23)$$

where C_u = unit cost of manufacturing operations.

C_{dm} = unit cost of material, including scrap and waste

C_{ot} = operation cost for tooling, dollars per unit

7.7 MAKE-VERSUS-BUY ANALYSIS

Unit costs are used in many important ways, such as pricing or make-versus-buy. Pricing is considered in chapter 8. In make-versus-buy the engineer determines if it is cheaper to self manufacture the part, or "make," versus purchase from an external source, or "buy." The objective is selection of the cheapest source of supply. The engineer compares the price as promised by the vendor's quotation to a calculated make value. If a purchased part is cheaper, then the firm will buy the design.

Vendor capability, quality of the vendor and buyer, schedule to complete and deliver, and future activity are factors beside cost that influence the subcontract decision. If those factors are more or less equal, it is to the firm's economic advantage to base the decision on the results of a make-versus-buy analysis.

The "buy" side of the comparison is supplied by a subcontractor. Table 7.9 is a summary of the cost elements estimated for the make side of the comparison. The inclusion depends on the level of plant capacity; that is, the heading "plant load" refers to whether or not the plant is operating at less than full or greater than 100% of fixed capacity. A plant below capacity will have fewer employees than normal. Idle equipment is apparent, and sales are less than in previous periods. If the plant is less than capacity, then fixed costs are not usually estimated for the make side. When the fixed costs are not included, the make value is less in value, allowing the factory to be more competitive when compared to purchased parts. A company will adopt this policy because it may want to keep its employees and plant operating.

TABLE 7.9. SUMMARY OF THE COST ITEMS TO CONSIDER IN THE "MAKE" SIDE OF THE MAKE-VERSUS-BUY COST ESTIMATING ANALYSIS

	Plant Load	
Cost item	Less than 100%	Greater than 100%
Direct labor	Include	Include
Direct material	Include	Include
Variable overhead	Include	Include
Fixed overhead	Omit	Include
Marginal cost	Include	Include
Sunk cost	Omit	Omit
Profit	Omit	Optional

Fixed costs are interest charges, depreciation, and those account items that must be paid but not necessarily from the production of the make part. Other activities of the business, it is assumed, will pay those fixed costs.

Direct labor, direct material, and variable overhead are always included in the make side calculation despite the level of plant load. An optional choice implies that management policy will dictate if the item is included. Omitting an item implies that the make side is less, giving the economic advantage to the engineer's company.

The marginal cost component is a special calculation that results from the decision to make or buy, which can be positive or negative. Consider an example where the plant is at less than 100% capacity. If the decision to buy results in employee terminations, then the unit cost approximation for marginal cost is positive to reflect increased unemployment insurance. Because the choice is relative between a price-versus-make value, there is little difference in the outcome if the marginal cost is subtracted from price, except that it is better estimating discipline to leave price unaltered.

A plant with greater than 100% capacity will have overtime, will have rushed work, and will be hiring employees. If the plant capacity is greater than 100% and the decision "make" will require overtime, new equipment, plant space, or additional shifts, then the marginal cost addition to the make side is positive. Problems are provided for make-versus-buy.

SUMMARY

Cost is the major objective for operation estimating, and material, labor, and tooling are the principal components. If the labor is designated and engineering performance data are available, then time estimates are selected for work elements. The total time is subsequently multiplied by the wage or the productive hour cost. Tools can be estimated and appraised for their economic justification.

In the next chapter we consider product estimating. Manufacturing operation estimates are a prerequisite for product estimating and provide information to cost and price products.

QUESTIONS

7.1. Define the following terms:

RFE	Speeds and feeds	Milling
Operation cost	Machining cost	Length of cut
Moderate production	Rough pass	Approach
Process	Tool life	Flat rates
Operation	Tool changing	Make-versus-buy
Setup	Amortization	Manufacturing work
Cycle	Operation sheet	Standard man hour

7.2. What role does the RFE have in the management of cost estimating?

7.3. Separate the types of production according to volume. How else can you classify production?

7.4. Why is setup, a fixed direct-labor cost, divided by the quantity of the lot?

7.5. Describe the origin of flat rates. Why are some labor hour estimates flat?

7.6. Name examples of nonrecurring initial fixed costs. How are those costs handled? Does the act of separating a tooling charge from the unit cost of the product for a buyer eliminate the amortization problem?

7.7. If a tool has been designed, constructed, and paid for, then would it be included as an item on the make side of a make-versus-buy analysis?

7.8. Define marginal costs for a make-versus-buy analysis.

7.9. List the pros and cons of trade-off policies dealing with the selection of quantity in the amortization of nonrecurring initial fixed costs.

7.10. "Make" estimates, with a variety of initial assumptions, are prepared for a product and are shown by Fig. Q7.10. The horizontal axis is the economic status of the plant, ranging from poor to normal to excellent. The vertical axis is cost per unit, ranging from the floor value of direct labor, direct material, and variable overhead (which are the minimum cost inclusions irrespective of the economic status of the business) to the maximum value (which is the price plus premium). An upper and a lower value for the cost per unit of a make-versus-buy analysis can be determined at any economic status of the plant by simply choosing different combinations of the cost elements. This results in an upper and a lower value of the cost per unit. The middle zone area is the possible range of the estimate that results from those different initial assumptions. Discuss the nature of the cost elements, assumptions, and the procedures that give a cost zone. Elaborate on the initial management and engineering policies that guide the selection of these cost elements.

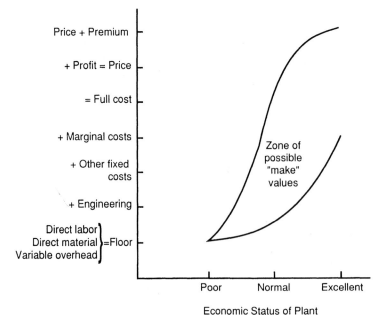

Figure Q7.10

PROBLEMS

7.1. (a) An operator earns $15 per hour, and handling and other metal cutting elements total 1.35 min. What is the cost for the element?

(b) The length of a machining element is 20 in. and the part diameter is 4 in. O.D. Velocity and feed for this material 275 fpm and 0.020 ipr. What is the time to machine?

(c) The Taylor tool life equation is $VT^{0.1} = 372$. What is the expected average tool life for $V = 275$ fpm?

(d) Tool changing time is 4 min, tool life equation is $VT^{0.1} = 372$, $V = 275$ fpm, $C_o = \$0.25/\text{min}$, $L = 20$ in., $D = 4$ in., and $f = 0.020$ ipr. Determine the tool changing cost.

(e) By using the information in part (d) and $C_t = \$5$, find the tool cost per operation. Would you recommend those machining conditions?

7.2. (a) Operator and variable expenses are $60 per hour, and handling is 1.65 min. Find the handling cost for this element.

(b) The length and diameter of a gray iron casting are 8.5 in. by 8.6 in. Velocity is 300 fpm, and feed is 0.020 ipr. Find the machining time.

(c) The Taylor tool life equation is $VT^{0.15} = 500$. Find the expected tool life for 300 fpm.

(d) The time to remove an insert and index to another new corner is 2 min, the tool-life equation is $VT^{0.15} = 500$, $V = 300$ fpm, $C_o = \$1/\text{min}$, $L = 8.6$ in., $D = 8$ in., and $f = 0.020$ ipr. What is the cost to change tools?

(e) By using the information in part (d), and that an eight-corner insert is $24, find the tool cost per operation.

7.3. (a) Stainless-steel material is to be rough and finish turned. Diameter and length are 4 in. × 30 in. Recommended rough and finish cutting velocity and feed for tungsten carbide tool material are (350, 0.015) and (350, 0.007). Determine rough and finish cutting time.

(b) Medium carbon steel is to be rough and finish turned by using high-speed steel-tool material. The part diameter and cutting length are 4 in. and 20 in. Determine the total time to machine by using Table 7.3.

(c) Gray cast iron is to be rough and finish turned with tungsten carbide tooling. Part diameter and cutting length are 8.5 in. and 8.6 in. What is the part rpm for the rough and finish? Find the turning time by using Table 7.3.

7.4. (a) The top of a square 250-mm block is to have a 65-mm slot machined on it. The end mill is 25 mm, and only one pass is traced over the slot. Calculate the length of cut if a safety stock is 5 mm.

(b) A slab milling cutter is 6 in. in diameter. The design length is 15 in., and two machining passes of $\frac{3}{8}$ in. and 0.015 in. depth of cut are necessary. Safety stock is $\frac{1}{4}$ in. If the cutter is required to move its vertical center line to the edge of work for roughing and completely leave the work for finishing, determine the length of cut for roughing and finishing.

(c) Find the length of cutting for a 25-mm drill in soft steel where the design length is 13 mm.

7.5. (a) The top of a 250-mm dia. circle avil is to have a 60-mm slot through the cross-hair center of the circle. Find the length of cut for a 25-mm dia end mill if the safety stock is 5 mm.

(b) Find the length of cut for a 2-in. slab milling cutter removing $\frac{1}{4}$ in. of thickness for a workpiece length of 10 in. The milling cutter is to mill completely off of the stock. Safety stock is $\frac{1}{16}$ in.

(c) Calculate the overtravel required for a $\frac{1}{2}$ in. drill in stainless steel; in cast iron.

7.6. (a) A stainless steel surface 1 in. × 10 in. is to be end milled with a 1-in. four-flute end mill for a depth of 0.0l5 in. The end mill will pass entirely over the material and have a safety stock of $\frac{1}{8}$ in. Velocity is 85 fpm, and chip load per tooth is 0.002. What is the length of cut? How much time is necessary for machining? What is the cutter rpm and feed rate in in. per min?

(b) A 1-in. slot is to be end milled in gray cast iron for a depth of $\frac{1}{4}$ in. for a design length of 20 in. A rough and finish pass is required for the high-speed steel four-flute cutter. The cutter must pass entirely over the surface, and a safety stock of $\frac{1}{8}$ in. is necessary. What is the machining length of cut for rough or finish? Determine rough and finish machining time. Use Table 7. 1 for machining data, where data are cutting velocity and tooth load. Find cutter rpm and cutting rate in in. per min.

7.7. (a) A surface 8 in. wide by 20 in. long is rough milled with a depth of cut of $\frac{1}{4}$ in. by a 16-tooth cemented carbide face mill 6 in. in diameter. The material is gray cast iron. What is the cutting length? Estimate the cutting time if $V = 120$ fpm and $f_t = 0.0012$ in./tooth-rev. Find the cutter rpm.

(b) A high-speed steel cutter vertical milling with 12 teeth is 100 mm in diameter and 125 mm long and is used to mill a soft-steel surface 75 mm wide by 225 mm long with a depth of cut of 20 mm. A cutting speed of 0.5 m/s and a feed of 0.25 mmpt are selected. What are the cutting time and cutter rpm?

7.8. (a) A stainless-steel part is to be drilled $\frac{5}{16}$ in. for a depth of 1 in. It is followed by a $\frac{3}{16}$ tap for a depth of $\frac{7}{8}$ in. Find the drilling length and the drilling and tapping time. Use Table 7.3. Repeat for steel.

(b) Steel is tapped drilled $\frac{3}{8}$ in. for a depth of 1 in. and is followed with a tapping element of $\frac{5}{16}$ 10 hole for $\frac{7}{8}$ in. Find the drilling length and the drilling and tapping time. Repeat for stainless steel.

7.9. (a) Find the cutting time for a hard copper shaft 2 × 20 in. long. A surface velocity of 250 fpm is suggested with a feed of 0.009 in. per revolution.

(b) An end facing cut is required of a 10-in. diameter workpiece. The revolutions per minute of the lathe are controlled to maintain 400 surface feet per minute from the center out to the surface. Feed is 0.009. Find the time for the cut.

(c) If the tool is K-3H carbide, the material is AISI 4140 steel, depth of cut is 0.050 in., and the feed is 0.010 in. per revolution, what is the surface feet per minute for a 4-in. bar and a 6-min life if the tool life equation $VT^{0.3723} = 1022$? Also rpm.

(d) Consider the Taylor tool life model, $VT^n = K$ for the following tool materials and work materials:

Tool	Work	n	k
High-speed steel	Cast iron	0.14	75
High-speed steel	Steel	0.125	47
Cemented carbide	Steel	0.20	150
Cemented carbide	Cast iron	0.25	130

What is the cutting velocity for a tool life of 10 min for each of these combinations?

7.10. Note the tool life curve in Fig. P7.10 for SAE 3140, feed of 0.013, depth of cut of 0.50, and a HSS tool material.
   (a) By using the curve find the parameters for $VT^n = K$.
   (b) What is tool life for 100 fpm?
   (c) What are the revolutions per minute for 2 in. bar stock and $V = 100$ fpm?
   (d) Repeat for 150 fpm.

Figure P7.10

7.11. A rough turning operation is to be performed on medium carbon steel. Tool material is high-speed steel. The part diameter is 4 in. O.D., and the cutting length is 20 in. The tool point costs $5. Time to reset the tool is 4 min. Part handling will be 2 min, and the operator wage is $20 per hour. The Taylor tool life equation $VT^{0.1} = 372$. Feed of the turning operation is 0.020 ipr for a depth of cut = 0.25 in.
   (a) Plot the four items of cost to find cost curve and select optimum velocity.
   (b) Determine the minimum velocity analytically.

7.12. Let the cost of $C_t = 0$ for problem 7.11, and plot time against velocity to find optimum velocity. Compute V_{max} and T_{max} and T_u.

7.13. Find the material cost and setup and cycle time for the operations of a stainless-steel pinion similar in all respects to Fig. 7.6 except for the following changes (work each subproblem separately; complete the operations process sheet and find the unit cost as demonstrated by Table 7.2). (Hint. Many work elements will be similar to the existing example.)
   (a) Let the 18.750-in. dimension be 8.750 in., the 1.100-in. dimension be 1.125 in., and no holes.
   (b) Let the 4.75-in. flat dimension be 8.00 in., and the raw material be 2 in. O.D. instead of $1\frac{3}{4}$ in. O.D.
   (c) Let the 4-in. O.D. milling cutter be 3 in. in width, and operation 10 will use high-speed tool material instead of tungsten carbide.
   (d) Let the material be medium carbon steel instead of stainless steel. Medium carbon steel costs $0.72/lb.

7.14. Let the estimated unit savings in direct labor be $0.30, burden on labor saved = 50%, cost of each setup = $100, interest rate = 20%, allowance for taxes and insurance = 10%, allowance for maintenance = 25%, and yearly allowance for depreciation and obsolescence = 50%.

(a) The cost of a fixture is $4000. With one run per year, how many pieces must be made per year to have the fixture pay for itself?

(b) Let depreciation be 100%, because the fixture must pay for itself within a year. How large must that run be?

(c) By using those data, how much money can we afford for a fixture for a single run of 15,000 units at an estimated savings of $0.30 per piece?

(d) How many years will a $4000 fixture require to pay for itself for an annual quantity of 20,000 units?

7.15. PN 8871 is to be analyzed for a make-versus-buy decision. The labor estimate is

Operation	Unit Cost
1	$0.0006
2	0.0130
3	0.0130
4	0.0007
5	0.0068

Direct material and material overhead = 0.0084 per unit. Variable overhead is 75% of direct labor, and fixed cost is $0.05 per unit. This company believes that the part deserves a profit of $0.025.

(a) If the company's plant capacity is underutilized, is the decision make-or-buy for a vendor's price = $0.075?

(b) For 100% plant utilization, what is the decision for a $0.118 price?

(c) If the plant chooses to make the article while at undercapacity, it will incur a 15% increase in direct wages owing to marginal costs of inefficient production. What is the decision for a $0.125 price?

(d) Evaluate the choices for a nonrecurring initial fixed price of tooling designed, manufactured, and paid for. Those initial costs for 2500 units were $25. What are the nonquantitive considerations of this sunk cost? Also, perform the analysis as if the $25 has not been spent.

7.16. Estimate the unit cost for Fig. P7.16 for material, productive hour cost, and tooling. Quantity is 7,500 units. Raw material is 3 in. O.D., 8 ft long, and is $1.25/lb. Density is 0.285 lb/in.^{3} Two operations are necessary: (1) facing, turning, chamfering, drilling, and reaming the 0.375 in. hole and cutoff with a lathe; and (2) drill and counterbore three holes. A drill jig is necessary for the drill and counterboring operation. Productive hour costs for turning and drilling are $40 per hour. Tool design is expected to cost $4000, and the PHC for tool building is $75 per hour. Use Tables 7.1, 7.2 and 7.3 to complete Table 7.5. Use Table 7.7 for tooling labor and opinion for tool material cost.

MORE DIFFICULT PROBLEMS

7.17. Construct individual cost curves similar to Fig. 7.5 for the following machining work. A gray iron casting having a diameter of $8\frac{1}{2}$ in. is rough turned to 8.020/8.025 in. for a length of 8.6 in. A renewable square carbide insert is used. The insert has eight corners suitable for turning work and costs $24. Operator and variable expenses less tooling costs are $60 per hour. The feed for this turning operation is 0.020 ipr. Taylor's tool life equation for part and tool material is $VT^{0.15} = 500$. The time for the operator to re-

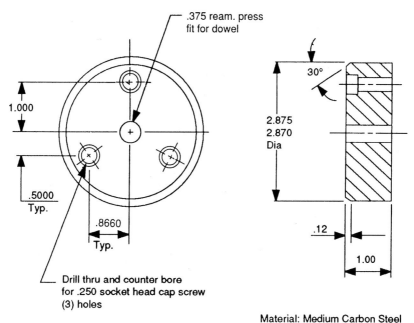

.375 ream. press
fit for dowel

1.000

30°

2.875
2.870
Dia

.5000
Typ.

.8660

Typ.

.12

1.00

Drill thru and counter bore
for .250 socket head cap screw
(3) holes

Material: Medium Carbon Steel

Figure P7.16

move the insert and install another new corner and qualify the tool ready to cut is 2 min. Part handling time is 1.65 min for a casting mounted in a fixture.

(a) If the y axis is unit cost and x axis is fpm, then plot the curves and locate optimum velocity.

(b) Determine optimum velocity analytically.

7.18. Assume that the cost of $C_t = 0$ for problem 7.17 and graphically find optimum time, V_{max}, and T_{max}.

7.19. A printed circuit blank is 300 × 450 × 1.6 mm in size. The surfaces are copper laminated to a pressed plastic core. The blank is eventually sheared into six 100 x 125 mm pieces. The remainder of the blank is necessary for margins and is material lost as waste. Production of the printed circuit pieces may require a setup for the operation, a setup for each blank, and a cycle time for each piece on the board or blank. Setup for the operation is one-time recurring for the lot. A blank setup is one-time recurring for the blank, and cycle time may be for the blank or piece.

Operation	Wage Rate ($/hr)	Operation Setup (hr)	Blank Setup (hr)	Cycle Time (hr per 100)
Clean blank	$15.25	0.5	0.2	1.677/blank
Photo resist	16.75	0.2	0	0.835/piece
Develop	16.75	1.0	0.1	3.345/blank
Etch	16.75	0.4	0.05	0.167/blank
Clean	15.25	0.1	0.1	0.085/piece
Pierce holes	14.50	0.1	0	0.250/piece
Shear	17.50	0.1	0	0.100/piece

A lot of 250 blanks is started at the clean blank operation, and, after shearing, only 95% will be satisfactory.

(a) Find the joint labor cost for the lot.

(b) Find the joint and unit labor cost for each operation.

(c) Determine the net unit labor cost for the piece.

(d) Develop an estimating formula for answering the previous three questions.

CASE STUDY:
STAINLESS STEEL PART

Estimate the unit cost of the design given by Fig. C7.1 for material, productive hour cost and tooling. The annual quantity is 500 units, which is produced in one lot. Stainless steel raw

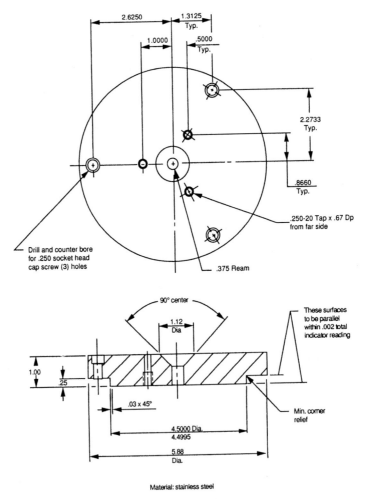

Material: stainless steel

Figure C7.1

material barstock is 6 in. O.D., 6 ft long, and is $2.25/lb. Density is 0.275 lb/in.^{3} Two operations are necessary, or (1) facing, turning, drilling, countersinking and reaming the 0.375-in. hole, turning the 4.500-in. diameter, chamfering, and cutoff in a lathe; and (2) drill, tap, and counterbore six holes. A drill jig is necessary for the drill, tap, and counterboring operation. Productive hour costs for turning and drilling are $90 and $40 per hour. Tool design is expected to cost $5000, and the PHC for tool building is $75 per hour. Use Tables 7.1, 7.2, 7.3 and 7.7 and equation 7.23. Use opinion for tool material cost.

8

Product Estimating

With design details at hand, the engineer begins the task of product estimating. Though providing information for the pricing step is one major object of the estimate, other documents are vital for management strategy. To permit the preparation of a product estimate, sales, marketing, and operation estimates must be concluded. En route to price setting, various procedures work on the task so as not to overlook the objectives of stockholders and consumers as well as management and engineering. The preliminary estimate provided early screening. Now a detail product estimate is necessary to help the pricing step and determine cash flow, rate of return, and profit and loss statements for a second look. *Concurrent engineering* tasks require timely and accurate estimates as well. The appraisal is made on new or continuing products and product lines.

8.1 PRICE

The product effort is ongoing, all inclusive, and basic to company survival. Care and stimulus for the product and its success rest on research, engineering, manufacturing, marketing, legal, and management. A product strategy calls for a wide assortment of decisions. This is a large undertaking for any firm. The hazard is high because of the several hundred fresh ideas only one will congeal into a successful new product. In view of those complications we now direct our efforts toward estimating aspects related to the product.

Even this portion of confined study is vast. In addition to price and cost of the product, problems exist concerning cash flow, rate of return, and meeting obligations to investors, owners, and shareholders.

288

Who is responsible for setting price? Practice varies. Is it an old or a new product? What precisely is the cash flow problem? Because products provide income, the realization of revenue from the sale of newly introduced or established products offsets costs of those same products.

Is the new or old product small in proportion to total income? If it is significant, then individual products and product lines constitute an important factor to the cash-flow problem. Some new products require a large outlay of cash for new plants and processing equipment in addition to engineering and construction costs. Investment analysis, called *profitability*, is one way to conduct an investigation. The capital obligations for this expansion come from profit, current depreciation, loans, or new issues of stock. Expansion may be necessary to produce the product.

The conclusion of a product estimate is *price*. Along the way to this result are a number of analyses, not the least of which is the cost estimate. There is a school of thought that asserts that "price is not related to cost." An argument begins by assuming that houses A and B are identical as to neighborhood, appearance, and the like. Price A is $100,000 and price B is $90,000. Now, owner A argues that the cost for house A is $ 100,000. The buyer, however, considers this as irrelevant and so would choose B. But what about owner B? If the cost is under $90,000, then owner B has made a profit, but if the cost is $90,000 or more, profit is weakened. What the buyer is willing to pay is influenced by the lowest competitive price and is determined at the point of sale, not in the factory. If price is less than cost, then the company must take steps to reduce cost or abandon the product. Of course, cost is not the only factor that sets a price, but it is a vital one. For long-term survival it is imperative to recover the full consumption of resources. A price strategy must accommodate this policy.

Older products for which an existing market is well defined present a different situation. Here, competition plays a larger role. Custom products for one customer require different treatment than do products for which a large market exists.

A cost estimate for a product undergoing redesign requires a different treatment than a new product. The components of the redesigned product are compared to the old design and classified as changed, added, or identical parts. The costs of the unchanged parts are found from records and may or may not be altered to reflect future conditions. Operation estimates are prepared for the altered and new parts. Methods of product costing and pricing vary in each of those several product conditions.

Figure 8.1 describes the estimating elements with price as the ultimate objective. The bottom layer, operation costs, consisting of direct material, direct labor and nonrecurring initial fixed cost, provides the grist for an operation estimate and is covered in chapter 7. Addition of the upper blocks is achieved by the several methods described in this chapter.

Administrative, marketing, distribution, and selling rates are found as in overhead rates. However, the denominator for this calculation is the *cost of goods manufactured*. Costs for administration and sales are structured and totaled, and finally the appropriate ratios are determined. Refinements such as separating divisional and corporate expenses or marketing from sales can be undertaken. Those

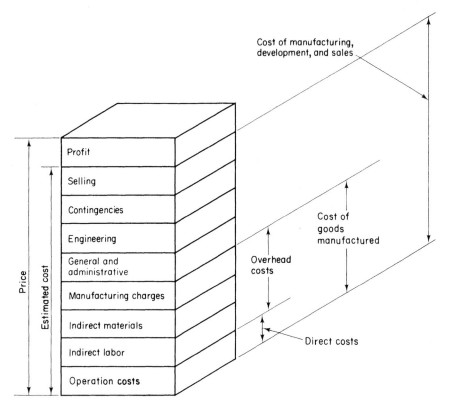

Figure 8.1. Elements in a product cost estimate.

refinements are usually worth the trouble because the costs can be more accurately absorbed to a product for its eventual recovery.

Contingencies are another category in Fig. 8.1. Uncertain costs may be estimated here. Although this is not a desirable category, there are circumstances where it becomes necessary. For instance, government-imposed safety requirements may be required for product reliability. The firm, uncertain of this future behavior and unprepared to introduce this feature until the government has provided the specifications, uses this category for a visible cost. The contingency identification has the advantage of providing special provisions for future legitimate costs. The use of the contingency, say, to cover careless detail-estimating practices, is not encouraged. Products having extraordinary research, development, and design problems are candidates for contingency.

The elements of Fig. 8.1 vary in importance depending on the business sector. Some industries may be labor intensive, and, thus, direct labor may be important. Material, such as standard purchased parts, raw materials, and direct utilities may dominate.

In capital intensive industries, the recovery of capital money is important. The relative importance of those major elements is seen by examining Fig. 8.2(a). An

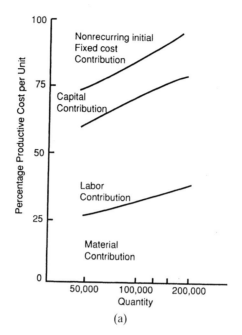

(a)

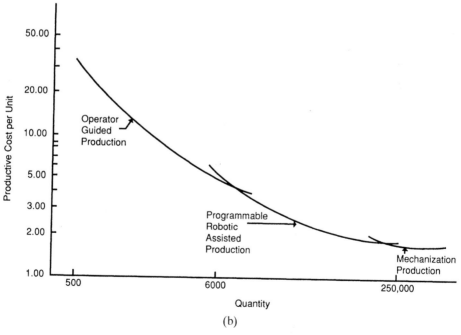

(b)

Figure 8.2.   (a) Proportion of major segments as volume varies for robotic-assisted production of products; (b) optimization of methods over range of quantity.

example for this figure is an expensive product. Figure 8.2(a) is a layer chart and is read by using differences between the chart lines to show the percentage importance of each category. For instance, at any quantity the percentage cost of capital is found by subtracting the higher y value by the lower y value. The estimating emphasis given to the basic elements depends on their percentage of total cost. The engineering department concentrates on the major contributors to cost, and those models identify the importance of the individual product situation. At the quantity 100,000, labor is the most significant contributor to the product cost.

It becomes possible to optimize the choice of the production system by using product estimating methods. A series of envelope curves can be determined for a specific product, that will indicate the best system and relate that to the regions of quantity. Note in Fig. 8.2(b) that "programmable robotic-assisted" production is desirable for the region of 6000 units to 250,000 annual units.

When is product estimating done? Figure 8.3 shows the location of this effort relative to design, production, and delivery. The figure is a simple time scale defining the point at which information is processed. The now time, or $E = 0$, is the moment when the estimate is made. Periods are denoted as positive or negative integers and may be days, weeks, months, or years. Some products will require only a few weeks from moment of estimate to delivery, but the majority are for several months' or many years' duration. Estimating methods must accommodate to those diverse requirements. The time scale in Fig. 8.3 shows the important events to be considered in product estimating.

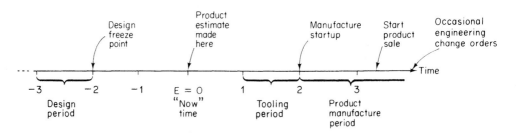

Figure 8.3.   Time scale line showing chronology of product estimate events.

A time compression of many of these activities exists in concurrent engineering efforts. In some cases, the estimating may be simultaneous to the design effort.

8.2 ESTIMATING ENGINEERING COSTS

In some cases, engineering costs for a product are not covered by overhead and are estimated separately. Those companion estimates include research and development and engineering. High-technology firms calculate engineering, development, and design as a separate line-item cost.

When products are similar to existing products, ordinary techniques useful in judging the *costs of engineering* are adopted. For those conventional situations, engineering does self-estimating on the basis of similar and historical records.

Engineering provides design of the product, specifications, bill of materials, and maybe physical models. In addition, preliminary engineering is performed for jobs that may not lead to orders. The engineering job may extend to production liaison, field and construction service, and maintenance engineering during the life of the product. For those specific functions, hours and a cost factor for hourly rates for wages and salaries of the group, charges for supervision, housing, light, and heat are estimated. Engineering costs to develop proposals that do not lead to new orders are additionally collected into the conglomerate hourly rate and used as the multiplier for the estimated number of hours. The total of this cost is amortized to the product by

$$C_e = \frac{\text{total engineering expenses}}{\text{projected product quantity}} \tag{8.1}$$

where C_e represents the engineering cost per unit.

Equation (8.1) is for exceptional charging not included in overhead practices. This model becomes a spread sheet such as in Table 8.1. The engineering estimate is

TABLE 8.1. ESTIMATING FOR DESIGN ENGINEERING

Design engineering estimate for __X-152__ products

Description __Product design__

Customer __Not known__ _____ Inquiry or quote no _____ Date _____
Based on quantity of __87 units__ _____ During period of _____

Type of labor	Hours	Rate per hour	Extended labor
Scientist, research			
Engineer, senior design	40	$33.00	$1,320
Engineer, design	800	$30.00	$24,000
Technician, electrical			
Designer/engr. aide	160	$18.75	$3,000
Tech. writer			
Illustrator			
Draftsman	80	$17.00	$1,360
Provisioning specialist			
Model shop			
Total design engineering labor	1080		$29,680

categorized by type of labor and rate per hour and is finally extended to total design engineering labor.

Cases exist where an engineering effort is significant in proportion to a contract. In special cases a firm will be hired to do the engineering and, say, manage a large-scale construction job. Estimating those costs may become a competitive bid among engineering firms. If major equipment is installed, say, as a result of this bid, and is in the range of $500,000 to $50 million, then engineering costs for complex pilot and chemical plants range from 20 to $7\frac{1}{2}\%$. In repetitive types of construction, engineering costs vary from $3\frac{1}{2}$ to 13% of total installed cost.

Engineering costs may be negotiated on a lump sum turn-key basis where the cost of engineering is included in the erection package for an entire plant. Sometimes the engineering contracts are negotiated on some cost plus basis. Variations include

1. Cost plus a negotiated fee or profit for the engineering contractor.
2. Cost plus a fixed fee contract with a guaranteed maximum.
3. A contract for engineering design manpower to be supervised by the client's engineering staff.

Contractual variations are discussed in chapter 12. No matter what the contract type, engineering contains these elements of total cost:

$$C_e = \Sigma S + \Sigma E + \Sigma OH + \Sigma F \qquad (8.2)$$

where C_e = sum of engineering costs
    S = salaries
    E = variable expenses such as travel, living away from home, communication
  OH = overhead, rent, depreciation, heat, light, clerical supplies, workman's compensation, and so forth
    F = fees paid to other specialists and engineers

In view of the proportion of salaries to the total cost, factors will multiply expected salaries to arrive at the estimate. Those factors vary from 1.8 to 3.0, depending on complexity, novelty, or secrecy of the work. In chemical and architectural work, the ratio of design drafting is of the order of two or three times other types of engineering. In electronics, this is reversed. Standards for engineering work have related size of drawing and design productivity. In tooling of dies, for instance, the size of the drawing can be associated to so many hours of design time. Other similar rules of thumb have been established. In any one industry there seems to be a tendency toward standardization in drawing size as well as the kind and quantity of information recorded on the print.

The cost of engineering is prorated to product charges like any other cost. Its impact may amount to very little, or it can be a major factor, particularly when a product, new or revised, calls for a new process or plant.

8.3 INFORMATION REQUIRED FOR PRODUCT ESTIMATING

A request for estimate (RFE) includes

1. Engineering bill of materials, specifications, and designs
2. Due date for completion of estimate
3. Quantity, rate of production, and schedule for product
4. Special test, inspection, and quality control requirements
5. Packaging and shipping instructions
6. Marketing information

The estimating of all labor and material cost and its extension by overhead calculations gives the important quantity *full cost.* This is increased for profit that gives price. Before that, however, there is the total bill of material cost for several or many parts and major assemblies. A *bill of material,* an important engineering document, is a listing of all parts required for a product.

The bill of material is a fundamental design document. Preparing and issuing bill of material lists may be on the corner of the drawing or may be separately listed on other documents. Whenever there is a separate bill of material, the list may be typewritten or appear as computer printout and contains information such as the number of the drawing, the name of the part or assembly, the material to be used for the part, and the quantity per unit of the final product.

A bill of material *explosion* is unnecessary if the manufacturer only sells single item parts. The cost estimate serves as the principal summary document for price setting. That is seldom found. When there are several or many parts and assemblies, cost estimates are organized into a bill of material. A graphical bill of material is shown in Fig. 8.4.

A tree is described by levels, three in the case of Fig. 8.4. The top level is the final product and is the sum of the costs of the components and assemblies below this top level. Each box of a tree contains a part number and part name. The tree gives the number of units required for the part in the next higher assembly. The number of units for the next higher assembly is the number shown at the left of each box. For example, one unit of final product (model X-152) requires one unit of the airframe product, one unit of SOHO, and eight units of a hatch assembly. Each hatch assembly contains eight machine screws, and so eight hatch assemblies require 64 machine screws.

Assembly of the second level boxes sums to the level 1 box or the final product. Note that the SOHO part does not have any lower parts.

Labor cost and material cost are indicated below each box. In some cases material cost (M) and productive hour cost (PHC) equals 0. A zero value for material could indicate consignment material, meaning that the material is supplied by the customer. A zero cost for labor indicates that the part is a purchased item.

Figure 8.4.   A costed bill-of-material tree.

Costs roll up from the bottom. For instance, the airframe assembly 50530, which has an estimated material cost of $10.41, is the sum of the labor and material costs adjusted for quantity of the lower levels. A labor cost of $0.24 is estimated to assemble the lower levels, which become the 50530 assembly. Level 2 labor and material estimated costs roll up to level 1 material cost after adjustment for quantity. But the airframe assembly, 50530, finds the material cost as the sum of its lower levels. For example, material cost is $4(0.04 + 0.10) + 4(0 + 0.04) + 1(3.01 + 6.68) = 10.41$. Note that lower-level productive hour cost (composed of gross hourly labor cost and machine hour cost) becomes the value added to the material to transform it to higher level material cost. The term $4(0.04 + 0.10)$ is the quantity 4 required for the next higher level, and $0.04 and $0.10 are productive hour costs and material costs. Those values are below the 50531 chassis box in Fig. 8.4.

The bill of material tree may show common materials. A purchased machine screw is used in two different locations in the third level. Each unit of final product uses four screws through the airframe assembly and 64 screws as a result of the hatch assembly. A total of 68 screws becomes necessary for one unit of final product, and 5916 screws are required for 87 units of final product.

Though the tree is a graphical illustration, more often the bill of material is provided as a table. The corresponding printout for Fig. 8.4 is given in Table 8.2. Level 1 is the final product. Level 2 indicates the next lower subassembly, and level 3 is the raw material or purchased parts for level 2. The heading for the costed bill of material indicates their purpose. The UM BOM provides the unit of measure for the bill of material that can have a variety of dimensions. Next, assembly quantity

TABLE 8.2.   A COSTED BILL OF MATERIAL

Level 12345	Part Number-RV	Part No. Description	UM BOM	Next Assm. Unit Quantity	Lot Qty	Total Lot Hours	O P	Unit Material Cost	Unit Labor Cost	Total Unit Cost	Next Assm. Material Cost
1	22245	X-152 Model	Ea	0	87		M	396.24	17.21	413.45	—
2	SOHO	Unknown	Ea	1.000	87	5.77	M	0	2.39	2.39	2.39
2	50530	Assembly	Ea	1.000	87	.84	M	10.41	.24	10.65	10.65
3	50531	Chassis	Ea	4.000	348	.89	M	.10	.04	.14	.56
3	Screw	Mach. Screw	Ea	4.000	348	0	B	.04	0	.04	.16
3	50532	Airframe Pt.	Ea	1.000	87	8.73	M	6.68	3.01	9.69	9.69
2	446754	Hatch Assm.	Ea	8.000	696	35.3	M	46.22	1.68	47.90	383.20
3	44656	Sheet Metal	Ea	4.000	2784	6.57	M	9.57	.52	10.09	40.36
3	Stiffener	Stiffener	Ea	2.000	1392	1.08	M	2.39	.17	2.56	5.12
3	50531	Sheet Metal	Ea	3.000	2088	.89	M	.10	.04	.14	.42
3	Screw	Mach. Screw	Ea	8.000	5568	0	B	.04	0	.04	.32

Courtesy: Penton Publishing

and lot quantity are important calculations. Order policy (OP) indicates either make (M) or buy (B), but other situations are possible.

The machine screw that appears in two places has different next-assembly requirements but has their quantity summed for the lot. It is this summed quantity used for the estimated quantity when calculating the lot for production or purchase. Total lot hours and unit material cost are values that estimates provide and are input information to the costed bill of material.

The bill of material (BOM) is an important list because other business documents use it as a central source of information. It is used to identify purchase orders if the item is purchased or to start production of the part; scrap, waste, and shrinkage are additional estimated values that contribute to the cost of material.

For new parts the engineer prepares new operation estimates. Thus, the BOM is a complete listing of parts, including standard commercial materials, parts, subcontract, and raw material. By using a BOM, the engineer is assured that all materials, both purchased and fabricated, will not be inadvertently overlooked.

The control and assignment of numbers for various engineering data such as drawings, bill of material lists, and engineering instructions is a function of design engineering. The design numbering system is comprehensive enough to serve engineering, manufacturing, and sales and yet unique enough to indicate readily specific conditions.

BOM provides an abundance of information leading to other existing documents, products, and cost estimates. New product designs incorporate some parts, subassemblies, or major subassembly units of other manufactured products, and the engineer uses the bill to find previously prepared estimates. Those older estimates are updated by using indexes or new productive hour cost rates and the like. It may not be necessary to estimate the parts of a product from scratch, because similar parts can reduce the estimating requirements.

8.4 FINANCIAL DOCUMENTS REQUIRED FOR PRODUCT DECISION

Old or newly introduced products require continuing analysis. What money is necessary for the venture, and what will come out of it? Management, stockholders, and money lenders are interested in this question. Operating capital, loans, and notes for new equipment or plant enlargement may be needed before there is any revenue from the products. Thus, it is mandatory to prepare several documents on products. Management takes action on the basis of these documents:

1. Product estimate
2. Cash flow statement
3. Rate of return analysis
4. Profit and loss statement

Product estimate. The product estimate indicates a full cost and selling price for the product. Almost all of this chapter is concerned with this document.

Cash flow statement. The cash flow statement considers the value of money in and out of a firm. It may be compared to a reservoir receiving a stream of water. At certain times more water is received than at other times. The demands placed on the reservoir fluctuate with a controlled quantity leaving. Money to a company behaves very much like this and thus is frequently called a stream.

If the cost of the product venture is small in proportion to the inflow or accumulated surplus, then the cash flow document may be unnecessary. If the venture is a big one, then the company evaluates its cash position to meet obligations. If the product requires a tooling-up period, long pilot runs, expensive equipment, and extensive engineering and preoperation break-in, then the construction of a cash flow document is vital. A small company may require operating capital before product revenue is received. It would find a cash flow document necessary.

A cost estimate involving price and quantity schedules, production rates, and marketing and sales rates, and capitalization costs for new equipment and plant enlargement is necessary to construct a cash flow statement. A sample cash flow statement is provided in Table 8.3.

For a cash flow statement we define

$$F_c = (G - D_c - C)(1 - t) + D_c \tag{8.3}$$

where F_c = total source of funds, dollars, year
$\quad G$ = estimated annual gross product income, dollars
$\quad D_c$ = annual depreciation charge, dollars
$\quad C$ = annual costs not estimated elsewhere, dollars
$\quad t$ = tax rate, decimal

TABLE 8.3. CASH FLOW STATEMENT

	Year 1	Year 2	Year 3	Year 4	Year 5	Year 6
Percent of capacity	25	50	75	100	110	116
Production, 10,000 lb/yr	12	25	37	50	55	58
Net profit after taxes	$51,000	$148,500	$188,500	$248,500	$269,500	$282,500
Depreciation	70,000	70,000	70,000	70,000	70,000	70,000
Total inflow of funds*	$121,000	$218,500	$258,500	$318,500	$339,500	$352,500
Startup expenses after taxes	75,000					
Fixed assets (from project estimate)	700,000					
Working capital per year (from operation estimates)	125,000	100,000	90,000	80,000	80,000	80,000
Total outflow of funds	$900,000	$100,000	$90,000	$80,000	$80,000	$80,000
Net cash flow, annual	−779,000	118,500	168,500	238,500	259,500	272,500
Cumulative cash flow	−$779,000	−$660,500	−$492,000	−$253,500	+$6,000	$278,500

*Equation (8.3) can be equivalently $F_c = (G - C)(1 - t) + tD_c$.

The *payback time and cash flow* is based on a nondiscounted basis (i.e., the face value of the cash flows for each year). Methods discussed in chapter 9 show how those values are discounted.

Operation of the chemical plant described in Table 8.3 increases up to 100%. It is then assumed that capacity grows owing to learning effects. Let us discuss the terms in the table.

Market quantity and price are known, obviously necessary to find net profits after taxes. *Depreciation* is a noncash tax expense, and a tax credit (a product of the depreciation times the firm's tax rate) helps to provide the inflow of funds. Depreciation is straight line over 10 years or 70,000 (= 700,000/10). *Preoperating expenses and investment costs* are first-year cash out. *Working capital* is made up of accounts receivable, raw material inventory, work in process, and finished goods inventory. Increases in inventory require immediate cash outlays that delay cash flow from generating sales revenue. Ordinarily, a higher requirement for cash on hand occurs during periods when operations are increasing. Determination of what constitutes working capital is usually meant to be incremental capital (i. e., differential capital between present and prior year).

The arithmetic for *net and cumulative cash flow* is evident in Table 8.3. The payback time calculates the number of years required to regenerate, by means of profits, depreciation, and tax credits, the total investment of the fixed assets and preoperating expenses required to launch the product. The payback time of 5 years illustrates that it will take this number of years to recoup the capital investment and pretax operation expenses.

Rate-of-return analysis.   A broad view of rate-of-return analysis is one that considers the net effective profitability of a product over its life span. Sometimes this analysis is called *profitability*. The rate-of-return analysis evaluates the return for a price volume situation over time. A product, including equipment and plant enlargement, capital costs and working capital, must, during the product's life, return to the firm a suitable fixed interest on the unpaid balance of outstanding cash flows. A company can value a product in terms of a constant annual rate of interest that will be produced on the unreturned balance of investment during a product's life. Thus, profitability is an analysis technique that measures the desirability of risking money for new products. Basic factors are capital, expense, revenue, and time. The end result of a rate-of-return analysis predicts the discounted net changes in the company's cash position. This topic is covered in the next chapter.

Profit-and-loss statement.   The profit-and-loss (P&L) statement is the final document important to making decisions on whether to go ahead with commercialization of a product. The P&L statement referred to here is a P&L statement specifically for the product and not the firm's P&L statement for its overall operation. Chapter 4 introduced this information. Important products merit P&L attention. Uneventful products that do not impinge significantly on the P&L statement can be evaluated by other criteria.

The P&L considers sales volume, manufacturing costs, and research and design costs. For the analysis of a proposed product we need an estimate of production level and an annual accrual of sales income. Direct manufacturing cost is a tabulation of the various cost components that are estimated. A P&L statement shows management the commitments that marketing, sales, and manufacturing operations must make to reach the predicted gross sales. P&L statements are made for several years ahead for significant products.

If the price for a product is known, then the cost estimate becomes a miniature P&L statement.

8.5 LEARNING APPLICATIONS

The theory of learning was introduced in section 6.6. Now our attention turns to the applications for product estimating. Applications are found in procurement, production, and the financial aspects of a manufacturing enterprise. In purchasing, a learning function negotiates purchase price or may be used for the make-versus-buy decision. In estimating, decisions related to cost are based in part on the concept. Contract negotiation is sometimes reopened after a first satisfactory model is produced. The contract for later units is based on learning reductions with time and cost known for the prototype unit.

There are explanations for this behavior, and verification has been uncovered by independent researchers and companies.

Principally, the reduction is due to direct labor learning and engineering programs. The direct labor learning process assumes that as a worker continues to produce, it is natural that he or she requires less time per unit with increasing production.

The engineering programs improve production, encourage quality, reduce design complexity, create technology progress, and foster product improvement. Those programs inspire time and cost reduction beyond direct labor learning.

This author suggests that the operator is responsible for approximately 15% of the total reduction, though engineering and their programs contribute the remaining 85%. In the manufacturing industries the 85% is broken down into 50% due to engineering endeavors, and manufacturing and industrial engineering activities are credited with the remaining 35%.

Ships, aircraft, computers, machine tools, and apartment and refinery construction have in common high cost, low volume, and discrete item production and can be treated by learning. Although the same principle applies to television production, for instance, the effects may take years to uncover because of the large production volume. Learning theory is usually not applied to high volume or low cost consumer products.

8.5.1. Follow-On Procurement

Costs for *follow-on production* are noticeably lower than are original costs. The learning function is defined if the number of direct labor hours required to complete the first unit is estimated and if the subsequent rate of improvement is specified. Alternatively, the learning curve is defined if direct labor man hours for a downstream unit and the learning curve rate are estimated. Other possibilities for defining the learning function can be selected. The number of direct labor hours required to complete the first production unit depends on these circumstances:

1. Experience of the company with the product: If the company has little or no experience, then the first unit time is greater than for a product having greater company experience.
2. Amount of engineering, training, and general preparations that the organization expends in preparation for the product: In some cases the first several units are custom made and tooling is not designed and constructed until more sales can be assured. This "hard way" production inflates first unit time.
3. Characteristics of the first unit: Large complex products are expected to consume more direct cost resources than something less complex.

Figure 8.5 describes follow-on estimating for procurement. The cumulative average is assumed to be the linear line. The follow-on estimating problem is stated as follows: Given that historical values are available for unit number and the associated direct labor man hours, find the value for the cumulative quantity of a follow-on estimate. This is shown by Fig. 8.6 for a 10-kW product, and a follow-on estimate is necessary at unit 18 for a 2-year bid. The follow-on bid quantity will be for 482 units (= 500 − 18). Because the estimate is for future conditions, history of the product's performance up through unit 18 is only one consideration.

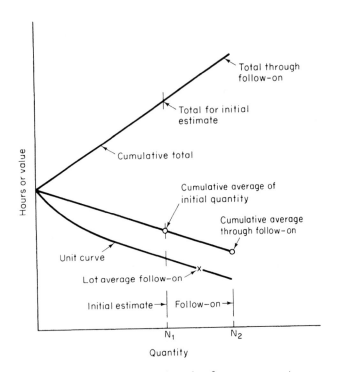

Figure 8.5. Follow-on learning for procurement.

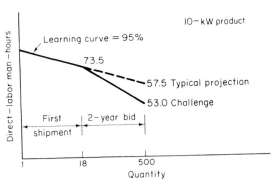

Figure 8.6. Follow-on learning.

If at unit 18 the hours are 73.5 and learning is 95%, then the first step finds K by using Eq. (6.14),

$$K = \frac{73.5}{18^{-0.074}} = 91.0$$

Finding the average time at unit 500, $T'_a = 91.0(500)^{-0.074} = 57.5$ average through follow-on. Those calculations can be confirmed with log-log paper and plotting those values. The lot time for the follow-on shipment is found by using Eq. (6.16), except that the subtraction is between shipments N_1 and N_2.

$$T_c' = KN_1^{S+1} - KN_2^{S+1}$$
$$= 91(500^{-0.074+1} - 18^{-0.074+1})$$
$$= 27{,}413 \text{ hours}$$

The follow-on estimate may indicate a value too high for competitive reasons. Engineering may change the 500th value from 57.5 to 53.0 hours, for example, to have a more competitively priced product. But then it is necessary to find the slope and assess whether the challenge is reasonable. By using Eq. (6.9),

$$s = \frac{\log T_i - \log T_j}{\log N_i - \log N_j}$$

we have

$$s = \frac{\log 73.5 - \log 53.0}{\log 18 - \log 500} = -0.0984.$$

Applying Eq. (6.6), $2^s = 0.934$. We have changed the slope parameter from 95% to 93.4%. Engineering will determine if the new slope is attainable and realistic.

8.5.2. Engineering Change Order

The effects of *engineering change orders* (ECOs) are evaluated by learning theory. ECOs may occur after the product is estimated, priced, and perhaps after a contract exists between a buyer and seller. Delivery may have started. Engineering change orders result from design improvements.

It is necessary to estimate an equitable adjustment to product cost and price if the contract terms allow. Now, consider Figs. 8.7(a) and 8.7(b), which show curves for the nonretrofit and retrofit cases. The early versions of the product are modified in the case of retrofit, but this is not the situation for nonretrofit. An ECO learning line could be above or below the follow-on line.

In the nonretrofit portion of the curve in Fig. 8.7(a), no adjustment is required for products supplied prior to ECO action. In Fig. 8.7(b), retrofit work is required for units before the ECO point. Additional time or value is also required after the ECO action, and an ECO learning line with different slope is shown. At the unit point of the ECO, the estimated ECO and actual learning lines do not intersect. The difference between the two lines at the ECO point is the unit time or cost to affect the ECO. It is simpler to estimate unit effects at the ECO point rather than average values. Thus the unit line is assumed linear.

Note Fig. 8.7(c), which shows average and total lines though still recognizing that the unit line is assumed linear. The convex portion of the average line is shown straight even though we know it is curved, as is shown in Fig. 6.7. The dashed lines in Fig. 8.7(c) represent the actual and projected time without an engineering change.

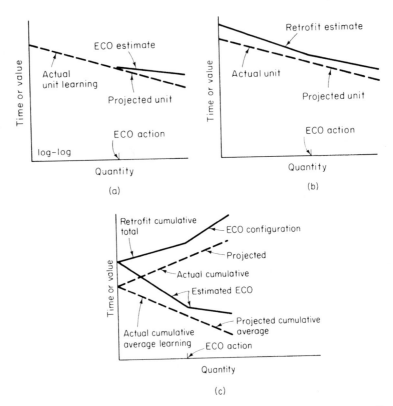

Figure 8.7. Engineering change order learning curves: (a) nonretrofit; (b) retrofit; (c) finding total cost for ECOs.

The solid lines represent average and total learning before and subsequent to ECO action. Note the change in slope for the retrofit and the ECO configured product. At the ECO point there is an incremental cost for the change. It is plausible that the ECO configuration product line could be less than the original projected line. But an ECO-causing retrofit would not be less than the original actual time.

Assume for Fig. 8.7(b) that the actual and projected learning line is $T_u = 1066N^{-0.1520}$ hours, similar to page 203. The average learning line is $T_a \doteq 1257N^{-0.1520}$ hours by using Eq. (6.13) and Φ is 90%. An ECO is planned for the 101st unit. Retrofit is scheduled for units 1–100, and units 101–500 are manufactured consistent with the ECO design. By using a similar experience the slopes of the ECO configured designs are estimated at 95% before the ECO point and 93% afterward.

Now we determine the amount of the contract change. This will be in two pieces (i.e., for the changes to the existing product from units 1 to 100 and for the configured ECO product from unit 100 to 500).

At $N = 100$, the retrofit line and the ECO design line meet but not at the original time. At $N = 100$, $T_u = 529$ hours and $T_a \doteq 624$. The time necessary for the ECO at only $N = 100$ is 75 hours. For units 1–100 and a slope of 95%, $s =$

log 0.95/log 2 = −0.074 by using Eq. (6.7). The retrofit intercept K for unit 1 is 105.5 hours (= 75/0.7112 = $K/T_{u=100}$). Note that the learning line is assumed linear because it is easier to estimate the unit difference at $N = 100$ than the average value. Cumulative retrofit hours = 8080 (= 105.5 × 76.5864 = $K × T_{c=100}$) by using appendix 3 for $\Phi = 95\%$. The 8080 hours is for the first piece of the retrofitting of previous manufactured products.

The slope for ECO configuration product is $s = $ log 0.93/log 2 = −0.1047 and the unit time at 100, $T_u = 529 + 75 = 604$. Its intercept $K = 978.2$ hours (= 604/$100^{-0.1047}$). Appendix tables are not provided for 93% learning, so other methods are required to find the total time for units 100–500. We use the essential idea of NT_a, in which the total is found with the product of the quantity and average time. For the unit line, the average time is approximated by $T_a \doteq KN^s/(1 + s)$. The approximate average time at unit 100 = 674.6 [= 604/(1 − 0.1047)]. For unit 500, the unit time is 510.3 [= $978.2(500)^{-0.1047}$] and the average is 570 [= 510.3/(1 − 0.1047)]. The total time for units 100–500 for a 93% learning with the ECO design configuration becomes 217,540 hours (= 500 × 570 − 100 × 674.6). This is the second piece of the time for the configured product.

The total time for the previous and new units are 8080 + 217,540 = 225,621 hours. This amount is multiplied by the Productive Hour Cost rate and becomes a cost addition to the contract.

The original design employed the 90% learning; thus, the ECO increases the cost. But by how much? The total time for units 100–500 is given by the product of the $K = 1066$ and the appendix. $T_c = 1066(228.7851 − 58.1410) = 181,907$ hours. The increase in time for the ECO is 43,713 hours (= 8080 + 217,540 − 181,907).

8.5.3. Break Even

Break-even analysis is discussed in section 6.10. Usually, those methods solve for the quantity where cost and price are equal. Those earlier methods did not consider learning effects. Now our study expands on those linear methods with single- and multiple-effect product learning break even models. We define

$$P = KN^s_{be} \tag{8.4}$$

where P = price of product, dollars
$\quad\quad N_{be}$ = break-even unit

Equation (8.4) assumes that the unit line is linear. The right side of the equation may be cost, and a profit exists when $N > N_{be}$. Note Fig. 8.8(a), which shows this idea. Let the price of a product be $4762, its cost $12,500N^{-0.2172}$, and $N_{be} = 86$. Profit develops if $N > 86$. The problem is more involved by having a value for P less contribution, margin, or other deductions.

Often, product learning is a composite of several separate learning models, as is shown in Fig. 8.8(b). Each of those models is identified with different major

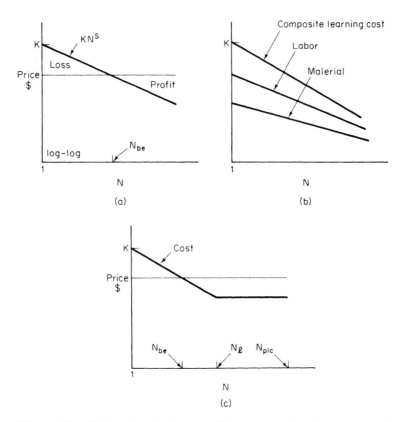

Figure 8.8.   (a) Learning break even; (b) composite learning constructed from individual learning models; (c) learning model with bottom-out effects.

assemblies, or the single model may be itemized into labor, material, and overhead. The former is a hardware classification, and the latter is functional. The chosen approach depends on the historical information. We assume knowledge of two learning curves for labor and material:

$$T_{ul} = 25 \times 106N^{-0.1405} = 2650N^{-0.1405}$$
$$T_{um} = 9850N^{-0.2435}$$

where T_{ul} = direct labor dollars per unit
  T_{um} = direct material dollars per unit

The 25 shown in the T_{ul} model is labor cost per hour. $K = 106$ hours in the equation.

A composite model finds T_{ul} and T_{um} at two points, say, 1 and 100, and adds their values. This is an approximation, and more points allow a new regression curve fitted to the data.

	$N = 1$	$N = 100$
$2650N^{-0.1405}$	2650	1388
$9850N^{-0.2345}$	9850	3209
Total	12,500	4597

$$s = \frac{\log 12{,}500 - \log 4597}{\log 1 - \log 100} = -0.2172$$

$$2^s = 2^{-0.2172}$$

$$\Phi \doteq 86\%$$

The composite learning model becomes $12{,}500N^{-0.2172}$, from which we find break even, as is indicated in Fig. 8.8(a).

Product learning may not continuously decline and may bottom or level out. This happens if engineering finds it uneconomic to continue cost reduction programs indefinitely. Cost levels out, and the product is manufactured until the termination of the product life cycle, which is an estimate of the number of units that will be ultimately sold. This is shown in Fig. 8.8(c).

Interest turns to uncovering cash flows or profit and loss over the duration of the product life. Loss cash flow for $1 \le N \le N_{bc}$ is given by

$$\text{loss cash flow} = \frac{KN_{bc}^{s+1}}{1 + s} - N_{bc}P \tag{8.5}$$

Gain cash flow for $N_{bc} < N \le N_1$, where N_1 is the limit of learning unit number and is given as

$$\text{gain cash flow} = (N_1 - N_{bc})P - \frac{K}{1 + s}(N_1^{s+1} - N_{bc}^{s+1}) \tag{8.6}$$

Level cost cash flow for $N_1 < N \le N_{plc}$, where N_{plc} is the product life cycle unit number, and is given as

$$\text{level cost cash flow} = (N_{plc} - N_1)(P - KN_1^s) \tag{8.7}$$

Recall the previous example having $N_{bc} = 86$ and price $= \$4762$. Marketing suggests that $N_1 = 125$ and $N_{plc} = 325$. The cash flows are found as follows:

$$\text{loss cash flow} = \frac{12{,}500(86)^{0.7828}}{0.7827} - 86(4762) = \$112{,}365$$

$$\text{gain cash flow} = (125 - 86)4762 - \frac{12{,}500}{0.7828}(125^{0.7828} - 86^{0.7828})$$

$$= 185718 - 177495 = \$8223$$

$$\text{level cash flow} = (325 - 125)(4762 - 4380) = \$76{,}426$$

The total cash flow over the product life $= -\$112{,}365 + \$8223 + \$76{,}426 = -\$27{,}716$.

8.5.4. Cost Application

Engineers need to know the recurring costs of production and the learning slopes for the several aspects of production (e.g., manufacturing direct labor, raw material, manufacturing engineering, tooling, quality control, and other indirect charges). In our next application we consider learning for direct labor and factory overhead. Material is assumed to be insensitive to the learning function.

$$C = \frac{KN^s}{s+1}(R) + \frac{KN^s}{s+1}(R)(R_{oh}) + C_{rm} \tag{8.8}$$

$$= \frac{KN^s}{s+1}(R_{dl})(1 + R_{oh}) + C_{rm}$$

where C = product cost per unit

R = direct labor hourly wage, plant average, dollars per hour

R_{oh} = overhead rate including engineering, tooling quality control, and other indirect charges expressed as a decimal of the direct labor hourly rate

C_{rm} = direct material cost

Our example considers the production of 200 units. R is $25 per hour, raw material cost is $250, and the overhead rate is 50%. The first unit is estimated to require 350 direct labor man hours, and a 90% learning curve slope with the unit line linear. The average cost per unit is

$$C = \frac{350 \times 200^{-0.152}}{0.8482}(25)(1 + .050) + 250 + \$7167$$

For a lot of 200, the total cost is 200(7167) = $1,433,400. We will return later to learning and show an important extension for estimating product cost.

8.6 METHODS

The product estimate determines the *full cost* of the product. Operation estimates, productive hour cost rates, and engineering and other companion estimates are available when the product estimate is computed. A preliminary estimate may have preceded a detail estimate, but that was a screening effort for feasibility.

The product estimate is a *pro forma* ("for the sake of form"), which is the company's official cost-estimating form. The *pro forma* summarizes total direct labor, material, and productive hour costs and it is the single most important document of the product-estimating effort. The product estimate is approved by the owner, president, or senior executive officer. The *pro forma* is contained on one sheet, and thus is a summary of other information and it provides the total of the manufacturing, development, and sales, which is the full cost.

Millions of worldwide firms exist that manufacture products. There are over 250,000 firms in the United States. Variations in product-estimating methods can be expected. Product estimates depend on availability of detail operational estimates, engineering data, and marketing information. We study two popular methods: the productive hour and the learning methods.

8.6.1. Productive Hour

This is the oldest of the product-estimating methods. Manufacturing operation sheets are prepared for each part, subassembly, and assembly. Those operation sheets provide the manufacturing procedure and give the operation setup and cycle standards, lot hours, productive hour cost rates, and material costs. The total unit cost for a part (and, hence, a part number) is given as

$$C_u = \Sigma C_{dlo}/N + C_{dm} + C_{ot} \tag{8.9}$$

where C_u = unit cost of manufacturing operations for one part or assembly on the bill of material.

C_{dlo} = operational labor and equipment costs, dollars per unit

C_{dm} = unit cost of material, including scrap and waste

C_{ot} = operation cost for tooling, dollars per unit

The symbols were previously defined. Practice dictates that setup and cycle standards for operations be determined from any of the methods of time measurement discussed in chapter 2. Each operation is estimated for setup hours and cycle minutes and lot hours are found by using $SU + N \times H_s/60$. If the lot quantity N is very large, C_{dlo} will not include setup time considerations, because SU/N becomes negligible on a unit basis. Under those circumstances, cycle time H_s is the only one considered. Setup costs for large volume is included in the overhead distribution basis. But C_{dlo} does include the direct labor, and productive hour for the equipment.

It is customary for the cycle H_s to be standard time, including allowances for personal, fatigue, and delay. Composition of this standard is discussed in chapter 2. On the other hand, performance against those standards invariably requires adjustment to actual time. The adjustment is either up or down and divides the operational cost by efficiency.

Note Table 8.4. The bottom line is found by using

$$C_p = \Sigma C_{phc} + \Sigma C_m + \Sigma C_e + C_c + C_s \tag{8.10}$$

where C_p = product cost of manufacturing, materials, engineering, contingency, and sales, dollars

C_{phc} = costs of direct labor, equipment, space, etc. (productive hour cost), dollars

C_m = costs of materials, dollars

TABLE 8.4. PRODUCTIVE HOUR METHOD OF PRODUCT ESTIMATING

C_e = engineering costs, dollars
C_c = contingency cost, dollars
C_s = sales cost, dollars

The summation sign includes all of the components that are a part of the product bill of material. Normally, the contingency cost is zero, unless extraordinary provisions exist.

Note, again, Table 8.4. This operation process sheet is the manufacture and planning of the final assembly of model X-152, the product given by the graphical bill of material, Fig. 8.4. Except for purchased parts, each item on the bill of materials will have a manufacturing process sheet. This operation process sheet is the assembly of the airframe assembly, SOHO, and the hatch assembly. The incoming material, shown in Table 8.4 and Fig. 8.4 as $396.24, is the value added of direct labor, machine, and assembly costs for all levels below it. This is given by the graphical bill of material and the operations process sheet.

The final operation to bench assemble the airframe assembly, SOHO, and the hatch assembly is estimated to require a setup of 0.15 hour and cycle minutes of 31.561. Those estimates are found from standard time data, the most accurate way to estimate. For 87 units, the lot hours of 46.51 [$= 0.75 + 87(31.561/60)$] is multiplied by the productive hour cost rate for bench assembly, $32.19. This rate is composed of the gross hourly direct labor cost (see chapter 2) and the machine hour cost (see chapter 4). The productive hour cost rate will vary, and more expensive manufacturing cells can be expected to have rates that reflect the cost of efficient operation of these assets. Multiplying the lot hours and the productive hour cost rate gives the cost for the operation, or $1497.27 (= 46.51 × 32.19). Unit productive hour cost for the operation is $17.21 (= 1497.21/87). The unit cost of the model X-152 product is $413.45 (= 17.21 + 396.24).

This estimating *pro forma* shows the material costs of the subassemblies, which are rolled up costs of direct material and productive hour costs. The material cost of $396.24 is the sum of the productive hour costs for labor and equipment and materials that have been rolled up from the lowest level. The engineering design cost is found from a companion estimate, Table 8.1, and listed in Table 8.4. Both a total quantity cost for 87 units and 1 unit are given.

There are many advantages to the productive hour method: accuracy; it combines the machine hour rate with gross hourly rate; some machine rates can be several times more costly, for example, heavy machining as compared to simple bench assembly; ability to consider the many systems of production, such as cell, programmable automation, flexible machining centers, and mechanization; and with this method it is possible to consider factories without direct labor, the so-called "automatic machines."

8.6.2. Learning

A sample of the learning *pro forma* is given by Table 8.5. The cost elements are operation estimates consisting of direct labor, direct material, and nonrecurring initial fixed costs. The estimate is made for the first unit, which subdivides the cost into fixed and variable portions. In this method direct labor and direct material are considered variable only. In chapter 7 the nonrecurring initial costs were considered fixed. Now we acknowledge that tooling and test equipment may have a variable component such as perishable tooling that varies with the quantity. Its proportion to nonrecurring initial fixed cost sum is small for Table 8.6. The learning slopes are estimated by using historical evidence. The 150th cumulative average factor is found in appendix 3 or by using Eq. (6.15). Each cost element total is found by using the following formula:

$$\text{cost element total} = C_f + C_v \times T_c' \tag{8.11}$$

where C_f = cost element considered fixed, dollars

$\qquad C_v$ = cost element considered variable, dollars

$\qquad T_c'$ = cumulative factor where cumulative average line is assumed linear

A plant overhead rate is applied to the total of the operation estimates. The sum of direct costs and plant overhead gives the cost of goods manufactured. General and administrative and selling overheads are applied on the basis of the cost of goods manufactured. Engineering costs require a separate estimate. The total of the

TABLE 8.5. LEARNING METHOD OF PRODUCT ESTIMATING

Cost element	First unit Cost estimate Fixed	Variable	Estimated learning slope	150-unit cumulative factor	Total
Direct labor					
Machining	—	$85,000	85%	46.44	$3,947,400
Sheet metal	—	170,000	90	70.04	11,906,800
Assembly	—	650,000	75	18.75	12,187,500
Quality	—	45,000	80	29.88	1,344,600
Direct material					
Raw	—	68,000	90	70.04	4,762,720
Standard	—	15,000	95	103.53	1,552,950
Sub contract	—	240,000	95	103.53	24,847,200
Nonrecurring					
tooling	175,000	500	95	103.53	226,765
Test	20,500	100	98	129.62	33,462
				Subtotal	$60,809,397

1. Plant overhead at 75%	45,607,048
2. Total cost of goods manufactured	106,416,445
3. General and administrative at 20%	21,283,289
4. Engineering	8,500,000
5. Selling at 5%	5,320,822
6. Contingencies	·0
Cost of manufacturing, development, and sales	$141,520,556

five major costs leads to the full cost, or the cost of manufacturing, development, and sales.

The learning method is required for major contractors selling to the agencies of the U.S. government and is also used for expensive low-quantity industrial products such as computers, turbines, boilers, and airplanes.

8.7 PRICING METHODS

An estimate provides information for price setting. In small companies the engineer is also charged with price setting. But in large or diversified or technological oriented companies, price setting is handled by sales or marketing, which is separate from engineering. The division of estimating and price setting functions is useful for other reasons. Engineers do not want to understate or overlook any costs. In price setting there is the urge to have price as low as possible to encourage sales, meet competition, and so on. If cost analysis and marketing functions are merged, then the cost estimate tends to become a "guesstimate" rather than a technically determined fact.

"Tension" between engineering estimating and sales can be useful. A constructive understanding between estimating and sales will lead to better cost estimates and prices if the functions remain separate.

There is seldom a price solution without complexities. The problems that appear first are those of competition and the consumer. Setting a price causes reactions by competition as well as a knotty evaluation of the kind of response to be expected from the consumer. Additionally, the law impinges on some price decisions. Prosecution is not uncommon for prices established through industrywide collusion. In times of fierce competition, sales people exert pressure to reduce prices.

A number of practices exist and are closely related to price setting and add to the complexities. For instance, promotional pricing, premiums, coupons, trade-ins, extras, fire-sale gimmicks, volume discounts, repeat discounts, geographic price differentials, lease-buy arrangements, and reciprocal agreements are deals that squarely affect the price choice.

The pricing situations presented to the price setter are diverse and may range from a price for a one-of-a-kind product to one where there are identical units offered to many buyers. The firm may have only a single product or it may have a multiple-product line. The product may be brand new to the firm or on rare occasions may be a product of research or invention. The product may have been manufactured for decades in the same form, or minor modifications may be introduced every so often.

Market distinctions do exist. There is the open market in which products are sold to an unknown buyer: Competition among products is based on price and nonprice factors. Then there is the bid or order market, in which manufacturers produce an item for a specific buyer. The type of contract, specification, and engineering design, in addition to price, are factors. In some cases, price is the single means of competition. In others, price may be relatively unimportant. In all those situations one pricing method is superior and others are not.

To make a rational price decision, we attempt to foresee the effects on the objectives of the company and on the several groups affected by the price. In view of those complications, we justify several concepts to determine a price. At this point we define price as the value of the economic want of a product design given, received, or asked in exchange. Of course, the value is monetary and expressed in the currency of the country (i.e., dollars, deutsche marks, yen).

We have deferred a rigorous definition of profit up to this point, allowing its general understanding. For the purpose of engineering, profit is the future monetary excess between price and the estimated cost by the producer of the product design. The actual or historical profit realized is of more importance to accounting than engineering. Estimated and historical profit may not be equal.

Pricing concepts follow:

1. Prices proportional to cost: If the concept produces the same percentage profit for all elements of cost, then it is full cost plus a markup. Another variation would have different markup percentages on the several cost elements.

2. Prices can be established proportional to conversion cost: This concept ignores the effects of the several kinds of material cost in its calculation. A conversion cost concept emphasizes value added or direct labor plus overhead.

3. Prices can be proportional to variable cost (i.e., that result in the same percentage contribution over variable costs): This concept can lead to different proportions for different designs; that is, fixed costs and profit. Direct labor and direct material (and variable overhead in some situations) are used as its base, and marginal cost of producing additional units is emphasized.

4. Prices can be systematically related to the stage of market and competitive development of the product: Price and cost-estimating relationships are possible features of the concept.

5. Prices can be established that depend on the elasticity of demand, a topic discussed in section 6.10.

Costs are stressed in the first four concepts. In the fifth concept, the theoretical desire to provide elasticity of information in time and quality for pricing meets with frequent failure. This concept seldom leads to practical methods. Four methods of product pricing are studied:

1. Opinion, conference, and comparison
2. Markup on cost
3. Contribution
4. Price estimating relationships

8.7.1. Opinion, Conference, and Comparison

Opinion, conference, and comparison are nonanalytic methods that involve the people who know the product's market and its price and understand the technical factors. Those people meet as a consequence of competitors' actions, or someone may notice that the price of an item may be out of line. Perhaps costs have gone up or down and adjustments are felt necessary. Discussion regarding the volume effect that each of the alternative prices would have and the product's profit as a result of alternative prices is undertaken. Data on competitors' prices are usually available as are comparisons of strengths and shortcomings of competing products. Cost estimates, past sales figures, and the history of price changes are discussed.

To assess the number of units that might be sold and their price, we try to imagine the customer response to each alternative. Further, the response of distributors and salespersons, changes in the sales of other products, and probable response of competitors, and the effect on the share of the market are evaluated. The predictions will be shaky, and we must recall their shortcomings. Though we discern the future actions of our customers and competitors, we can be certain that reactions in the future will differ from those in the past. Analytic methods to determine a price are overlooked. Despite the absence of such methods, opinion, conference, and comparison remain essential to price setting.

8.7.2. Markup on Cost

Markup or cost method, variously called the cost plus or markup, sets prices proportional to cost and is probably the most popular method because it identifies the types of the costs from the *pro forma* and adds an additional percentage of those costs as a markup. If the total cost of manufacturing, development, and selling is used, then it is a full-cost base. The formula for a full-cost markup is

$$P = C_t + R_m(C_t) \tag{8.12}$$

where P = unit price, dollars

C_t = total cost of manufacturing, development, and sales, dollars

R_m = markup rate on cost, decimal

Note Table 8.2, where C_t = \$413.35 and R_m 13%. Then, P = \$467.09.

Simple as this appears, there are literally dozens of variations. For instance, the C_t term can be broken down to material, labor, burden, and engineering with each term having its own markup. This is found in government contracting where limitations on the markup of the cost components are sometimes required. Some companies use the same add-on percentage year after year. Others use markups reflecting the preceding year's actual percentages. For the most part, those percentages vary with business conditions, and this feature along with its ease of understanding are its best features. As business falls the add-on is reduced and vice-versa. When large companies use those procedures, it is usually as a starting place for a price decision. In some companies the sales managers in the territory eventually decide what price they will actually quote, and the markup percentage serves as background.

Conversion cost pricing emphasizes value added or direct labor plus overhead in the base for a markup calculation. In some cases, material will be provided cost free to the manufacturer, because it may not be competitive to add markup on materials. We can define a price by using the following formula:

$$P = \Sigma C_{dlo}(1 + R_{oh})(1 + R_m) + C_{dm} \tag{8.13}$$

where C_{dlo} = direct-labor cost on operations, dollars

R_{oh} = general overhead rate, decimal overhead dollars to direct labor dollars

R_m = markup rate, decimal

C_{dm} = direct material cost, dollars

In some cases, C_{dm} may equal zero. By using a plan like Eq. (8.13) shifts the emphasis to products with high material costs and economizes on company labor and overhead. If the cost base is direct labor and direct material, then the price objective is the incremental cost of additional units, or

$$P = (\Sigma C_{dlo} + C_{dm})(1 + R_m) + C_{oh} \tag{8.14}$$

where C_{oh} denotes the costs of overhead, dollars.

In some cases, C_{oh} can equal zero. Equation (8.14) requires a larger markup on a smaller base than in the case of the full-cost base.

It is basic that if a new venture or a major improvement is desirable then the net return from it must exceed the cost of capital required. This philosophy provides a pricing method that depends on invested cost and requires that the cost of operation and the consumption of fixed assets and an acceptable rate of return be estimated. The acceptable rate of return on investment varies with numerous economic factors, but overall cumulative values have emerged and range from a low of 10% to over 50%. A simplified markup pricing model based on investment cost is

$$P = \left(\frac{iI}{N_Y} + C_f + C_v N \right) \Big/ N \tag{8.15}$$

where i = desired return on investment, decimal

I = investment, dollars

N_Y = number of years for payback of investment

C_f = product fixed costs, dollars

C_v = product variable cost unit, dollars

N = number of units sold

In this model the return on investment substitutes for the markup rate. Another cost-based approach to pricing includes methods to achieve a markup on sales value. It may be formalized as

$$P = \frac{C_t}{1 - R_s} \tag{8.16}$$

where C_t = total cost of manufacturing, development, and sales, dollars

R_s = markup on sale values, decimal

Refer to Table 8.2, again. If C_t = $413.45, R_s = 13%, then P = $475.23. This compares to $467.09 for a markup of 13% of total cost. Terms found in sales have discount synonymous with margin, list with selling price, and net with cost.

8.7.3. Contribution

As we have said, some costs vary with changes in production quantity, and others do not. Pricing is related to this variable cost as a percentage of the full variable cost. In this development we find the principal items of variable cost, including labor, material, marketing, and administrative costs. Upon this C_v value, we find the contribution based on sales price, or

$$P = \frac{C_v}{1 - R_c} \tag{8.17}$$

where P = list price for manufacturer, dollars
$\quad\quad C_v$ = full variable cost, dollars
$\quad\quad R_c$ = contribution rate, decimal

If the variable cost of a product is $125 per unit and a contribution percentage of 35% is assumed, then P = $192.31 (= 125/0.65).

Contribution is the amount left over from revenue after paying the variable costs and is used first to pay the fixed expenses. Any overage is a contribution to profit.

Obviously, the percentages for the several kinds of pricing rates (i.e., markup, contribution, etc.) are not comparable. The percentages have meaning only in relationship to a pricing method.

8.7.4. Price-Estimating Relationships

Price estimating relationships (PERs) are mathematical models or graphs that estimate price. Similar in construction to cost estimating relationships (section 6.6), a price driver is the independent variable. Time is usually chosen, although quantity and other variables are possible. A variety of functional forms is available, and the reader can refer to chapter 6 for details.

Now consider PERs in the application of a price ceiling and a price floor. The marketplace determines the price at which products will sell and sets a price ceiling. The cost, profit, and price determined by the firm establishes a price floor. In this context we define an opportunity margin offered by the market. Opportunity margin is the difference between the ceiling and the floor. If the policy is to accept the marketplace as determining the price at which products will sell, then, when the two meet, the firm may choose to stop manufacturing of the product or reduce price or cost. Price for a product will remain static in constant dollars if production and demand remain in equilibrium. Excess demand tends to increase prices, and excess supply suppresses prices. If profit and opportunity margin are high, more competitors are apt to enter the market.

We define

$$\text{opportunity margin} = P_o e^{-k_m t} - P_f e^{-k_f t} \tag{8.18}$$

where P_o = price of the initial unit at inception of production, dollars
$\quad\quad k_m$ = decay experience for the product, decimal
$\quad\quad t$ = time, typically years
$\quad\quad P_f$ = price floor initially, dollars
$\quad\quad k_f$ = decay experience of the price floor, decimal

A typical graph for a product selling price is as shown in Fig. 8.9. The floor price is the point at which the most efficient producer will make a reasonable profit. With a

Figure 8.9. Opportunity margin, projection for the model $P_0 e^{-k_M t} - P_f e^{-k_f t}$.

fairly stable floor price the margin over this floor price decays rapidly as more competitors attempt to capture the market.

Here is an example: A new product has been developed. Novelty will sustain early sales but long-range estimates have concluded that competing designs with a constant retail price of $9 will provide a floor. The initial price of the firm is $21, and a decay of 0.25 is expected on the basis of experience. The number of years for the intersection is

$$9 = 21e^{-0.25t}$$

$$t = 3.4 \text{ years}$$

8.8 SPARE PARTS

Spare parts are defined as product for separate production and procurement required for maintenance or repair. This excludes end items that have redundancy designed into the product to avoid failure. Spare parts range from piece parts to minor assemblies to end items.

Production scrap and waste are not included in spare parts. Cataloged and off-the-shelf replacement spare parts occurring in a purchase action are trivial and are overlooked. Failure rates and the number of spare parts required are not studied, except to point out that historical averages, mean-time-between-failure types of statistics for similar equipment, or opinion are employed. An estimate of the number of replacement spare parts is required before a cost estimate can be made.

Usually, spares are quoted after a product has been sold. Actual cost may be available to allow a basis for making the new estimate. When using actual costs, material and labor costs can be updated. If changes in setup quantity are noticeably different, then adjustments in this, too, are necessary. For companies who deal in selling spares, either on a single or lot basis, many of the techniques described in this chapter are used.

Figure 8.10 graphically describes policies for spare part and regular production. Figure 8.10(a) is constant production followed immediately by spare parts production but at a lower rate. In Fig. 8.10(b), spare parts are concurrent to regular production. For Fig. 8.10(c), the production of spare parts is prior to regular production. In some situations spare-parts production may continue during regular production. Figure 8.10(c) is found in the auto industry, where spares are in the hands of car dealers before introduction of a new model. Figure 8.10(d) shows a period of time before spare parts production is started. In Fig. 8.10(e) we see spare parts production starting after initial production and concluding with the regular production.

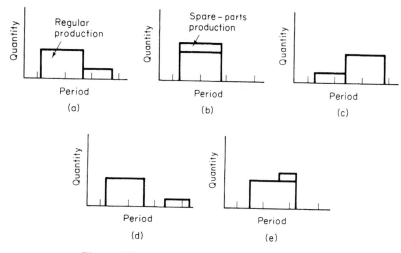

Figure 8.10.   Production plans for spare parts.

According to our definition, spare parts require separate production. Figure 8.10(b), though it has provision for spare parts beyond a basic production number, would not qualify for special cost-estimating treatment because the spare parts are piggyback to the production requirement. The spare parts in this case would be identified as spare parts certainly, but cost-estimating procedures and formulas do not recognize them as a distinctive group if joined to the basic group. Special estimating treatment is given to spare parts scheduled apart from basic production. In Fig. 8.10(a), the spare parts are produced after regular production, which may not be the same as spare parts production. In Fig. 8.10(d), there is a significant period between regular and spare parts production. Figure 8.10(c) shows a change between regular and spare parts production and indicates a lower throughput rate for spare parts. Another nuance could be imagined, which would be a combination of Figs. 8.10(b) and 8.10(c) or a delayed start for a slower scheduled rate for spare parts.

Spare parts are estimated by using the methods in section 8.6. Regular production is considered first, and modifications are examined later. Adjustments by setup quantity, indexing, rates of labor, material, or overhead are possible depending on the *pro forma* method. Inasmuch as the nonrecurring costs are amortized by regular production, those costs do not appear in spare parts. Contrariwise, an increase in quantity broadens the base for amortization of fixed costs, permitting more economical production methods. But better practice will acknowledge that spare parts are dependent on *full variable cost*.

A learning theory approach for assembly or major items is possible. The learning theory approach concentrates on the variable cost per unit and any effects of nonrecurring cost are progressively diminished. Figure 8.11 gives the thinking for a learning theory approach. Figure 8.11(a) is similar to a follow-on estimate. Figure 8.11(b) demonstrates the situation where there is a significant time period between regular and spare parts production. In Fig. 8.11(c), spare parts production is at a lower scheduled rate. Those practices are demonstrated by chapter problems. Spare parts pricing and profits are often attractive. But buyers who purchase spare parts are aware of those potential lucrative prices and may negotiate smaller profit markups.

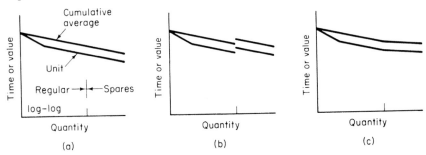

Figure 8.11.   Learning curve approaches for spare parts.

Repair estimating is unlike the spares problem. Until the product is examined, the work needed to restore the product to an acceptable condition cannot be fully identified. A number of practices are found in repair estimating. The product may be stripped at the site and estimated by a person experienced in identifying the work necessary, or a time, material, and profit contract between the original supplier or repair company and the owner may be specified. The product may be shipped back to the factory and a repair inspection undertaken there. The time is estimated, and a unit cost per hour is applied.

8.9 DESIGN TO COST

For the most part, engineers "build up" an estimate starting with operation estimates and concluding with either product cost or price. Most detail estimating is "bottoms up." Sometimes a reverse procedure is required. Beginning with a competitor's market price for a given product the engineer works downward to find total cost and the cost for various design assemblies and components. This practice is called *design-to-cost*. Those design-to-cost targets become goals for engineering, procurement, and production.

Design-to-cost is the early estimating of the product before design. So, actual hardware is unavailable. Sometimes a competitor's product is "reverse engineered," which implies that the product is de-manufactured and estimated.

What is meant by "early" cost estimating? If the intention is to provide a cost incentive for better design, then estimating before those actions is early. If we wish to encourage design engineers to be cost conscious, then early may be before their design. Design-to-cost programs supply a target for direct labor, direct material, and productive hour costs for significant hardware. A reward system to encourage design and other engineers to meet target can be associated with its attainment. This program strongly assumes that design engineering plays a prominent role in the cost of hardware.

Major and minor cost elements are uncovered by estimating or by the use of factors. In the design-to-cost procedure, and when detail exists, learning theory, tooling philosophies, quantities, rates of manufacture, and escalation or de-escalation of costs are introduced into the process.

The apportioning of cost is in accord with a logical bill of material (BOM) structure for the product even though the designs and a BOM are unavailable. Each design is given a cost goal. Thus, the designer knows that he or she controls the design and is able to meet the goal.

The design is again estimated after the drawings become available, and it is compared to the design-to-cost goal. The essential purpose of a design-to-cost program is to ensure that the product meets a designated price to allow product competition.

The process starts with the market price. Retail markup is removed, leaving the manufacturing cost. Profit and other overhead items are removed, many times by using ratios. Eventually, the cost is stripped down to the product parts and subassemblies. It is at this point that the cost-to-design targets are applied. Usually, those goals are not applied for marketing, overhead, and other management operations. Those designs are costed by opinion, comparison, and other preliminary methods. Remember, design-to-cost is before a bill of material and designs are available.

Note Fig. 8.12, a power supply. This is one subassembly of an electronic consumer product. The design-to-cost team determines that a power supply cost goal of $25 is necessary. This $25 is distributed to lower level hardware. Thus a case has a goal of $4.75 for direct labor and material.

Naturally, it is important that the total goal be distributed fairly and that no subassembly or component be favored at the expense of another. As the design effort progresses, the usual pattern of preliminary and detail product cost estimates is undertaken. A cost and price will ultimately be found by using methods given in this chapter. Obviously, it is important that the final price harmonize with design-to-cost goals.

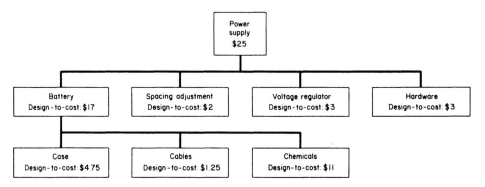

Figure 8.12. Estimating typical subassembly and component design-to-cost goals.

SUMMARY

In this chapter we considered product-estimating procedures. Important as other motives may be in price setting, the cost of a product must be fully recovered with profit to assure that the firm will ultimately survive. Financial documents are important in the appraisal of the worth of a product to a firm. Learning theory, useful for large-scale production goods, impinges on the product estimate in important ways. Methods of product costing and strategies of pricing lead to the information required for various financial documents and help to improve the product's success.

QUESTIONS

8.1. Define the following terms:

Price	Machine hour method
RFE	Learning price method
Cost of goods manufactured	Markup
Amortization	Investment price
Cash flow	Margin
Follow-on estimate	Contribution
Engineering change orders	Opportunity margin
Bill of material	Payback
Operation method	Design to cost

8.2. What are the ways a product estimate will be used? List others not included in the chapter.

8.3. What kinds of information are required for a product estimate? Which is internally determined by the product engineer?

8.4. Relate the flow of information between the various financial documents. At what point would those documents be required? When would they be unnecessary?

8.5. Describe the purpose of learning for a cost estimate. In addition to production quantity, what other things show learning correlation? Where in the cost-estimating process is the learning theory applied?

8.6. How may the parameters of the learning theory be defined? What factors contribute to this definition?

8.7. Segregate the methods of product costing. Outline those differences.

8.8. Indicate the complexities of pricing. Rank several objectives of a pricing policy.

8.9. When is the full-cost method of pricing appropriate? What are its failings? Devise a new model that uses the full-cost and investment method of pricing.

8.10. Is contribution pricing a marginal method of analysis? How important is it to have a contribution? What are the problems involved in determining fixed costs and associating them with various product lines?

8.11. Describe the distinctions in spare-part estimating.

8.12. What is design-to-cost? How does this procedure aid the control of costs during the design process? What features of this management procedure bear watching?

8.13. Each sketch of Fig. Q8.13 is defined by a horizontal axis of time and a vertical axis of dollars per unit. Market and engineering studies have calculated a price and a cost estimating relationship. Indicate the type of mathematical functions and their parameters that give these curves. Discuss the engineering and business implications of the product estimate that leads to these curves.

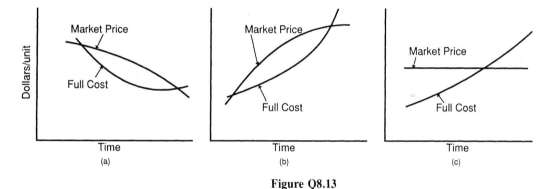

Figure Q8.13

PROBLEMS

8.1. An investor is considering an invention, and costs and revenue have been estimated. Calculate the net cash flow representing this investment. Disregard taxes and discounting.

Year	Cost	Revenue
0	$50,000	0
1	5,000	$ 2,500
2	5,000	10,000
3	2,500	25,000
4	0	25,000
5	0	20,000

8.2. The cash flow for a product entering production is given as follows:

Year	Cost	Revenue	Amount for Depreciation
0	$60,000	0	0
1	5,000	$ 2,500	$9,000
2	5,000	10,000	9,000
3	2,500	25,000	9,000
4	2,500	25,000	9,000
5	2,500	20,000	9,000

Fixed assets are $45,000, and initial operating expenses are $15,000. This is the first-year cost. If the company is in the 25% tax bracket, then determine the net cash flow. When is the payback point based on a nondiscounted cash flow? *Hint:* Use carry-forward depreciation when there is insufficient income for a depreciation tax expense.

8.3. Buffalo Chip, a new product from Buffalo Engineering Services, is considering a cash flow for one of its new inventions. A projection, in 10^4 units, is provided.

Year	Cost	Revenue
0	$50	$ 0
1	4	2
2	5	3
3	6	8
4	10	25
5	15	40
6	20	45
7	21	46

The tax rate for this company is 25%. The initial cost of $50 is composed of $40, which is a nonrecurring initial fixed cost subject to a straight-line depreciation schedule of 5 years. The $10 is ordinary start-up costs. When is the payback period anticipated? Assume that this is the only producing income opportunity for this firm.

8.4. Construct a cash flow document. Investment and preoperating charges are $100 and 25×10^5 for the first year. Production is 10 million pounds and is expected to increase by 2 million pounds per year for the next 5 years. Net profit after taxes amounts to 20×10^5 for the first year, depreciation is straight line for 10 years, and working capital is 8×10^5 dollars for year 0. Costs are assumed constant over the period. When is the payback period?

8.5. If the 2-year bid in Fig. 8.6 is estimated to decline to 52 hours, then what is the slope? What is the reduction in total hours for the 2-year bid as compared to 53 hours?

8.6. A company that produces engine-generator sets is planning follow-on bidding with the learning model. At the 980th unit a 5-kW product required 68 man hours of direct labor with a slope of 95% learning. The company adopts the cumulative average learning line as linear.

(a) Plot the cumulative average line between units 1 and 980 on log-log paper.

(b) Follow-on will involve 2460 units. Extend the constructed line to include this second shipment. What is the value of unit 3440 graphically? Repeat analytically.

(c) Now engineering believes that the 5-kW product must reduce to 56.3 hours at the 3440th unit. Find the new slope and the lot time for the second shipment. What is the average time per unit for the follow-on lot?

8.7. A company uses the learning concept to cost products. A summary of one product is given below. The average hour is assumed linear, and its best-fit equation is $T_a' = KN^s = 365N^{-0.256}$, which is the regression equation for columns 6 and 7.

Lot	Lot Quantity	Recorded Total Hours	Average Lot Hours	Cumulative Hours	Cumulative Quantity	Average Hours
1	5	1300	260	1300	5	260
2	22	2600	118	3900	27	144
...	...	...	...	...	...	...
8	55	3520	64	24020	267	90
9	45	2760	61	26780	312	86

(a) How many hours will be required for lot 10 if 42 units are to be built?

(b) What is the average unit estimate for the lot?

(c) If the full cost per hour is $35.00, then what is the total lot cost?

8.8. For the engineering change (EC) proposal A, the cost at unit 14 is $8000 and the learning rate is 80%. For EC proposal B, the unit 14 cost is $6000 and the learning rate is 90%. Which proposal is best at unit 15 or 30? Assume the unit line as straight.

8.9. The actual and follow-on learning equation is $T_u = 1000N^{-0.322}$, where the unit-hour line is assumed linear. An engineering change order is planned for $N = 4$, and the slope parameter $\Phi = 75\%$ is estimated. Find the total hour effect for units 4 and 5. Retrofit is not required. Repeat for $\Phi = 85\%$. (*Hint:* Units 1, 2, and 3 are unchanged.)

8.10. The actual and follow-on learning equation is $T_u = 1000N^{-0.322}$, where the unit-hour line is considered linear. At $N = 4$, an engineering change is planned with a slope of 85%. At $N = 4$, an additional 50 hours will be necessary. The slope for the retrofit portion is 90%. Find the total time for the ECO for units 1–5.

8.11. The equation $T_u = 35,000N^{-0.322}$ is actual and projected hours for a design, and the unit line is considered linear. An ECO is planned at the 50th unit. Retrofit is scheduled for units 1–50 and units 50–100 supplied according to ECO configuration. From a similar experience, 90% and 85% are estimated as learning before and subsequent to the ECO action, and 150 hours are estimated at the ECO point. Find the number of hours resulting from the ECO for the ECO and projected units 50 to 100. Use appendix 3.

8.12. A new product will sell for $6250 each. Unless a learning curve approach is adopted, the product will be dropped because the cost summary indicates that the full cost exceeds the potential price initially. Engineering wishes to determine the break-even point N between price and full cost. The following facts are gathered:

Selected Cost	Value at 500 Units	Cost per Hour	Percent Learning
Direct labor	105 man-hours	$17.76	95
Purchased Materials			95
Engine	$ 850		95
Semifinished	250		90
Raw	700		80
Other	650		100
Total	$2450		

Overhead is at 100% of direct-labor cost. Find the break-even point N_{be} to recover full costs. Use an approximation for a composite learning curve with units 1 and 500. What is the unit profit available at 1000 units? Use the unit line as the linear line.

8.13. A product, PN 8871, has these operational times summarized:

Operation	Setup	Hours per 100 Units	PHC
Shear	0.1	0.0048	36.50
Punch press	0.4	0.150	49.00
Punch press	0.4	0.150	49.00
Tumbler	—	0.010	26.05
Degreaser	0.1	0.100	29.45

Material unit cost = $0.095. The 2500 quantity is to be shipped in one order. We assume no learning for this product. Find the full cost for manufacturing, development, and sales by using the productive hour cost method. Find the price if the markup is 20%.

8.14. A single component product is to be estimated for cost and price for a lot of 175 units. The operation estimates for this product are as follows:

Operation	Setup Hour	Cycle Hours per Unit
Shear	0.1	0.001
Pierce	0.5	0.038
Countersink	0.3	0.043
Brake	0.6	0.010
Weld corners	0.2	0.035
Grind weld	0.1	0.020
Deburr	0.4	0.006

Average shop labor is $17.25 per hour. Factory overhead is 75% of direct labor. Material cost and overhead unit is $0.075, and G&A is 25% of the cost of goods manufactured. There is no selling or contingency cost. The percentage price on full cost is 20%.

(a) Develop a system of cost estimating, and find the unit and lot cost and price.

(b) Which operation is most costly? If a 50% reduction can be achieved in this operation, what addition to profit would be expected at the same selling price? (Assume that profit is increased only by direct-labor costs reduction.)

8.15. A universal projection screen mount, Fig. P8.15, is to be estimated. This product provides a means of suspending a screen from the ceiling and eliminates the tripod support. The device permits rotation of 360° and tilting of a maximum of 30°. Various vendors supply parts to the company. An indented bill of material is used to summarize cost facts.

No. Req'd				Part Name	Material Cost per Unit	Labor Estimate (hours/ units)	Gross Hourly Cost
1				Complete assembly		0.01	$20.05
	1			Gear pivot assembly		0.03	20.05
		1		Pivot, geared	$2.25	0.005	18.75
		1		Bolt	0.25	0.004	16.00
		2		Nut	0.10	0.002	15.00
	1			Adjustment shaft	0.75	0.01	20.05
		1		Worm	1.15	0.01	18.75
			1	Shaft	0.60	0.01	16.00
		1		Connector	0.50	0.02	17.75
			2	Thumbscrews	0.20	0.01	15.00
			2	Rod	0.40	0.03	16.00
	1			Housing		0.02	20.05
		1		Cover plate	0.60	0.005	16.00
		2		Side plate	0.30	0.015	16.00
		1		End plate	1.40	0.010	16.00
		1		Top plate	1.75	0.010	16.00
		1		Bottom plate	1.65	0.020	16.00

(a) Find the direct costs for labor and material.

(b) Define the cost of manufacturing as labor and material. Let the overhead costs for the plant be at 200% of the cost of manufacturing. Find the full cost.

(c) If markup is 25% of full cost, then find the profit and price.

(d) Repeat part (c) if the margin is 25% of the sales price.

8.16. Electronic components must be protected during manufacture. For example, spurious electric charges can destroy microelectronic chips. To avoid those problems, a tote-box liner is designed that uses a static-free plastic material. An assembly sketch is given in Fig. P8.16. A vendor has provided a quotation for items 3 and 4 as follows:

	Item 3	Item 4
Quantity		
500	$4.10	$5.22
1000	3.51	4.47
2500	3.28	4.17
Material	0.28	0.33

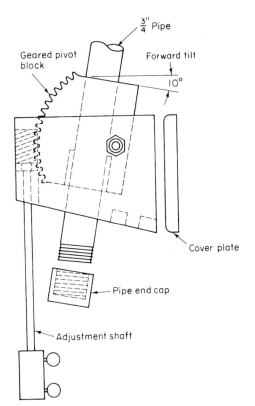

Figure P8.15. Universal projection screen mount.

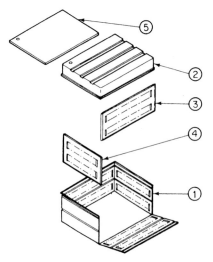

Figure P8.16. Tote-box liner.

Items 1, 2, and 5, assembly, and packaging are estimated by the company. Additional cost factors are as follows:

	Unit Material Cost	Tooling Cost	Direct Labor Setup (Hours)	Direct Labor Hours/Unit	Gross Hourly Cost
Item					
1	$1.22	$3500	6.5	0.025	$19
2	0.92	1750	4.0	0.013	19
5	0.67	1200	3.0	0.012	19
Assembly			0.5	0.005	19
Packaging	0.20		0.5	0.005	19

General administrative and sales costs are at 100% of operational costs. Develop a *pro forma* system to estimate full cost, and then use it for quantities of 500, 1000, and 2500.

8.17. A company uses the learning-curve approach to estimate its product. The average line is assumed linear.

Cost Item	Prototype Cost Estimate Fixed	Prototype Cost Estimate Variable	Estimated Learning (%)
Direct labor	—	45,000 hr	90
Direct material	—	$20,000	95
Manufacturing support	25,000	$5,000	95

Labor costs $18 per hour, fixed and miscellaneous overhead is 100%, distribution and administrative overhead is 20%, and selling costs are at 2% of direct labor and manufacturing support. The company uses the full-cost method with profit at 10%. Construct a method of estimating. Find the price for the 10th unit. Find total price for 10 units.

8.18. **(a)** If the margin on sales is 20% and cost is $160, find the selling price. **(b)** If the margin on sales is 30%, then find the markup percentage.

8.19. An engineer has found the following costs per unit:

	Cost
Direct labor	$0.10
Direct material	0.20
Overhead	0.06
Total	$0.36

(a) Find the price if the markup of full cost is 50%.

(b) If the markup of direct labor and material is 100%, then determine the price for an objective of incremental cost of additional units.

(c) Disregard the cost of materials for a markup of 200% on conversion cost. Find the price.

(d) Find price for a margin of sales as 50%.

8.20. An estimate is as follows: direct labor, $4.50 per unit; direct material, $8.00 per unit; and overhead, $12.50 per unit. Find the price for a margin on sales as 25%.

8.21. A firm desires a return on investment of 30% before taxes. A project estimate indicates that the required investment is $100,000, fixed costs = $20,000, and variable costs are $2000 per unit. The firm desires that its investment is to be paid back within 5 years. Sales are anticipated as 500 units per year. Find the price per unit.

8.22. A product is priced at three levels:

Unit Price	Projected quantity
$4	40,000
5	30,000
6	15,000

The full variable cost is $3. For each unit price find (a) contribution and contribution percentage, (b) total revenue and contribution, and (c) which price level is preferred.

8.23. A revolutionary product has been developed. It is believed that novelty will sustain early sales, but long-range estimates have concluded that competing designs with a constant retail price of $30 will provide a floor. An initial price of $70 and a decay of 0.25 is expected. How many years will elapse until the new product intersects the competing price of $30 per unit?

8.24. A price dictated by Eq. (8.18) has the following information: initial price = $10, decay = 0.25, floor price = $6, and decay = 0.15.

(a) By using those estimates, at what year will there be an intersection of the selling price with competitive floor price?

(b) At a floor price of $3.50, what is the opportunity margin? At what year will this level be reached?

(c) Discuss what your pricing policy should be after the intersection of prices.

8.25. Regular production is defined by $T_a' = 38{,}204 N^{-0.152}$ dollars for 50 units.

(a) If units 50 to 100 are expected as spare parts, then find their cost.

(b) A period of 1 year will elapse at the conclusion of the 50th unit before spare-part production is started. An increase of 10% over the 50th unit is estimated for the 51st unit. If there is no change in slope from regular production, then find the cost for units 50–100.

(c) Spare-parts production is expected to change to 95% at the 50th unit. Find the cost for units 50 to 100.

MORE DIFFICULT PROBLEMS

8.26. A single-product firm is bringing a new design to market. A cost estimate indicates that the full cost will be $142.86 per unit. Capital expenditures, at the startup time, will be

$250,000, and straight-line depreciation period of 5 years is IRS allowable in this case. Initial working capital is $50,000. Price is figured on markup of full cost, or 40%. All taxes on this startup company are 20%. The number of expected units per year is given as (year = 0; N = 0), (1; 1000), (2; 2000), (3, 4, 5 and 6; 3000). Find the cash flow payout time.

8.27. **(a)** Buffalo Engineering Services has information on the average time for the second and the 35th units as 125 and 90 hours. What is the average time for the 300th unit from a plot of the data? Find the cumulative time.

(b) Construct a factor table for 87% for units 1, 2, and 3.

(c) A product is estimated with a learning equation as $T_u = 1000N^{-0.322}$ where the unit line is considered linear. But after the initial design an engineering change is required and will be effective for the fourth and any follow-on units, which is estimated to have a slope of 85%. It is estimated that 50 hours will be necessary for retrofit at N = 3. The slope for the retrofit line will be 90% for retrofit units 1, 2, and 3. Unit 4 will require 47.9 hours additional to the original product-learning equation. Find the total time for the retrofit and new production owing to the ECO for units 1–5.

8.28. **(a)** A company's experience in building gear transmissions indicates that the first unit will consume 100 hours of fabrication and assembly time. A learning rate of 75% is anticipated. A contract is being estimated in which 40 units will be supplied. Labor cost of $24 per hour covers labor time, and indirect manufacturing expense and material expenses should be $25 per unit. Material is not subject to learning. If a 20% add-on for profit is historically applied, then what should the bid price be? How much must the time estimates be off to consume the profit? Assume that the unit line is linear.

(b) A prototype unit has been constructed with 45,000 hours of work recorded on job tickets. The direct labor hourly average rate is $24, and the overhead rate on the basis of labor is 100%. Raw material costs charged to the job were $20,000. The company, after a period for design changes, will build units 2, $\cdots$, 10 for sale. Customizing for specific customers is negligible. They anticipate a learning rate of 90% for direct labor and 95% for material. What are the total estimated costs for this product, the average, and the 10th-unit cost? The unit line is linear.

8.29. A multinational company designs and manufactures products, both complete and semi-finished, for sale to international divisions scattered throughout the world. Though specific cost-estimating practices depend on which two trading companies are dealing, consider the following points: A non-U.S. firm wants to buy a product for use as a standard purchased material for its own finished product. The U.S. international division buys the product from the manufacturing division by using a variable pricing formula, or variable cost plus 35% for the firm's contribution [price = variable cost $\div$ (1 − 0.35)]. The international division adds 20% for its own overhead based on the manufacturing division's price. This transaction is F.O.B. (free on board) at the U.S. plant and requires the consignee to pay all transportation costs from factory to destination. Export and import duty at the border is 0 and 20% of the *ad valorem* (invoice price at the port of shipment). Freight cost from the U.S. plant is $0.18 per unit. If the manufacturing variable cost = $1 per unit, then what is the price that the importer ultimately pays? Recall that during production of this product the importer will add labor, material, overhead, and contribution to the standard purchased material. Discuss what happens to the trade of this exported product if the rate and efficiency of production are approximately equal between those trading partners.

8.30. A multinational company is considering a "twin-plant" project where two plants are adjacent in the sense of low-cost transportation (separated by a national border, or the material is nonbulky and transportation costs both ways are negligible on a per unit basis). On the U.S. side, labor rates and productivity conform to typical standards, and in the foreign country labor rates vary from one-tenth to three-quarters as much. The U.S. plant will process the material, transport the semifinished material to the twin plant, where the foreign plant then adds labor value to the product and returns the product to the U.S. plant. The U.S. manufacturing unit cost for the semiprocessed material is $1. At the border, custom fees amount to 5% and return custom fees are 5%. An *ad valorem* of 20% of value added is assessed. In the foreign plant the labor work value is $0.25 for an equivalent $1.25 per unit of U.S. work. Once back in the United States, a 35% contribution is added to give final price. Find the product price for a twin-plant and a single-plant operation. Discuss the implications of a policy that sends goods for intermediate processing to other countries. What must the labor cost of the non-U.S. labor be to make the decision indifferent between the twin plants?

CASE STUDY:
UNIJUNCTION TRANSISTOR METRONOME

Ray Enterprises is releasing a new design and hopes to market it through a well-established chain of catalog houses. President Art Ray says that the key to the quality of the product lies in its new electronic circuitry, shown in Fig. C8.1. The bill of material is given by the parts list, and other materials cost $1.25 per unit, which covers the case, plug, and vector boards. Marketing, accounting, finance, and sales have been asked for information and the following preliminary suggested price-volume (P-V) is received:

Annual Volume	Potential Market Price
200,000	$95.90
210,000	93.30
235,000	90.40
260,000	89.20

New investment: $20,000 in tooling; depreciation policy, 2 years. Current tax rate: 40%. Standards for production are determined from operation estimates as follows:

	Hours per 100	Rate per Hour
Finishing	18.00	$14.60
Machine shop	7.00	15.00
Assembly	6.25	15.00
Inspection	2.80	16.20

The learning curve for finishing and inspection is estimated at 90%, and all else is 75%. The manufacturing burden is broken into variable and fixed with a rate of 75% and 25%. Ad-

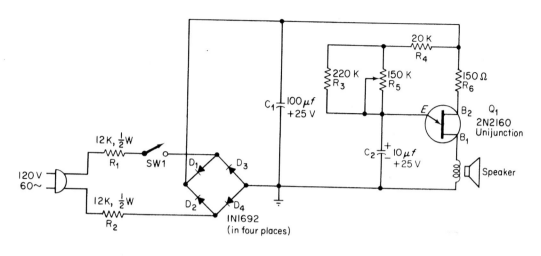

Figure C8.1. Unijunction transistor metronome.

	Qty	Unit price	Extended	Remarks
12–KΩ, $\frac{1}{2}$W resistors	2	0.12	0.24	Any
20–KΩ, 150Ω, 220KΩ, $\frac{1}{4}$W resistors	3	0.13	0.39	Any
150–KΩ, log taper potentiometers	1	1.02	1.02	Mallory U42
Signal diode	4	0.49	1.96	Newark
SPST switch	1	0.66	0.66	Cutler Hammer 7580K4
Capacitor, 25 V, 100 μf	1	0.81	0.81	Cornell–Dubilier Electrolytic BR100–25
Capacitor, 25V, 10μf	1	0.60	0.60	Cornell–Dubilier Electrolytic BR10 –25
Unijunction transistor, 2N2160	1	1.49	1.49	Allied
Speaker, 3.2Ω, 2 W	1	1.85	1.85	Quam 30A05

ministrative and marketing burden costs amount to 50% fixed only. Engineering development costs have amounted to $24,000. Manufacturing startup expenses will be about $15,000.

The distributor's charge for a product of this sort is usually 40% less than list price. A full-cost practice of adding 25% to all costs has been practiced before but is now used as a guide. A 20% return on appropriate investment is a minimum desired level. Ray Enterprises uses quality assurance techniques, and for products like this one has empirically determined that about one failure in 50 is expected. Its warranty policy is adamant: "Replace with new model if the old unit fails during the first year." The company reimburses the distributors for a new unit.

Construct a cost estimate and determine a price based on the full-cost method. Provide a cash flow statement over the next 2 years and a profit and loss statement for this product. Should this product be made and sold? In view of the preliminary marketing data on price and volume and estimating data on cost, what recommendations on price and cost can you make to President Art Ray?

9

Project Estimating

Project design is concerned with investment and is dissimilar to other designs because of the need for money appropriation. If the money is immediate expense rather than a capital cost, then we are concerned with an operation design. Otherwise, we have a *project design*. A project design is for a *one-of-a-kind end item*. Appropriations are lumped-sum or first-cost type. The principal concern about investment evaluation is due to its long-range impact on the financial health of the firm.

Examples of project designs are numerous: plant, turbine, high-voltage transmission line, and major equipment. Those things are physical. When we consider a task-dominated design to improve paperwork flow through an office, it is a one-time design, certainly, but, in the absence of capital expenditures, we call this an operation design. A product design is time dependent on production of units and is of different character.

To clarify the discussion for the remainder of this chapter, it is assumed that the technical feasibility of equipment, plants, and other physical services has been determined but that the project cost estimate is not yet revealed.

Project engineers are frequently the first to recognize the need for new equipment, processes, plants, or their replacement. As the preliminary engineering plan is originated, and as a matter of good practice, the engineer contributes information to the budget defining the costs that may be required. Should the preliminary estimate call for additional planning, a detail estimate is made. If the estimate looks encouraging, then the question to spend money becomes an executive-coordinated decision, particularly if the capital money is large when compared to readily available resources. Thus there is a special responsibility that rests within the project engineering function.

Of the decisions that executive engineering makes, few affect the financial stability and the future earnings of the firm more than those pertaining to capital investments. Those decisions commit the firm to manufacture or distribute certain products, to construct plants at certain locations, and to utilize certain materials processes, methods, machines, or groups of machines and to establish the structure within which the organization will operate for years to come. Those decisions involve thousands if not millions of dollars to any one firm.

The design decision has a substantial impact on cost. Although engineering is able to influence the efficiency of the transformation of the design into the actual product or service, the approximate level of cost is nominally fixed after the plans and functional engineering concepts have been finalized. This holds true because those concepts and designs determine the limits and cost of the processes, materials, and labor used. Each of the various original design concepts results in a different final cost. It is important that the design concept be initially chosen in the light of the cost of the processes and the methods that it dictates and the capital investment requirements that result from those processes.

Although project designs calling for evaluation differ, the economic techniques are common despite seemingly large differences in the engineering design. Nor are there distinctions in methods for estimates ranging from several thousand dollars to several billion dollars. An equal intellectual challenge awaits the owner's or contractor's engineer. Ideas in this chapter are germane to either an owner or contractor and all sizes of projects.

9.1 PROJECT BID

The project engineer prepares an estimate eventually required for a bid. This preparation responds to a request for estimate (RFE) from engineering if the engineer is employed by the owner. Similarly, the request for proposal (RFP) is the response by the contractor's engineer to an owner, major contractor, or other business opportunity. In either case, the effort leads to a bid. Very simply, a *bid* is the cost and profit that a project requires for labor, materials, and overhead. Sometimes a bid is known as a quotation, proposal, investment, price, or even cost, because terms are used loosely in practice. Jargon for a complete estimate of this type is known as a "grass roots" or "greenfield" estimate.

Cost elements of a project vary widely. Their selection differs between a plant, electrical transmission line, or turbine even though the design is for a single end item. One approach recognizes that those costs can be divided into fixed capital and working capital. *Fixed capital investment* is the amount of money required from concept to the finished end item. But once the plant is ready to operate, working capital is required.

Net working capital is the difference between current assets and current liabilities. Sometimes working capital is calculated as the allowance in a capital estimate for necessary operating inventory (raw, in-process, and finished materials), cash, and net receivables. Working capital requires immediate cash outlays not realized until the sales revenue is generated. Thus, timing is involved.

What follows should be tabulated and totaled to obtain total working capital: changes in current payables and accounts receivable, change in inventory of raw materials and supplies, changes in cash-on-hand balances, and changes in current liabilities and nondepreciable items.

Increases in working cash are required to support the operations resulting from the investment. A higher requirement exists during periods of expansion, and a lower requirement exists during contracting periods.

Working capital is important because of the time in its recovery. At the termination date of the project, a credit for working capital is estimated to offset, although not equally, the initial requirements for working capital at the start of the project.

But classifying project cost into fixed and working capital diminishes the complexity of a project estimate. Instead, we symbolically define project cost as given by the layers in Fig. 9.1. Not all the layers may be required. It depends on the design.

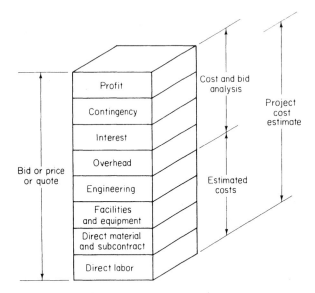

Figure 9.1. Components of project cost estimate.

Operation and product estimates are available as information in the preparation of a project estimate. In operation estimating, the emphasis is on man hours and man days, which were defined as standard, and are estimated for tasks. Chapter 8 deals with methods of estimating product designs. A project design could be a turbine that has thousands of turbine blades. Those blades are estimated by using principles of product estimating. The project engineer in gathering cost facts may estimate the man days for an operation or the price for product, or these may be provided from an RFP.

Direct labor is work associated with the materials required for an engineering project design. Alternatively, we can define labor as the initial or ongoing labor required for plant operation and is for working capital or operating requirements of a plant once constructed. The work of pouring concrete footings for a turbine base is direct, and indirect labor would be the guards or clerks on duty. Turbine operation and turbine maintenance require direct and indirect labor.

Initially, direct labor is expressed in time units, man hours, man months, or years. An individual or crew wage multiplies the units of time, leading to direct cost. Previous chapters have dealt with this computation. If those costs are identified as *allowed,* or *standard*, then an adjustment by productivity becomes necessary. The *productivity factor* adjusts allowed time to actual time. The productivity factor varies with time of season, location, worker experience, and so on. See problem 9.1 for an example.

Direct materials are subdivided into raw, standard commercial items, and subcontract items and appear in the end item. The amounts must be increased for losses stemming from waste, scrap, and shrinkage. Indirect materials, for example, are supplies, lubricants, and small tools necessary for the construction of direct materials. Standard commercial materials are hardware or items generally selected from designer catalogs or materials having common specification and design. Those materials are broadly available. Subcontract materials are custom designed, and a Request for Quotation (RFQ) instructs an offeror to submit a bid to the project engineer.

Depending on the particular plant process, raw materials can constitute a major portion of operating costs. A list is developed from the process flowsheet. Information obtained for each raw material would include units of purchase, unit cost, sources of supply, quantity required per unit time and unit of output of product, and quality (concentration, acceptable in purity level, etc.) of raw material. Naturally, the quantity is increased for waste and yield. Fuels involved in the process or catalysts and other processing material are raw materials. The initial fill of fuels and catalysts is a part of the capital cost, and refill materials at the conclusion of the first year is ongoing direct material. Process by products, including wastes and pollutants, must be considered for the operating cost estimate of direct materials. Every input and output stream must be considered in the operating cost estimate. Those costs may be credits (for salable or usable by-products) or debits (for wastes or unsalable by-products).

Consider a simple flowsheet for processing raw milk. A dairy is designed to convert raw milk into 2% milk and 40% cream. The percentages refer to fat content. A flowchart is shown in Fig. 9.2. Input materials are raw milk and vitamins. A charge of $0.50/lb is required for this milk plant to dispose of its cream to a butter plant. Vitamins are added to the 2% milk. There is 98% yield, or 2% shrinkage, for three items. A 4-lb waste is a sludge removed to a landfill dump. This raw milk problem is not a joint cost problem, as discussed in chapter 3, because there is only one marketable product. Cream is a result of the process, but this plant has no equipment to process or market its cream. Because it is not feasible to drain the

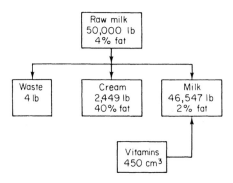

Figure 9.2. Simple flow chart for converting raw milk to 2% vitamin-fortified milk.

cream, $0.50/lb is a charge for tanker truck removal to a butter plant. A tanker truck supplies 50,000 lbs of raw milk daily to the dairy.

Material	Cost per Unit	Quantity	Cost
Raw milk	$32/100 lb	50,000 lb	$16,000
Vitamins	$20/450 cm^3	450 cm^3	20
Processed cream	$0.50/lb	2449 lb	1,225
Total			$17,245

The output cost of 46,547 lb is $0.37/lb, or about $3.19 per gallon if 1 gallon = 8.61 lb. This considers the cost on the output amount.

In addition to operating materials, an expanded flowsheet provides a listing of the equipment. This raw milk processing is handled by equipment that must be estimated. If this equipment is standard, then a quotation can be obtained. Or, it may be conceivable that the process is novel and that special equipment must be designed and manufactured.

Facility and equipment cost is a term that can be broadly defined. For a uranium-ore–processing plant, the pieces of capital equipment, such as conveyors, tanks, and rod mills, are examples. A high voltage transmission line requires a field office or facilities. Equipment for the construction of the transmission line is used for other lines. In this case, equipment is an overhead charge. Indirect materials and labor are other charges conveniently handled by overhead.

Engineering costs are the costs incurred for design, drawings, specifications, or reports. Included are the salaries and overhead for engineering administration, drafting, reproductions, and cost engineering. This was discussed in section 8.2, and differences for project design are minor.

Contingency costs are for those situations having no experience or prior data. Projects requiring extraordinary research, development, and design are the best candidates for contingency. Unfortunately, contingency is sometimes a cover up for poor estimating practices.

Because projects are a huge financial undertaking, interest charges are usually charged against the contractor or owner during construction progress. Those interest charges can be substantial and are a part of doing business.

The layers in Fig. 9.1 from direct labor to interest are the elements of a project cost estimate. Profit is calculated on those items. The sum of cost and profit constitutes a bid.

The remainder of this chapter uses those terms and definitions.

9.2 WORK PACKAGE

Steps for project estimating are given in Fig. 9.3. A precise sequence is not intended. The likely way is a simultaneous maneuver, but a project engineer develops this style only after experience.

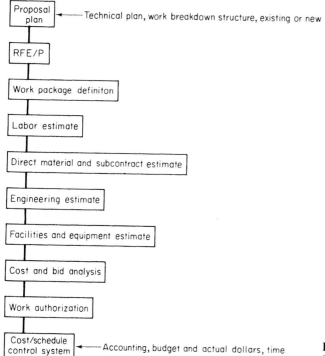

Proposal plan ◄── Technical plan, work breakdown structure, existing or new

RFE/P

Work package definiton

Labor estimate

Direct material and subcontract estimate

Engineering estimate

Facilities and equipment estimate

Cost and bid analysis

Work authorization

Cost/schedule control system ◄── Accounting, budget and actual dollars, time

Figure 9.3. Flow chart for integrated project.

The proposal plan gives the technical statement, preliminary designs, and preliminary estimate. An encouraging plan will lead to additional effort. This results in the RFE/P, which initiates the work. Bills of material and designs are the principal technical information for cost-estimating products. A project, however, may have significant nonhardware costs. The project design, by itself, is insufficient for estimating. Other necessary information is consolidated with the project designs.

The collected project information is called a *work package*. Forms included within a work package are the definition, designs, work breakdown structure, schedule, and estimates. If the project is large, then the amount of information in a work package is extensive. Even small projects require a thick notebook.

A *project definition* is a planning form that identifies what is to be done, when it is to be done, and by whom it is to be done. There is a numbering system that relates the definition to other forms and shows a baseline initiation date. The project definition gives a workable scheme to achieve the end item. The design engineer is concerned with a design that will satisfy the specifications at least cost. The results of this effort become the definition that translates performance objectives into a project design.

A project engineer will take an overall project definition and assign items of the work breakdown structure to cognizant engineers of the proposal team.

For the remainder of this chapter, we assume that the definition phase has been completed but that the cost estimate remains.

Data collection for the definition can follow one of two approaches. It is possible to gather data for a specific estimate as the estimate unfolds. This is accomplished by contacting vendors for quotations for each part of the estimate, and the data are re-collected for each job. A second way is to identify, collect, analyze, and divulge data that are more-or-less permanent, and, though adjusted for technology and price changes, are standard once initially collected. Indexes change, but the base standard information remains consistent. Standard information is then used over and over.

The definition subdivides the end item into large-scale tasks and eventually operations for physical items. For example, a reinforced concrete wall may be one task, or the task may be divided into erect outside forms, tie-reinforcing steel, erect inside forms and bulkheads, pour concrete, remove forms, and clean.

The work breakdown structure (WBS) is a graphical display of the project and results from reduction of an end item into logical components that are arranged into a treelike figure to allow visibility and analysis of a single component. Components can also be grouped into larger tasks or the end item itself. The WBS is used for estimating, planning, and performance measurement and control. In some cases accounting charges to the design by means of WBS are linked and can be arranged to show manpower.

The WBS is widely accepted by contractors and U.S. Department of Defense.

The WBS differs greatly from the bill of material (BOM). The WBS will show nonhardware costs. The WBS also shows hardware but will not indicate all items, which a BOM is required to do. The BOM is identified with part numbers of the hardware.

Subdivision of a large complex end item into smaller, less complex, and more manageable tasks is not new and has been basic to production for a century or more. The work breakdown structure is composed of hardware, software, services, and other work tasks.

Levels are used to specify the WBS. Level 1 is the entire project. Level 2 elements are the major elements, such as land.

A contract WBS is the complete WBS for a contract developed and used by the contractors and may contain three or more levels as necessary. Five levels are usually considered adequate.

Figure 9.4 is an example of a six-level WBS. Each lower level adds another digit, and 1322 is a fourth-level transformers and control panel. The numbering of the WBS corresponds to the definition and the estimates. The definition is the source of the original WBS number system. Though the bill of material deals only with hardware, the WBS also deals with nonhardware costs.

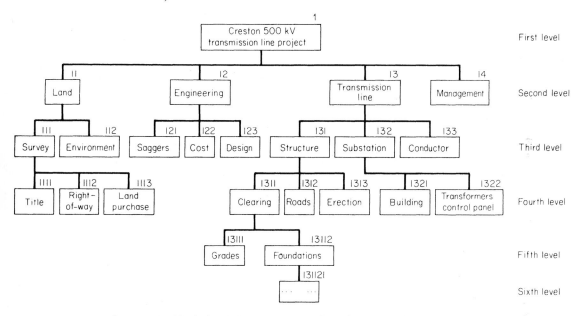

Figure 9.4. Work breakdown structure (WBS) of a project to the sixth level.

After the project summary WBS is formulated and numbered, individual contract WBS are used for procurement actions for vendors. The vendor may extend the WBS to lower levels as the basis of an RFP.

When attached to the project summary WBS, the project WBS is formed. The project summary WBS and its derivatives are used throughout contract definitions, construction, design, and operation for technical and engineering activities. Reporting of progress, performance, and financial data is often based on the project WBS. In some instances it is possible to have specifications and drawings conform to the numbered WBS.

Schedules are an important part of the work package. Though the definition is the planning document and notes what must be done, scheduling determines the calendar dates for the start and conclusion of the WBS activity. Details of scheduling like CPM (critical path method), PERT (program review and evaluation technique), or other network methods are considerable, and we refer the reader to the many excellent texts available for that.

Instead, refer to Fig. 9.5, which symbolically shows the integration of the WBS, schedule, and costs. The fourth level of the WBS is estimated at $5, $10, $5, $10, and $5 for each task. The third-level summary is $35. Two other third-level tasks are

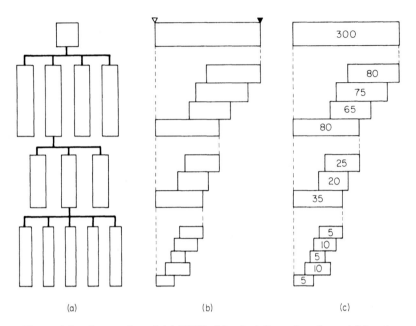

(a)   (b)   (c)

Figure 9.5.   Integration of (a) WBS, (b) scheduling of work, and (c) estimating tasks.

estimated at $20 and $25 for a second-level summary of $80. The first-level summary is $300. In Fig. 9.5(b), open and darkened triangles indicate the start and conclusion of the major task. Each task can be similarly scheduled. The length of the bar is an indication of its time requirement and not its cost.

If we add period designation to the horizontal axis, we can match the costs for levels of the WBS as shown by Fig. 9.6. In Fig. 9.6(a), the $150 bid is broken down for the WBS. In turn, Fig. 9.6(b) shows the baseline defined as the time-scheduled out-of-pocket cash flow for the estimate. If the engineer works for the owner, then this will represent the owner's cash flow. Similarly, it may represent the estimated cash flow for a subcontractor or major contractor. The contract cost of $150 is noted. The baseline estimate is similar to an ogive curve and is useful for estimate assurance and cost control.

Commitments for a project vary with the type of material. Note Fig. 9.7 where the axes are period and dollars per period. E indicates the moment of the estimate, M is the elapsed time for construction, fabrication, assembly, or test, and D is the moment of delivery or startup of operation. Open and darkened triangles are the starting and ending milestones. The division of material into four types depends on the project. Seldom will any two projects have the same mix of materials.

In Fig. 9.7(a), facilities and equipment are committed earliest after the point of the estimate. In this example, facilities and equipment make up 20% of the value committed. The long-lead-time components and subcontract materials, as in Fig. 9.7(b), are committed next, and they account for 30%. In Figs. 9.7(c) and 9.7(d) the stan-

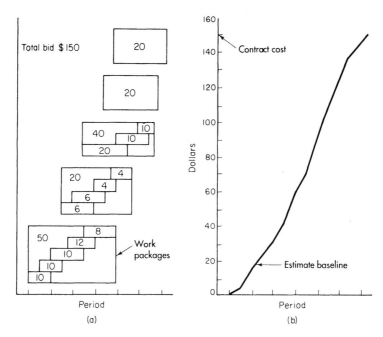

Figure 9.6. Definition leading from (a) estimated tasks to (b) scheduled cash flow called the estimate baseline.

dard commercial materials and raw material and stock items are committed successively. By using addition of areas, the actual area is summed for all four types in Fig. 9.7(e). An approximate triangle is shown with the peak at midpoint. The data for those commitments are determined from the work package.

Commitment dates are not the same as delivery or payment dates. The commitment date allows a time lag for a subcontractor for delivery and payment.

Scheduling of expenditure differs from scheduling commitment. A reverse order is sometimes necessary when time staging expenditures to commitments.

Note Fig. 9.8, where expenditures of those items previously committed are now examined. The factors that affect scheduling of expenditure for a major program are initial funding, progress payments, type of contract, and billing practices (such as 2% discount if paid within 30 days).

Obviously, expenditures are discrete and lump sum, yet a continuous line and area is assumed. For major projects this smoothing assumption is acceptable, but for smaller projects discrete payments cannot be ignored.

The order for *expenditure scheduling* implies several ideas. For purchase commitments, the center of gravity of the area of the cash flow triangle is before the midpoint of the time span of the period axis. For *expenditure scheduling,* the center of gravity is after the midpoint of the period axis. Also, *raw material* is required to allow construction or production an earlier start. *Standard commercial materials* and *long-lead-time components* are required and should be stocked and accounts

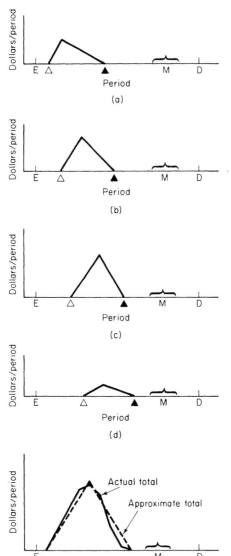

Figure 9.7. Purchase commitment for a project: (a) facilities and equipment, 20%; (b) long-lead-time materials, 30%; (c) standard commercial materials, 40%; (d) raw materials and stock items, 10%. (e) Sum of the actual areas.

paid as work in progress. Payments on long-lead-time items often start early owing to required progress payments. Payments for *minor raw materials* may be on a letter contract or a voucher system. Thus, scheduling of expenditure relates to the commitment yet has different time staging and a cash flow triangular shape.

An isosceles triangle is used to approximate the actual total area. The midpoint of the isosceles triangle is halfway between the initial and final expenditure milestone.

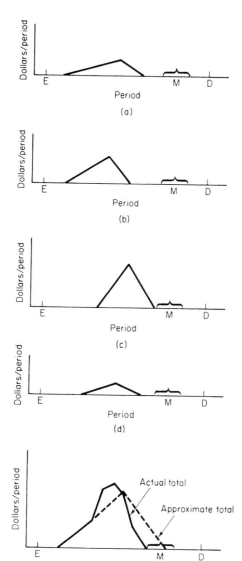

Figure 9.8. Expenditure scheduling for a project: (a) facilities and equipment, 20%; (b) long-lead-time materials, 30%; (c) standard commercial materials, 40%; (d) raw materials and stock items, 10%. (e) Sum of the actual areas.

The approximation allows simpler and earlier estimating of the points or periods of expenditure. If commitments and expenditures can be estimated and scheduled, engineering is able to measure performance against project goals. (Those ideas are extended in chapter 11 and deal with cost/schedule control.) But the isosceles triangle can be divided into periods and the percentage determined for each period.

Selection of the isosceles shape and midpoint is arbitrary. The period length is longer than the actual project length, and ending points are chosen based on policies of self-funding by the contractor and subcontractor.

Consider Fig. 9.9, which shows the approximating isosceles triangles for commitment and expenditure. The *y* axis is designated dollars per period. The numbered *x* axis is labeled "period" with the origin E set equal to zero. The numbered period axis implies fiscal half years, but other calendar periods are also suitable.

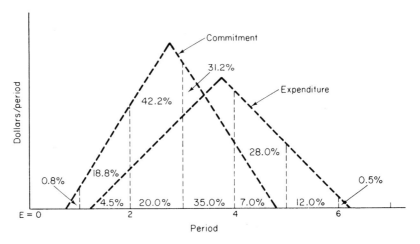

Figure 9.9.　Matching commitment and expenditure to period.

Some organizations define a scheduling month as 4 weeks (or 160 hours) and an estimating year as 13 months to avoid the problem of short and long months. A half year or quarter period does not proportionately have those scheduling period differences.

Note Fig. 9.9, again, where the commitment triangle is five periods long and is divided into fiscal half-year percentages by using geometry. Similarly, the expenditure triangle is six periods long, and its amount can be synchronized to the fiscal half-year percentages. Those half-year percentages shown on the figure are repeated in the following table:

Fiscal Half Year	Percent Commitment	Percent Expenditure	Contractor Dollar Commitment	Owner Dollar Expenditure
0	0.8	—	$　94,000	—
1	18.8	4.5	2,209,000	$　528,750
2	42.2	20.0	4,958,500	2,350,000
3	31.2	35.0	3,666,000	4,112,500
4	7.0	28.0	822,500	3,290,000
5	—	12.0	—	1,410,000
6	—	0.5	—	58,750
Total			$11,750,000	$11,750,000

Total project funds can be distributed by using those percentages. If the project is estimated as $11,750,000, then the committed and expenditure dollars for each half year are given by the product of the percentage and the total project funds. With those ideas and continuity of periods, it is straightforward to find monthly commitment and expenditure percentage and dollar amount.

Curves approximating the actual commitment and expenditure other than the isosceles triangle are used. Rectangular, trapezoidal, and bell shapes are possible.

If the *committed or expenditure dollars* as found in the previous table are progressively accumulated for the periods and a smooth line drawn through the points, then an ogive curve results. This is shown in Fig. 9.10. A third line, *available funds*, is included.

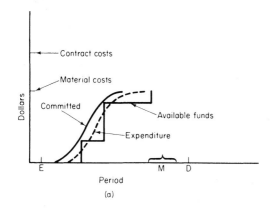

(a)

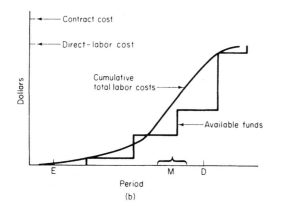

(b)

Figure 9.10. Collection of many cost estimates to form (a) smooth committed, expenditure curve, and available funds and (b) labor costs and funds available.

Available funds represent the money received periodically from the owner and used by the major contractor to pay subcontractors or themselves.

The *funds available* line is herky-jerky as shown in this figure because discrete payments are received.

Those three lines are important in financing a project. At the midpoint, commitments lead available funding dollars, and if the graph had divisions, then this lead time could be measured.

Expenditures lead available funds by so many periods, although less than committed. Current liabilities represent the difference between the expenditure and available funds line. Current liabilities for material can be defined as total dollars that must be assumed if the contract were terminated. Unliquidated commitments comprise the unpaid balance between commitments and expenditures. Those three lines are used in cost/schedule control. Now we turn to cost scheduling of direct labor.

The discussion so far in cost scheduling has been with materials, but many of those ideas are effective with direct labor. There are several ways to schedule direct labor cost. The distinction is whether or not the work package has been concluded.

The labor estimate is used for time scheduling direct labor cost. Recall that the labor estimate form provides entry opportunity for labor craft, period and year, total hours, average hourly rate, and total dollars. The labor hours are also totaled for the period. For any modest or major project, those data from the labor estimate are already period-scheduled. This scheduling is either in hours or dollars. Hours are more useful for work planning, though dollars are necessary for financial planning and control. Hours are not subject to escalation because of inflation effects, although dollars are.

The labor estimate integrates the elements of the estimate, learning theory if it exists, schedule, and proper lead times to accomplish the tasks in their order. After summing the labor estimates for various periods, Fig. 9.10(b), an ogive curve can be drawn. During the construction, assembly, and test phase, the curve is steeper, indicating more manpower.

Estimates are another major part of the work package. Estimating is done concurrently with the work package development, and the estimates are the most important documents relating to the project bid.

9.3 ESTIMATING

A *project estimate* is composed of labor, direct material and subcontract, facilities and equipment, and engineering forms. Procedures vary because of application or dollar amount. The larger valued projects may have addendums to those forms. Distinctions exist between construction, either building, road, plant, and so on. An owner or contractor has needs that call for variety. But for instructional purposes, these four forms are sufficient.

The *labor estimate* deals with the bidder's labor or work force. This assumes that labor will be direct hire for the duration of the project rather than a subcontract arrangement.

Direct-hire labor is hired and paid on a per hour basis to the worker. Consequently, the labor estimate may involve effective gross hour wages, payroll overhead, and productivity. In those situations where union labor is used, the term "off the bench" implies that the hiring is though a union-controlled hiring office and that the labor is waiting for a "call." Where union labor is not used, the firm will conduct its own hiring.

Whether the labor is union or nonunion, the company pays the workers directly. Subcontract labor works for a subcontractor, and the costs are estimated and quoted on a fixed unit price or lump sum, often including material costs.

TABLE 9.1. LABOR ESTIMATE

Project *High-voltage transmission line.*

WBS	Description	Total hours	Periods 3	4	5	6	Gross wage	Cost
1311	Clearing	4,000	4,000				$30	$120,000
1312	Roads	8,000		4,000	4,000		35	280,000
1313	Tower erection	55,743	8,862	22,797	16,723	7,861	40	2,229,720
1321	Substation building	24.285		5,000	15,000	4,295	35	850,000
								$3,479,720

An abbreviated labor estimate is shown in Table 9.1. The labor estimate is cross referenced to the WBS and the definition. Information can be listed such as start date, duration, conclusion and labor types, hours, and wages. Some estimates will include a check-off that indicates how hours were estimated (i.e., opinion, comparison, standards, cost estimating relationship). The project engineer determines if the work is recurring or nonrecurring. *Recurring work* for project estimating is cyclic; that is, it is done and estimated with repetition in mind, such as the erection of transmission towers.

The estimated time may be adjusted for *productivity.* The same job may require more time in Alaska than Texas because of weather conditions. Though a labor standard for a job is consistent, effects of location, crew skill, native or green, and so on, are factors that the engineer weighs.

Productivity for the bidder's and subcontractor's work force can vary, and some experience suggests that subcontractors have a more favorable productivity.

The preferred approach is to estimate separately the allowed man hours (or man days) and their productivity, rather than posting a lump sum that is the product of the two quantities. Those features are a part of the usual labor estimate, but are not demonstrated in Table 9.1.

The *direct material and subcontract estimate* requires similar title block information as the labor estimate. It is cross identified to the WBS and labor estimate.

There is a difference between direct material and subcontract. Direct materials end up in the project design and are usually installed by the bidder's work force and include bulk materials or structural steel, piping, concrete, wiring, and so on. We have consistently referred to those materials as raw or standard commercial materials. Those direct materials do not include the contractor's work force in their cost.

On the other hand, the WBS and the definition may indicate that a subcontractor will be hired. *Subcontract materials* are specially designed or standard commercial materials. Especially important is the notion that subcontract materials use subcontractor labor, and this dual nature makes it different from direct materials, which uses the contractor's work force. A subcontract may include both the material and the subcontractor's labor.

A basic quantity of the item is posted. Additions for scrap, waste, shrinkage, and spares can be included for direct material. The subcontract materials may be item estimated or handled as a lump sum.

The source of the direct material estimate may be opinion, comparison, take-off for shape and rate, or statistical relationship. The information for a subcontract may be similar, but a quotation may be additionally available. Selection of the subcontractor may be based on low bidder, technical competence, or best delivery schedule. Table 9.2 is an abbreviated example of the direct material and subcontract estimate.

TABLE 9.2.   DIRECT MATERIAL AND SUBCONTRACT ESTIMATE

Project *High-voltage transmission line*

WBS	Description	Direct material or subcontract	Periods				Cost
			2	3	4	5	
1312	Road	Direct material	15,000				$15,000
1313	Tower	Direct material	1,000,000	940,625			1,940,625
1321	Substation building	Direct material			90,000	90,000	180,000
1322	Substation transformers, control	Subcontract		200,000	150,000	50,000	400,000
133	Conductor	Subcontract		195,000	195,000	195,000	785,000
							$3,215,625

Facilities and equipment are also direct materials but of a different character; that is, delivered and erected equipment, such as large storage tanks or field-fabricated vessels involving both material and labor as a single lump sum. Facilities and equipment are specified in detail and are custom manufactured by vendors for the project and may be produced and estimated like a product. Those job shop manufacturers are sensitive to demand and adjust prices to accommodate demand. Price fluctuations can be expected in those costs, especially for external suppliers.

It is also necessary to estimate any raw or bulk materials that may be necessary with the facilities and equipment. For example, an electric transmission tower will require concrete foundations. Will the foundations be estimated with the transmission towers? Usually not, and a cost connection must be made between the WBS, definition, and the estimate. The concrete foundations are estimated as direct material by using the contractor's work force.

Because facilities and equipment are custom designed, a quote becomes necessary. Often a predesign estimate is made. Thus, the source for those data can be external to the estimating team or may be based on internal information.

Facilities and equipment estimates can include land and building and process-ing equipment. Equipment supporting a subcontractor is included in the subcontrac-tor's quotation. Rental charges for construction support equipment can be included in the labor estimate but usually are included in project overhead. Table 9.3 is an ab-breviated example of a facilities and equipment estimate.

TABLE 9.3.   FACILITIES AND EQUIPMENT ESTIMATE

Project   High-voltage transmission line

WBS	Description	Type	Period 1	2	3	4	5	6	Cost
111	Land, Survey	Facility	75,000	11,575					$86,575
112	Environment	Facility		6,425					6,425
1321	Substation building	Facility						50,000	50,000
1322	Substation transformer	Equipment				400,000	400,000		800,000
									$943,000

Engineering estimates have been previously discussed in chapter 8. Those costs are significant for a project design, and their separate consideration points to this importance. Table 9.4 is an example of an engineering estimate for project designs. Productivity can vary and should be forecast. Overtime, job size, and specific work-ing conditions can affect productivity.

TABLE 9.4.   ENGINEERING ESTIMATE

Project   High-voltage transmission line

WBS	Description	Hours	Hourly rate	Applied overhead	Period 1	2	3	Total cost
12	Engineering	750	$44.61	25%	35,000	5,000	1,875	$41,875
121	Saggers	150						
122	Cost	80						
123	Design	520						

Those estimating procedures have similar advantages. The uniformity encour-ages consistency for estimate assurance. The source of information, auditing, a stan-dardized communication format, cross reference, and the central WBS engineering document make those estimating methods of value throughout the project.

9.4 COST AND BID ANALYSIS

Four kinds of estimates provide the factual basis for *cost and bidding analysis*. If the estimates are poorly done, then no amount of superficial analysis will improve the estimates. But we separate the task of estimating from its later analysis. It is important that the bid be in line competitively as well as compatible to the firm's ability.

More information becomes available to the engineer during the estimating and analysis period. Tips may be found in the local newspapers, budget disclosure, or the owner or major contractor may even indicate boundaries for the bid. Rebidding may occur in large projects, and first bids are known. By law, past winning bids are open knowledge for public works. Business magazines and trade newspapers publish information regularly. Even rumors are sought.

Competitive bidding is usual for projects. In the simplest case, the offeror will announce a deadline date for the bid, and sealed envelopes containing the bid and other information are opened and the winner announced. In technical projects, the bidder provides a design along with the bid. Evaluation may take a long time before the winner is selected.

When formal advertising and competitive bidding are impractical, a bargaining process begins between the parties, each having its view and objective. This is termed *negotiation*. Cost and bidding analysis is different for each of those situations. The estimating procedures should be identical, however. Eventually, the bid is "laid on the table," so to speak, and its acceptance or denial depends on many factors.

The bid is the sum of the estimates, overhead, contingency, interest, and profit. As a *pro forma* document, the bid is the center of much interest. If the project estimate is a public document subject to audit by an owner, major contractor, or the government, and this depends on the contract, then analysis is handled by a contracting officer, negotiator, or engineer on an "arm's-length" basis (i.e., each side having a competitive and self-serving interest). The audit is from several vantage points: engineering, accounting, purchasing, and estimating. Thus, the cost and bidding analysis must satisfy many objectives.

9.4.1. Overhead

Overhead costs were discussed in section 4.9, which pertained to operation and product overhead. Many of those principles apply to projects. Overhead for projects is of two types: office and job. Overhead costs exclude direct labor, direct and subcontract materials, and facilities and equipment. Items appearing in overhead must not be included in those estimates. However, for practical reasons, some minor costs that could be treated as direct are classified and handled as overhead.

Office overhead includes general business expenses, such as home office rent, office insurance, heat, light, supplies, furniture, telephone, legal expenses, donations, travel, advertising, bidding expenses, and salaries of the executives and office employees. Those charges are incurred for the benefit of the owner's or contractor's overall business. In office overhead, the final cost objective is multiple (i.e., several projects) and cannot be isolated as specific estimating amounts. Though variety is possible in the calculation, one approach would use

$$R_o = \frac{C_o}{C_p} \times 100 \tag{9.1}$$

where R_o = office overhead rate on basis of direct costs, percentage

$\quad\;\; C_o$ = overhead charges summed for office activity of contractor, dollars

$\quad\;\; C_p$ = cost of direct labor and direct and subcontract materials, dollars

A company has an annual value of direct cost of construction as \$60,000,000 and a *general office overhead* of \$2,400,000. $R_o = 4\%$, which is applied against those estimated future costs, provided that the future will be similar to those historical costs. Though the rate is guided by historical patterns, those computations are for future periods, specifically budgeted for the duration of the project. It is necessary that both the base C_p and the overhead charges C_o be forecast and computed for identical periods.

Job overhead pertains to the project. If a firm has only one field project in mind, then a consolidation of office and job overhead is a convenience. On the other hand, improved accuracy and other advantages become apparent if separate job overheads are found for one or all projects. Job expenses are those costs that do not become an integral part of the construction and are directly chargeable to the contract and must be separated in the accounting journals from overhead expenses, which are general.

Typical items of job overhead are listed below. Each project requires a special analysis to determine its own items.

Permits and fees	Insurance
Performance bonds	Electricity at job location
Job office expense	Water at job location
Office salaries	Barricades
Cost clerk	Badges
Timekeeper	Survey
Supplies	Parking areas
Working foremen	First aid
Depreciation	Storage and protection

Those overhead charges do not include mandatory contributions for direct labor such as FICA, workmen's compensation insurance, unemployment insurance, or any of the union contract contributions. Employee-related costs are included in the gross hourly cost of direct labor.

$$R_j = \frac{C_j}{C_p} \times 100 \qquad (9.2)$$

where R_j = job overhead rate on basis of direct cost, percentage
        C_j = overhead charges summed for project, dollars
        C_p = cost of direct labor and subcontract materials for project, dollars

As a rule of thumb, if the base C_p increases, then the overhead rate R_j should be decreasing. Conversely, if the base is decreasing, then the overhead rate generally will be increasing.

Those short-term fluctuations depend on cost control and estimate assurance. Comparison of rates between projects may make no sense, remembering that the rate is a ratio of two cost sums. A low rate may indicate a project involving manual labor. As a result, the direct labor base is high. The overhead account is less because it includes little or no charges for depreciation or rental of equipment.

The overhead is applied to the project by using

$$C_{op} = C_p'(R_o + R_j) \qquad (9.3)$$

where C_{op} = overhead charged to future project, dollars
        C_p' = cost of future direct labor and direct and subcontract materials estimated for forthcoming project

This value is posted to the estimate summary.

9.4.2. Contingency

Contingency is another cost element. Sometimes this element is called *engineering reserve,* and uncertainties in developmental projects are included by this self-insurance. Especially in the early stages of a project where technology detail and its cost are absent, contingency estimating is preferred to careless estimating or unjustified padding of cost elements. *Contingency cost estimating* is practiced for projects albeit by various techniques.

As the project design matures and as more information becomes available, the contingency estimate becomes less in absolute value and percentage of the total proposal value. In large-scale project cost planning, contingency amounts are more important during early estimating. Eventually, a detail estimate is determined by using a work breakdown structure. Thus, contingency dollars follow the rule: more information and less contingency.

In high-risk high-dollar projects, contingency is usually present and can appear as a line item in the bid. As the project spends money for materials, direct labor, subcontracts, and so on, the dollars of budgeted contingency are available for the payment of unexpected problems that arise.

Two methods of assessing contingency are considered. The first recognizes the importance of experience and opinion, and the second treats it analytically

In discussing an informal method of assessing contingency, we assume that dollar amounts for the major WBS levels are already estimated. By using opinion the engineer estimates an additional dollar amount for major elements. For instance, laser glass is risky development, and prudent evaluation would determine an incremental cost for possible overrun. This increment is contingency.

Contingency analysis can also be associated with the estimating of the major elements, as is shown by a second method. It is an extension of range estimating discussed in section 6.7.3. Note Fig. 9.11 for the steps.

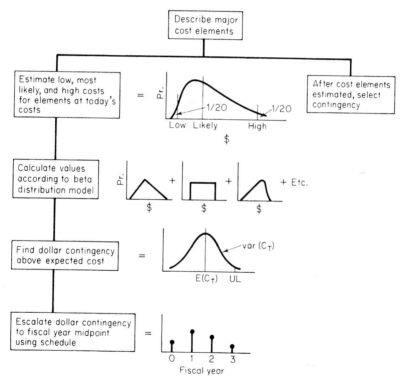

Figure 9.11. Contingency cost analysis.

Contingency cost estimating begins with estimating a lowest value (L), most likely value (M), and the highest cost value (H) of the cost elements of the WBS. The most likely value is the modal or most common value that would be repeated in an unlikely repetition of this cost element. Similarly, the lowest and highest cost values are not to be exceeded either downward or upward with a probability of $\frac{1}{20}$.

Seldom, if ever, are those instructions exactly executed, but three values are necessary to allow the use of the beta probability distribution. The form of this general distribution has properties that make for simple calculation, which is important for early estimates of the project. It is unnecessary to specify the probability distribution of the individual cost elements, which is fortunate, because their behavior is non-normal and unknown. After many elements are added, according to central limit theory and practice, the resulting distribution is normal and can be manipulated by using conventional probability rules.

The three values for each major cost element are estimated at today's costs even though it is future expenditure that is desired. If information is available, then it is often at current value. Further, in conference and round-table estimating, current values are easier to estimate intuitively than future values.

Now, assume that the expected cost of Table 6.1 $E(C_t) = \$10,600$, and its variance $\text{var}(C_t) = 22,300$ dollars^2 are to be analyzed for contingency. Recall Eq. (6.28):

$$Z = \frac{UL - E(C_T)}{[\text{var}(C_T)]^{1/2}} \tag{9.4}$$

where Z = standard normal variable

   UL   = upper limit cost, arbitrarily selected, dollars

   $E(C_T)$ = expected cost of range estimating model, dollars

   $\text{var}(C_T)$ = expected variance of range estimating model, dollars^2

The square root of the variance is called the standard deviation. $E(C_T)$ and $\text{var}(C_T)$ were calculated by using Eqs. (6.26) and (6.27).

Values of UL are selected by the project engineer and cover the potential range of overrun or underrun cost of the project. The range of those values depends on the risk of the project. With several values picked, Table 9.5 is completed by using Eq. (9.4) and appendix 1, standard normal distribution tables. The appendix provides probabilities as measured from the reference $Z = 0.00$. Overrun and underrun are defined with respect to this reference. At this symmetrical location, 50% of the probability is above and below $Z = 0.00$. For example, if $Z = 1.00$, then the table reads 0.3413, which is the area or probability from $Z = 0.00$ to 1.00. If we want $P(Z \geq 1.00)$ and upper tail area is given by $0.5 - 0.3416 \geq 1.00$, then an upper tail

TABLE 9.5 CALCULATION OF PROBABILITY GIVEN THAT A COST UPPER LIMIT EXCEEDS A MEAN VALUE

UL	Z	$P(\text{cost} \geq UL)$	UL	Z	$P(\text{cost} \geq UL)$
$10,300	−2.00	0.98	$10,750	.99	0.16
10,400	−1.35	0.91	10,800	1.32	0.09
10,500	−0.68	0.75	10,850	1.66	0.05
10,600	0	0.50	10,900	2.00	0.02
10,650	0.32	0.37	10,950	2.33	0.01
10,700	0.66	0.25	11,000	2.67	0.004

area is given by $0.5 - 0.3416 \geq 0.16$. If we are interested in the probability that cost exceeds \$10,300, then we would have

$$Z = \frac{10,300 - 10,600}{(22,300)^{1/2}} = -2.00$$

Because we are interested in $P(\text{cost} \geq 10,300)$, the appendix would be used as $0.5 + 0.4772 = 0.98$, which is observed as the first entry in Table 9.5.

Note Fig. 9.12, which gives the probability of exceeding the expected project cost. The expected value, \$10,600, gives a 50% probability. The probability that cost will assume values less than \$10,600 is greater than 50% naturally. From this analysis the engineer gains an opinion for the risk with this curve. Several contingent maximum costs are considered, keeping in mind the uncertainty, design, technology, and construction conditions. Eventually, one upper limit cost is selected as the contingent maximum cost. This selection is made with engineering approval. For example, suppose that engineering selects \$10,700 as this value, which will be exceeded with a 25% probability. The next step is to schedule the increment of contingency of \$100 ($= 10,700 - 10,600$) over the project periods. The contingent maximum is at current cost. If inflation or deflation is suspected, then this value is adjusted by means of indexes relating to the project. Indexes such as those provided by Table 5.6, which assumes declining costs owing to active technology, are now

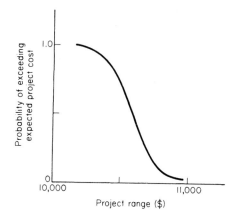

Figure 9.12. Risk graph of contingency.

TABLE 9.6 SCHEDULING CONTINGENCY AMOUNT OF \$100 FOR PERCENT EXPENDITURE AND INDEXES

	Period				
	1	2	3	4	Total
Expenditure (%)	15	25	40	20	100%
Period cash flow ($)	15	25	40	20	$100
Index	1.000	0.941	0.891	0.876	
Contingent cash flow ($)	15	23.5	35.6	17.5	91.6

used. Additionally, expenditure scheduling given in percent terms is known. This information is shown as Table 9.6. The scheduled contingent amount for period 2 is $23.5(= 100 \times 0.25 \times 0.941)$.

The total of \$91.60 is posted to the project summary as a line item for contingency. Similarly, the \$10,600 is entered as the project estimate amount. Note that contingency is shown for the total project rather than single cost elements. Even though it is element variability that gives project contingency, it is as likely that other cost elements may be overlooked and this contingent amount becomes available for unspecified requirements.

Opinion and technical experience are inescapable in estimating contingency amounts. On the other hand, algebraic refinements such as those presented do not substitute for effective methods in project estimating. Unfortunately, in practice, contingency is a device that replaces qualified cost-estimating methods.

Depending on the contract type, if unjustified contingency is added to the cost, then the bid becomes noncompetitive. In some cases, should a surplus of contingent dollars be available at the conclusion of the project, they are returned to the owner, according to provisions of the contract.

9.4.3. Interest

Projects may involve periodic payments between the owner and contractor or between the contractor and subcontractors. Those payments are known as *partial or progress payments* and reimburse for work.

Often the owner will hold back an amount and not immediately pay for work concluded. This amount is called *retainage*. The contract may require that retainage be placed in an *escrow account* with a bank, thus assuring the contractor of its availability. An escrow account removes any unfair advantage and the motivation is contractual. Eventually, the owner will pay all monies, but with a sufficient delay to assure that the project design is concluded according to the terms and specifications of the contract.

Projects may exceed cash-on-hand, and bank loans are necessary to meet the ongoing cash flow requirements. Before interest is calculated for the estimate *pro forma*, it is necessary to establish a prepayment plan. The difference between payment for work concluded and available funds is the amount required for bank loans and is the principal on which *short-term construction interest* is charged. Both an owner and contractor are required to establish this cash flow stream.

Figure 9.13 shows project cash flows where the x axis is the contract time scaled as a percent. The y axis is baseline dollars per period. The triangle approximation has been previously discussed. For example, a rectangular cash flow approximation may result from a pipe laying project. A trapezoid is the more common approximation. Those sketches assume continuous cash flow. Let

t_i = initial time when baseline dollar per period is maximum as a percentage of contract time

t_f = final time when baseline dollars per period is maximum as a percentage of contract time

C_a = average baseline dollars per period and equal to the total value of contract divided by frequency of progress payments

C_m = maximum baseline dollars per period

Geometrically, for the trapezoid, Fig. 9.13(c),

$$\frac{C_m}{C_a} = \frac{200}{100 + t_f - t_i} \tag{9.5}$$

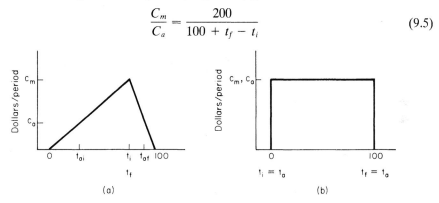

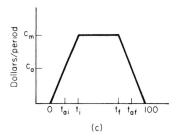

Figure 9.13. Cash flow models: (a) triangular; (b) rectangular; (c) trapezoid.

Values of C_m/C_a can be calculated as follows:

			t_f	
t_i	50	60	70	80
20	1.54	1.43	1.33	1.25
40	1.82	1.67	1.54	1.43
60	—	2.00	1.82	1.67

where $0 \le t_i < t_f \le 100$.

An early step in the analysis is to assume the geometrical approximation t_i, t_f, and C_m/C_a. Those assumptions may be based on experience.

The periods of contract time during which the baseline dollars per period is equal to the average value can be shown by similar triangle geometry to be

$$t_{ai} = \frac{t_i}{C_m/C_a} \frac{N}{100} \tag{9.6}$$

and the final instance for an average cash flow rate is found when

$$t_{af} = \left(100 - \frac{100 - t_f}{C_m/C_a}\right) \frac{N}{100} \tag{9.7}$$

where N = number of project payments

t_{ai} = period when average is initially reached

t_{af} = period when average is finally reached

A construction estimate is \$150,000 and has a contract time of 10 months and payment is monthly. If the cash flow assumption is trapezoidal and $t_i = 40\%$ and $t_f = 70\%$, then we have

$$C_m/C_a = \frac{200}{100 + 70 - 40} = 1.54$$

as the ratio of maximum to average cash requirements.

where C_a = 150,000/10 = \$15,000 per period average cash flow

C_m = 15,000 × 1.54 = \$23,100 per period from t_i to t_f

t_i = 0.40 × 10 = fourth month, when maximum cash flow rate starts

t_f = 0.70 × 10 = seventh month, when maximum cash flow rate concludes

t_{ai} = $\dfrac{40}{1.54} \times \dfrac{10}{100}$ = 2.6th month, when initial average cash flow rate is reached

t_{af} = $\left(100 - \dfrac{100 - 70}{1.54}\right) \dfrac{10}{100}$ = 8th month, when final average cash flow rate is reached

As the project progresses, the cumulative cash flow appears as an *ogive or S curve*. These cumulative curves are the integral of the curves given in Fig. 9.13. But it is also possible to have the cumulative baseline value as a percentage of the total contract amount, y axis, to the percentage of the total contract time, x axis. This is shown in Fig. 9.14. Both the triangular and trapezoid of Fig. 9.13 give the S-shaped curve. The rectangular model, Fig. 9.13(b), will give a straight line. For a trapezoid cash flow model we have after integrating

$$\text{CBV} = \frac{C_m t^2}{2 C_a t_i} \qquad 0 \le t \le t_i \tag{9.8}$$

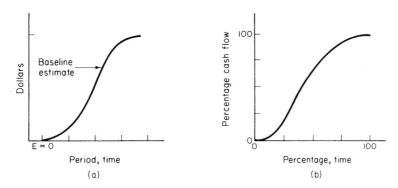

Figure 9.14.   Cumulative baseline curve plotted as (a) period versus dollars and (b) percentage time versus percentage cash flow.

$$\text{CBV} = \frac{C_m}{C_a}\left(t - \frac{t_i}{2}\right) \qquad t_i \leq t \leq t_f \tag{9.9}$$

$$\text{CBV} = \frac{C_m}{2C_a}\left[2t_f - t_i + (t - t_f)\left(1 + \frac{100 - t}{100 - t_f}\right)\right] \qquad t_f \leq t \leq 100 \tag{9.10}$$

where CBV denotes the *cumulative baseline value,* percentage. Application of those formulas requires t to be a percentage number. With those equations and $t_i = 40\%$ and $t_f = 70\%$, we have what follows:

t (%)	CBV (%)	t (%)	CBV (%)
0	0	60	61.52
10	1.92	70 (t_f)	76.90
20	7.69	80	89.72
30	17.30	90	97.53
40 (t_i)	30.77	100	100.00
50	46.15		

The cash flow requirements for this project need additional information. For example, there may be a delay from work concluded to money paid. Now assume that for the $150,000 project there is a 1-month delay, and profit is 10% of the project bid by the major contractor. A different profit would result if the 10% were a markup rate on cost. Refer to Table 9.7. In column 1, each period is a progress payment and begins with zero and ends with the total number of periods of contract time plus those periods before all monies, including retainage, will be received. Column 4 is the product of column 3 and the project contract value. Profit, column 5, is found by multiplying column 4 by 10%, the markup profit rate. Expenditure is the difference between columns 4 and 5. Available funds depend on contract terms. For our example, there is a 1-month delay for the owner to certify that the work is done and is paid on the basis of 85%. For example, at $t = 2$, column 7 is found by

TABLE 9.7   DETERMINING PROJECT CASH FLOW, RETAINAGE, AND EXPENDITURE

(1) Month	(2) Contract Periods (%)	(3) CBV (%)	(4) CBV	(5) Profit	(6) Expenditure	(7) Available Funds	(8) Retainage	(9) Net Cash Flow
0	0	0	0	0	0	0	0	0
1	10	1.92	$ 2,880	$ 288	$ 2,592	0	$ 2,880	$ (2,592)
2	20	7.69	11,535	1,154	10,382	$ 2,448	9,087	(7,934)
3	30	17.30	25,950	2,595	23,355	9,805	16,145	(13,550)
4	40	30.77	46,155	4,616	41,450	22,058	24,097	(19,482)
5	50	46.14	69,210	6,921	62,289	39,232	29,978	(23,057)
6	60	61.52	92,280	9,228	83,052	58,820	33,451	(24,223)
7	70	76.90	115,350	11,535	103,815	78,438	36,912	(25,377)
8	80	89.72	134,580	13,458	121,122	98,048	36,532	(23,074)
9	90	97.53	146,295	14,630	131,666	114,393	31,902	(17,273)
10	100	100.0	150,000	15,000	135,000	124,351	25,649	(10,649)
11	110	100.0	150,000	15,000	135,000	127,500	22,500	(7,500)
12	120	100.0	150,000	15,000	135,000	150,000	0	15,000

multiplying the cumulative baseline one period earlier by 85%, or 2448 (= 2880 × 0.85). Retainage, column 8, is found by subtracting available funds from the cumulative baseline value for each period. For period 2, column 8 is $9087 (= 11,535 − 2448). The last column describes contractor's net cash flow and is found as the difference between columns 7 and 6 or 5 and 8. For period 2, (7934) denotes a negative value, and a loan equal to this amount is required. It is only the last period, $t = 12$, when the apparent profit of $15,000 is available to the contractor. Note that column 9 is a cumulative deficit or surplus. Values in this column are a guide to the amount to which the contractor must finance the project with a bank loan. Interest must be paid for this interim financing.

The net cash flow column when plotted against the month provides the S curve. Note Fig. 9.14. The y axes can be identified in dollars per month in actual or percentage terms.

Though labor scheduling can be done directly by using the work package, it is also possible to schedule with geometrical models. Both the trapezoid and the S curve would deal with manpower in this case. We slightly change the example, and we add information as follows. Assume that the general contractor's markup for overhead and profit is 20% of cost and he subcontracts all work. Subcontractor's markup for overhead and profit is 25% of cost. Labor accounts for 60% of cost and is paid $20 per hour. Now let the contract total be $1,500,000 and $t_i = 40\%$ and $t_f = 70\%$.

The cost to the general contractor is 1,500,000/1.2 = $1,250,000. Cost to the subcontractors is $1,250,000/1.25 = $1,000,000. Total cost of labor is 1,000,000 × 0.60 = $600,000. Manpower requirements = 600,000/20 = 30,000 man hours. Man months = 30,000/173.3 = 173.1, where 173.3 (= 52 × 40/12) is the available hours per month. The average number of direct labor workers is 30,000 ÷ (10/12 ×

2080) = 17.3. If the maximum on the job manpower is 154% of the average manpower, then the maximum number of direct labor employees is 27 (= 1.54 × 17.3). A manpower curve is shown in Fig. 9.15(a).

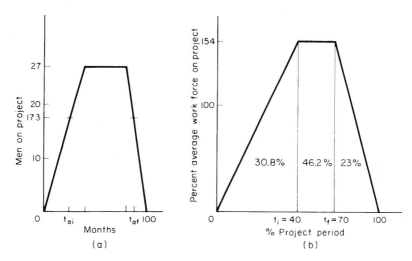

Figure 9.15.   Project trapezoidal models for (a) manpower requirements and (b) percentage requirements with respect to timing.

Another variation to labor scheduling can be approximated by Fig. 9.15(b). Assumptions are that the maximum on-the-project manpower is 154% of the average. The maximum first occurs after 30.8% has been expended and when 40% of the project time has elapsed. The period of maximum force accounts for 46.2% of total manpower. The cumulative manpower curve for time would appear as an S curve, as typically shown in Fig. 9.10.

Because projects involve large dollar undertakings, the owner or contractor may have insufficient cash on hand to meet all financial requirements. Borrowing may be necessary. Once the project is concluded, the contractor is paid, but the owner has incurred a long-term debt.

For project analysis there are two components to owner's interest. The first is during the period of the construction phase and the second is following the conclusion of construction. The owner deals with both, but the contractor is concerned with project construction cash flow requirements only.

The owner will use a cash flow analysis such as in Table 9.7. In particular, the owner will pay attention to column 7. Initially, there will be some cash on hand available for progress payments. If we assume that the owner has 25% available money to finance the project, then from period 5 on a loan for the remainder is required. This is *construction financing* and terminates at project end. It is usually more expensive (i.e., concerning interest) than ordinary long-term debt for which collateral and other securities are pledged. The short-term loan is converted into long-term debt.

In a similar way, the contractor needs money to meet obligations. Table 9.7 was prepared from an owner's viewpoint, because the owner held back contractor's profit in retainage. Realistically, the contractor may need short-term financing measured by the difference between columns 6 and 7 up to period 10, or column 9. The money out of pocket is the same as indicated by column 9. Repayment by the contractor to the lender would be concluded following final payment by the owner. The amount of paid interest by a contractor to a lender also depends on initial working capital.

Consider Table 9.8, which demonstrates the calculation of interest for short-term financing by a contractor. In this evaluation we assume total financing by the contractor and that column 9 of Table 9.8 reflects the net cash flow requirement to be financed monthly. Line 1 is the total loan requirement at the start of the period, say, a month in our case. Line 1 is the balance of the loan at the start of the period. Line 2 is the money needed for construction during the upcoming month. It is found from Table 9.7, column 9, and is the difference between the current and previous month. Deficit quantities are considered positive for the purpose of a loan requirement. The total loan is the sum of lines 1 and 2 and is the quantity on which interest is charged. For this example we use a 1% monthly interest rate. Line 4 is the product of the total loan and the interest rate. The outstanding loan is the sum of lines 3 and 4, which becomes line 1 in the next period.

Note in Table 9.8 that the outstanding loan is paid off in month 12. The total of the monthly interest, $1831, would be posted to interest on the *pro forma*. In this situation it reduces the apparent profit of $15,000 to $13,169.

The owner would use a similar procedure, and interest is based on column 7 of Table 9.7. This quantity is converted to long-term debt at the conclusion of the project construction.

9.4.4. Pricing and Bid

Pricing for project designs differs from product in several ways. Many products are sold on the *open market* to an unknown buyer. In the *bid market,* however, a manufacturer or contractor produces or constructs an end item for a specific buyer.

Projects are procured by contract. Methods of pricing depend on the type of contract selected by the owner or buyer and pricing is in accordance with the contract. For products, price adjustments are routinely based on market or technology or cost movements. This choice is made every so often for the same or similar products. A bid requires a unique analysis where opportunity for repetition is unlikely. Some product engineers develop many prices, though project engineers submit bids infrequently.

The winning or losing of a bid award may give no hint about how well the project was estimated and priced. A competitor may "low ball" a bid to keep a project work force busy. Contrariwise, the winning of a bid may be against uninterested competition who submit high bids. In the product market a 3% increase may reduce the number of sold units by 5%, for example. A similar increase in the bid may lose the contract.

TABLE 9.8 SHORT-TERM FINANCING OF LOAN BY CONTRACTOR

	Month											
	1	2	3	4	5	6	7	8	9	10	11	12
1. Loan, start of month	0	2,618	8,040	13,792	19,921	23,731	25,146	26,563	24,503	18,889	12,388	9,331
2. Monthly net cash flow	2,592	5,342	5,616	5,932	3,575	1,166	1,154	−2,303	−5,401	−6,624	−3,149	−9,331
3. Total loan	2,592	7,960	13,656	19,724	23,496	24,897	26,300	24,260	18,702	12,265	9,239	0
4. Interest amount	26	80	137	197	235	249	263	243	187	123	92	Σ = $1,831
5. Outstanding loan	2,618	8,040	13,792	19,921	23,731	25,146	26,563	24,503	18,889	12,388	9,331	0

Net profit = $13,169

Contracts can be broadly divided into two groups. In the first, risk is borne by the contractor, and the contract is called *firm price*. The contractor may or may not determine a contingency, which is added to the estimate. Second, an owner or buyer may assume all economic risk, and the contract is called *cost plus*. In this type, costs are reimbursed by the owner. A profit percentage or a fee is usually agreed to between the parties, which increases the total amount.

In view of those contract complexities we defer their discussion to chapter 12. We recommend that estimates be prepared similarly for either type of contract. Preparation of the price and bid analysis will depend on the contract type, however.

The procurement objective of an owner or buyer is to buy the end item from a responsible source for a reasonable bid. We define bid (also price and quote) to be the value of the economic want of a project design given, received, or asked in exchange. The value is monetary and expressed in the currency of the country. The bid does not include costs of operation and maintenance for the end item. Procurements are usually competitive and may result from formal advertising or from a request for proposal to an individual firm. In bidding competition, the quote, which may be lowest, may not be selected. Quality, specification adherence, delivery, and performance may be more important than price. Contrariwise, the lowest price from a responsible bidder may be selected.

Many view project pricing as an interpretive art: "Price can be anything you want to make it." Others would change this art to a routine calculation where price is found by formula. In engineering and estimating, though opinion is unavoidable, several pricing strategies are important but none guarantee success. Pricing success, when achieved, is a consequence of technical and estimating factors of moderate importance. Seldom is it a result of brilliant strategy. But to evaluate potential bid prices, analysis and the use of formulas are required. Because the spread between cost and price is an indication of value, there is a temptation to base cost on price alone. We, however, stress that pricing proceeds once cost is estimated. The estimated cost is the most significant and recognizable component of price.

The most common approach is to use cost plus a markup. A formula would be

$$P = C_t + R_m(C_t) + C_m \tag{9.11}$$

where P = project bid, dollars

C_t = total cost of direct labor, direct and subcontract materials, facilities and equipment, etc., dollars

R_m = markup rate on cost, decimal

C_m = miscellaneous costs that may be inappropriate for markup (i.e., contingency), dollars

Another variation would use various markup factors on the several cost elements.

A less frequent approach is to base the markup on the price. For instance, see Table 9.7, where the calculation of profit was found by multiplying a rate with the contract value; that is, $0.10 \times$ cumulative baseline value.

Table 9.9 is the bid summary for a 500-kV voltage transmission line project. Information from the estimates, Tables 9.1–9.4, are removed and posted. The esti-

TABLE 9.9.   PROJECT BID SUMMARY

Project *High-voltage transmission line*
Customer *Bonneville Power Administration*

Direct labor	$ 3,479,720
Direct material	2,035,625
Subcontract items	1,180,000
Facilities	143,000
Equipment	800,000
Engineering	41,875

Overhead

1. Office, applied to direct cost
   $R_o = 4\%$, direct costs $= \$6,695,345$ 267,813

2. Job, applied to direct costs
   $R_j = 5\%$, direct costs $= \$6,695,345$ 334,767

Subtotal	$8,282,800
Contingency, $\frac{1}{2}\%$ of subtotal	41,414
Interest with retainage	103,500
Total	$8,427,749
Profit 8% (on cost less contingency)	670,904
Bid	$9,098,653

mating analysis for overhead, contingency, interest, and profit is conducted after this posting. Eventually, a bid value of $9,098,653 for 5 miles of a high-voltage transmission line is found.

9.5 ENGINEERING ECONOMY

The purpose of converting money into plants and equipment is to return an amount of money that exceeds the investment. This statement assumes that capital is productive and earns a profit for the owner of this capital. In efficiency terms productive capital is related to a ratio of output to input. But unlike physical processes, the economic efficiency of capital, assuming long-term success and a capitalistic society, must exceed 1. The productivity of capital comes from the fact that money purchases more efficient processes for making goods and supplying services than consumers could employ themselves. Those products are then offered to the public at attractive prices that pay a profit to the manufacturer.

In earlier chapters, stress is placed on cost and price as the measure of importance. For purposes of capital investment, however, the owner uses a term called *return*. Like cost and price, return can be expressed in several ways. Among those are total dollars, percent of sales, ratio of annual sales to investment, or return on investment, which is favored by most project engineers.

The methods for calculation of return on investment are (1) average annual rate of return, (2) payback period, and (3) engineering-economic rate of return. Naturally, the selection of a method must be consistent with engineering's goals of profitability.

The goal of any project estimate is to predict the net change in the company's overall cash position. The engineer makes studies of alternatives, looking for the change in cost and revenue that must be considered. Those studies provide factual quantitative data to measure the interaction of future events.

In this text we favor the compound-interest–based investment computation that considers the time value of money. Many other fine texts expand on engineering economy. However, as a background let us look at methods used because of simplicity.

9.5.1. Average Annual Rate-of-Return Method

In some economic studies, return on investment is expressed on an annual percentage basis. The yearly profit divided by the total initial investment represents a fractional return or its related percent return. This recognizes that a good investment not only pays for itself but also provides a satisfactory return on the funds committed by the firm. There are several variations, of which this is one:

$$\text{percent return} = \frac{\text{earnings per year}}{\text{bid value}} \times 100 \tag{9.12}$$

Earnings are after tax, and deductions for depreciation usually represent some average future expectation. For instance, consider the following example: The bid for new equipment is $175,000, salvage will provide $15,000, and an average earnings of $22,000 after taxes is expected.

$$\text{percent return} = \frac{\$22,000}{160,000} = 13.75\%$$

Now, assume that a private investment opportunity of $25,000 has been brought to your attention. This is broken down to $20,000 for the investment and $5000 for initial working capital (cash, accounts payable, and so forth). Annual operating and other expenses are estimated at $10,000 and income at $15,000 per year. This investment is analyzed as follows:

	Amount
Income	$15,000
Expenses	10,000
Net income	$ 5,000

$$\text{percent return} = \frac{\$5000}{25,000} = 20\%$$

Another variation is expressed as

percent return

$$= \frac{\text{average earnings} - (\text{total investment} \div \text{economic life})}{\text{average investment}} \times 100 \qquad (9.13)$$

The earnings in the formula are the average annual earnings after taxes plus appropriate depreciation charges. The original investment is recovered over the economic life of the proposal by subtracting the factor of total investment divided by economic life from average earnings. This difference denotes the average annual economic profit on the investment.

The average investment is defined as the total investment times 0.5, acknowledging that the life of an investment for tax purposes and its true economic life are not the same.

The original investment is based on the normal physical life or as legally defined by the Internal Revenue Service. Average investment represents the profitable life of the investment, which is frequently a different period. If engineering desires, it may incorporate a risk element by further shortening economic life.

For example, an average after-tax earning of $22,000 is expected from an investment of $175,000 with an economic life of 10 years. Straight-line depreciation is assumed for 12 years and salvage is $15,000.

$$\text{percent return} = \frac{22,000 + (160,000/12) - (175,000/10)}{160,000 \times \frac{1}{2}} = 22.29\%$$

9.5.2. Payback Period Method

The payback method is easy and widely adopted. Essentially, the method determines how many years it takes to return the invested capital return. The formula as normally given is

$$\text{years payback} = \frac{\text{net investment}}{\text{annual after-tax earnings}} \qquad (9.14)$$

The payback method recognizes *liquidity* as the basis for the measure of economic worth of capital expenditures.

Payback separates proposals of doubtful validity from those that call for additional economic analysis. Obviously, the payback signals the immediate cash return aspect of the investment, which may be desirable for corporations where a high-profit investment opportunity and limited cash resources exist. In some situations, the payback is used for those investment situations where the risk does not warrant earnings beyond the payback period.

Let us use an example to illustrate this. The installed cost for new equipment is $175,000, and old equipment will be sold for $15,000. Better productivity of the new equipment will return $40,000. For a composite 55% corporate tax rate earnings amount to $22,000.

$$\text{years payback} = \frac{175,000 - 15,000}{22,000} = 7.3$$

Now, consider two investment opportunities:

	Equipment A, $60,000	Equipment B, $60,000
Revenue		
Year 1	20,000	30,000
Year 2	20,000	30,000
Year 3	20,000	30,000
Year 4	20,000	—
Year 5	20,000	—
Total annual after-tax earnings	$100,000	$90,000
Payback period	3.0 years	2.0 years

In this case, equipment B is preferred over A because of the smaller payback period. If sufficient resources were available, then an engineering fiat could allow any investment that was under an arbitrary level such as 5.

The average annual rate of return is acknowledged to have faults and assumes equal distribution of earnings throughout the economic life of the asset. Even if this were true, there is a significant difference between the value of the dollars earned in the first year and those earned in later years. The time value of money is ignored here. A project yielding savings in early years of its life is more beneficial, because these funds become available for additional investment or for alternative use and often are subject to less risk than savings projected many years ahead. The differences are overlooked in salvage values and their relation to the time element. Nor is interest on borrowed money in any way reflected.

The payback method suffers similarly. The life pattern of earnings is ignored in payback formulas. In the example, equipment B had a shorter payback period than equipment A, yet A will return $10,000 more. New equipment may not be profitable during the early part of the payback period. On the other hand, new equipment may be quite profitable in the future. Payback does not provide for a system of ranking with other investment possibilities. Nor does it take into account depreciation or obsolescence or the earnings beyond the payback period. For example, the payback method does not recognize that one investment with an earning of $10,000 the first year and $2000 the second year is more desirable than another that earns $6000 in each of the 2 years. The situation for which payback is suited, and then only provisionally, is as a rough measure of evaluation.

The *engineering-economic method* of determining return overcomes those shortcomings. This method is applicable to every possible type of a prospective investment and can yield answers that permit valid comparisons between competing projects.

There are many methods of capital investment analysis, and, unfortunately, only a few will be studied in detail here. It is the analysis that leads to a strategy that in turn leads to the decision.

The time value of money is the application of compound interest formulas to the additional cash flow produced by the investment. This concept enables engineering to place a value on the money that will become available for productive use in the future as well as for the money available today. Fundamentally, the time value of money begins with simple interest, or

$$I = Pni \qquad (9.15)$$

where I = interest earned, dollars
P = principal sum, dollars
n = number of compounding periods
i = interest rate, decimal

This formula can be restated as the amount including principal and simple interest that must be repaid eventually, or

$$F = P + I = P(1 + ni) \qquad (9.16)$$

where F is the principal and interest sum collected at some future time. In the payment of simple interest, the interest is paid at the end of each time period or the sum total amount of money is paid after a given length of time. Under the latter condition there is no incentive to pay the interest until the end of the contract time.

If interest is paid at the end of each time unit, then the lender could use the money to earn additional profits. Compound interest considers this point and requires that interest be paid regularly at the end of each interest period. If the payment is not made, then the amount due is added to the principal and interest is charged on this converted principal during the following time unit.

An initial loan of \$10,000 at an annual interest rate of 5% would require payment of \$500 as interest at the end of the first year. If this payment were deferred, then the interest for the second year would be (\$10,000 + \$500)(0.05) = \$525, and the total compound amount due after 2 years would be \$10,000 + \$500 + \$525 = \$11,025.

When interest is permitted to compound, as in the following computation, the interest earned during each interest period is permitted to accumulate with the principal sum at the beginning of the next interest period. This compounding is shown in Table 9.10. The resulting factor, $(1 + i)^n$, is referred to as the single-payment

TABLE 9.10   DERIVATION OF BASIC COMPOUND INTEREST FORMULA

Year	Principal at Start of Period	Interest Earned During Period	Compound Amount F at the End of Period
1	P	Pi	$P + Pi = P(1 + i)$
2	$P(1 + i)$	$P(1 + i)i$	$P(1 + i) + P(1 + i)i = P(1 + i)^2$
3	$P(1 + i)^2$	$P(1 + i)^2 i$	$P(1 + i)^2 + P(1 + i)^2 i = P(1 + i)^3$
	$\cdots$		
n	$P(1 + i)^{n-1}$	$P(1 + i)^{n-1}i$	$P(1 + i)^{n-1} + P(1 + i)^{n-1}i = P(1 + i)^n$

compound-amount factor. The total amount of principal plus compound interest due after n periods is

$$F = P(1 + i)^n \tag{9.17}$$

where F is the future amount, dollars.

The single-payment compound-amount factor may be used to solve for a future sum of money F, the interest rate i, the number of interest periods n, or a present sum of money P when given the other quantities.

Engineering economic methods are preferred because they depend on time-value-of-money concepts. Do not conclude that all methods employing interest computations are useful for all occasions. Some have limited applicability. We present here four distinct variations. When these methods are given correct information and properly understood, their answers are equally valid:

1. Net present worth
2. Net future worth
3. Equivalent annual cost
4. Rate of return

Each of those methods measures a different factor of the investment. They give different evaluations. Nonetheless, they lead to the same recommendation for consistent decision making.

Each method is demonstrated with the same standard problem. Cents are dropped from calculations for ease of understanding.

Year	Cost	Revenue
0	$1025	$ 0
1	0	450
2	0	425
3	0	400

The annual compounding and end-of-year convention are used to simplify understanding. It is assumed that the nonuniform revenue is instantaneously received at the end of year.

9.5.3. Net Present Worth Method

Net present worth is also known as *net present value* or *venture worth*. The present worth of future revenue is compared with initial capital investment, assuming a continuing stream of opportunities for investment at a preassigned interest rate. The procedure compares the magnitude of present worth of all revenues with the investment at the datum time 0.

One way of defining net present worth is as the added amount that will be required at the start of a proposed project by using a preassigned interest rate to produce receipts equal to, and at the same time as, the prospective investment. For a given interest rate of 10%, the net present worth of the previously given problem is computed by discounting all revenues to year 0 at this rate and subtracting the proposed investment, or

Period	$\dfrac{1}{(1 + i)^n}$ Present Worth Factor at 10%		Amount
Year 1 to zero	450×0.9091	=	$ 409
Year 2 to zero	425×0.8264	=	351
Year 3 to zero	400×0.7513	=	301
Total			$1061
Less proposed investment			1025
Net present worth			$36

The $36 is the amount that must be added to the $1025 to set up the amount that would have to be invested at 10% to achieve receipts equal to and at the same time as those predicted for the recommended investment, or

$$(\$1025 + \$36) \times 1.1 = (\$1061) \times 1.1 = \$1167$$

$$\text{less payment} \qquad \underline{450}$$

$$\$717$$

$$717 \times 1.1 = \$789$$

$$\text{less payment} \qquad \underline{425}$$

$$\$364$$

$$\$364 \times 1.1 = 400$$

$$\text{less payment} \qquad \underline{400}$$

$$0$$

9.5.4. Net Future Worth Method

Assets and revenues can be invested at the preassigned interest rate where there is a continuous exposure of investment according to opportunities $[F = P(1 + i)^n]$. A comparison of investment of the original sum plus reinvestment of revenues at the preassigned interest is made against the standard alternative of investing only the original asset value. The calculation results in that added amount obtained at the end of the project economic life if the project's anticipated revenues were invested instead of the proposed investment. A common comparison uses the same stipulated interest rate.

For the sample problem and 10%, the *net future worth* is computed by compounding future revenues to the terminal year and then subtracting from this the amount that would have resulted from the other alternative of investing the original asset at the same preassigned interest rate to the terminal year:

Period	$(1 + i)^n$ Compound Amount Factor at 10%		Amount
Year 1 to 3	450×1.10^2	=	$ 545
Year 2 to 3	425×1.10	=	468
Year 3	400×1.00	=	400
			$1413
Less disbursement compounded to terminal year at 10%			
Year 0 to 3	$1025 \times (1.10)^3$	=	1364
Net future worth			$ 49

The calculations point out that if the project is funded and if the revenues materialize as estimated, then a surplus of $49 will be expected over the simple alternative of investing only the asset of $1025. The same period of time and equal interest rates are parts of this comparison.

9.5.5. Net Equivalent Annual Worth Method

Management often wants a comparison of annual costs instead of, say, present worth of the costs. Here, we refer to net costs; that is, the net difference between any cost or revenues or credits. This method considers a supply of opportunities for investment of both assets and receipts at the predetermined interest rate plus a supply of capital at the same interest rate.

Now, the sample problem does not have uniform annual receipts, and the receipts first must be converted to total present worth and then to annual equivalents. The total present worth at time zero is $1061 (from section 9.5.3). The annual equivalent is found by dividing by the sum of the present worth factors, or

$$\text{annual equivalent} = \frac{1061}{(0.9091 + 0.8264 + 0.7513)} = \frac{1061}{2.4868} = \$427$$

$$\text{net annual equivalent worth} = 427 - \frac{1025}{2.4868} = \$14$$

This $14 is the amount by which the anticipated revenues from the proposed investment, rated at 10% interest, exceed the annual equivalent of the proposed investment:

	Amount
Anticipated equal annual receipts	$ 427
Less equal annual equivalent worth at 10%	14
Equal annual receipts to be generated by investing	$ 413
1025 × 1.10 =	$1128
Less payment	413
	715
715 × 1.10 =	$ 787
Less payment	413
	374
374 × 1.10 =	$ 413
Less payment	413
	0

Starting with the investment that earns interest and subsequently subtracting payments leads to a balance of zero dollars at the end.

9.5.6. Rate-of-Return Method

Rate of return evaluates the rate of interest for discounted values of the net revenues from a project to have those present worths of the discounted values equal to the present value of the investment. *Rate of return* thus solves for an interest rate to bring about this equality. Other titles, such as *ROI* (return on investment), *true rate of return, profitability index,* and *internal rate of return,* also exist. The adjective "true" distinguishes true rate of return from other less valid methods that have been labeled rate of return [i.e., Eqs. (9.1.3) and (9.14)]. For this method there is no assumption of an alternative investment and no predetermined interest rate. We define this interest rate at which a sum of money, equal to that invested in the proposed project, would have to be invested in an annuity fund in order for that fund to be able to make payments equal to, and at the same time as, the receipts from the proposed investment.

The solution for the interest rate is by repeated trials or by graphical or linear interpolation. For the sample problem

Year, n	Revenue		$\dfrac{1}{(1 + i)^n}$ PW Factor at 15%		Discounted Amount
1	$450	×	0.8696	=	$391
2	425	×	0.7561	=	321
3	400	×	0.6575	=	263
Total					$975

Year, n	Revenue		$\dfrac{1}{(1 + i)^n}$ PW Factor at 5%		Discounted Amount
1	$450	×	0.9524	=	$ 429
2	425	×	0.9070	=	385
3	400	×	0.8638	=	346
Total					$1160

The two trial values bound the initial asset value of $1025. Either graphical or interpolating methods can be used to locate the rate of return, or from which

$$\frac{0.05 - i}{1160 - 1025} = \frac{0.05 - 0.15}{1160 - 975}$$

it follows that $i = 12.3\%$. This rate of return is the interest rate at which the original sum of $1025 could be invested to provide returns equal to, and at the same time as, the receipts of the prospective investment:

	Amount
$1025 \times 1.123 = $1151	
Less payment	450
	$ 700
$700 \times 1.123 = $ 787	
Less payment	425
	$ 362
$362 \times 1.123 = $ 400	
Less payment	400
	0

This shows that the earning rate is true and is the actual return of the invested money. It has the important advantage of being directly comparable to the *cost of capital*.

9.5.7. Comparison of Methods

A summary of the four different methods follows:

	Amount
Net present worth at 10%	$36
Net future worth at 10%	$49
Net annual equivalent worth at 10%	$14
Rate of return	12.3%

We can demonstrate that the first three are commensurable answers. For instance, the net present worth of $36 can be compounded to the terminal year by using Eq. (9.17) or

$$F = 36(1.10)^3 = \$49$$

The preassigned interest rate, 10%, gives the net future worth or $49. The net annual equivalent worth of $14 can be uncovered by dividing the net present worth value of $36 by a sum of present worth values, of 10% or

$$\frac{36}{2.4867} = \$14$$

On the other hand, the rate of return cannot be calculated from the foregoing answers and can be found directly from the data.

Equivalency among the first three can be found for any arbitrary interest rate, and if those are equivalent to the rate of return, then it is equivalent at only one preassigned interest rate.

The first three methods are only differences based on the choice of an interest rate. This arbitrary selection of an interest rate makes the three methods little more than decision tools for comparing projects.

The rate of return is the only method that can provide a consistent measure of the extent of the economic productivity of prospective investments. The answer can be compared directly with the cost of capital.

However, no one single method or criterion of profitability analysis is preferred for all situations.

9.6 EXAMPLE

A 500-kV transmission line is to be estimated. An owner provides an RFQ to the project engineer stating that a double circuit electrical transmission line will be about 5 miles in length from a power station bus wall to the Creston substation. "Double circuit" implies two parallel transmission conductors and hardware that are constructed side by side. The aluminum conductor is 3 × 1.6 in. O.D. per circuit. Performance is 0.28 voltage drop per mile at 525 kV. Figure 9.16 describes the route and tower. Figure 9.4 has already provided the WBS. The second, third, and fourth levels of the WBS need to be determined for a *pro forma* bid. Those higher-level estimates are preliminary, and, if lower levels are evaluated, then we are dealing with a detail or takeoff bid.

The RFQ will give the design, schedule, location, and specifications and will indicate that the estimate is preliminary and discuss the purpose, which may be for study, budget, or bid. Because the transmission line is a *single end item* and will require significant money, a project estimate is clearly needed.

The project engineer first establishes the definition and WBS. Organizational policy may require that other company units assist the project engineer in this work.

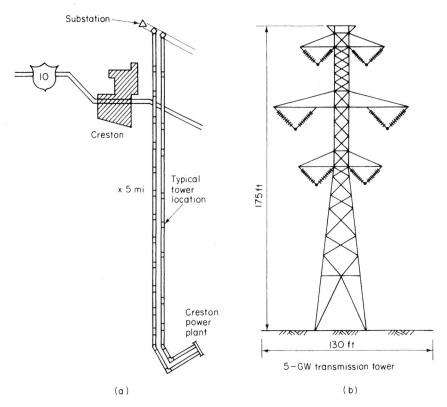

Figure 9.16. Preliminary design of Creston 5 GW transmission line: (a) double circuit transmission; (b) transmission line tower.

Complicated projects require a team. There is no single approach, because many engineering styles can be successful.

We show several methods for making estimates. Those methods have been introduced previously. Table 9.11 is a summary of the data, methods, and references. Figure 9.17 is a summary of curves useful for the transmission line. Although the problem is a simplification of real data, it demonstrates the ideas necessary for project estimating.

Note Table 9.11: Entry is WBS and type of estimate. This summary is instructional only. Estimating data, curves, and equations can fill many notebooks. Similar information may also be obtained from nationally published data manuals. The application of the data depends on design parameters.

Note WBS 111, land survey, which uses a unit estimate. Because the average cost is $17,315 per mile, and the distance from the power plant to substation is 5 miles, $86,575 is entered in the facilities estimate, Table 9.3.

Road clearing, WBS 1311, is a comparison and opinion estimate, and a value of $120,000 (= 4000 x 30) is posted on the labor estimate, Table 9.1. Road clearing

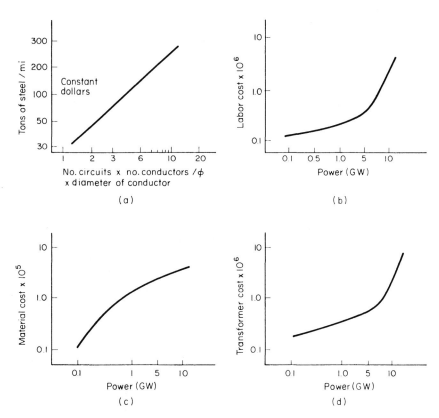

Figure 9.17. Typical curves for high-voltage transmission line project: (a) WBS 1313; (b) WBS 1321; (c) WBS 1321; (d) WBS 1322.

does not require materials beyond that of overhead supplies, and the road equipment is covered by the job overhead. This labor is the contractor's work force. Construction equipment necessary for clearing is covered by job overhead. Construction equipment is different from equipment installed in the substation.

WBS 1312, Roads, has two entries in Table 9.11. The first deals with direct labor, and the second provides for the material base, asphalt, and so on, for the road. Note that those two results, $280,000 and 15,000, are entered in two different estimates, Tables 9.1 and 9.2.

WBS 1313, Erection of Tower Structure, shows a labor estimate by using the average learning theory. Parameters are listed by Table 9.11. For 26 towers, the entry of 55,743 hours (also $2,229,720 = 55,743 × 40) is posted on the labor estimate (Table 9.1).

Direct material, WBS 1313, is found by using Figure 9.17(a). The cost driver for this curve is "no. circuits × no. conductors/circuit × diameter" or 9.6 (= 2 × 3 × 1.6), and the curve value of 225 tons of steel per mile is read. When multiplied by the unit cost of material and welded fabrication for tower manufacture, we have

TABLE 9.11 SUMMARY OF ESTIMATING DATA FOR HIGH-VOLTAGE TRANSMISSION PROJECT

WBS	Type of Estimate	Method of Estimating	Data, Curve, Equation	Reference
111	Facilities	Unit	$C_a = \$17,315/\text{mi}$	Eq. (6.3)
112	Facilities	Unit	$C_a = \$1285/\text{mi}$	Eq. (6.3)
12	Engineering	Linear regression	$a = 25,000$; $b = 1700/\text{mi}$; rate $= \$44.67$ overhead $= 25\%$	Eq. (5.3)
1311	Labor	Comparison, opinion	4000 hours; \$30/hr	Secs. 6.2, 6.3
1312	Labor	Comparison, opinion	8000 hours; \$35/hr	Secs. 6.2, 6.3
1312	Direct material	Opinion	\$15,000	Sec. 6.2
1313	Labor	Learning	$K = 3,518$; $\phi = 90\%$; average line linear, $N = 26$	Sec. 6.6.1
1313	Direct material	Curve	See Fig. 9.17(a); $I_c = 1.15$; cost/lb $= \$0.75$	Eq. (5.53) Eq. (6.3)
1321	Labor	Curve	See Fig. 9.17(b)	
1321	Direct material	Curve	See Fig. 9.17(c)	
1321	Facilities	Historical quote, comparison	\$50,000	
1322	Subcontract	Power law and sizing model	$Q_r = 1.2$ GW; $C_r = \$200,000$; $m = 0.9$; $I_r = 1$; $I_c = 1.18$	Eq. (6.18)
1322	Equipment	Curve	See Fig. 9.17(d)	
133	Subcontract	Cost estimating relationship	$22,800x_1^{2.38898}x_2^{1.15675}$; $x_1 =$ diameter of conductor, $x_2 =$ number of conductors per phase	Eq. (6.22)

$1,940,625 \ (= 225 \times 5 \times 2000 \times 0.75 \times 1.15)$ for the direct material estimate (Table 9.2). Since Fig. 9.17(a) is for constant dollars of a bench-mark year, the value is multiplied by an index 1.5 to inflate the material cost to the period of out-of-pocket cash flow.

WBS 1321, Substation Building, involves a labor, direct material, and facilities estimate. Figures 9.17(b) and 9.17(c) provide \$850,000 and \$180,000 and are entered for Tables 9.1 and 9.2 for a value of 5 GW. A minor facilities value of \$50,000 is judged on the evidence of a past quote and comparison for Table 9.3.

Together with other information, the four estimates, Tables 9.1–9.4, are concluded. With various overhead rates, contingency, construction interest, and a profit markup, a bid value of \$9,098,653 is found in Table 9.9.

This value of \$9,098,653 is the investment charge for a utility. The life of this transmission line is 50 years, which is approximately perpetuity because of the nature of the time-value-of-money tables. It is possible that this investment will be subjected to an engineering economy analysis to determine its profitability.

SUMMARY

The project engineer has the job of forecasting capital investment and operating expenses. This aids the engineering process in choosing and evaluating one-of-a-kind end items. There are several kinds of estimates for projects. Once those estimates are concluded, a bid and pricing analysis is conducted. Eventually, alternative designs and their estimates are analyzed by using engineering-economy methods. Accurate estimates are crucial to any type of time-value-of-money computation.

QUESTIONS

9.1. Define the following terms:

RFP	*Pro forma* document
Fixed capital	Job overhead
Working capital	Contingency
Standard commercial items	Optimistic cost
Facilities cost	Retainage
Work package	Baseline
WBS levels	Markup
Commitment	Time value of money
Subcontract estimate	Payback
Cost spreading	Equivalent annual cost
Bid summary	Venture worth

9.2. What is the purpose of a project estimate? Why are project estimates important?

9.3. What kinds of information are necessary to undertake a project estimate?

9.4. Illustrate fixed capital, working capital, and direct and indirect costs.

9.5. Distinguish between direct material and subcontract in a project bid.

9.6. Give the purpose of a work package. What is the definition part?

9.7. How do commitment and expenditure differ? What causes this difference?

9.8. Describe and construct a new labor estimating form.

9.9. Why is the bid document important for project designs?

9.10. Why is overhead separated into office and job?

9.11. Is there a difference between the interest calculated for project funding and the interest included in time-value-of-money concepts?

9.12. Describe two pseudo-return methods and indicate why they are faulty. What advantages do they serve?

9.13. What is meant by "equivalent or equality" with methods of net present worth, net future worth, and net annual equivalent worth?

9.14. What criteria do you suggest for making project decisions? Will a shortage of capital influence your decision? Do you believe that a successful company is cash poor and bank mortgage rich?

9.15. Suppose that at a certain date an organization had a listing of engineered projects, each showing the amount of capital required and the rate of return and that those can be plotted as shown in Fig. Q9.15 Given a limited amount of equity capital, the organization asks the bank to cover these capital ventures, and you show them your curve. Now the bank, knowing your financial credit, offers its own curve with increasing interest because of their risk as more capital is loaned. What is the favorable aspect about this approach to the firm? Unfavorable? What does this method say about the marginal interest to be earned? At what point should you stop borrowing?

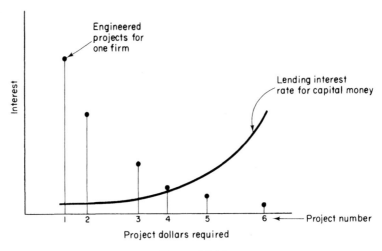

Figure Q9.15

PROBLEMS

9.1. A task requires 17,000 allowed hours. Location of the work is above the Arctic Circle, and the productivity factor is estimated as 45%. If the gross hourly cost is $45, then what is the estimated cost? If the work is in Texas, hourly cost is $25, and a productivity factor of 85% is used, then what is the cost of direct labor?

9.2. Five similar bridges are to be constructed under the direction of one field office. Facility costs for the field office are $250,000. Overhead on the basis of direct labor and material cost is 45%. Additional data for each bridge are as follows:

| | Hourly Rate | | Estimated | Estimated | Professional |
Task	Wage	Equipment	Time (months)	Material	Estimate
Masonry and concrete	$25	$40	15	$30,000	
Forms, scaffolding	20	30	6	20,000	
Asphalt	20	80	3	15,000	
Grading	15	40	2		
Surveying	30	10	3		
Engineering					$50,000

Each month is a 173.33-hour period. Develop a simple project cost estimate by using items in Fig. 9.1. Find the project cost and the cost for one bridge.

9.3. The flow chart for the manufacture of printed polyethylene food bags is given in Fig. P9.3. Costs for this process are

	Costs
Polyethylene	$3/kg
Color chips	$5/kg
Additives	$4/kg
Ink	$9/liter
Waste removal	$1.50/kg

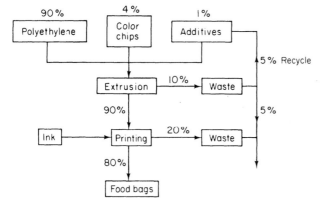

Figure P9.3

Find the cost of 1 million good bags if 1 kg = 500 bags, and 1 liter of ink will print 10,000 bags. (*Hint:* Work from good bags upward to requirements. Assume that ink has a 20% loss.)

9.4. A flow chart for a hydrocracker is proposed. A simplified version is illustrated in Figure P9.4. There are a large number of physical materials, such as gas-oil input, catalyst, hydrogen, fuels, and so on.

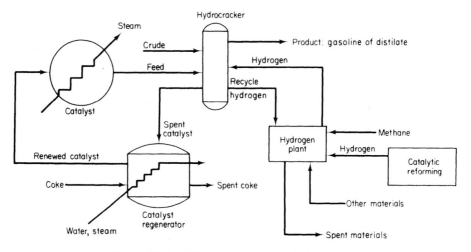

Figure P9.4. Schematic model for hydrocracker.

(a) Develop a listing of the materials.

(b) Tabulate a qualitative material balance.

(c) What input and output material costs do you identify?

9.5. From an experience of yours, provide a work breakdown structure showing and describing the principles used in its development. Decompose a project to the fifth level, where one block in each level is further reduced to a lower level. Indicate the numbering and title for each block.

9.6. A satellite repeater link has the following definition:

WBS	Definition Title	WBS	Definition Title
1	Satellite repeater	15	Equipment
11	Management	16	Training
12	Engineering	151	Preamplifier
13	Documentation	152	Mixer
14	Integration	153	IF, amplifier, filter
		1511	Other hardware

Construct a graphical WBS. How many levels are used?

9.7. A project is defined as follows:

WBS	Cost	Starting Milestone	Ending Milestone
1		0	11
11	$165	6	11
12	225	0	8
13	140	7	10
14	160	6	9
121	55	3	8
122	95	0	6
123	75	3	7
1221	15	3	6
1222	25	2	5
1223	15	2	3
1224	25	1	4
1225	15	0	3

Prepare a graphical work breakdown structure. Construct a graphical schedule of work. Plot the cumulative baseline costs. What is the total cost?

9.8. (a) Prepare a cash flow plan for $6.5 million, matching dollars commitment to expenditure.

	Period					
	0	1	2	3	4	5
Percent commitment	10	18	36	28	8	0
Percent expenditure	0	8	25	39	21	7

(b) A project costs $200,000 and the isosceles triangles in Fig. 9.9 are an approximation for the cash flow plan. If the periods now represent quarters, then prepare a cash flow plan matching dollars commitment to dollars expenditure. When is the approximate midpoint of the commitment and expenditure triangles? What is the difference in money at this time?

9.9. A project definition for conversion of a plant, and the design, manufacture, and assembly of a kerosene heater product line, is given.

WBS	Definition Title	Start Period	Ending Period	Cost Estimate (x 10^3)
1	Kerosene project	0	10	
11	Design	0	2	$1,000
12	Model assembly	2	8	750
13	Plant conversion	1	3	8,000
14	Reliability	4	10	1,600
121	Body assembly	2	6	400
122	Tank assembly	2	6	200
123	Burner assembly	2	5	150
1211	Lower body	2	5	300
1212	Upper body	3	5	100
1221	Upper tank	2	4	150
1222	Lower tank	3	6	50
1231	Glass cylinder	2	3	100
1232	Net	3	4	50

(a) Prepare a graphical work breakdown structure.

(b) Construct a schedule of work similar to a bar chart.

(c) Plot the cumulative baseline costs. Determine project cost.

9.10. A contractor has the following historical costs, expressed in 10^6, for the last 4 years:

Year	C_p	C_o	C_j
1	$240	$12	$7
2	320	14	5.6
3	280	13	8
4	300	15	7.6

(a) Calculate office and job overhead rates, and determine the rates for year 5.

(b) If in year 5 a project is estimated to have a direct labor and material cost of $150 million, then calculate the overhead amount.

9.11. Historical costs for one project have provided the following data:

Equipment: $600,000

Direct and subcontract materials: $10,000,000

Direct labor:
    100 employees at $25 per hour for 525 hours each
    500 employees at $20 per hour for 875 hours each
    200 employees at $15 per hour for 1050 hours each

General expenses include office rent, $250,000; insurance, $160,000; furniture and supplies, $150,000; telephone and computers, $50,000; and salaries, $500,000. Project expenses include permits, $25,000; superintendence, $100,000; storage and protection, $75,000, and other project expenses, $400,000. Find the office and job overhead rate. (*Hint:* consider equipment as an overhead cost.)

9.12. Find direct cost and job overhead rate and amounts from the following information: direct materials, $200,000; labor hours for project = 2000 hours and labor wage without fringes and mandatory contributions = $20 per hour; fringes and mandatory contributions = 40%; office overhead rate = 50% on basis of full direct labor and materials; permits and fees = $2500, bonds = $5000, job office expenses are salaries = $18,000, supplies = $1500, and telephone = $1200; insurance = $15,000; utilities = $8000; surveys = $28,000. What is the full cost of the job?

9.13. A developmental project has an expected cost of 25×10^6 and a variance of 9×10^{12} dollars^2.
  (a) Plot a risk graph of contingency.
  (b) Find the upper limit cost for 25% overrun; for 25% underrun.

9.14. **(a)** A preliminary estimate is considered uncertain. Engineering directs a contingency plan as insurance to the conceptual estimate. Mean cost and standard deviation are $100,000 and $50,000. Construct a risk graph of contingency for the current time.
  (b) Project life is 4 years, and money flow is scheduled as 25, 50, 15, and 10%. Escalation indexes for those years as related to bench-mark year are 1.000, 1.015, 1.050, and 1.100. Find the total contingency cash flow amount necessary to add to the project estimate where the contingent maximum cost is selected at 25% risk.

9.15. An owner requests the contractor create a bias contingency in his estimate; that is, artificially calculating a high estimate to avoid asking for additional funding from a bank and thus giving the impression of a project in control. Assume an expected cost of $100,000 and a standard bias contingency of 10%. Project cash flow and indexes are identical to problem 9.14. Construct a risk graph for the standard deviation of $50,000. Assume a contingency for a current period as well as inflated periods. Use a 25% overrun risk level. Discuss the pros and cons of this manipulation.

9.16. A construction estimate is $200,000 and has a contract period of 5 months. The cash flow is trapezoidal, and $t_i = 40\%$ and $t_f = 80\%$.
  (a) Draw the cash flow trapezoid similar to Fig. 9.13(c). Find average and maximum cash flow and when they occur.
  (b) Earned profit is 10%, and there is a 1-month delay for payment by the owner. Retainage is 20%. Construct a table similar to Table 9.7.

9.17. A construction bid is $8 million and has a contract period of 5 months. Cash flow is trapezoidal, and $t_i = 20\%$ and $t_f = 60\%$.

    (a) Draw the cash flow trapezoid. Find average and maximum cash flow and when they occur.

    (b) Profit earned by the contractor is 20%. Retainage is 10%, and there is a 1-month delay in payment by the owner. Find the project cash flow table.

9.18. A contract of $150 million is estimated. A trapezoid model of cash flow is assumed with $t_i = 30\%$ and $t_f = 80\%$. Project life is 15 months. Other assumptions are as follows: profit $= 10\%$, owner retainage $= 5\%$, and money as paid from owner to contractor is 1 month late.

    (a) Find the net surplus or deficit for each period from the contractor's viewpoint.

    (b) Plot the cumulative baseline dollars against the months.

    (c) Plot the cumulative baseline percentage against the cumulative contract percent. When does the cumulative baseline percentage equal the elapsed contract percentage?

9.19. Schedule and geometrically model the labor for problem 9.16. The general contractor's markup for overhead and profit is 20% of cost, and he subcontracts all work. Subcontractor's markup for overhead and profit is 25% of cost. Labor accounts for 60% of cost and is paid $20 per hour.

9.20. A construction contract has an overall value of $8 million and a contract period of 5 months. Cash flow is trapezoidal, where $t_i = 20\%$ and $t_f = 60\%$. The general contractor's markup for overhead and profit is 10% of cost and he subcontracts all work. The average subcontractor's markup for overhead and profit is 20% of cost. Labor accounts for 50% of cost and is paid $30 per hour. Plot manpower requirements for the project and as a percent of average work force. For how long does the maximum work force last as a percentage?

9.21. A project that costs $150 million total is scheduled to be constructed in 15 periods. The major contractor markup on labor for overhead and profit is 20% of cost. All work is considered subcontracted. The subcontractor's markup for overhead and profit is 25% of labor cost. Labor accounts for $50 million and is paid $40 per hour (includes wage and fringe costs). Maximum on-the-job manpower first occurs after 30% of the total manpower requirement has been paid. The period of maximum labor occurs for 50% of the project duration. Consider only the $50 million.

    (a) Plot the percent of average work force on project versus percent elapse time of project.

    (b) Plot people on project versus project time, periods.

    (c) Provide an ogive curve for labor cost versus periods.

9.22. Let the interest rate be 1.5% per month for Table 9.8. Determine the interest and net profit.

9.23. Calculate the interest and net profit amount where interest is 1% per period: (a) for problem 9.16 and (b) for problem 9.18.

9.24. Calculate the interest amount, where interest is 2% per month for the net cash flow given in millions of dollars.

Month	1	2	3	4	5	6
Net cash flow	(1.92)	(2.10)	(1.83)	(1.05)	(0.64)	1.60

(*Hint:* The net cash flow is cumulative. A bank loan of $2,100,000 − $1,920,000 = $180,000 is the amount necessary for month 2.)

9.25. **(a)** A new capital equipment will cost $225,000, salvage is $25,000, and an average earning of $20,000 after taxes is expected. Find the average annual rate of return. If earnings are doubled, then what is the rate?

    (b) Consider part (a). The investment has an economic life of 10 years. Straight-line depreciation for 8 years is used. Find the percent return.

9.26. **(a)** Investment for new equipment is $100,000, and salvage will be $10,000 eight years hence. Earning from this equipment will be $15,000 on the average after taxes. What is the nontime value of money return? What is the payback?

    (b) Two investment opportunities are proposed:

	Process A	Process B
Total investment	$60,000	$60,000
Revenue (after tax)		
Year 1	20,000	30,000
Year 2	20,000	30,000
Year 3	20,000	30,000
Year 4	20,000	
Year 5	20,000	

For an interest of 10%, which has the least period of time before capital recovery is complete?

9.27. Two bids are compared for performing the same design:

	Bid A	Bid B
Total bid	$100,000	$150,000
Revenue (after tax)		
Year 1	25,000	25,000
Year 2	30,000	40,000
Year 3	35,000	55,000
Year 4	30,000	40,000

    (a) Determine the payback.

    (b) For an interest of 5%, which bid has the least period of time before the bid value is returned?

9.28. **(a)** What is the principal amount of interest at the end of 2 years on $450 for a simple interest rate of 10% per year?

    (b) If $1600 earns $48 in 9 months, what is the nominal annual rate of interest?

    (c) An investment of $50,000 is proposed at an interest rate of 8%. What is the future amount in 10 years?

    (d) What is the present worth of $1000 for 6 years hence if money is compounded to 10% annually?

    (e) What is the compound amount of $3000 for 15 years with interest at 7.25%?

    (f) Find the annual equivalent value of $1050 for the next 3 years with an interest rate of 10%.

    (g) How many years will it take for an investment to triple itself if interest is 5%.

    (h) An interest amount of $500 is earned from an investment of $7500. What is the interest rate for 1 year? For 2 years?

9.29. (a) Find the principal if interest at the end of 2.5 years is $450 for a simple interest rate of 15% per year.

(b) A loan of $5000 earns $750 interest in 1.5 years. Find the nominal annual rate of interest.

(c) In 1626 Native Americans bartered Manhattan Island for $24 worth of trade goods. Had they been able to deposit $24 into a savings account paying 6% interest per year, how much would they have in 2000? At 7% interest?

(d) What payment is now acceptable in place of future payments of $1000 at the end of 5, 10, and 15 years if interest is 5%?

(e) What is the compound amount of $500 for 25 years with interest at 15%?

(f) Calculate the annual equivalent value of $1000 for the next 4 years with interest rate at 5%.

(g) How many years will it take for an investment to double itself if interest is 10%?

(h) An investment of $10,000 earns an interest amount of $750. Find the rate if the amount is earned over 1, 2, or 3 years.

9.30. (a) Reconsider the cash flow of cost and revenues provided on page 372. Provide the net present worth, net future worth, and net annual equivalent worth if interest is 11%.

(b) Repeat for an interest of 9%.

9.31. (a) Reconsider the cash flow of bid A in problem 9.27. Find the net present worth, net future worth, and net annual equivalent worth if interest is 4%. Find the rate of return. Present a summary of the four methods.

(b) Reconsider the cash flow of bid B in Problem 9.27. Repeat and present a summary of the four methods. Which bid is preferred, A or B?

9.32. Reconsider the cash flows of equipment A and B given on page 370. Present a summary of the four methods. Assume an interest of 20%. Is A or B preferred?

9.33. A project estimate has the following cost and revenue cash flows. The cash flows are end-of-period.

Year	Cost	Revenue
0	$800	$ 0
1		450
2		425
3		400

(a) If interest is 10%, find the net present worth, net future worth, and net annual equivalent worth.

(b) Find the rate of return.

(c) Present a summary of the four methods.

9.34. A prospective venture is described by the following receipts and disbursements:

Year End	Receipts	Costs
0	$ 0	$800
1	200	0
2	1,000	200
3	600	100

For $i = 15\%$, describe the desirability on the basis of present worth.

9.35. **(a)** Evaluate the cash flow diagram, Fig. P9.35. Arrows pointing down represent end-of-the-year costs and upward arrows are revenues. Determine a present sum, an equivalent annual payment, and a future sum. Use $i = 25\%$.

   (b) Repeat with $i = 10\%$.

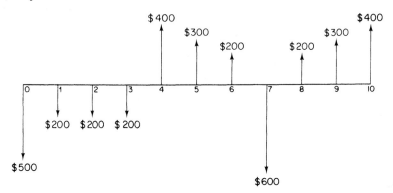

Figure P9.35

9.36. **(a)** Evaluate the cash flow diagram, Fig. P9.36. Downard-pointing arrows are costs and upward arrows are revenues. Find present value, annual value, and a future value. Use $i = 15\%$.

   (b) Repeat with $i = 10\%$.

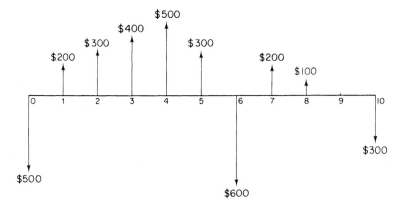

Figure P9.36

9.37. By using Tables 9.1–9.4, construct the cumulative baseline cost ogive curve plotted against period. In what period will roughly 50% of the estimated costs be spent?

9.38. Assume that a transmission line is to be estimated. Apply the data by using Table 9.11 and the following assumptions: length of transmission, 8 miles; 525 kV, single circuit, 1.5 GW capacity, 3 × 1.3 in O.D. conductor per phase, 30 towers, and other data the same. Determine the labor, direct material and subcontract, facilities and equipment, and engineering estimates. Conclude by finding the bid value similar to Table 9.9.

9.39. Estimate the following data for 735-kV double-circuit transmission line. The transmission line length is 25 miles. The conductors are 4 × 1.38 in. O.D. per phase. Capacity is 5 GW, and there are 90 transmission towers. Other data of Table 9.11 remain the same. Find the four project estimates and the bid similar to Table 9.9.

MORE DIFFICULT PROBLEMS

9.40. You, as a contractor, find that monthly estimates for payment increase uniformly for the first quarter, decrease uniformly for the last quarter, and are level for the middle half of the job. The job will require a $2,400,000 contract lasting 12 months. The first monthly estimate will be paid 2 months after the start of the job, less 10% retainage, and monthly thereafter. It is necessary to pay all expenses promptly. Profit is figured at 15% of direct and indirect costs. Prepare a monthly cash flow schedule of the project. Determine timing and amount of debt. When does the percentage of work in place equal the percentage of elapsed contract time?

9.41. Consider a project where materials are divided into four categories: (1) facilities and equipment, (2) long-lead-time materials, (3) standard commercial materials, and (4)

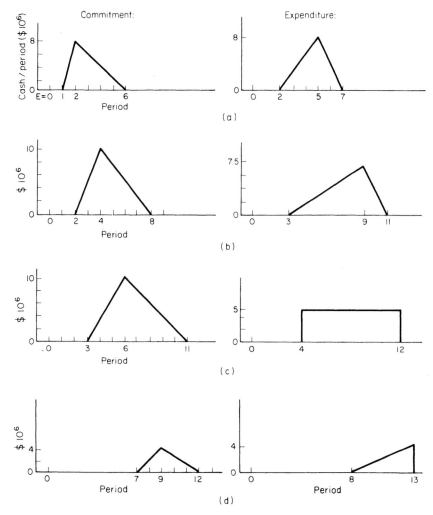

Figure P9.41. Commitment and expenditures for (a) facilities and equipment; (b) long-lead-time materials; (c) standard commercial materials; (d) raw materials and stock items.

raw materials and stock items. The engineer is required to set levels of commitment. Ultimately, subcontractors and vendors supply the materials to the contractor and demand their money. The commitment and expenditure cash value per period can be geometrically modeled against period number. $E = 0$ is the point of the estimate, and 16 is the period at which manufacture will be started. Assume that cash flows are instantaneous because the project is large and there are many suppliers, so a continuous cash flow model is considered reasonable. The four categories, graphical value per period and period numbers, are given as Fig. P9.41.

(a) Plot the total commitment dollar ogive beginning at period 1 and concluding at period 12. Describe the method.

(b) On the same curve, plot the total expenditure ogive beginning period 2 and ending period 13. Show the method.

(c) Graphically or analytically determine the effective retainage for the major contractor for periods, 2, 7, and 13. In those curves, title the commitment and expenditure axes as budgeted cost material committed and budgeted cost material expended. To the contractor all material costs are money-out-of-pocket.

(d) Using the same curves, redefine the range from 1 to 100% for both the x and y axes.

CASE STUDY:
URANIUM ORE PROCESSING

Chamberlin Mining Ltd. has discovered a rich ore deposit in Saskatchewan, Canada. Prospector Chamberlin hires consultant Ralph Light to provide him with preliminary estimate facts and flowsheet data.

Light reports that three process methods are used: acid-leach countercurrent decantation solvent type, acid-leach resin, and an alkaline leach. Operating costs are affected greatly by the grade of ore, mill capacity, and reagent consumption for a particular operation. Although sulfuric acid costs are decreasing, labor costs are not, and the general trend is upward. Construction costs are increasing with new plants requiring pollution controls.

Because Chamberlin is unwilling to disclose his proprietary facts such as profits above costs, transportation, depreciation rates, overhead rates, and assets, Light is forced to submit only preliminary data for his client:

Type of Plant	Plant Capacity (tons of ore feed per day)	Cost Installed ($ per ton)	Operation Cost ($ per ton)	Uranium Purity
Acid-leach CCD	500	$110,000	$57.00	0.90
	1000	75,000	45.50	0.93
	2000	55,000	38.00	0.95
Acid-leach resin	500	90,000	53.50	0.88
	1000	60,000	46.00	0.90
	2000	42,500	38.00	0.94
Alkaline leach	500	120,000	61.00	0.93
	1000	80,000	49.50	0.95
	2000	60,000	41.50	0.96

Now the prospector determines that a plant runs 300 days per year and figures a model that incorporates the full cost of capital and operation will be good for a starter. On the basis of those preliminary estimating data, what will be the direct operating dollars per recovered pound of U_3O_8 (uranium) if the ore grade of feed is 0.20% U_3O_8? What conclusion as to capacity and flow sheet do you recommend on this basis?

From a separate exploration report Chamberlin knows that the ore body is limited to 10–12×10^6 tons before primary grades are exhausted. New processing equipment will then be required. What is the capital investment cost for the nine plants? Ignoring depreciation, which type of plant minimizes total capital and operating costs over the life of the ore? Which flowsheet should be analyzed by detail estimating methods?

10

System Estimating

In this chapter we outline methods for applying system estimating to system designs. Unfortunately, a procedure cannot be set out beforehand in some recipe fashion. If that were possible, and after we have learned those procedures and by strict adherence to them, then we would be guaranteed absolute confidence of result. The preparation of each estimate is unique, and, to a large extent, the methods are adapted to the design. Each system estimate is a result of the skill, experience, and resourcefulness of the engineer. Nevertheless, there is a methodology and a set of principles that aid the construction of system estimates.

We shall show that the measure of system estimating is *effectiveness*. Previously we considered cost, price, and bid as the appropriate measure for operation, product, and project types of designs. To make a cost effectiveness evaluation, it is necessary to contrive and then construct an information flow structure. Given the information flow structure and the corresponding model, it becomes possible to use various analytical aids to arrive at system effectiveness. Finally, several small-scale system estimates are provided.

Concepts inherent in system estimating have been applied successfully to a broad listing of problems, such as water resources, health systems, social welfare programs, hospital planning, space systems, community relations interaction, industrial growth, weapons development, and production. The techniques have been applied at levels ranging from preliminary to detail.

What is meant by the word *system?* Three classifications are normally suggested: executive, operational, and physical. For *executive* we have in mind procedures and organizational forms that connect organized effort. In *operational systems* the central thought is about responsibilities, authority, and aspects of functional relationships as they deal with minor or major problems. *Physical* concerns engineering and the social system.

Because this is a book about cost analysis, our system deals with a *configuration* of operations, products, and projects. Thus, scrutiny of those divisions is possible by the means of the previous three chapters. A system design will require operations, products, and projects, and those estimates become a part of the larger system estimate.

10.1 EFFECTIVENESS

In system estimating the word "effectiveness" is the focus of a great deal of attention. *Effectiveness* is something similar to efficiency. In the traditional sense, efficiency is nondimensional, and as a quantity it approaches one. In engineering the object is to have the highest efficiency as possible. But effectiveness means more than efficiency. For one thing, it may have dimension.

Efficiency reports usually express performance in terms of a percentage of a predetermined norm or standard. If the daily standard for a production operation is 100 units and actual production is 97 units, then efficiency for that day's operation is 97%. If the standard is properly determined, the efficiency indicates this operation close to expectation, but there is room for improvement. Even though the efficiency may be close to 100%, it does not necessarily prove that the operation is effective.

Efficiency describes the ability to do things correctly. Effectiveness is the ability to do the correct thing efficiently.

The finding of system effectiveness measures is not easy. Although many effectiveness measures are possible, only a few ever serve a practical purpose.

For any meaningful evaluation, analysis cannot be conducted on lofty levels because of the lack of detail inherent in the general terms. Consequently, more specific measures are desirable.

To choose between two energy systems by listing the things we value, such as satisfaction, dependability, and lack of pollution, is not very helpful. However, this does not indicate that we should not form such a list early in the analysis.

Effectiveness measures such as cost, heat loss, waste heat recovery cost, and equal marginal fuel cost are more specific in meaning. With narrow measures of effectiveness, there is a greater chance of finding information to allow analysis.

While we could associate system effectiveness with a narrow meaning, this would restrict the actual character of the concept. There is no one best measure of effectiveness. With the present state of knowledge that obscures perfection, we settle for something less. For cost analysis, that something less implies a dimension in monetary units such as dollars. This is our basic dimension of effectiveness.

There are ways to overcome the philosophical problem of effectiveness: For instance, make effectiveness more specific. A major design can be reduced to elemental designs. For each of the elemental designs, assume that a special effectiveness measure can be constructed that permits individual attention. A firm or governmental agency cannot have one system engineer examine all the designs simultaneously and

pick each course of action in light of all the decisions. The magnitude of the task requires that the design be broken into its elements. Thus, high-level engineers should
make broader policy choices while other engineers are delegated to lower levels.

This piecemeal analysis makes it possible for more attention to detail. However, dangers are inherent in piecemeal analysis, because lower-level effectiveness
may be unrelated to the higher-level designs. If the chosen effectiveness provides
only approximate results for smaller suboptimal designs, then a hierarchy of crude
effectiveness measures would be considered simultaneously and potential inconsistencies become abundant.

10.1.1. Cultivation of Alternatives

A system engineer is expected to consider a broad range of alternatives in a study
and is encouraged to visualize most alternatives in the realm of practicality (denying
acts such as perpetual energy, for instance). Although ideal opportunities are appealing, they are really not worth the effort. The engineer rules out other less re-
warding possibilities due to the shortage of time, effort, and money. The remainder
are reasonably exhaustive. For some situations it is necessary to initially describe a
list of possible alternatives that is at first quite large. Preliminary estimating methods
abridge the list to something more valuable. The converse is also true. As the analysis continues, opportunities may present themselves that were not known at the start.

Selection of appropriate alternatives for further study is guided by several
rules. For instance, the scope of the system to be compared in conjunction with the
selection of the effectiveness tends to deny alternatives of doubtful value. When
called on to narrow the selection of alternatives, the engineer will logically state restrictions that had only been verbalized. This prevents certain alternatives from
creeping into the evaluation process. Naturally, a good rule is familiarity with overall system objectives.

10.1.2. Intangibles

An engineer may be unable to commit some aspect of a design into ordinary monetary units. Function, beauty, safety, quality of life, and ease are difficult to evaluate
and are intangible or irreducible. Although special and nonfundamental units for a
scale of measurement or ranking might be forced on the intangibles, the engineer
usually believes that it is not worth the effort and that it is truly inscrutable in terms
of ordinary units.

If intangibles are not treated in monetary units, then there are other ways to
consider them. If the stakes are not high, then it may be convenient to ignore them.
Though "intangibles" may be advantageous to some, there may be a contrasting dis-
advantage to others. A mere listing of the intangibles, both pro and con, may be a
sufficient examination. Despite the ability to render, although superficially, various
intangibles to a deterministic scale, the preponderance of practice chooses to accept
most intangibles as closed to numerical estimating. Thus, it leaves the engineer in
the position of considering what is tractable in dollars and what is not.

10.1.3. Long-Term Uncertainty

The cost estimates in chapters 7, 8, and 9 can be called average or expected outcomes. When viewed as a random variable, those cost estimates may be off their mark. For this present discussion there are two types of uncertainty. First, the uncertainty may be about the state of the world in the future. This is called long-term uncertainty and includes factors such as technology uncertainty or strategic uncertainty by competitors. Second, there is statistical uncertainty, which is different from long-term uncertainty and results from purely chance elements. Statistical uncertainties exist even if long-term uncertainties are zero. Statistical uncertainties are usually less troublesome to handle in system studies. Attention to detail, meticulous care, and other techniques deal with statistical fluctuations. Discussion about random statistical variation is found in chapters 5 and 6.

Probabilistic techniques and sensitivity analysis are tools for long-term uncertainties. Analysis to recognize those long-term uncertainties can be involved or simple. If the design warrants attention for this reason, then the analyst may choose to provide a low, modal, and high value for critical input variables. This range estimating would determine how sensitive the results are to variations in certain information. The number of combinatorial variations rises at remarkable rate for a statistical evaluation of this kind. Policies dominant irrespective of minor or major variations in input data are desirable solutions. However, it is an unusual system design that is so simple.

10.1.4. Time Horizon

Most decision making is a part of the history of decisions. Previous choices have affected the now time, and current decisions will influence future actions. Viewed this way, all system estimates can be classified as having no time limit. Sometimes, dynamic models have an unbounded horizon and others have a terminating date. Project-estimating models are mostly terminating, such as time-value-of-money policy actions. Obsolescence and depreciation are two factors that influence systems significantly. As a rule, the particular system design is posed in a dynamic context, and the engineer considers time explicitly.

With time phased costs from the cost streams of different alternatives, the irregular amounts may be discounted by using an appropriate interest rate. Methods dealing with this idea were introduced in the last chapter, and they fit system estimating problems.

10.1.5. Policies

Discussion has been leading up to the manner in which system studies are undertaken. The two approaches are either fixed effectiveness or fixed cost. First, for a fixed effectiveness system, various alternatives for achieving a prescribed capability or level of effectiveness are studied to determine how such effectiveness can be attained with the least resources. In other words, there is a given objective, and the question is to find that alternative or feasible combination of alternatives to achieve the objective. Second, fixed cost analysis deals with a fixed budget or specified cost. Given a fixed resource, the engineer looks for the most efficient manner to achieve the level of effectiveness.

In either of those approaches there are two ways to conceive the effectiveness measure. The analyst considers either relative or absolute effectiveness. For *relative effectiveness,* the measure is viewed more as an index, and only differences are considered important. For comparisons between competing systems, it is simpler to use relative differences to make the decision. For *absolute effectiveness,* it is the total cost measure. In addition to the selection among alternatives, the absolute measure indicates the magnitude and is an additional bit of wisdom for the decision maker.

10.2 DEFINITIONS

A *system design* is a configuration of operations, products, and projects in any combination. There are characteristics that separate system design from other engineering practices. There seems to be an urgency to design configurations based on performance, general requirements, equipment specifications, and their compatibility. Interwoven in system design are the engineering requirements directing an integrated effort. Every design is a new combination of preexisting knowledge that satisfies an economic want. The system engineer provides the measure of the economic want. Symbolically, the measure is *cost effectiveness,* where cost is in dollar terms and effectiveness is a design measure.

To a large degree a system estimate is determined by the information that can be obtained. In this respect the system estimate is similar to other types of estimates, because accuracy, precision, and its general goodness are no less dependent on the available information.

There is one distinction, however. In system estimating the kind, quality, and amount of information is usually established after the start of the estimate. With an operation, product, and, to a lesser extent, project design, information is reused, and agencies outside of engineering are in the habit of supplying raw data, performance figures, and the like. This is not true for the system estimate. Because this type of estimate is unique, the information is determined after the estimate is started. Usually, it is gathered by the system engineer.

A *request for proposal* (RFP) initiates the estimating procedure. Internally, the requestor may be systems engineering, or some governmental activity may spur it. The system engineer is not much concerned with comparisons between competing designs. For example, a transportation system for a growing metropolitan area may include concepts of bus, monorail or light rail, subway, subway-bus, and so on. Each of those concepts is to be estimated. Engineers find the economic want and leave optimization, comparison, and trade-off techniques to other members of the system engineering team.

The *system estimate* is composed of elements as shown in Fig. 10.1. We define terms to prevent semantic difficulties because system estimating is different from a

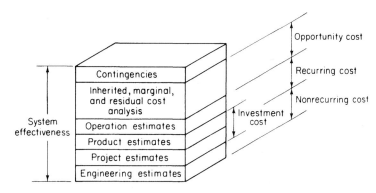

Figure 10.1. Elements of system estimate.

general business understanding. Our systems are large scaled, and the estimating practice is from a governmental or not-for-profit viewpoint.

The first difference is the absence of a specially defined profit element. Certainly, the components include profit for the manufacturers, contractors, and so on, but a system estimate does not include an explicit profit entry. Naturally, the estimates are future values, and, thus, accounting values are unacceptable because future costs will differ from historical values.

Some costs are *nonrecurring,* that is, occur only once for a system design. This type of cost is not the same as *fixed cost,* which implies a constant periodic cost such as annual debt, rent, or depreciation. Some costs are *recurring,* which is not quite similar to variable cost. Recurring costs "recur" relative to a performance specification of the system.

The system estimate is measured in money terms, dollars. The four estimates are figured considering inflation, deflation, technology, and system performance effects. It is an inviolate principle to this author that estimate cash flows are *out of pocket,* or *real.*

Though there is no specific order in considering the elements, a skilled engineer begins with the engineering estimate. Those costs occur early and include research, development, design, production or construction, testing, service, and so on. Test models, research laboratory, and facility support are included. *Engineering costs* are basically nonrecurring. Chapters 8 and 9 discuss techniques for calculating an engineering estimate.

Product estimates deal with several or many end items, such as airplanes for an aircraft carrier or people carriers for a subway. Key to the logic of product estimating is the idea of replication of the design and the production of several or many units. The measure of the economic want of a product is *price.* Chapter 8 deals with methods of product estimating.

A *project cost* is for a single end item given by a unique and special design, such as an aircraft carrier and subway construction. The measure of the economic want is called a *bid.* A project estimate is dominated by thinking on "first cost," which is only a part of the system cost. Chapter 9 provides methods for project estimating of single end items.

The total of the project and product estimates is referred to as *investment* in system-estimating jargon. Construction and products are simultaneously necessary for a system (i.e., people movers and the subway).

Operation costs include the costs associated with distribution, operational use, and the sustaining life-cycle logistical support of the system in the field. Maintenance of the investment, training, and utility costs are typical of many thousands of specific costs. Knowledge of whether the costs are borne by the seller or buyer (government in the case of the system estimate) is necessary. Fringe benefits, insurance, medical, retirement, and other overhead charges are a part of this cost pool. Both direct and indirect operational costs are included. Techniques provided by the discussions of operation estimating (chapter 7) and overhead (section 4.9) are necessary to understand this component. Those costs, operation and maintenance (O&M), are figured on an annual basis, and the yearly distinctions required by the system are considered.

Those estimate types are sufficiently understood and offer no new difficulties when incorporated into system estimates. However, a cost category unique to system estimating is inherited, marginal, and residual cost.

If a new system is able to adopt an existing subsystem, then we have an inherited cost. We assume that the inheritance is already paid for and thus is noncharge-able in the sense of future out-of-pocket monies. An example is existing missile silos for new missiles. The missile silo does have economic value provided that a missile system will be used.

The term *avoided cost* suggests a contradiction of definition, except in the sense that the silo is single purpose and will not be used unless the decision is made to design a contemporary weapon system compatible with the silo.

Can a dollar value be assigned to an *inheritance* for which no commercial purpose is likely? The answer is yes if the silo prevents or displaces other monies to achieve the same function. The answer is no if substitution is unlikely, or if an equivalent amount of money will be spent anyway despite the incorporation of the missile system.

However, the dilemma is partially resolved if the value of the inheritance is assessed to the system and is placed in a fictitious account titled "opportunity."

Opportunity cost is another concept peculiar to system estimating and is better understood if we recognize that a shortage of resources is ever present. Governments are no different from the governed. There are greater opportunities for spending than there are taxes or income. As a consequence of the scarcity of taxes and income the options for their use are limited. If we have $10 we could spend it proportionately on systems A and B or invest it entirely on C. In either case, one choice prevents the other.

There are two sides of an opportunity cost. We could spend the money on one system now and prevent its use for a choice now or later. The choice may be greater or smaller, and an opportunity cost is created. Perhaps, urgency for some reason requires that we take immediate action even though it does not compare favorably with other actions.

Opportunity cost is a true cost that must be assessed to the system alternative selected and is germane to the selection. On the other hand, opportunity cost is not money out of pocket. For example, inheritance cost of a missile silo is relevant to the system design choice and is a positive quantity that is reckoned in the special account of opportunity. Remember that the missile silos have been built and the money is spent. As another example, consider the relocation of a road for straightening purposes. The bypassed road has residual value and is assessed to the choice of the system that straightens the road in an opportunity account.

Bookkeeping records do not keep track of those opportunity costs. Those estimates of the opportunity are listed in a category separate from the other estimates and *marginal costs*, which are real. The purpose of the opportunity category is to complete the total requirements for resources. Additionally, opportunity costs that have been recorded avoid the "hidden" cost for a system. Conversely, if opportunity costs were considered equal to the estimates, a possible distortion exists.

Opportunity cost should not be confused with sunk cost. *Sunk cost* is money spent that cannot be recovered by a current or future decision. If the missile silo is abandoned, then its cost is sunk and unrecoverable. In fact, there may be a real *residual* cost for reclaiming the land.

Marginal costs are additional costs due to an alternative design. Section 6.10 already defined marginal cost in the context of production quantity, which is the usual variable. On the other hand, the marginal costs can be defined with respect to system design change, policy selections, or effectiveness level. Note Fig. 10.2, which shows the system cost level for I. A change in design, policy, or effectiveness level indicates that marginal cost for II is the difference from the base level. It is also possible to have a negative marginal cost from a base level. Marginal costs are real costs and can be estimated by the chapters already discussed.

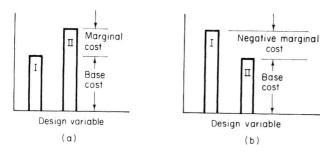

Figure 10.2.   Base cost and marginal cost for design levels I and II.

Residual costs or salvage costs are future values of an investment. Examples are portable machines that may have commercial value or an automobile that has trade-in value or materials that can be sold. Residual value has a different meaning. The initial value is determined by project and product estimates. Like *salvage value*, residual values are negative, positive, or zero. It may be necessary to detoxify or reclaim a land area. Contrariwise a sale may be possible, thus providing revenue on disposal of the asset. In some cases the value may be meaningless: for example, a highway that is bypassed by a newer road. It is also conceivable that a road may have residual value 20 years hence. Residual value is useful only if the function of the road persists. Thus, residual value deals with governmental system estimates and may not have portability or a conventional market may be unavailable for its disposal. However, the dilemma for an evaluation is partially resolved if the residual value is assessed to the system and placed in a fictitious account titled *opportunity.*

10.3 ANALYTICAL AIDS

We now examine popular devices used to analyze system proposals. These are brief encounters, and the reader is encouraged to examine the references for greater detail on the mechanics of special methods.

Often, analytical aids for system estimating are necessary. Those aids may be subroutines to the model or may be freestanding techniques useful for system estimating.

10.3.1. Discounting

Chapter 9 considers aspects of discounting as they relate to project estimating. Those aspects work here, too, in our discussion on system estimating. The central conclusion that the time-value-of-money concepts can be adopted for time dynamic system models is valid. Discounting allows a systematic method of comparing streams of costs and returns that have differing monies and times of payment. Generally, the discounting or present worth model is used. If the annual rate is 5%, then a dollar received 1 year from now is really worth $(1 + 0.05)^{-1}$ dollars and in n years $(1 + 0.05)^{-n}$ dollars right now. Or if you presently had $(1 + 0.05)^{-n}$ dollars and lent it at 5% interest compounded, you would have back 1 dollars n years from today.

Regardless of the costing period chosen it is desirable to *time phase system cost* throughout the total life period. Thus, the annual cost model is not as appropriate as *life-cycle costs* or the present worth of system design and development, acquisition, and operation costs.

This raises the point: What is the life cycle? In public works 30 to 100 years can be used. In weapons system the horizon may terminate within a period of months to 20 years. In commercial enterprises, a system may logically extend between a fad period to an enduring life of 20 to 40 years.

All kinds of priorities, political and social, affect the length of time. As the system life cycle becomes longer the actual length becomes less crucial, because the discounting factor drops rapidly if either i or n or both increase.

The total system life cycle is used for another reason. A *trade-off analysis* between design, investment for construction and acquisition, or operating cost becomes possible if the birth retirement or death horizon concept is used. In this event it is possible to time phase all costs and see if, say, additional costs in design might be warranted over excessively large investment and operating costs.

When considering *present value,* the higher the interest rate i the smaller the *discounted value.* The interest rate pertinent for a system estimate is an important subject in its own right and is a lively topic among scholars and practitioners. An unrealistically low or high rate can give a distorted value to the proposed design. In addition, what are the rates for public and governmental units as compared with commercial enterprises? Inasmuch as the government is concerned with social good, there is an undeniable mixture of politics associated with economics. The general effect of an increase in the rate of discounting is not only to make it undesirable to engage in some investment but also to change the character of those system designs adopted, making them less capital-intensive and more labor-intensive.

10.3.2. System Boundaries

Though discussion has been devoted to the effectiveness issue, little has been discussed about boundaries or constraints that limit the effectiveness. Constraints, whether implied or unimplied, are always existent. Those bounds may be in the form of a budget limit in terms of a fixed amount of cash or a mathematical statement.

The finding of well-expressed constraints is not an easy task. Usually, the process starts by verbalizing a known constraint situation. Creating a notation and devising an algebra that translates the verbal statements may be the next step. Or the constraints may be well known, and empirical evidence may have substantiated its statistical behavior. The constraint may be theoretical, resting on mathematical hypothesis. However the constraint is derived, it is of major importance to system estimating.

Sometimes the constraints are obvious and thus easily overlooked. For instance, a constraint type of importance is nonnegativity (i.e., $x_i \geq 0$, $i = 1$, $2, \cdots, n$), which denotes that the x variable cannot assume negative values. Negative production, meaning that the sunk cost of production could be recovered by a backward process such as disassembly of a product, is prevented by the statement that production $x_i \geq 0$.

10.3.3. Probability

The application of the *probability art and science* is commonplace in system estimating. In making system, product, project, or other design decisions, *risk* and *uncertainty* cloud future events. We deal with the future about which we cannot know, but certain things are inferred from what we know about the past. Time series models, discussed in chapter 5, are typical. It is information gained from past events used to predict future events. This involves the study of probability. To a small extent, probability has been discussed piecemeal throughout the previous chapters (see especially chapters 5 and 6). But depth and rigor are necessary for system estimating.

Because future events are random, the insistence on single valued determined data for input and output is misleading. It is frequently more meaningful to compare system alternatives in terms of a probability of being attained rather than by comparing mean values. As was shown in Fig. 6.9, cost estimates can be shown as probability distributions. The mean and variance are important in selecting between two system alternatives. The dilemma is of total system cost of an alternative with a lower mean cost but a higher standard deviation: How much higher is naturally germane to the selection.

Given that neither system cost nor effectiveness can be calculated precisely, the determination of the amount of uncertainty is accorded a most-probable cost effectiveness value. This requires finding the probability distribution of the major units of information. A cost estimate using this reckoning would be given as $250,000 ± $75,000 for a probability of 0.95 of being in this range.

One way to handle probability system cost is by means of *Monte Carlo analysis.* A computer simulation of the system elements is an effective way to model those problems. Many conditional combinations of random costs can be contrived to simulate the system.

10.3.4. Sensitivity

Analyses undertaken for system problems are rarely confined to single numerical values of the optimal solution. How much can the input parameter values vary without causing alterations in a computed optimal solution or the composition of some policy? An investigation of this sort is termed a *sensitivity analysis,* where fluctuations of the unit profit, item cost, item horizon, product demand, calamities, or indirect secondary benefits are permitted. Sensitivity questions are sufficiently complicated to require electronic computation.

In the simplest case, say, for straightforward computation of system data, each parameter is varied in turn to determine its effect on the model. Those parameters shown to have little or no effect may be treated as constant, or the analyst could be tolerant of variation in those data. Those parameters that when incremented show large variation in effectiveness and alter the optimal solution or reverse policy decisions are called sensitive and bear watching. Those kinds of data should be examined in detail since they are important to system effectiveness.

10.4 METHODS

System estimating is relatively new. Even so, engineers have favorite practices. Though widely divergent in detail, the thrusts of the practice are known as (1) benefit cost, (2) life-cycle cost, and (3) system budgeting.

10.4.1. Benefit Cost

Benefit cost is a practical way of assessing the desirability of system alternatives where it is important to take a long time and broad view. It is an effectiveness measure in its own right, because it implies the enumeration and evaluation of all or nearly all the selected costs and benefits.

Benefit cost has a long history, initially applied to the federal improvement of navigation back in 1902. Earlier it was used in France. For the most part, benefit cost (B/C) is used by governmental agencies. B/C analysis is applied for water improvement and related land use, and its application is controlled by federal statute.

We begin the discussion about *B/C analysis* with a small illustration on highway analysis. Costs have long been a factor in decisions made by highway engineers. From the many considerations of highway transportation, an economical road is achieved whenever the total cost is a minimum consistent with convenience, safety, transportation, and the ability to pay. The benefit cost ratio method is a comparison of the difference in annual cost to highway users when there are vehicles using an existing road in one case and an improved road in another with the annual cost of making the improvement. The equation is

$$\text{benefit cost ratio} = \frac{r_o - r_i}{(I_i - I_o) - (M_o - M_i)} \tag{10.1}$$

where $r_o - r_i$ = decrease in road user costs after improvements per year

$\quad\quad I_i - I_o$ = increase in investment costs per year

$\quad\quad M_o - M_i$ = decrease in maintenance costs per year

This would simplify to a ratio of decrease in user cost per year divided by net increase in investment costs per year. A similar approach is found for irrigation, recreational development, dams, and so forth. Benefit cost ratios are calculated for the logical alternatives and are compared with the basic condition. Now we leave this highway illustration for a more general discussion on B/C analysis.

The current objective is to show how benefit cost estimates are made and methods of analysis that start after conclusion of the estimates. Criticism against the B/C method usually deals with the methods of analysis and overlooks the schemes of estimating, their definition, and logic.

The same input data, if applied consistently, will lead to similar decisions between alternatives. The benefits and costs are analyzed by present worth and equivalent annual cost. The interest rate, sometimes called minimum attractive rate of return, is preestablished. Those methods were discussed in section 9.5, and many texts on engineering economy provide tables to ease calculations. For simplicity, we adopt a present worth approach and confine discussion to that.

One of the confusing choices in benefit cost analysis is that the benefit to some is a disbenefit to others, or a cost to a governmental agency serves as income to other firms. For purpose of this discussion, we rule that relevant consequences to the governmental units are classified as costs.

Two methods of calculating the benefit cost ratio exist. The first involves subtracting annual benefits from annual costs to establish a *net annual benefit*. Annual net benefits are discounted back to the date of the program's inception and summed to establish a present value of discounted net benefits. The benefit cost ratio is then formed by relating this figure to the capital cost of the program. This approach is vaguely similar to a business calculation of the rate of profit that can be earned by capital. The second approach is to establish the *gross benefits and costs* for a typical year. The costs include annual operating costs and amortization of investment. No discounting is used.

Of the several ways to estimate a benefit cost ratio, the one chosen here uses discounting. Principally, the B/C ratio has in the numerator the present worth of all benefits, and the denominator is the present worth of all costs. General notation is as follows:

B_n = future benefits in year n, dollars
C_n = future costs in year n, dollars
i = interest rate, decimal
n = year, $n = 0,1, \cdots, N$

We adopt an end-of-the-year convention, meaning that benefits and costs, however they may occur, are assumed to be instantaneously received or paid at the end of the year. An investment or first cost occurs at time 0. The excess of the present worth of benefits over the present worth of costs is

$$\sum_{n=0}^{N} \frac{B_n}{(1 + i)^n} - \sum_{n=0}^{N} \frac{C_n}{(1 + i)^n} > 0 \tag{10.2}$$

Investment costs are already at time 0 and do not require discounting. If the difference in Eq. (10.2) is not positive, then we presume that the contemplated action is unfavorable.

A present worth approach to benefit cost ratios is determined by

$$P_b = \sum_{n=0}^{N} \frac{B_n}{(1 + i)^n} \tag{10.3}$$

$$P_c = \sum_{n=0}^{N} \frac{C_n}{(1 + i)^n} \tag{10.4}$$

$$\text{B/C} = \frac{P_b}{P_c} \tag{10.5}$$

Consider the stream of cash flows for project X, and let $i = 8\%$.

	Year			
	0	1	2	3
Costs, C_n	6	3	4	5
Benefits, B_n	0	10	12	15

The analysis for project X would be

Year	$(1 + i)^n$	$\dfrac{B_n}{(1 + i)^n}$	$\dfrac{C_n}{(1 + i)^n}$
0	1	—	6.
1	1.080	9.529	2.778
2	1.166	10.292	3.431
3	1.260	11.905	3.968
Total		31.726	16.177

The value of the discounted cash flows of benefits and costs is \$31.726 and \$16.177. The excess of the present worth benefits over costs is 15.549 (=31.726 − 16.177), and the B/C ratio is 1.96 (=31.726/16.177). In this example we assume that the benefits were amenities such as might develop from increases in irrigation, recreation, and fish and wildlife revenues and assume that they were nonexistent before.

In another situation, such as the straightening of a road, the benefits accrued to the user are reductions in road-user costs and are a favorable consequence to road users. We then look on the difference of the improved and the original road as a benefit to the road user.

Consider now projects X and Y, and determine their benefit cost difference and ratio. Project X is in bad condition, and reconditioning is planned. Project Y is new and involves higher investment but lower costs to the user. Operation and maintenance costs of the project also differ.

	Project X	Project Y
Investment	6.000	9.000
PW of maintenance	10.177	13.115
PW of user costs	31.726	25.267

By using present worth (PW) calculations and by using $P_b = P_{bx} - P_{by}$ as the difference in benefits

$$P_b = 31.726 - 25.267 = 6.459$$

and the difference in costs as $P_c = P_{cy} - P_{cx}$.

$$P_c = (9.000 + 13.115) - (6.000 + 10.177) = 5.938$$

$P_b - P_c > 0$ can also be expressed as the ratio 1.09 (=6.459/5.938). The conclusion is to adopt project Y, but barely, on the basis of the marginal B/C ratio.

As with other effectiveness measures, the B/C ratio has limitations that we cannot cover fully. It is a ratio; that is, simply benefits divided by costs. At that point, however, complications set in. Which costs and whose benefits are to be included? How are they to be valued? At what interest rate are they to be discounted? What are the relevant constraints? There is bound to be arbitrariness in answering those questions.

In most cases the scope and nature of the system to be analyzed are clear. There is a wide class of costs and benefits that accrue to organizations other than the one sponsoring the system and an equally wide issue of how the parent agency should consider them. For instance, a hydroelectric dam can be costed and benefits determined, but what about the recreational amenities, water for farming, and improvements in scenery? The net rise in rents and land values is a result of the benefits of hydropower. Those *secondary benefits* may be more important to one governmental agency than to another, and calculations can impute those *financial spillovers* in a more or less favorable way.

Some critics suggest that a B/C analysis side steps the issue by requiring public agencies to operate on a commercial basis, leaving resource allocation to be resolved through an artifice of the pricing system. But welfare economics, the well-being of people, income redistribution, market imperfections, and the like make a reasonable demonstration of B/C in a commercial environment difficult. Thus, a B/C philosophy has to have a *comprehensive public viewpoint.* When constrained by laws and appropriations, the B/C ratio is best used as a means of ranking various systems. Those higher are considered better from the B/C viewpoint.

As an example, consider the following. A project is planned largely to provide water for industrial, municipal, and domestic use in connection with anticipated development of coal and oil shale reserves. It would increase irrigation supplies for production of livestock feeds and also would benefit recreation, fish and wildlife, and flood control.

Note Table 10.1, which is a summary of project and operation estimates and benefits. Project estimates determined by using rules in chapter 9 are listed for each of the separate projects. In turn, operation estimates that use the advice in chapter 7 are averaged for the span of life, 100 years in this example. Project costs are nonrecurring. Operation estimate provides for the labor, materials, and overhead to operate and maintain the investment. Occasional replacement of minor capital requirements are, for convenience, included in the operation estimates.

The benefits are determined by using price schedules of similar opportunities. For example, irrigation water has a market value and is found by using comparison estimating methods. In circumstances where there is no market advice, opinion estimates will lead to imputed values. It may be possible that some benefits reduce financial gain. For instance, the impounding of water into a reservoir will reduce the grazing pasture for cattle, and those efforts cause disbenefits, or benefits lost. It is a practice that benefits lost are charged against the benefits and not added to the denominator cost term.

TABLE 10.1   SUMMARY OF BENEFIT COST ESTIMATES

Costs	
Projects	
Dam 1	$11,850,000
Dam 2	6,400,000
Dam 3	5,100,000
Canals and diversion dams	23,870,000
Laterals and drains	5,350,000
Operating and equipment	330,000
Fish and wildlife equipment	610,000
Recreational facilities	1,246,000
Constructional total	$54,756,000
Operational costs	
Operation, maintenance, and replacement for 100 years, annual	$ 147,200
Benefits	
Average annual for 100 years	
Industrial, municipal, and domestic water use	$ 4,590,900
Irrigation	869,800
Recreation	245,800
Fish and wildlife	137,700
Flood control	11,000
Less benefits lost	9,900
Average yearly net benefits	$ 5,845,300

With the benefits and cost posted to the *pro forma* summary (Table 10.1), it is possible to determine the net present worth of the cash flow and the B/C ratio. We use an interest rate of 6%.

$$P_b = \sum_{n=0}^{100} \frac{B_n}{(1 + i)^n} = \$5,845,300 \ (16.618) = \$97,137,195$$

$$P_c = \sum_{n=0}^{100} \frac{C_n}{(1 + i)^n} = \$54,756,000 + 147,200 \ (16.618) = \$57,202,170$$

The $P_b - P_c > 0$ and B/C = 1.70.

Once the tangible benefits and costs are recognized, classes of *reimbursable* and *nonreimbursable cost allocations* are made. We have seen the problem of cost allocation before. In a B/C situation though, it is the proper distribution of the costs of the features that serve several purposes that are the problem. This problem does not arise in the cost of a single-purpose project nor when national policy has determined in advance that the purpose to be served outweighs all costs. The costs of a multiple-purpose project are composed of the costs of individual project features, such as irrigation canals, power houses, or navigational works, which serve only a single purpose. A dam, of course, serves those several single purposes, and its cost must be allocated to both reimbursable and nonreimbursable services.

Broad principles of cost allocation are possible. Each purpose should share equitably in the savings resulting from multiple-purpose construction within the limits of maximum and minimum allocations. The maximum allocation to each purpose is its benefits or alternative single-purpose cost, whichever is less. The minimum allocation to each purpose is its specific or its separable cost. *Joint costs* are apportioned without regard to the ability of any particular purpose to pay.

In an oil shale water project as major as Table 10.1, it is necessary to recognize that many of the costs can be repaid by various public or private entities that will benefit. Land rents, payment for irrigation water, and recreational amenities by the general public are typical opportunities. Laws require repayment, and practices have established the categories of reimbursable and nonreimbursable costs.

Reimbursable costs for Table 10.2 have a period of 50 years for repayment. Often, interest of a few percent may be added to those costs. Some expenses are considered nonreimbursable and are borne by the federal government, such as the expenses for project investigations. Note, however, that the total of both costs is used in determining the B/C ratio.

If there are several designs requiring a decision for one system, invariably the cost and benefits are different, and a marginal method must be used to find the best system. Several guidelines are necessary. First, the same interest rate should be used to figure costs and benefits. The same period of life, N, is necessary for the system alternatives. We initially calculate the B/C ratio for each system, such as that given in Table 10.3. In this table there are four mutually exclusive choices. System choices that were less than 1 would be rejected.

Inspection of those B/C ratios suggests that alternative B is chosen because its ratio is the greatest. This is an improper selection. The proper choice depends on the principles of interest rate methods of engineering economy.

Table 10.4 shows the required calculations. The system choices are considered in order of increasing cost. System D is used as the initial base since it requires the minimum present worth cost. The first row of this table repeats row D from the previous table. Insofar as this first calculation is concerned, the decision is to accept D. Next, the marginal increase in cost and benefits is determined by using the next alternative above the least costly alternative. This would be C, and the C minus D values are indicated for the row. The marginal B/C ratio is 0.46, which is less than 1, and C is rejected. We proceed by considering the alternatives in order of increasing costs. Now choice B is compared to D and the B/C ratio exceeds 1 and design B is preferred to alternative D. System design B is the current best choice. Last, the marginal gain and loss of design A minus design B is computed. The ratio of marginal present worth benefits to marginal present worth costs is greater than 1, indicating that design A is preferable to B. The final choice is A, and it assures that the equivalent present-worth benefits will be less than the equivalent present-worth costs and the system return is maximized. Choice A is contrary to the initial selection B.

Thus, B/C analysis deals with the first step of estimating the costs and benefits. If this is successfully achieved, then the next step is to rank alternatives in order

TABLE 10.2.   COST ALLOCATIONS AND REPAYMENT SUMMARY

	Project Costs	Annual Operation, Maintenance, and Replacement Costs
Reimbursable costs*		
Industrial, municipal, and domestic	$39,330,000	$ 61,200
Irrigation	10,160,000	40,100
Recreation	166,500	17,800
Fish and wildlife	375,000	
Subtotal	$50,031,500	$119,100
Nonreimbursable costs		
Recreation	$ 2,765,500	$ 27,500
Fish and wildlife	1,823,000	400
Flood control	136,000	200
Subtotal	$ 4,724,500	$ 28,100
Total	$54,756,000	$147,200
Repayment		
Industrial, municipal, and domestic use†		
Prepayment‡	$ 474,000	
Water conservancy district	38,856,000	$ 61,200
	$39,330,000	$ 61,200
Irrigation		
Prepayment	$ 126,000	
Water conservancy district	6,720,000	$ 40,100
Apportioned to others	3,314,000	
	$10,160,000	$ 40,100
Recreation, fish and wildlife†		
Nonfederal interests	$ 541,500	$ 17,800
Total	$50,031,500	$119,100

*Reimbursed over 50 years.
†Repayment rate at 3.5% annually.
‡Nonreimbursable expenses for project investigations.

TABLE 10.3.   BENEFIT-COST RATIOS FOR FOUR MUTUALLY EXCLUSIVE SYSTEM CHOICES

System Design	Present Worth Benefits	Present Worth Costs	B/C
A	$120,000	$60,000	2.00
B	112,000	53,000	2.11
C	75,000	58,000	1.29
D	64,000	34,000	1.87

of increasing investment, such as the present worth, and then check to see if the marginal investment is effective. This is tested by checking the differences between successive pairs of B/C alternatives against a status quo condition, or do nothing. A

TABLE 10.4. MARGINAL BENEFIT-COST RATIOS

System Design	Marginal Present Worth Benefit	Marginal Present Worth Costs	Marginal B/C Rates	Decision
D	$64,000	$34,000	1.87	Accept D
C−D	11,000	24,000	0.46	Reject C
B−D	48,000	19,000	2.52	Accept B
A−B	8,000	7,000	1.14	Accept A

conclusion is reached whenever the last alternative is compared with the last acceptable alternative.

10.4.2. Life-Cycle Cost

Life-cycle cost (LCC) is the summation of all estimated cash flows from concept, design, construction and manufacture, operation, and disposal of the system at the end of useful life. The design is generally a system, but a product or project can also be evaluated. Intuitively, individuals have used LCC principles for economic evaluation of cars when they concern themselves not only with initial cost (sticker price) but with operating and maintenance expenses (gas mileage, worn parts, insurance, license) and residual value (resale price).

LCC attempts to estimate all relevant costs, both present and future, in the decision-making process for the selection among various choices. Figure 10.3 illustrates engineering, product, project, and operation costs as separate cash flows determined by using the estimating methods previously discussed. However, in LCC analysis, the estimates are scheduled as period cash flows starting with $E = 0, +1, +2$, where $E =$ moment of the estimate, $N_c =$ end of costing period, and $N_l =$ end of life cycle. The periods are year designations. Note in Fig. 10.3 that design precedes project and product, which precedes the operating costs. It is not necessary that the curves be symmetrically shaped. Costs conclude at the end of cycle, of course.

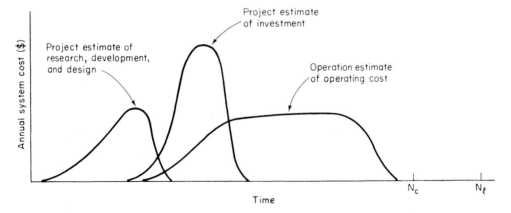

Figure 10.3. Cost time phasing for a system life cycle.

Operating cash flows can easily be greater than the original R&D or the investment. Moreover, a system with higher engineering and investment costs but lower operation and maintenance costs may, depending on service life, be a least LCC system. It has been shown that for military hardware systems approximately two-thirds of life-cycle costs are unalterably fixed during the design phase. LCC encourages trade-off analysis between one-time costs and recurring costs. This is described in Fig. 10.4, where either system A, B, or C is selected on the basis of minimum LCC and operating years. The analysis suggested by Fig. 10.4 is important in trade-off selection.

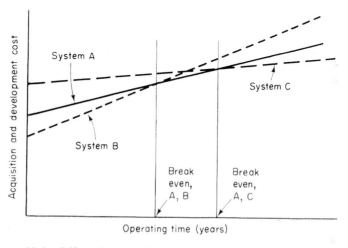

Figure 10.4. Life-cycle cost where operations over a period of years determine selection.

We shall consider a case study of LCC applied to an environmental chamber used to simulate high-altitude pressures for electronic equipment experimentation. (See the case study in chapter 1 for additional background.) Figure 10.5 is a preliminary design of the product. The system is a retractable chamber that opens for placement of various electronic equipment. Various vacuum pressures are possible. The design has a life cycle of six periods because the experiment will conclude then. The length of life cycle is sensitive to wear-out, casualty or destruction, economic, or technology factors. For our case study, the life cycle is set by the contract period for the experiment of electronic equipment. In preparation for the LCC system estimate, the following data are required:

1. Engineering estimate
2. Product estimate
3. Operation estimate
4. Operating profile
5. Maintenance schedule

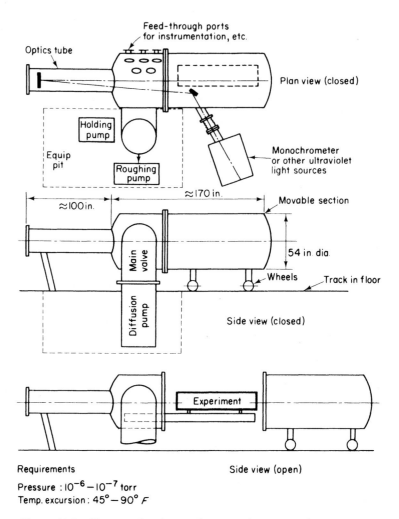

Figure 10.5. Vacuum chamber environmental conceptual test system.

The operating profile has a repetition time and contains all the operating and nonoperating modes of the equipment. It is sometimes possible to have operating profiles internal to other operating profiles. For trade-off studies, candidates are evaluated with the same operating profile. The operating profile says when or in what way the equipment is operating. For example, the diffusion pump will run 60% of the time in the environmental chamber. It will not run in a shutdown mode. Instead, a roughing pump holds nominal vacuum.

Operating costs incurred during the life of the equipment are found by using the profile. Two unusual cost parameters are mean time between failure (MTBF) and mean time to repair (MTTR). Warranty costs are customarily included in the product estimate and depend on MTBF information. Time between overhaul, power consumption rate, and preventive maintenance routines such as cycle and the preventive maintenance rates are required information.

For our case study, installation costs $20,000 and is a one-time nonrecurring cost. The aerospace firm estimates its operating profile and determines that a continuous three-shift operation for about one-third of a year, or 2920 hours per year, is necessary. By using the schedule and a labor rate, 2920 hours per year $\times$ $28 per hour $\times$ 1 person per machine = $81,760.

Labor for preventive maintenance (PM) actions is calculated by using

$$\text{number of PM actions} = \frac{\text{scheduled operating hours}}{\text{PM cycle time}} \qquad (10.6)$$

The number of preventive maintenance actions is

$$\frac{2920}{160} = 19 \text{ actions}$$

where each maintenance cycle is 160 hours. Since each maintenance action requires 4 hours, we have 19(4) = 76 hours of PM time for a total yearly cost of $1976 (= 76 hours $\times$ $26 per hour).

An aerospace firm has reliability studies that the system will fail every 500 hours. The 500 number is the mean time between failures. The cost of corrective maintenance is found by using

$$C_{cm} = \frac{\text{SOH}}{\text{MTBF}}(\text{MTTR})C_m \qquad (10.7)$$

where C_{cm} = cost for corrective maintenance per year
   SOH   = scheduled operating hours
   MTBF = mean hours between failures
   MTTR = mean hours to repair
   C_m = cost of maintenance labor

We have

$$C_{cm} = \frac{2920}{500}(40)(26) = \$6074$$

Power consumption cost is found by multiplying input power in kilowatts by total hours of operation and the cost per kilowatt hour.

power cost = 2920 hours per year $\times$ 10 kW $\times$ $0.20/kW = $5840

Spare parts, discussed in chapter 8, are also estimated by a historical percentage factor of the original product cost. For $\frac{1}{2}\%$, we have

spare parts cost = 0.005 $\times$ (125,000 + 210,000 + 65,000) = $2000 per year

Those costs are summarized in Table 10.5. Their sum of $656,200 is the undiscounted life-cycle cost. However, LCC analysis discounts all costs to a time zero, sometimes considered to be the moment of the estimate at $E = 0$. A discount rate of 10% is applied to the data, where each estimate is assumed end of period. The discount factor is found from $1/(1 + i)^n$. Total discounted cash flow sums the product of the discount factor and cash flow for each year, or $470,237.

TABLE 10.5. DESIGN, PRODUCT, OPERATION, AND MAINTENANCE CASH FLOW ESTIMATES SCHEDULE OVER LIFE CYCLE OF ENVIRONMENTAL VACUUM CHAMBER

	$1000 Cash Flow for Period					
Cost elements	1	2	3	4	5	6
Design	25	11	5			
Product						
manufacture		125	210	65		
Operation						
Installation				20		
Manpower					81.8	81.8
Preventive						
maintenance					2.0	2.0
Corrective						
maintenance					6.0	6.0
Power					5.8	5.8
Spare parts					2	2
Discount factor, $\dfrac{1}{(1.1)^n}$	0.909	0.826	0.751	0.683	0.621	0.564

Life cycle is determined from wear-out, casualty, economic, or technical obsolescence factors. When obsolescence is a factor, opinion is required, because the life of the equipment may suddenly be terminated by a change in company policy, buying habits, government legislation, competitive pressures, or new designs.

Another bewildering prediction is salvage value. An experienced appraiser may give opinions of future land and factory values. If the life is not expected to be great, then the engineer is in a position to trust this source of information. For the longer-lived equipment, information may be unavailable. Despite the fear of distant predictions, errors in evaluating salvage value are not normally serious. Error in the present or the near future should be given more concern or study because the effects of those conclusions are greater. Another prediction for equipment and plants is the efficiency of utilization. This efficiency is important because it controls cost to some extent. Errors in the degree of efficiency of equipment are considered serious, and predictions of this nature should be studied closely.

In forecasting the life of an asset there are two major concerns: *annual deterioration* and *obsolescence*. Where physical life or annual deterioration establishes the life, statistical data from past records become the basis for future prediction. If life is regarded as economic, then statistical methods find the probability that the economic life of a proposal will terminate during each year of its service life. In either of those two cases, deterioration or obsolescence, the life of a particular proposal terminates because it is worn out or because the product or service is no longer profitable.

In the first instance, the physical condition has deteriorated and does not produce the desired quality, or the cost of maintenance exceeds the cost of replacement. In the absence of statistical data, reliance is placed on the opinion of people having experience such as engineers, operators, and the people producing the equipment. In the case of obsolescence, competition introduces substitute products, processes, or machines with better prices, qualities, or services.

Once the estimate is concluded, continuing analysis becomes possible. The estimate allows for visibility of "tall poles," a jargon implying significant cost elements. For the vacuum chamber, the tall poles are product manufacture and operating manpower.

10.4.3. Budgets

Summing of various cost elements is a popular method. System estimating is no different. Aggregation of cost may be handled by a general summation model of the form

$$C = \Sigma E + \Sigma M + \Sigma L + \Sigma OH + \cdots \tag{10.8}$$

where E = sum of engineering cost

M = sum of direct materials

L = sum of direct labor

OH = sum of overhead

Budgets are *pro forma* methods that provide the summing of costs, although not as formal appearing as intended by Eq. (10.8). Its importance is made clear by its popularity, for the budget is indispensable for planning and traceability.

We have previously mentioned fixed, variable, operating, and appropriation types of budgets in chapter 4. Minor modifications for concepts of opportunity cost, inherited (sunk or residual) cost, and marginal cost make those standard procedures suitable for system budgets.

From the viewpoint of system estimating, cost continues to be broadly interpreted and can be considered to be the amount paid or given for anything whether labor or self-denial to secure a benefit. For system engineers, cost is only one element of *value forgone* to secure a benefit. In short, cost is a *negative benefit*. In those terms, cost includes money, time, performance, consumption of scarce resources, and ordinary human skills.

It is relatively easy to determine a value scale that relates the relative worth of one resource to another if an interchange is indexed by the dollar. The money value of inputs is not difficult to establish, because the market place provides a mechanism for assessing those costs. For a commercial system the dollar value measure system applies, and competition aids in establishing the price. In other situations the market mechanism may be unavailable; for example, weapons system or river basin development. Electrical power, one component of a river basin development, may have market value, but recreation or flood risk does not have this advantage. Whenever a competitive market action and reaction is nonexistent, the analyst determines an opportunity cost function for those components that require a value. This is equivalent to imputing a price that might be comparable to a market price.

Inasmuch as resources are always scarce, the options for their application are limited. The selection of one system design precludes another. Thus, an opportunity cost arises from the fact that the expenditure of money on one design preempts its use for another. Failure to take advantage of the other opportunity may result in foregoing a profit or benefit that otherwise might have been obtained. This is a *true cost* that can be assessed against the alternative selected. In this context it is called opportunity cost. This indirect cost measurement approach evaluates the resource, which may not have been clearly identified and measured for the selected alternative.

Inherited and marginal costs cannot be separately treated in matters of system budgeting. Inherited or residual cost is a value of earlier resources committed to the system, and marginal costs are the additional costs resulting from a change in objectives or level of the decision. In earlier chapters we suggested that a policy can be made based on marginal analysis irrespective of inherited values. For instance, an optimum operating point n existed when $dC_T/dn = 0$. It is not really this simple. Not all decisions can be made separate of inherited costs and values such as marginal cost theory seems to imply.

Certainly, the logic to ascertain a marginal decision is based in part on inherited values. To remove or ignore that base and to deny its importance for future decision making overlooks its value as a base representing existing capability. In determining how many additional resources are needed to acquire some specific capability or, conversely, how much additional effectiveness will result from some additional cost is a marginal cost budgeting problem.

How are those concepts incorporated into the budget? Opportunity cost can be a line item in the accounts of the budget, because it is certainly germane to the design selection. A clarification must be made as to what it precisely means on the budget form to avoid misleading interpretations.

Two budgeting extremes for inherited and marginal costs are total of inherited values plus marginal cost, or full, and marginal costs only. Because of their extreme position, both can be faulty or correct at various times. In the full method, the overstatement of fixed values may cause insensitivity. One means to overcome a disproportionate statement is to identify that which is fixed inheritance from that which is variable or proportional to the decision. In the latter case, an allocation of a part of the inherited system value tends to reduce the invariant part of the fixed budget.

If we use an absolute budget, then full inherited values and their *amortization* are indispensable methods. If a relative general budget is to be employed, marginal cost values can be used. Of course, there is the type of budget that is in between those two. Caution in the preparation and understanding must be exercised.

SUMMARY

Because system designs are configurations of operations, products, and projects in any manner, a *standardized treatment* is largely illusory. An estimating method is devised for the design, and the principal elements deal with operations, products, projects, and engineering. The methods have the rubric of benefit cost, life-cycle cost, or budget. A measure, called symbolically effectiveness and having a dollar monetary unit, is the usual one for cost analysis. The task of determining the kinds and sources of information is next. As always, data are vital, and careful methods of analysis can extend the usefulness of the information. If the system is long range and complex, then the effort for a system estimate is formidable.

QUESTIONS

10.1. Define the following terms:

System	Opportunity cost
Effectiveness	Sunk cost
Efficiency	Residual value
Intangible	Discounting
Tall poles	Sensitivity
Recurring costs	Life-cycle costs
Nonrecurring costs	Benefit cost ratio

10.2. Two approaches for system studies are fixed effectiveness and fixed cost. Describe how they work, and contrast their differences. What are the advantages and disadvantages of specific effectivenesses?

10.3. How is salvage cost different from residual cost?

10.4. For a real-life problem, write some system boundaries, and then find their equation.

10.5. State some disbenefits for water navigation projects. What makes the benefit cost ratio like an effectiveness measure?

10.6. When is a relative effectiveness superior to absolute effectiveness? If accuracy of data is a problem, then how does this affect your discussion?

10.7. For an imaginary system design, list several intangibles. Indicate the pros and cons.

10.8. Why are budgets important in system estimating?

10.9. Ever since World War I, stockpiles of military chemical weapons have been stored in remote beehive locations throughout the world. These ten million weapons (rockets, artillery shells, and land mines) are composed of explosives and noxious chemicals consisting of nerve gases, mustard gases, etc., which are fatal. The chemical compositions have probably altered due to the many years of their aging. The United States will demanufacture these weapons. But how to do it at minimum cost and safety is an engineering systems problem. For example, shall the weapons be decomissioned at one location or at their point of storage? To have one location is cheaper for demanufacturing, but more expensive and perhaps more dangerous in terms of transporting the weapons to the central depot. Two concepts of demanufacturing are (1) cryogenic baths that neutralize the action of the explosive and reagent because of the temperature of the liquid nitrogen, and (2) disassembly of the weapons with automatic mechanical presses and robots in chambers followed by furnace combustion of the components, explosives, and chemicals. In system one, a cryogenic plant is necessary, and once the weapon is at cryogenic temperature, the weapon is struck with the ram of a press, breaking apart the weapon. Explosion of the charge is considered unlikely. In system two, even if the weapon explodes and releases the chemicals, a chamber confines the explosion and gases to a restricted space. Disassembly of the weapon is automatic. Both systems have advantages and disadvantages. There are also political factors, which superimpose on the technical and engineering considerations.

Depending on the assumptions, either system can dominate the other in terms of cost. The economic measure of the system analysis is total discounted life-cycle cost. Because there is a finite number of weapons to decommission, and the effort can be shortened or extended over a number of years, both systems are cost sensitive to years and a minimum discounted LCC can be found such as given by Fig. Q10.9(a). Different discount rates provide curves given by (b), where i_g is the greater discount rate as compared to the smaller i_s. Levels of initial investment where I_g is greater than I_s, the smaller investment, are given by (c).

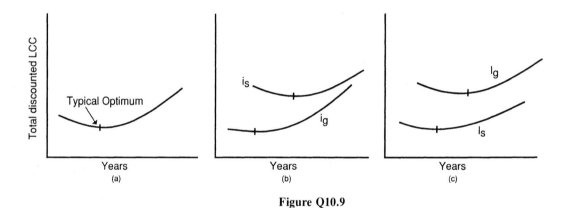

Figure Q10.9

In a report, list the technical assumptions that you believe are significant for the two systems, identify the economic assumptions that will guide your analysis, and assess the strengths and weaknesses of the discounted LCC for this actual problem. Speculate on the shape of the LCC curve, and draw two discounted LCC curves for the systems focusing on your assumptions. Be prepared to defend your conclusions regarding the best alternative.

PROBLEMS

10.1. There is no uniform method for the analysis of system problems. If one were available, and after you have learned the procedures and strictly adhered to the method, then you would be assured of a good chance for a successful result. Considering the system concepts of (1) effectiveness and system models, (2) cultivation of alternatives, (3) intangibles and tangibles, (4) time horizon, and (5) fixed, variable, inherited, and marginal cost streams, how do you propose to analyze a technical problem that you would be concerned with? State your own problem, give its ramifications, and provide the system procedure for its evaluation.

10.2. A system is conceived in terms of three goals. Engineering considers each desirable and of equal importance. The available alternatives are mutually exclusive and are measured by a relative effectiveness consistent for the goals and alternatives. Higher effectiveness is considered more desirable.

	G_1	G_2	G_3
A_1	80	22	19
A_2	50	40	43
A_3	35	70	21
A_4	39	47	48
A_5	25	25	75

(a) Which is the preferred alternative for each goal?

(b) Assuming that goals G_2 and G_3 are equally desirable, what weighting would goal G_1 require to favor alternative A_1?

10.3. Your company is attempting to sell a system design to a client. The design will provide a profit of $50,000 once it is concluded. So far, your company has spent $20,000 promoting the sale. There is a sense of confidence that the sale will be assured by an added expense. What is the maximum amount of resources (over what your firm has already spent) you should spend to secure the sale? Discuss.

10.4. (a) In a period of rising prices, a contractor is maintaining an inventory at a constant level. A subsystem is purchased by a contractor for $20,000 and then resold as a part of a system for $30,000. It is replaced for $40,000. Comment on their unit cost estimating and pricing policies.

(b) Now assume a period of declining prices. A contractor buys a subsystem for $40,000 and then sells it as part of a system for $30,000. It is immediately replaced for $20,000. Discuss those policies.

10.5. Suppose that a building owned by a business can be rented for $10,000 per year to another firm or used for the manufacture and construction of a system. After an initial cost for equipment, the new system will provide a net equivalent annual income of $20,000. However, the analyst did not charge for expenses for self-use of the building because "it is owned by the company." Discuss this action considering opportunity cost. Is there a more meaningful measure? What if there were no interested renters or if the space were unsuitable for rent. Would this affect your measure? What if there is only one renter who desires the space for 1 year but the system has a potential market for 2 years? Contrariwise, the renter will lease for 2 years though the market will last only 1 year. Will those situations require a time-dependent weighting factor? Discuss the point: "Once a decision is made, the advantage of the alternative is foregone and further consideration is useless."

10.6. There are two possible system designs. System 1 is refurbished, and system 2 is new. The second system is located away from the older system. Both systems are identical in meeting performance requirements. Following Fig. 10.1 we select symbols that represent the discounted sum of the cash flow and their comparison is as follows: Engineering $(E_1 < E_2)$; product $(P_1 = P_2)$; project $(W_1 < W_2)$; operation and maintenance $(OM_1 > OM_2)$; inherited $(I_1 > 0, I_2 = 0)$, residual $(R_1 = 0$ if choice 1 or 2 selected, and R_2 is unavailable if choice 1 is selected, or $R_2 > 0$ if choice 2 is made); and contingency $(C_1 < C_2)$. Construct a total symbolic effectiveness measure for choices 1 and 2. Find their marginal difference. Indicate nonrecurring and recurring cost. Give their investment costs. What is the opportunity cost in the context of the circumstances of selecting options 1 or 2? Does opportunity cost differ before and after the decision of selecting the system?

10.7. (a) Cash flows for a project A are given as follows:

	Year				
	0	1	2	3	4
Costs, C_n	20	5	5	6	8
Benefits, B_n	0	25	20	15	10

For an interest rate of $i = 8\%$, find the present worth of costs and benefits. Also calculate the net present worth and benefit cost ratio.

(b) Cash flows for project A are given as

	Year				
	0	1	2	3	4
Costs, C_n	20	8	6	5	5
Benefits, B_n	0	10	15	20	25

By using an interest rate of $i = 8\%$ find the present worth of costs, benefits, their difference, and the benefit cost ratio.

10.8. Two projects, B and C, have the following cash flows:

		Year			
		1	2	3	4
Project B	Maintenance	5	5	6	8
	User Cost	25	20	15	10
Project C	Maintenance	4	4	4	4
	User Cost	20	15	10	5

(a) Investment costs for projects B and C are $20 and $30. The interest used in the analysis is 9%. Find the present worth of cash flows. Determine the benefit cost difference and ratio between the two projects.

(b) Repeat for 10%.

(c) Repeat for 15%.

10.9. An engineer is considering a new project. Possible choices and estimates of costs and benefits are shown in the table. Each project has a 5 year life. Interest is 9%.

	Project Value ($\times 10^5$)		
	A	B	C
Initial cost	$144	$50	$380
Annual net benefits, 5 years	40	16	100
Residual value	0	0	10

(a) For each project find the net present worth. Which project should be chosen? Find the benefit cost ratios.
(b) Find the marginal benefit cost ratios. Which project should be chosen?
(c) Reconsider with an interest of 15%.

10.10. A governmental agency that uses an interest of 5% is able to select one of three projects:

	Project Value		
	A	B	C
Initial cost ($ $\times 10^6$)	40	50	60
Annual benefits ($ $\times 10^6$)	15	20	25
Project life (years)	4	4	4

(a) Find the net present worth. Determine the B/C ratios. Rank the projects according to method of evaluation. Determine the marginal benefit cost ratios. Rerank the methods. Which one do you prefer?
(b) Reconsider with 10%.

10.11. A welfare program costs $350,000 with a $20,000 annual operating cost. Benefits are anticipated to be $40,000 in the first year and to increase by $30,000 each year for 4 years. It will decline to no benefit in 2 years. Interest = 5%. What is the benefit cost ratio with and without discounting?

10.12. A sociopolitical system has a first cost of $1 million and an annual maintenance cost of $25,000 each year over a 50-year life cycle. Benefits average out as $50,000 per year.
(a) At 5% interest, what is the system net present worth?
(b) What is the benefit cost ratio?

10.13. The government tells potential contractors that a product will be evaluated according to this LCC model:

$$\text{LCC} = \text{unit operating cost}$$
$$= (\text{unit price} + \text{logistic cost}) \div \text{service life}$$

The selected contractor demonstrates service life in a post award reliability acceptance test. If the reliability test does not meet the level stated by the contract, then a penalty function deducts from the unit price as (1 − test value MTBF/quoted MTBF) × (unit price + logistic cost). The logistic cost is $200.

Company	Unit Price	Hours MTBF
A	$350	1000
B	400	1200
C	700	2100

Find the winning company. Now suppose that the contractor failed to meet the quoted MTBF by 10%. What is the penalty and the final price?

10.14. The government establishes a LCC model to evaluate tires. The model is given as LCC = quantity (unit price + shipping cost + maintenance cost per unit). Three tire manufacturers are invited to bid. Each is asked to provide a sample for simulated landing tests to determine the number of landings per tire. The government determines the number of tires for each company's quote based on a tire landing index, which is given by number of required landings divided by best performance.

$$\text{number of required landings} = 1,200,000$$

Shipping costs are evaluated from the manufacturer to a central inventory. Maintenance cost to change a tire is $47.50. Make a bid evaluation to determine LCC price. Which company do you select?

Company	Landings per Tire	Shipping Cost per Tire	Bid Price per Tire
A	110	$13.50	$1380
B	105	7.50	1280
C	95	13.00	1360

10.15. A purchasing agent announces that a system will be evaluated according to the following LCC model:

$$C_m = \frac{\text{SOH}}{\text{MTBF}}(\text{MTTR})C_m + \text{Unit Price}$$

The following bids were received. Which bid do you choose?

Company	Unit Price	SOH	MTBF	MTTR	C_m ($/hour)
A	$200	5000	1000	1.5	$25
B	280	4500	1500	1.4	25
C	260	4000	2000	1.3	25

10.16. The B/C test has been applied to welfare and job training for underprivileged youth. In this case an undiscounted B/C ratio is given as

$$\text{benefit cost} = \frac{B_p - B}{C_a - T_n}$$

where B_p is graduate earnings, B is original earnings of student, C_a is annual amortization payment, and T_n is taxes on net increased earning of student. The following items are estimated:

	Estimated amount
Direct program training cost	$ 725,000
Allocation of center overhead based on planned enrollment	950,000
Subtotal	1,675,000
Capital investment cost at 5%	83,750
Job corps cost at 25%	418,750
Total cost	2,177,500
Number of graduates	400
Total cost/graduate	5,443
5-year amortization cost (C_a)	1,089
Average starting salary	4,222
5-year average salary (B_p)	5,026
5-year average taxes (T_n)	562
Original earning power of students (B)	1,040

(a) Find the undiscounted and discounted benefit cost ratio.
(b) Let the interest rate be 5%. What happens to the B/C ratio as the interest rate increases? Decreases?

10.17. Construct a personal B/C ratio for your own education along the lines of the one in problem 10.16.

10.18. An aerospace firm has received a second bid for an environmental chamber. Design cost is estimated as $30,000 and $10,000, which occur in the first two periods. Installation costs $30,000 at $n = 4$. Product manufacture is quoted as $90,000, $150,000, and $25,000 for $n = 2, 3,$ and 4. The new bidder says that operation will require two operators. The preventive maintenance cycle time is 200 hours, and each maintenance action requires 5 hours. MTBF = 650 hours and MTTR = 25 hours. Power requirements are 23 kW. Spare parts are 0.25%. Find the LCC. Determine the discounted value of LCC where the opportunity interest rate = 10%.

MORE DIFFICULT PROBLEMS

10.19. An existing highway, ABC (Fig. P10.19), originally constructed in 1944, is 8.5 miles long. Average daily commercial traffic is 10,000 vehicles with 10% trucks. Now requiring reconstruction, the unit estimated cost of improvements to existing highway is $2 million per mile. The right-of-way will cost $550,000 per mile. A 7-mile supplemental

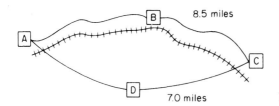

Figure P10.19

location, ADC, can be constructed for $3 million per mile. The right-of-way costs $220,000 per mile. ADC requires two viaducts for $500,000 each. Long-term mainte-nance cost of either road is an average $10,000 per mile per year. This maintenance includes occasional resurfacing. Trucks cost $1 per mile, and cars are evaluated to cost $0.25 per mile. If ADC were constructed, then it is estimated that 40% of the traffic between points AC will move to ADC and the remainder would use ABC. Describe the alternative plans.

(a) Based on a project estimate of capital cost only, which plan do you advise?

(b) What are the discounted and undiscounted traffic costs for 20 years for ABC and ADC? Use $i = 5\%$.

(c) Let $i = 5\%$. Construct a B/C analysis for this total problem, and advise a solution.

10.20. Four mutually exclusive designs have their benefits and costs estimated.

	Present Worth	
Design	Benefits	Cost
A	$48,000	$38,000
B	35,000	24,000
C	37,000	31,000
D	45,000	34,000

Determine the individual B/C ratios, and analyze the four choices on the basis of mar-ginal yield to find the best one.

10.21. A nonrepayable electronic component is up for bid and an LCC approach is deemed mandatory. The purchasing agent advertises that an LCC model to select the winning bid will be based on the model

$$\text{cost per unit} = \frac{\text{unit price} + \text{unit stocking cost}}{\text{bid MTBF}}$$

The stocking cost of $110 per unit is the total cost for storage, installation, and dis-mantling. Each bidder supplies this information to the purchasing agent.

Bidder	Unit Price	Bid MTBF
1	$1000	800
2	1250	615
3	1175	917

Which bidder wins the contract?

CASE STUDY:
LIFE CYCLE OF PUMPING STATION

A new pump station is being designed by Mr. Jarrett, a municipal planning engineer. Jarrett secures the following information on the new pump:

	Amount
Installation cost (1 year from design)	$22,000
Preventive maintenance cycle	2190 hours
Operating hours per year	2190 hours per year
Preventive maintenance action	4 hours
MTBF	10,950 hours
MTTR	8 hours
Power demand	60 kilowatts
Power cost	$0.07 per kilowatt hour

Jarrett estimates design cost as $40,000 for year 1. Maintenance labor costs $26 per hour. In addition to the consumption charge of $0.07 per kilowatt hour for power, there is a monthly demand charge of $390 for power. Find the LCC for the first 10 years of pump operation. What is the discounted LCC assuming an 8% interest rate?

11

Estimate Assurance

The engineer's interest in his or her estimate continues beyond the conclusion of the estimate. The estimate accuracy, reliability, and quality are important. If the estimate leads to a sale or winning bid, then there is an opportunity to verify its general goodness. This chapter describes the analysis of estimates when compared to a so-called *actual value*. There are lessons to be learned in, for instance, improving the capture rate, understanding the importance of human behavior, and in various techniques.

It is an important dogma of this book that the cost estimate is more useful to the design than the actual or standard cost. Each of the four kinds of estimates (operation, product, project, and system) requires modifying features to assure that reported costs concur with the estimate.

11.1 ANALYSIS OF ESTIMATES

Cost estimating is seldom done without an effort to check its success. A common method of verification is to find so-called actual costs and compare those with the original estimates. That was done in a study of 157 cost estimates for tools in a manufacturing plant, as shown in Fig. 11.1.

An error is measured as

$$E = \left(\frac{C_e}{C_a} - 1 \right) \times 100 \tag{11.1}$$

where E = percentage error of estimate

C_e = estimate of cost, price, bid, or effectiveness

C_a = actual cost, price, bid, or effectiveness

428

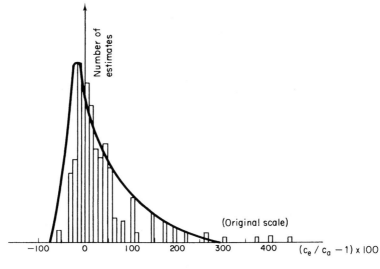

Figure 11.1.   Distribution of elements for various tooling jobs.

Now, if a firm determined this percentage error for a large number of estimates and plotted those findings on a histogram, then they could, by assuming an infinite population of estimates, form a *density distribution curve* to those *error estimates*. In this actual study the deviations ranged from a low of 50% to a high of 450%, but the sum of the estimated amounts is less than 4% above the total actual cost.

Obviously, all the jobs that were estimated low did not yield as much as expected. The others brought in more than expected because they actually cost less than estimated.

A natural question then is, What can be learned from an analysis of this sort to help improve the cost estimating practice? Should some amount be added to each estimate to reduce the losses from the low ones? If so, then how much? Obviously, to do so would raise the high estimates and diminish their chances in competition. The low estimates cannot be identified beforehand or there would not be any low estimates.

Though analysis of this sort is useful, it does not give information about jobs that did not meet competition. The actual costs of unsuccessful bids were never determined. Study of estimating policies must consider all estimates, not just the ones that have produced orders.

One objective for the estimating function is to produce estimates that are exact. While this is a commendable purpose, it is more realistic to say that the goal is to have the estimate value fall within some acceptable range. This advances the notion of a *tolerance,* which is open to analysis and consideration. Factors such as cost of preparation, time available, impact on the organization, and data requirements bear on the selection of an estimate tolerance. Inasmuch as estimates precede the fulfillment of the design, a passage of time exists between the estimate and the historical determination of the actual value. In some situations, data may never be gathered to allow even a nominal comparison. At the other extreme, an abundance of data may be on hand. Due to accounting effects a tidy comparison may be impractical. In practice a *reconciliation* between estimate and actual measure, though very desirable, is difficult to bring about.

There is another problem in assuring the value of the estimate. Engineers estimate, but it is management that controls the elements of cost. Certainly, engineers are a part of the general management team. However, the prerogatives of cost control, workers, budgets, and so forth are the responsibilities of nonengineers. We may argue that an engineer is responsible for forecasting the peccadilloes of management. Engineers should try to be aware of those factors when costing designs. Nonetheless, the responsibility of cost control is either delegated or shared with others.

In further discussion, the estimated cost C_e is understood to be the total price for a job. This includes the usual items of cost, direct and indirect, overhead, contingencies, and an expected profit. The *contingency* allows for unassignable extra costs that may occur. The total price is for the market, a quotation, or a bid. The actual cost C_a includes those same factors of cost and contingencies because they are inherent in the performance of each job.

For simplicity the deviations in cost estimates will be expressed by the ratio C_e/C_a. This does not change the shape of Fig. 11.1 to which a new scale may be applied to show the deviations in terms of C_e/C_a. Obviously, this ratio equals one where actual cost equals the estimated cost. This is the important break even point. Those points to the left on the horizontal axis are operationally undesirable because the firm will be operating at a loss. Depending on bidding policies and objectives, the distribution of estimates is to the right of the break-even point. Long-range survival depends on this.

But curves such as Fig. 11.1 are seldom, if ever, found because actual costs may not match neatly with estimates. Those nonsymmetrical data can be normalized for additional analysis, but those techniques are beyond the scope of this text.

Our purpose in this chapter is not to deal with "cost control." Instead, more important is the principle that estimate assurance is the critical factor. *Cost control* places emphasis on cost containment. But estimate assurance recognizes that the value of the estimate is the best value, even better than actual costs, and the objective becomes one of regulating costs to match the estimate.

Costs are identified as estimate, actual, or standard. *Actual costs* are incorrectly assumed to be the more important. However, it is seldom that actual costs are accurately known. Further, actual costs are not ordinarily indicative of future values. Actual costs are more expensive to determine than estimates and are not available until after the operation, product, project, or system is complete. Then actual costs are too late except for analysis and re-estimating. As is suggested throughout the chapter, actual costs are seldom known with precision.

Standard costs are hypothetical and in a sense are "should be" costs. Standard costs are fashioned on accounting principles and are useful for income and expense recording, tax finding, and other financial reports. Of course, standard costs are never "true" except in the sense of a definition.

It is a dogma of this author that the *estimate value,* properly determined, is superior to either actual or standard values. It may be surprising to the reader that actual costs may never be determined, but it is interesting to speculate on the deviation with estimates that would be revealed if actual costs were known. This difference is due to the error of the estimate. Three kinds of errors are mistakes, policy, and risk.

Mistakes result from imprecision, blunders such as $2 + 2 = 5$, and omissions. Typically, mistakes pass unnoticed, but if they occur, "nature" may be kind due to compensating effects of off-setting mistakes. Prevention results from uncompromising arithmetic and strict attention to methods that inspire faultless computation. The use of a computer is a popular solution for overcoming mistakes. As a tireless machine, it removes the burden of routine calculation. But most estimates are not manipulated by a computer but by pad and pencil. Cross checks by other engineers or the stapling of a calculator's paper tape to the estimate or row and column arithmetic simultaneously agreeing for the final value are simple ways to reduce mistakes. Preventing omissions in cost elements is encouraged by *pro forma* procedures, cross talk within the estimating team, and checklists. Unfortunately, most management views errors as implying only mistakes.

Policy errors are errors of belief made through ignorance or inadvertence. Simplified illustrations include failure to recognize material price breaks for quantity purchase or by overlooking a planned contractual increase in direct labor cost. Excessive or low values for overhead ratios, for instance, or first cost or cost of operation are typical. A cost-estimating relationship may have higher statistical correlation with a nonlinear relation instead of a linear model. Errors of policy are prevented by well thought-out policies, practices, and self-learning.

Risk errors are the least understood. Assume that in a competitive bidding situation your value is "perfect" and yet a competitor submits a lower value. The lower value may be the result of their more productive business, and, thus, your failure to win is a consequence of your unproductive business. On the other hand, the competitor's bid may be a "low-ball," a bid simply to keep the work force busy. Thus, your bid still lost. Contrariwise, your bid may win because your business is more productive or competitors are not really interested and submit high bids or they make mistakes and over bid. Maybe your bid is set intentionally low to keep your work force busy. Profit margins may be overlooked also. Risk error is the difference between the estimate and the winning value and is uncontrollable by the firm's engineer.

The winning bid for public works may be known, where the law dictates open disclosure of information. Some competitors' estimates may be proprietary and are unknown. On the other hand, some of those values can be determined by simply asking. Thus, the deviation between the estimate and actual cost for estimates not realized may never be known.

Consistency is another quality of estimates and engineers. There is the criticism that engineers are seldom consistent one day to the next and that different engineers will estimate the same design differently. Yet the nature of estimating requires a pinch of judgment and experience. Technical work, by its very nature, has variances between competent and trained engineers. Estimating is no different. One model, useful for evaluating consistency, is to construct the signal-to-noise ratio or C_a/SD where the C_a is the average estimate of several engineers of the same design under the same procedures. The SD is the standard deviation of the number of estimates. Usually, if there is a data base, and rules are followed in making estimates with the data base, then the signal-to-noise ratio is smaller than for those situations where judgment has a greater role.

Equation (11.1) finds the error by comparing actual costs to estimates and overlooks policy and risk errors. Finally, the comparison is not available until after the actual costs are reported.

An improved error measure uses a preliminary and a detail estimate and is given as

$$E = \left(\frac{C_p}{C_d} - 1\right) \times 100 \qquad (11.2)$$

where C_p = preliminary estimate of cost, price, bid, or effectiveness

$\quad C_d$ = detail estimate of cost, price, bid, or effectiveness

Naturally, the preliminary and detail estimate are made for an identical design. The preliminary estimate is made quickly and independently of the detail estimate. A detail estimate uses a *pro forma* procedure and emphasis is on comprehensiveness. There are pitfalls to avoid. It is necessary to prevent the preliminary estimate from becoming a self-fulfilling value where a detail estimate is guided to match a preliminary estimate. Two groups or individuals or alternative methods may make those estimates separately. The value of C_p should be restricted on a need-to-know basis so as not to influence the value of C_d.

Ultimately, estimates either win or lose in the sense that business is obtained. The purpose of a detail estimate is to determine an economic value of want that is desired by competing selfish interests.

Contractors and job-shop manufacturers frequently submit cost estimates to potential customers. In the case of products, estimating is done to set cost and price. Estimates may win or lose in this case as well. Engineering may drop a proprietary product because it may not be economically successful, or a vendor may not be selected. Winning and losing business can be functionally related to cost-estimating performance. This analysis is called capture rate and is defined as

$$\text{capture rate} = \frac{C_w}{C_t} \times 100 \qquad (11.3)$$

where C_w = cost estimates won, number

$\quad C_t$ = total cost estimates, number

This capture rate may be determined monthly or yearly. Some firms may make hundreds of estimates per month, though for projects only a few are made yearly. The capture rate may differ between new and repeat business. Some firms adjust their costs or margins depending on the direction of the capture rate. If the capture rate falls, then profit margins are reduced, and vice-versa. Some firms operate successfully with 1% and unsuccessfully with a 90% rate. This measure alone does not give the entire picture of successful estimate assurance.

Actual costs deviate from estimates because of inefficiency, for instance, or superior efficiency. Though it is a goal to have actual costs approach the estimate, the usual term describing the reasons for this departure is productivity.

Adjustments to the cost estimates are made by using an overall productivity factor. Engineers usually think of productivity as analogous to efficiency, or output/input, which is a number less than 1. We recommend the reverse ratio. But an output/input factor would divide the new estimate to indicate a "realized" estimate, or "adjusted estimate."

The productivity factor is found by using

$$PF = \frac{\Sigma C_a}{\Sigma C_e} \tag{11.4}$$

where PF denotes the productivity factor, a dimensionless number. For instance, if ΣC_a = \$5157.58 and ΣC_e = \$4715.07, then PF = 1.094. If those data were considered representative of future work, the total estimated cost for a future design would be multiplied by 1.094 to anticipate the cost actually reported. The productivity factor may be found from a single sample or a time series of experiences. It is doubtful if a constant productivity factor is achievable. Ultimately, the engineer will use opinion and experience to state if a historical value can be used for the future.

Despite a variety of possible ways to make an estimate, resources for the preparation are always restricted because time, money, and the technical staff are limited. Eventually, a point is reached where the objectives must be satisfied with what time, money, and intelligence are available. The ideal policy would have the estimate coincide with the reconciled actual cost.

Actual cost is, perhaps, never known. No procedure, mathematical technique, or policy employed in the engineering world is without its flaws and shortcomings or is able to guarantee perfect estimates. Although flaws in estimating may be obvious, those procedures and techniques are used for the simple reason that they are the best means at hand. Imperfection seldom deters usage.

11.2 BEHAVIORAL CONSIDERATIONS

The primary goal of estimate assurance is to develop and sustain business systems and to inform management about the extent, type, and status of expenditures. The

computer has given impetus to those systems to enable those systems to be more effective. Though computerized business systems should lead to improvement of estimate assurance, it is recognized that this is only the routine part. Though those
enamored features are traditionally thought to be the essence of estimate assurance,
it may be surprising to the reader that despite their development there remains a
general lack of estimating success. Little heed is given to the opportunity that those
systems operate in an organization. Behavioral principles should be considered by
the engineers since they encourage the better functioning of estimates.

The general "control" model consists of objectives and standards, measurement
of actual results, comparisons of actual results to standards, and engineering action.
Figure 11.2 describes a simple network. The "sensor" measures the output transmit-
ted for comparison with a standard. If an unacceptable difference is noted, then action is taken by means of an "actuator." Those control models that do not respond to
external social or economic influences are known as closed. Systems that, for example, respond to poor labor efficiency, late material delivery, engineering changes,
competition, and price changes are known as open. Open systems do not lend
themselves to automatic regulation and are more difficult and hence challenging.
Less management time and finesse are necessary for a mechanistic or closed system.
(A detailed discussion of behavioral principles is beyond the scope of this text, but
considerations are amplified throughout this chapter.)

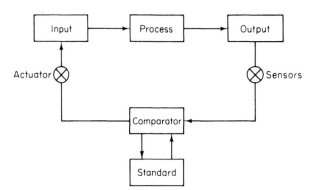

Figure 11.2. Simple control model
for cost assurance.

Let us relate those model components to engineering. Objectives and standards deal with performance, time, and cost. To be meaningful, unambiguous, and
useful, objectives and standards must be quantitatively stated in measurable operational units. An objective could be the cost estimate, but there are many other objectives. Measurement of results records actual cost and relates to the cost code,
work break-down structure, operation sheet, product task, and so on.

This measurement may be supplemented by special reports and audits. Comparison of actual results to standards is done by means of reports. A simple example
is the side-by-side comparison of estimate and actual costs for material, labor, and
overhead, broken down in as fine detail as required. Much of this is done routinely
in conjunction with results management. Trending and charting of milestones are
other techniques for this purpose.

Engineering actions encourage performance to match the objective. If there is agreement, then action is unnecessary. A match of this kind is unlikely. Because estimates are "open" systems, change is commonplace, and management is necessary to encourage conformance. This is the most crucial component in the cost-estimate assurance program.

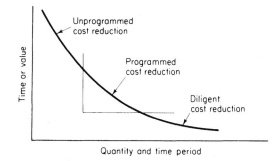

Figure 11.3. Reactions to match cost estimate.

A policy issue may determine whether a cost estimate should be made tight or loose. Behaviorally, it can be argued that tight estimates will motivate attainment and lower cost, as Fig. 11.3 suggests. A counterargument is that loose cost estimates easily attained will probably generate good feelings. Here, psychological motivation is needed for self-esteem. Research findings are biased toward tight estimates for productivity attainment. How tight the cost estimate should be is unresolved because if set too tight then the cost estimate may demotivate and cause lower performance. The cost estimate should reflect the anticipated *money out of pocket* that the element will require. Thus, if a diligent cost-reduction program is planned, lower cost is anticipated by the cost estimate.

Engineering must be aware of what cost elements are significant in a cost-estimate assurance program. A law, often attributed to Pareto, is the application to objective identification. It identifies that 20% of the designs, for instance, contribute 80% of the cost. Whether 20:80, 30:70, or another ratio is the correct rule of thumb is unimportant, but the rule relies on an analysis to determine the significant contributors to cost.

11.3 OPERATION ESTIMATE ASSURANCE

An *operation* is the conversion of direct material by direct labor into various shapes. Indirect costs such as factory overhead are necessary for the conversion. Operation estimates are for the immediate future period. Because of this brief time, comparison of actual to estimate values is sometimes possible.

"Standard" is often used synonymously with "estimate" for operation work. Accountants are involved with standard costs. The engineer, when referring to standards, thinks in terms of a rigid specification, but analysts from other fields have dissimilar viewpoints. A standard cost as discussed here provides a dollar amount that is a "should be" amount and is not an immutable natural law.

An attitude is necessary in viewing standard costs. Standard costs may be classified as perfection level standards, and engineering encourages attainment of those standards as goals. Some businesses contend that perfection standards are preferable to attainable standards because perfection standards provide a stimulus to workers and engineering to achieve the best possible performance.

A standard unit cost of a labor operation, part, or product is a predetermined cost computed even before operations are started. In constructing the standard unit cost of an item it is necessary to study the kind and grade of materials that should be used, how each labor operation ought to be performed, how much time each labor operation should take, how the indirect services should be best administered, and the entire specifications for the complete and total operation. The aim, of course, is to specify the most effective method of making the item and then, through adherence to the specifications in the actual operations, achieve the lowest practical unit cost.

A system of this type provides a calculated and anticipated cost of all products by cost elements; comparisons of the anticipated cost with actual results and the reasons for any differences; the effectiveness of all cost elements, including material, labor, and overhead; and measurement of departmental or individual performances against accepted standards. This is, of course, the method of estimating.

Despite the care in establishing standard costs, the reported actual costs are likely to deviate from the standard. These differences are known as variances and are expressed as dollar amounts or percentages. They are favorable variances when the actual costs are less than the standard costs and unfavorable when actual costs exceed standard costs. It should not be interpreted that excess of actual cost or of standard cost is nonbeneficial to the firm. Similarly, not all favorable variances represent actual benefits to the company. The terms favorable and unfavorable when applied to variances indicate the direction of the variance from standard cost.

Some businesses use *variance analysis* to understand inflation, deflation, schedule delays, engineering change orders, or rework. Those variances are the result of a buyer or seller action or a governmental regulatory agency. Other causes are possible. A consistent approach allows for reconciliation and explanation of the total cost increase.

The elements used to estimate *direct material cost* are quantity, shape, and the raw material cost. Consequently, actual material cost may differ from the estimate because of a quantity variance or raw material cost variance or both. For instance,

$$\text{Estimate 200 units at } \$17.38 = \$3476.00$$

$$\text{Actual 210 units at } \$17.05 = \$3580.50$$

An unfavorable variance of $104.50 resulted in part from an excess of 10 units and in part from a deficit unit cost $0.33. The analysis is described further in Fig. 11.4. There is a simple way to analyze the difference. The figure starts with actual material cost per unit times actual units. We change one of the factors (cost or quantity) to the estimate. The difference gives a variance due to the factor that

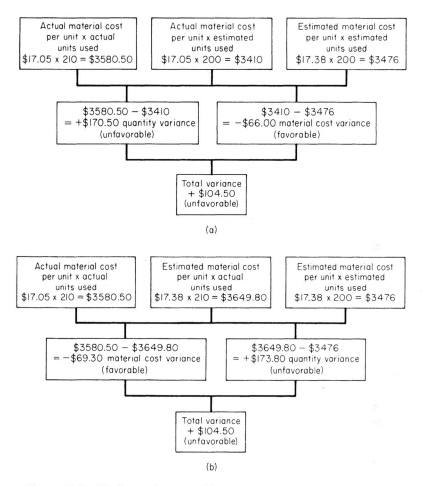

Figure 11.4. Finding variance on (a) quantity first and (b) material cost first.

was changed. The remaining difference is the variance for the other factor. In Fig. 11.4(a) the quantity factor was changed first, and for Fig. 11.4(b) the material cost factor was changed first. Those procedures result in a difference in the quantity or cost variance, depending on the order of calculation.

This ambiguity can be significant in owner-contractor or contractor-subcontractor disputes. Terms of the contract may stipulate the manner in which variance analysis is undertaken. A typical resolution is as follows:

$$V_m = (N_a - N_e)C_e$$

$$V'_m = (C_a - C_e)N_a \qquad (11.5)$$

$$\text{net material variance} = V_m + V'_m$$

where V_m = dollar variance for material owing to quantity change

      V'_m = dollar variance for material owing to material cost per unit change

      N_a = actual quantity, number

      N_e = estimated quantity, number

      C_e = estimated material cost per unit

      C_a = actual material cost per unit

Those calculations are shown in Fig. 11.5(a). Both favorable and unfavorable variances are possible. By using Eq. (11.5) the solution to Fig. (11.5) becomes V_m = \$173.80 and V'_m = \$69.30. The net material variance is an unfavorable \$104.50 (= 173.80 − 69.30).

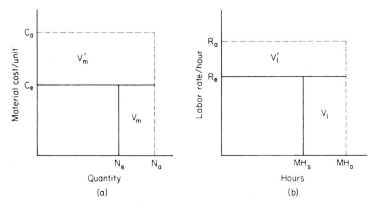

Figure 11.5.   (a) Variance for material; (b) variance for labor.

Labor costs can be analyzed for both man hours and the labor dollar rate. Figure 11.5(b) illustrates the notation.

$$V_l = (MH_a - MH_s)R_e$$

$$V'_l = (R_a - R_e)MH_a \qquad (11.6)$$

$$\text{net labor variance} = V_l + V'_l$$

where V_l = dollar variance for labor owing to a difference from estimated hours

      V'_l = dollar variance for labor owing to a difference from estimated hourly rate

     MH_s = estimated total standard hours for operation

     MH_a = actual total hours for operation

      R_e = estimated labor hourly rate, dollar/unit

      R_a = actual labor hourly rate, dollar/unit

Consider the example where the estimated wage rate = \$18.76 and hours are 283. The actual wage rate and hours were \$17.32 and 325. The unfavorable variance owing to changes in hours is \$784 [= $(325 - 283)18.67$], though the favorable rate variance is $-\$439$ [= $(17.32 - 18.67)325$]. The unfavorable net labor variance is \$345 (= 784 − 439).

Man hours are the product of a labor standard and the number of units. A three-dimensional variance analysis is possible for labor operation if we consider quantity, labor hourly rate, and the allowed or standard unit time. Define

$$V_{dl} = (N_a - N_e)H_s R_e$$
$$V'_{dl} = (H_a - H_e)N_a R_e \qquad (11.7)$$
$$V''_{dl} = (R_a - R_e)N_a H_a$$

net direct labor variance $= V_{dl} + V'_{dl} + V''_{dl}$

where V_{dl} = dollar variance for direct labor owing to a difference from estimated quantity

V'_{dl} = dollar variance for direct labor owing to a difference from estimated standard hour rate

V''_{dl} = dollar variance for direct labor owing to a difference from estimated wage rate

H_s = estimated standard hours, hours per unit

H_a = actual hours, hours per unit

For a standard rate of 1.415 hours per unit and an estimated quantity of 200, the actual rate of 1.548 hours per unit and 210, and an estimated wage rate of \$18.67 and actual of \$17.32, the following facts are uncovered: $V_{dl} = \$264$, $V'_{dl} = \$520$, and $V''_{dl} = -\$439$. The unfavorable labor variance is \$345.

Indirect expenses are important for operations. Methods of their regulation and assurance are not as straightforward as direct material and direct labor, where variances are the focus of attention. Indirect costs, since they are in support of direct costs, are assumed to vary proportionately. A ratio of indirect to direct costs is one way to monitor those costs. Overhead can be analyzed for variances also.

Behavioral considerations are important for operation cost of labor and materials. Labor cost reduction can be observed with the learning theory, as is shown in Fig. 11.6. Reduction of direct labor hours is due to working harder, improved methods of design, or material changes. Contrariwise, slow down, morale problems, unforeseen design changes, or destabilizing personnel reactions can alter the slope, and unlearning may occur. For the most part, labor estimates are lowered because of improvements caused by technical productivity.

Direct materials are sometimes the major part of an operation. Figure 11.7 describes the percent increase for material. In this actual history, the cost of a material ran away for 1 year before engineering efforts were applied to regulate the material cost. The darker line is the actual purchase cost of raw materials, and the dashed line is the compensating price. The difference is lost revenue. Figure 11.8 is a learning theory application for a subcontract material, diesel engines in this circum-

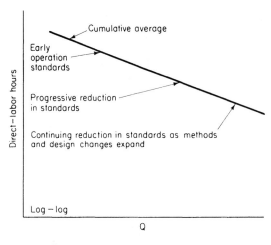

Figure 11.6. Typical effect of operation learning for standards. Note: Standards are lowered because of methods, design, or material changes only.

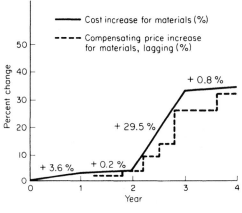

Figure 11.7. Effects of material cost increases.

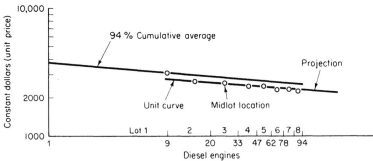

Figure 11.8. Learning curve application for purchase of subcontract materials.

stance. Those diesel engines were adapted for a special design and thus are a subcontract material. The constant dollar cost of the engines declined because of the estimate assurance program.

11.4 PRODUCT ESTIMATE ASSURANCE

A product estimate deals with material, labor, and overhead cost elements for a design that is repeated and changed. This notion, introduced in chapter 1, suggests a similarity between subsequent designs. Model 2 is like model 1, and we apply that principle to guide the logic of product estimating. This similarity is primary to the practice of product estimating. Product quantity may be anywhere from a few to millions and is another important factor. Seldom is a product design so novel that the firm has virtually no experience. Estimate assurance programs for the product depend on similarity in design and quantity performance.

Methods for estimating products were introduced in section 8.6. There are corresponding techniques for product assurance. Table 11.1 is an example dealing with a variable cost method for product estimating. This method is first discussed

TABLE 11.1 VARIANCE ANALYSIS OF INCOME AND EXPENSE STATEMENT FOR MODEL 1 OF SPECIAL PRODUCT

	Amount
Net sales (20,000 units at $0.965)	$19,300
Variable costs	
Labor, 20,000 units at $0.20	$ 4,000
Material, 20,000 units at $0.384	7,680
Manufacturing expenses, 20,000 units at 0.02	400
Material overhead expenses, 20,000 units at 0.007	140
Administrative and marketing expenses,	
20,000 at 0.017	340
Total variable costs	$12,560
Profit contribution at standard	$ 6,740
Percent of net sales	35%
Variances on variable costs	
Labor	$ (450)
Material usage and scrap	(1025)
Manufacturing expenses	25
Material expenses	45
Administrative and marketing expenses	(20)
Purchase price	(1250)
Total variance	$(2675)
Profit contribution at actual	$ 4,065
Percent of net sales	21%
Period costs	
Budgeted costs	4,060
Variances on period costs	(140)
Total period costs	$ 4,200
Total cost	$19,345
Net profit	$ (135)
Percent of net sales	0.7% loss

with marginal cost implications, and Fig. 6.15 serves as a planing attempt to understand marginal cost analysis. The pricing model (8.17) uses the notion of contribution. The student will want to review this background to understand Table 11.1.

At this point we assumed that 20,000 units of a special product model 1 have been produced, sold, and all cost facts have been historically determined. Observe in Table 11.1 that net sales are $19,300 and variable costs for labor, material, and various overhead are determined. A total variable cost $12,560 is found. Contribution at standard is 35%. At this point, variances exist, both favorable and unfavorable, on the variable costs and purchase price. Profit contribution at actual reduces to $4065 and 21%. Period costs, which are of a fixed nature occurring within the model 1 production period, are identified with budgeted and variance values. Now, total costs are estimated as $19,435, showing a loss of −0.7%. Apparently, model 1 was not as successful as was initially hoped.

It is at this point that a product estimate assurance program is necessary. Little can be done with the results of Table 11.1, because the money is already spent except that the results can lead to better operation methods, design, sales, and so on. Marketing and design engineering changes may lead to models 2 and 3. Note Table 11.2, where an analysis of models 1, 2, and 3 is summarized. Clearly, model 2 is preferred because its percent of net sales is the highest.

Behavioral considerations, while not a root solution alone to product estimate assurance, lead to achieving desired results. Learning theory is appropriate for products having significant aggregate value and low volume. Figure 11.9 is an example of an actual history and is evidence that the learning theory applies to this circumstance. Once this fact is established, learning can be used for estimating this or similar products produced by this company. Successful estimating with learning is a consequence of achieving results: Learning should not be used as an estimating technique unless company history has demonstrated its success.

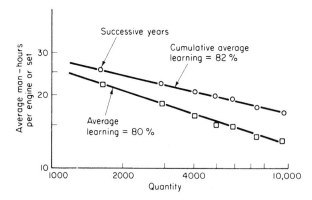

Figure 11.9. Historical analysis of learning effects for engine-generator project.

Design effort is necessary for learning theory reduction. Figure 11.10 is an example of the effect of an early or later estimate assurance engineering program. The right vertical axis, engineering change notices (ECNS) and drawing change notices (DCNS), are related to a cost reduction impact. This axis scale increases downward. Those engineering design programs are the investment in slope reduction.

TABLE 11.2.   ANALYSIS OF INCOME AND EXPENSES FOR MODELS 1, 2, AND 3 OF SPECIAL PRODUCT

	Total	1	2	3
Sales	$97,300	$19,300	$36,000	$42,000
Variable costs				
Labor	$23,000	$ 4,000	$ 7,000	$12,000
Material	38,330	7,680	11,000	19,650
Manufacturing expenses	2,200	400	850	950
Material expenses	740	140	250	350
Administrative expenses	1,850	340	560	950
Total variable costs	$66,120	$12,560	$19,660	$33,900
Profit contribution	$31,180	$ 6,740	$16,340	$ 8,100
Percent of net sales	32%	35%	45%	19%
Variances on variable costs				
Labor	$ (830)	$ (450)	$ 120	$ (500)
Material usage and scrap	(1075)	(1025)	$ 750	$ (800)
Manufacturing expenses	(100)	$ 25	$ (75)	$ (50)
Material expenses	(90)	45	(85)	(50)
Administrative expenses	(45)	(20)	0	(25)
Purchase price	(2500)	(1250)	0	(1250)
Total variances	$(4640)	$(2675)	$ 710	$(2675)
Profit contribution at actual	$26,540	$ 4,065	$17,050	$ 5,425
Percent of net sales	27%	21%	47%	13%
Period costs				
Budgeted costs	$18,060			
Variances on period costs	(800)			
Total period costs	$18,860			
Total costs	$89,620			
Net profit	$ 7,680			
Percent on net sales	8%			

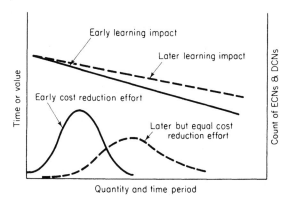

Figure 11.10.   Early and later engineering effort to assure product estimate.

Some products that have low value and high quantity may be inappropriate for learning theory. Other techniques, such as design to cost, can be successfully used. Note Fig. 11.11, which shows a product work breakdown structure. Some of the boxes have been completed and cost has been closed out, and costs are in progress

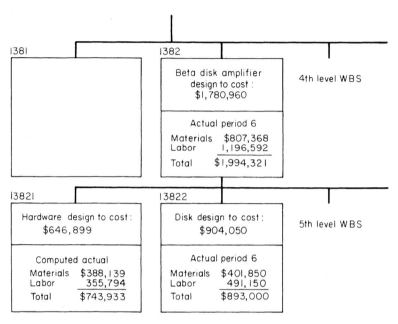

Figure 11.11.　Reporting actual costs with design to cost methods.

for other boxes. The S curves of chapter 9 can be used to note the trends. For long-term products, it may be possible to reduce the level of expenditure and assure the value of the estimate.

11.5 PROJECT ESTIMATE ASSURANCE

The capital investment for a single end item is determined by a *project estimate.* The money required for a project is large compared to the available resources. Because of the importance of projects to the long-term financial health of the firm, considerable effort is devoted to assuring the success of estimate. Projects are scheduled over a long duration, and opportunities exist to regulate the actual costs. Both the owner and the contractor desire those objectives. However, an estimate assurance program is more than a mere objective: It defines details, provides a strategy, and indicates the steps. Owners sanction the project because of a profitable time value of money concept, and money and time (or schedule) are the major factors of concern.

The project engineer in preparing the work breakdown structure (WBS) package provides details, strategy, and steps. The WBS package is discussed in chapter 9, which developed the scope and definition, designs, work breakdown structure, schedule, and estimates. The WBS is used for estimating, planning, and performance measurement and to assure that actual costs achieve the estimate. Now we extend those principles for estimate assurance.

The elements of the project estimate are direct labor, direct material and subcontract, facilities and equipment, engineering, overhead, interest, contingency, and profit. Contingency allows for unassignable extra costs that occur, and its eventual outcome is not known beforehand. Actual cost includes those contingencies because contingencies are inherent in the performance of the cost element. The purpose of cost assurance is to guarantee the project profit. Though estimate assurance deals with overall project cost, it is defined as the procedures where actual costs are controlled to the level of the elements. It makes little difference whether the budget is a *scheduled commitment* or a *scheduled expenditure* for the elements. Nor do values of dollar magnitude alter the approach.

The S curves in chapter 9 are developed first by plotting cumulative expenditure against the planned period. The subjects for the plot are the major elements of the project. The baseline values were determined by using trapezoidal or other models or by using the WBS schedule. This ideal baseline expenditure is known as the *budgeted cost for work scheduled* (BCWS) and is available at the moment of the project start. This vertical axis may also be titled BCWS%, which is equivalent to cumulative baseline percentage. The contract period percentage is an alternative *x*-axis for cost/schedule performance reports.

Once the project is under way the contractor's reporting methods will provide the *actual cost of work performed* (ACWP), and by cross checking actual work to the WBS will find the derivative *budgeted cost for work performed* (BCWP). Additionally, the cost/schedule status report will provide a forecast cost at completion and budgeted cost at completion. Those concepts are illustrated in Fig. 11.12.

At the close of each period the engineer graphically forecasts the remainder of unfinished WBS elements for ACWP and BCWP. The extension depends on a

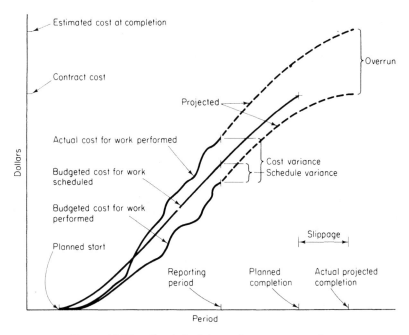

Figure 11.12.   Cost/schedule performance reporting.

graphical ability and, more important, that extra knowledge and opinion about delivery promises, potential strikes, bad weather, and so on.

There is a problem, however, with work in process tasks having a long time period. If the reporting cutoff period slices into work in process for some task, then the percentage of completion must be subjective. Estimating the percent of task complete must be done by the person responsible for the task.

Work packages vary. For example, fabrication work packages tend to be short and discrete. Engineering work packages are difficult to plan since the work is variable, making it difficult to judge for percentage completion. Contrariwise, work completion can be better judged over several reporting periods, because single-period judgment of percentage completion is tricky.

A related problem deals with reporting cost and schedule variance of fixed-price subcontracts, since some subcontractors may not report internal progress on work. For those subcontractors who do not report progress, the engineer will do "vendor talk" and learn informally of their progress and use that to keep the cost/schedule system informed.

Performance measurement is a comparison of budgeted versus actual accomplishment. Comparing the budgeted cost for work scheduled to the budgeted cost of work performed produces a dollar schedule variance. If the BCWP exceeds BCWS (i.e., is higher on the vertical axis), then more work is accomplished than was planned, and the favorable variance reflects the dollar (or percentage value) of the extra work. If BCWP is less than BCWS, then less work is accomplished than was planned, and an unfavorable schedule variance is indicated. The horizontal difference between BCWS and BCWP is the schedule variance expressed as periods or period percentage.

Contract cost performance is related to work done versus work planned. The BCWP when compared with BCWS indicates a schedule variance, but when compared with ACWP a cost variance is obtained. If the BCWP exceeds ACWP, then a favorable variance is noted. If ACWP is greater than BCWP, then a negative cost variance, or an overrun, is noted. The cost and schedule variance are indicated in Fig. 11.12.

The dashed projection of BCWP to the contract cost level indicates the conjectured schedule completion. Project slippage is the number of calendar days or periods between the actual and planned project completion. The dashed projection of ACWP in Fig. 11.12 indicates the projected dollar overrun when compared with BCWP.

If the estimate assurance program reduces the magnitude of the actual expenditure, then project attractiveness is enhanced. If the project decision is based on optimistic cost expectations, then a cost assurance program should reveal this fact early enough for reassessment. Cost information must be available so that significant items of the project cost may be watched. The project budget is tabulated to have this record correspond to the WBS, definition, and estimates. It has happened that projects that had overly optimistic estimates, poor performance, and scheduled delays have been canceled after reassessment.

Consider a construction bid of $8 million having a contract period of 5 months. Cash flow is trapezoidal, where $t_i = 20\%$ and $t_f = 60\%$. Profit earned by the contractor is 20%. The estimated cost total is $6,400,000. The owner has a retainage policy of 10% with a 1-month delay in the owner's payment. Table 11.3 provides the expenditure calculation as BCWS. Actual data have been provided for periods 1–4 as the work progresses. At period 4 there is a current unfavorable cost variance of $2,500,000 (= 7,000,000 − 4,600,000) and a project time delay of 1 month. The BCWP is extended to the contract cost. A horizontal difference between BCWS and BCWP gives a project slippage of three-quarters of a month. The extended ACWP and BCWP indicates a project overrun of $1,350,000. The cost/schedule performance is shown as Fig. 11.13.

TABLE 11.3.   COST/SCHEDULE PERFORMANCE REPORT WHERE COSTS ARE $1000

Period	Period (%)	Cumulative				Variance		Forecast		
		BCWS (%)	BCWS	ACWP	BCWP	Schedule	Cost	BCWP	ACWP	Variance
1	20	14.3	$9150	$ 500	$1250					
2	40	42.9	2745	2000	2250					
3	60	71.5	4575	5500	3500					
4	80	92.9	5950	7000	4600	1	$2400			
5	100	100	6400			1.1	2250	$5600	$7500	−$1900
6	120	100	6400			3/4	1800	6400	7750	−$1350

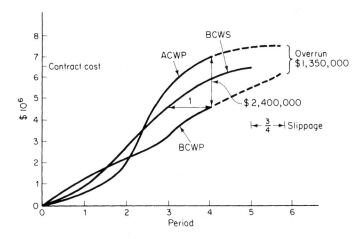

Figure 11.13.   Cost/schedule performance for contract of $6,400,000.

11.6 SYSTEM ESTIMATE ASSURANCE

The system estimate is composed of operation, product, and project estimates. The collection of those estimates leads to a measure of the design called *effectiveness,* generally implying a benefit-cost ratio, life cycle cost, or total cost for a large-scale design. System estimates are regularly made, for instance, for weapon systems, water and navigation systems, and hospital and health care planning. Preparation of a system depends on design circumstances, and usually no two are alike. The time horizon from estimate start to life cycle is from several to 25 years or so. Those ideas were first discussed in chapter 10.

A spectrum of time relationships is possible for the estimates and can vary from immediate or short-term operational requirements to the long-range planning of conceptual products and projects not even in the research and development stage. Many of the explanatory variables range from simple to complex. The crux of the difficulty in assuring the value of a system estimate cannot be routine or short term as is found with operation estimates. Effects of inflation or deflation, design changes, or definition alterations make a cost assurance program unlikely when we consider the complexity of system estimates.

Cost estimating relationships are used to evaluate system design. Regardless of their complexity or simplicity, they can be derived only from historical data. The past may prove to be unreliable as a guide to the future. It is usual that some design characteristics are outside the range of the historical sample in dealing with advanced hardware systems. Those dilemmas are not entirely resolved despite statistical procedures.

Not insignificant is the matter of legislative budget approval and eventual appropriation authority. Political pressures to shorten or lengthen schedules have an important role in whether the estimate is actually ever compared to the actual costs. A side-by-side comparison made between the estimate and actual costs for system estimates is seldom found.

Estimate assurance programs appropriate for operations, products, or projects are only partially successful when applied to a system design. As a consequence, emphasis shifts to increased mathematical analysis of the estimate. Techniques such as simulation, probability analysis, and sensitivity are used.

The preponderance of system estimates are used to make selections between alternative designs. The most popular technique is frequently identified as sensitivity. Once the "tall poles" are identified for a design, the values of the significant design factor are ranged over a wide magnitude to determine if a competing design becomes more cost effective.

Note Fig. 11.14. Each alternative has a range for system cost at any value of a design factor. The region of preference for a alternative 2 or 3 is clearly dominant for small or large values of the design factor. There is a region of overlap where one system's domination is a matter of probability. Those circumstances are described by Fig. 6.9. Once this overlapping region is discovered, the system-estimating team will enlarge its analysis to reduce the zone of uncertainty by improved cost and design studies. Continuing analysis, it is reasoned, will make the alternative choice

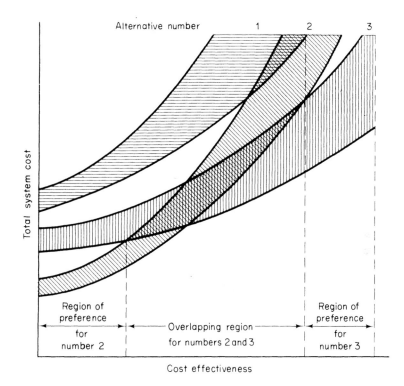

Figure 11.14. Sensitivity analysis with respect to system effectiveness.

clearer. Thus estimates are assured by greater attention to detail prior to the spending of money.

SUMMARY

An important purpose of an estimate is to improve business opportunity. Not all do, but those that are successful are matched against a comparison. The comparison could be an actual or preliminary value. Actual costs suffer from inaccuracy or lateness and do not always provide a valid future lesson. Nonetheless, a variety of techniques, including behavior considerations, variance, productivity, trend charting, and sensitivity, guide the actual cost values to become estimated values.

QUESTIONS

11.1. Define the following terms:

Estimate error	True value
Contingency	Mistake error
Policy error	Standard
Risk error	Variance

Capture rate	Unlearning
Closed estimate	ACWP
assurance systems	ACWP
Pareto rule of thumb	

11.2. Why are estimates compared to actual cost or standard cost?

11.3. Describe the philosophical differences between an estimate assurance and a cost control program.

11.4. Engineering usually stresses the abolition of mistakes. Name practices to achieve this objective.

11.5. Describe management actions for (a) a declining capture rate, and (b) for an improving capture rate. Why is a corrective policy based on a capture rate strategy alone defective?

11.6. Develop a list of behavioral principles that would be useful for cost estimating.

11.7. Give different formulas for variance calculations. Can ambiguity be completely overcome?

11.8. Why are system estimates difficult to cost track? Cite some cost overruns from local or national circumstances. List the factors that cause these overruns.

11.9. Define a capture rate based on dollars won versus dollars estimated. What advantages does this have over Eq. (11.3)?

PROBLEMS

11.1. (a) A cement contractor bids a job of steps and risers for an apartment. His estimate is 125 hours at $15.25 per hour, material takeoff of 32 yd^3 at $43/yd^3, other materials are $145 for the job, and an overhead rate of 50% on the basis of direct labor. His pricing policy is 20% as a markup on sale value. Determine his bid for the job.

     (b) Reconciliation of all costs by the bookkeeper revealed the actual cost of $5250. Determine the error of the estimate. What is his final profit for the actual costs?

11.2. (a) A job-shop firm receives a confirmation order of 20,000 1018-CRS manufactured parts. Material cost per unit is $0.0293. Direct labor setup is 1 hour, and cycle time is 0.009 hour per unit. The machine hour rate, which includes labor and overhead, is $20. Pricing policy is a 25% markup on full cost. Determine the price for the part.

     (b) After the job is closed, an audit showed that material, labor and overhead, and tooling actually cost $6000, $3600, and $5500. Find the error of the estimate. What is the effective profit margin?

11.3. (a) A prototype unit is being constructed with 45,000 direct labor hours and $20,000 of direct material. The direct labor rate is $18. The overhead rate is 100% of direct labor. The estimated learning rate is 90% and 95% for labor and material. The company has a policy of using a linear unit line. Pricing policy is 15% of total cost. Each unit is estimated and sold separately. Determine the 10th unit price.

     (b) Examination of the actual cost records showed a value of $1,250,000. Find the error of the estimate. Give the net profit and effective profit margin.

11.4. (a) The tooling vendor submitted 298 quotations for the winning jobs of Fig. 11.4. What is the capture rate?

(b) Of the winning jobs reported by this survey, 63 were for tools required by an engineering design change resulting from 109 quotes. The tooling vendor had supplied the original tools, which were not reported by this survey. What is the capture rate for new and repeat business?

(c) Discuss ways to improve his repeat business considering profit margins that are 20% on repeat business and 2.5% on new business, versus profit margins that are 2.5% on repeat business and 20% on new business.

11.5. A contractor reviews last year's performance of bids left on the table. What is the capture rate for 78 attempts? Because winning bids were openly announced, the contractor is able to analyze the bidding strategy.

Bids Left on the Table	Percentage above Winning Bid
2	0–2.0
1	2.1–4.0
3	4.1–6.0
7	6.1–8.0
10	8.1–10.0
10	10.1–12.0
5	12.1–14.0
2	14.1–16.0

(a) Discuss a strategy for no change in the future bidding environment.

(b) Assume that each quotation is about $1 million and includes $50,000 for profit. Discuss the strategy with this additional information. Are there adjustments that you can suggest?

11.6. An analysis of the estimating operation has been made for the previous fiscal year quarters:

Quarter	Estimates Made	Estimates Won	Cost for Jobs Won ($\times 10^6$) C_p	C_d	C_a	Business Value Lost ($\times 106$)	Estimates ($\times 10^6$)
March	216	112	$195	$191	$193	$380	$17.4
June	237	118	201	198	202	426	17.22
September	293	138	219	224	233	544	16.8
December	338	149	221	227	244	625	10.2

Find quarterly and yearly estimating error on the basis of actual and preliminary estimates, capture rate, and productivity. What trends do you spot? Advise engineering on cost-estimating policy for the next quarter. Evaluate the performance of estimating over the last year. Engineering has a goal of bidding 250×10^6 from detail estimates. What can be expected?

11.7. Repeat the variance calculation for Fig. 11.4 if actual material is $18.25 and quantity usage is 195.

11.8. Estimates for a material are 650 lb at $17.15/lb. The record shows am actual usage of 665 lb at $17.10/lb. Find the variance for material cost, quantity, and total.

11.9. A 1018 CRS material is estimated to cost $0.0293 per unit for 20,000 units. Actually, material costs $0.032 per unit and 20,500 units were necessary to allow for greater scrap and waste. Determine the material and quantity variance.

11.10. The estimates for labor are 72.6 hours at $13.76 per hour. The accountant's record indicated Kathy Holthaus, 20.2 hours at $13.50 per hour; Radon Tolman, 14.3 hours at $13.75 per hour; Louis Roth, 22.7 hours at $13.80 per hour; and Jim Morrison, 15.8 hours at $13.85 per hour. Find the labor rate variance, hours variance, and total variance.

11.11. A labor estimate is 1 hour setup and 0.009 hour per unit for 20,000 units. The shop direct labor was expected to cost $15. In reality, total time for setup and cycle required 208.15 hours, and shop labor was $14.50. Find the variance for the labor rate, hours, and total.

11.12. A time study demonstrated assembly production of an electrical home outlet as 101.8 pieces per hour (see Fig. 2.3). Subsequent reports revealed an average of 1.062 hours per 100 units. An estimated job anticipates an order of 16,000 units, but actually 15,900 units were stocked. The labor rate was $14.75 instead of the expected $14.50. Determine the individual and net variances. Find the productivity factor. A future quantity of 20,000 units is anticipated.

(a) If operation improvement is not possible, then what do you recommend as the future standard hour?

(b) If competitive reasons require a new labor time estimate, then what percentage reduction is necessary for a methods engineering goal?

11.13. An aluminum part weighing 2.7 lb is drilled in 37 locations with a computer directed drilling machine. The part is $\frac{7}{8}$ in. thick and hole diameters are $\frac{1}{2}$ in. and $1\frac{1}{2}$ in. Standard cost data expressed as setup dollars for the lot and cycle dollars for 100 units are given as follows:

$$\text{setup} = 1.90 + 0.78(\text{number drill sizes})$$

cycle dollars/100 units $= 8.70 + 1.14(\text{number lb})$

$$+ 1.10 \text{ (number holes)} + 1.15(\text{number sizes}) + 13.0(\text{each hole depth})$$

The first term of the setup and cycle estimating linear relationship is the constant mandatory for estimating. The lot quantity is 250. Actual costs and quantity, as eventually determined, were $950 and 255. Find the variances and productivity factor.

11.14. A free abrasive lapping machine is estimated to require 0.5 hour for setup and 1.721 minute per unit for the cycle. Estimated labor wage is $12.58. Actual man hours and wage were 225 and $12.61.

(a) Find the dollar variances for hours and hourly rate.

(b) Calculate the net labor variance and productivity factor.

11.15. A dip-brazed assembly is estimated to require 0.15 hour for setup and 0.96 minute per unit. The estimated quantity is 625 units, but actual history is 9.85 hours for a lot of 620 units. Labor wage as estimated and as actually determined is $16.80 and $17.05. Determine the variances and productivity factor

11.16. An 18-in.-long SAE 1020 steel bar weighs 83 lb. A lot estimate is required for 40 parts. Material cost is estimated as $0.73/1b. An invoice shows that 43 parts were consumed for $2960. Setup and cycle were estimated as 0.94 hour and 7.33 minutes. Records show that 6 hours were needed, and the labor rate was $17.25 instead of the planned $16.90. Find the material, labor, and net variances for material and labor.

11.17. An actual cost audit reveals that labor has an unfavorable 20% variance, material is favorable by $0.001, and overhead is reported unchanged. If standard part cost is $0.35 for labor, $0.075 for material, and 200% on the basis of prime cost for overhead, then determine the total standard cost and actual cost. What is the net variance?

11.18. An estimate results in $5.28 for labor, 150% for overhead on the basis of direct labor, and $17.38 for material. An actual fact-finding study reveals that labor had an unfavorable variance of 10%, material is favorable by $0.45, and overhead percentage is unchanged.

    (a) Determine total estimated and actual cost. Find the net variance.

    (b) Repeat if the overhead amount remains unchanged.

11.19. **(a)** Assume sales income for Table 11.1 is $20,300, and repeat calculations to find the net profit percent of net sales. All other values remain unchanged.

    (b) Assume sales income for Table 11.2 is $20,300, $32,000, and $40,000, and repeat calculations to find the net profit of net sales. All other values remain unchanged. Which model suggests future emphasis?

11.20. **(a)** Sales income for Table 11.1 is $23,250, and repeat calculations to find the net profit percent of net sales. Other values remain unchanged.

    (b) Sales income for Table 11.2 is $23,250, $36,000, and $38,000, and repeat calculations to find net profit percent of sales. Which model should be emphasized in the future?

11.21. Find the overrun and slippage for the following data and current and forecast variances.

	Period								
	1	2	3	4	5	6	7	8	9
BCWS	3	14	22	45	68	92	115	134	142
ACWP	4	10	19	60	83	113			
BCWP	1	4	15	28	52	82			

11.22. Refer to Table 9.7. Plot period number and period percent versus BCWS, BCWP, and ACWP, and BCWS% for the following additional data. Extend the data to completion and estimate cost at completion and project slippage. Determine schedule and cost variance and budget, project, and difference until completion.

	Cumulative	
Month	ACWP	BCWP
1	$ 2,000	$ 3,000
2	5,000	12,000
3	13,000	18,000
4	28,000	27,000
5	41,000	32,000
6	54,000	52,000
7	92,000	78,000
8	122,000	108,000

Interpret the early periods. What action do you think might have been taken on the basis of the early periods?

11.23. Refer to problem 9.41. Now assume that a project is under way and a material cost/ schedule procedure reports on budgeted cost material expended. Determine schedule and cost variations for periods 4 and 5. Find project overrun and slippage. Use graphical analysis.

	Period						
	1	2	3	4	5	6	7
BCWS	915	2745	4576	5946	6400	6400	6400
ACWP	250	1500	5000	6500			
BCWP	500	2500	3500	500			

MORE DIFFICULT PROBLEMS

11.24. A mining equipment manufacturer compiled the following data on three models of drill steel it produces:

Machine Center	Hourly Rate	Hours Required Per Unit for Model			Cost per Unit for Model		
		1700	1600	1500	1700	1600	1500
Cutoff	$14.20	0.01	0.01	0.01	$ 0.142	$ 0.142	$ 0.142
Upsetting	15.65	0.03	0.04	0.05	0.47	0.626	0.783
Machining	16.11	0.05	0.04	0.03	0.806	0.644	0.483
Heat treat	10.40	0.20	0.18	0.16	2.08	1.87	1.664
Bench	13.90	0.5	0.5	0.5	6.95	6.95	6.95
Finishing	15.02	0.1	0.1	0.1	1.50	1.50	1.50
		Unit conversion cost =			$11.95	$11.74	$11.52
		Unit material cost ($0.20/lb) =			34.80	31.60	27.99
		Unit selling cost (5% price) =			3.75	3.20	2.50
		Total variable cost =			$50.50	$46.54	$42.01
		Selling price =			75.00	64.00	50.00
		Contribution/unit =			24.50	17.46	7.99
		Contribution percent =			33	27	16

Variance or variable costs are as follows: manufacturing labor and expenses, 3% unfavorable; materials and scrap, 5% unfavorable; selling and distribution, 1% unfavorable.

(a) Make an income and expense statement if the manufacturer produces 200 1700-series, 100 1600-series, and 170 1500-series drill steel.

(b) If the total period costs (budgeted amount plus variance) were $10,500, then what are the net profit and percent of net sales?

(c) What advice can you provide management for future production of those three products?

11.25. Determine the schedule and cost variance, graphically forecast BCWP and ACWP, and find overrun and slippage assuming differing values for Table 11.3 as given by

(a)

	Period			
	1	2	3	4
ACWP	250	1500	5000	6500
BCWP	500	2000	3500	5000

(b)

	Period			
	1	2	3	4
ACWP	100	750	4000	5000
BCWP	750	2250	3500	4250

CASE STUDY:
COST SCHEDULE PERFORMANCE REPORTING (CSPR)

When a contractor provides a proposal to an owner, the contractor promises to perform work at a cost within a time limit and to meet design requirements. For the contractor to fulfill a

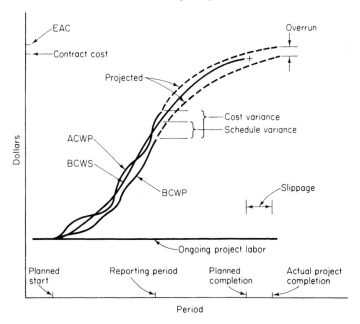

Figure C11.1.   Cost/schedule performance reporting with ongoing project labor.

contract, specific work is scheduled at various times of the project. Some contracts require periodic preparation of cost schedule performance reports that tell the owner what has been done and when and additionally what needs to be done and when. A graph such as Fig. C11.1 is often a significant part of the report. This figure differs from other CSPR figures in that a certain amount of ongoing project labor is considered fixed.

Create the specifications for report system that builds on Fig. C11.1. Consider what follows as necessary to your reporting procedure:

1. Alert engineering to conditions that might affect either cost or schedule.
2. Establish an orderly method of compiling supporting data.
3. Permit timely and corrective action to minimize negative effects or maximize positive breakthroughs on project cost and schedule.

12

Contract Considerations

If the cost estimate and the resulting price are adopted by the customer, or if the firm chooses to produce its product, then the business process continues on to contract considerations. Legal requirements are important, and that is why the engineer needs an understanding about contracts.

A variety of contract types exists. Fixed price and cost reimbursable are the two major families. Stipulations regarding contract changes, claims, disputes, and many other factors become possible. An intermediate step to a completed contract may be negotiation. Auditing and ethics are other considerations for the engineer.

12.1 IMPORTANCE

By now the student has recognized that cost estimating is the *para-business partner* with design engineering. The relationship involves understanding contract law, ethics, and responsibilities to the employer, client, or customers and professional obligations to society at large and public law. Though the engineer may not be a lawyer, he or she needs a fundamental understanding of the law because preparation of the estimate often consists of legal considerations.

Operation, product, project, and system estimates have differing *legal requirements*. Contractual distinctions are based on other factors. Because the estimate is a measure of the economic want, a handshake or amount of money in a legal contract may be the means of agreement between two parties. The notion of wants, initially described in chapter 1, requires two people, companies, buyer and seller, vendor and contractor, and contractor and owner or government. Self-interest leads the parties to consent to an exchange that may be recognized by a handshake, money, or a formal contract. *Negotiation* may or may not be involved. In disputes, engineering changes, patents, warranty claims, and so on, litigation may follow, and the estimate may be a document of evidence.

We use the term *contract* to describe a variety of agreements or orders for the procurement of services or materials. A modification of this contract may be an alteration to the specification, delivery point, rate of delivery, contract period, price, quantity, or other provision of an existing contract accomplished under one of the clauses, such as change order, notice of termination, or supplemental agreement.

There is a variety of contract types. They refer to specific arrangements used for the employment of work or the supply of materials or compensation arrangements.

12.2 BASIC CONTRACT TYPES

There are two basic contracts to measure the value of the estimate: firm fixed price and cost reimbursable. Firm fixed-price arrangements have in common that one party, a vendor or contractor, is to deliver a product, project or system, or perform an operation in accordance with the terms and conditions of the contract, and an agreement by a buyer, firm, agency, or government to pay a price equal to that specified by the contract. In contrast, cost-reimbursable contracts pay the vendor or contractor for money spent, subject to restriction and special negotiated understanding.

The element of risk, for instance, or willingness of the parties, competition, advanced or ordinary technology, complexity of design, urgency, and product cycle life influence which general type of contract is selected. Highly complex development uses a cost-reimbursable type of contract. Low technology, for instance, or short time periods, well-developed production consumer products, and large volume, use firm fixed-price contracts. Cost-reimbursable contracts transfer the economic risk to the buyer or customer, and fixed-price contracts place the economic risk on the contractor, supplier, or vendor.

The fixed-price contract requires that the design (operation, product, etc.) be delivered as described on a predetermined schedule. The price is fixed for the life of the contract precluding changes allowed by the contract. However, the terms and conditions may allow for adjustment. Though the fixed-price contract provides the greatest risk to the seller, it also offers incentive and opportunity to realize the greatest profit. The vendor or contractor fully recovers savings due to cost reductions. Thus, if the actual costs are less than estimated costs, then greater profit materializes.

Almost all public project contracts, as well as a large portion of private construction projects, are selected on the results of competitive bidding. Competitive bid contracts for projects that are fixed price have interesting variations. A unit-price construction contract is based on the placement of certain well-defined items of work and the costs per unit amount of each of those work items. For example, a price per linear foot of pile or cubic yard of excavation allows a reasonable variation to be made in the driven length of the individual piles or quantity of excavation. Thus, a contractor submits a bid on the basis of number and depth of the piles, which are verified by the engineer working in the field. Some metal-machining shops do much the same thing by providing a "time" service; that is, by charging a productive hour cost rate that will include a portion for profit.

The fixed-price or lump-sum contract is popular from an owner or buyer's viewpoint, because the total project cost is known in advance. If the project cannot be accurately estimated, then the fixed-price contract may not be suitable.

The cost-reimbursable contract places the risk on the buyer and is used whenever research, development, design, or urgency is necessary. The buyer has to be assured that the contractor is reputable in quality, delivery, and design. Inherent in a cost-reimbursable contract is the best effort by the contractor to complete the work. Even so, the costs are passed on from the seller to the buyer.

It is as necessary to estimate accurately for cost-reimbursable contracts as it is for fixed-price contracts.

Cost-reimbursement contracts have provisions for payment of allowable, allocatable, and reasonable costs incurred in the performance of the contract. Certain formulas, to be described shortly, permit adjustments for fees, incentives, and penalties.

Operations involve direct labor and direct material in the estimating process. Often, the raw and standard commercial materials are listed in a catalog, price list, or schedule regularly maintained by a manufacturer or vendor and may be published or made available for inspection by customers. The information may state prices of current interest or give prices to buyers. Though those values may not be formal contracts, they are established in the usual course of business between buyers and sellers either free or not free to bargain and agree. Competition establishes the prices and thus is a sufficient standard for a contract. Those catalog prices, or the value of the estimate, say, from $0.05 to $50 million, do not diminish the importance of the estimating, negotiating, or contractual procedure.

Labor costs may be contracted between the company and the union or between the company and an individual willing to work for the conditions as specified. Negotiations may be a factor. At the lower wage level, federal laws may dictate the value of the wage. (Labor laws were discussed in chapter 2).

Some vendors bid for work on the basis of a productive hour cost rate. For example, the manufacturing company specifies the following schedule:

Heavy machining: $113.75 per hour
Light machining: $43.50 per hour

and then conduct the work. This is a form of fixed pricing. Note Table 4.11, where rates were determined, and formula (8.12), which adjusted the cost on the basis of 15% profit markup. Obviously, the buyer needs to be satisfied that a fair and reasonable time is charged against the contract. Some companies maintain an open purchase contract allowing this arrangement between the buyer and seller.

A time and material contract is used between a buyer and seller for work at a fixed and specified rate (hourly, daily), that includes direct labor, indirect costs, profit, and materials. The materials may be at cost or cost plus profit. This contract is suitable for operations where the amount or duration of work is unpredictable or insignificant. Repair work is often handled on a time and material contract. The labor is considered a fixed part, but the material is cost reimbursable since the nature of repair materials is unknown at the time of the contract.

12.3 FIXED-PRICE ARRANGEMENTS

The parties agree to the price before a firm fixed price (FFP) contract is awarded. The price is firm for the life of the contract unless revised according to the change clauses given in the contract. An example for a $1 million contract under varying consequences is given as

Contract price	$1,000,000	$1,000,000	$1,000,000
Actual cost	900,000	1,000,000	1,100,000
Realized profit	$ 100,000	$ 0	$(100,000)

The contractor either gains or loses based on performance.

In another fixed-price contract, a negotiated pricing formula can be agreed to that motivates and rewards the contractor for performance. In those fixed-price incentive (FPI) contracts, the process involves an estimated cost, target profit, target price, ceiling price, and profit sharing for costs incurred above or below the estimated cost. The example above with the $1 million contract price is a 0/100 sharing arrangement. This means that the buyer does not share and the contractor accepts 100% of the difference between price and cost. Figure 12.1 shows the basic fixed price with 0/100%.

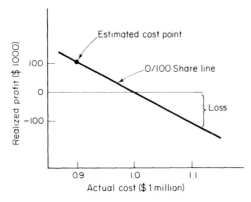

Figure 12.1. FFP arrangement with 0/100 sharing line.

The fixed price incentive is written to have a target cost, and a target profit percentage along with a contract ceiling price and a price adjustment formula.

Estimated cost	$ 900,000
Target profit (11.1%)	100,000
Target price	$1,000,000
Price ceiling (125%)	$1,125,000
Cost sharing	80/20 above estimated cost
Cost sharing	90/10 below estimated cost

The owner will not pay above the ceiling price.

Assume that a contractor experienced an overrun of 15% of estimated cost, or $135,000 $(= 0.15 \times 900,000)$. The reimbursement would be as follows:

Estimated cost	$ 900,000
Plus 80% of $135,000	108,000
Total reimbursement cost	$1,008,000
Target profit	$ 100,000
Less 20% of profit	20,000
Total profit	$ 80,000
Cost plus profit	$1,088,000

Thus, the owner covered 80% of the increased cost, but the contractor lost 20% of profit. Instead of a 11.1% profit $(= 100,000/900,000)$, profit declines to 7.9% $(= 80,000/1,008,000)$. But the profit return can be computed another way. On the basis of actual and received costs, the profit $= 5.1\%$ $(= (1,088,000 - 1,035,000)/1,035,000)$ where $1,035,000$ $(= 900,000 + 135,000 + 80,000)$.

Suppose that the contractor completed the job, but costs were $820,000, or $80,000 less than estimated. Under the terms of cost sharing of 90/10 below estimated cost, what follows is found:

Estimated cost	$900,000
Less 90% of $80,000	72,000
Total reimbursed cost	$828,000
Target profit	$100,000
Plus 10% of $80,000	8,000
Total profit	$108,000
Cost plus profit	$936,000

Instead of 11.1% profit, the profit increases to 13% $(= 108,000/828,000)$. Note that profit is expressed as a percentage of the cost of the contract.

The FPI represents joint responsibility for ultimate cost. This arrangement shares in any dollar difference between the estimate and final cost. In an 80/20 example, the contractor is responsible for 20% of the difference either as an addition to or a deduction from target profit. Though shares are always a total of 100%, the proportions vary because of uncertainty, amount of target profit, and the spread between estimated cost and ceiling price. Expressions of the owner/contractor or government/contractor shares are 60/40, 75/25, 50/50, and so on.

The share line can be straight both above and below estimated cost. But an 80/20 above estimated costs can be matched with a 90/10 below estimated costs. Slope changes such as those depend on the design and negotiation.

Assume a 70/30 sharing proposition for the following cost facts:

Price ceiling	$11,500,000
Target price	$10,850,000
Estimated cost	$10,000,000
Final cost	$ 9,600,000
Difference	$ 400,000 underrun

The contractor receives $120,000 (= 0.30 × 400,000) as an increase in profit. This is added to a target profit.

Target profit	$850,000
Contractor's share	120,000
	$970,000

The owner or government receives 70%, or $280,000 (= 0.70 × 400,000) difference, as a reduction in price.

Final cost	$ 9,600,000
Final profit	970,000
Final price	$10,570,000
Target price	10,850,000
Price reduction	$ 280,000

Now, assume a final cost of $10,500,000.

Estimated cost	$10,000,000
Final cost	10,500,000
Difference	$ 500,000 overrun

The contractor receives 30%, or $150,000 (= 0.30 × 500,000) as a decrease in profit.

Target profit	$850,000
Less 30% of overrun	150,000
Final profit	$700,000

The owner receives 70%, or $350,000 (= 0.70 × 500,000) as an increase in price.

Final cost	$10,500,000
Final profit	700,000
Final price	$11,200,000
Target price	10,850,000
Price increase	$ 350,000

If the final cost is $12,000,000, or $500,000 in excess of the ceiling, then the ceiling of $11,500,000 is the final price.

A contract arrangement of fixed price incentives with successive sharing ratios for early development, design, and production is possible.

Fixed price with redetermination (FPR) is another arrangement. The redetermination may be prospective or retroactive. In prospective, negotiation is undertaken for a fixed price in a future period, and then successive fixed prices are renegotiated periodically. Data available are past costs, performance, and estimate assurance.

The retroactive type of FPR contracts provides for adjusting contract price after the work is completed. A ceiling price is initially determined and actual audited costs are the starting point for negotiation. The terms of the contract do not provide for mutual sharing. The retroactive type requires an opinion of the contractor's performance by the owner or government.

Quote or price in effect (QPE) types of contracts are also known as escalation contracts (and were introduced in section 3.3). Long-term uncertainties that result from inflation or deflation effects are the reason for their use. The contractor will estimate costs and then establish a mutually agreed on bench mark or index. Adjustment is mostly up, but downward adjustment is possible. The price adjustment clause must identify a base or benchmark period and one or more indexes to measure changes in price level in relation to the reference cost and reference period. If the contract provides for economic price adjustment, then the contingency element of the cost estimate is ignored if it deals with inflation effects. Remember that contingency may be for considerations other than inflation.

12.4 COST-REIMBURSEMENT ARRANGEMENTS

The cost plus incentive fee (CPIF) develops an incentive sharing formula and is used in lieu of cost reimbursement with a 0/100 share. CPIF uses a ceiling price, and costs that comply with contract terms are reimbursed. The total of reimbursed costs is the final cost of the contract. In FPI plans the reasonableness and necessity of costs are established, and cost is finally found after negotiation. However, under a CPIF, the maximum and minimum fee are limited by negotiation. A cost level above and below estimated cost is negotiated for minimum and maximum levels. Contract sharing ceases, and the contract reverts, in effect, to a cost plus fixed fee

100/0 sharing plan. In contrast, the ceiling price in a fixed-price contract establishes a point over target cost where the owner ceases to share and the contract becomes a FFP with a 0/100 share model. The parties to a CPIF contract believe the cost risk is too great for a realistic price ceiling. CPIF arrangements encourage incentive over a greater variation from estimated cost than would be expected with an FPI contract. What follows is an example of a CPIF plan.

Estimated cost	$10,000,000
Target fee	750,000
Maximum fee	1,350,000
Minimum fee	300,000
Share formula	85/15

Assume a trial cost of $9,000,000.

Estimated cost	$10,000,000
Final cost	9,000,000
Difference	$ 1,000,000 underrun

The contractor receives 15%, or $150,000 (= 0.15 × 1,000,000), as an increase in fee.

Target fee	$750,000
Share	150,000
	$900,000

The owner receives 85%, or $850,000, between estimated and final cost as a reduction in price.

Final cost	$ 9,000,000
Final fee	900,000
Final cost plus fee	$ 9,900,000
Estimated cost plus fee	10,750,000
Reduction in price	$ 850,000 underrun

In this example the incentive is an effective overrun of $7,000,000, an underrun of 40% and an overrun of 30%. The contractor share of a $4,000,000 underrun is 15%, or $600,000. The share of a $3,000,000 overrun is 15%, or $450,000. Those adjustments are added to or subtracted from a target fee of $750,000. The maximum and minimum fee ranges from $1,350,000 to $300,000. If the actual cost is greater than +$3 million or −$4 million from the estimated cost, the plan fixes the fee at either the maximum or minimum level.

Another cost-reimbursement plan is the cost plus award fee and provides a base fee and an additional fee that may depend entirely or partially on performance. The cost plus award fee plan does not include targets or automatic fee adjustment. The amount of the fee is subject to unilateral judgment by the government.

The cost plus a fixed fee (CPFF) plan pays the contractor a fixed fee above reimbursable costs. The fee may change when the work breakdown structure changes. Conceptually, a CPFF is the opposite of a FFP arrangement. The FFP has a 0/100 sharing and CPFF has a 100/0 share.

Estimated cost	$15,000,000
Fixed fee	900,000
Estimated CPFF	$15,900,000

The fee is unalterable with respect to actual costs. CPFF is used when the risk is great as when effort is inconsistent with performance.

12.5 CONTRACT CLAUSES

Organizations have procurement policies expressed as *standard* or *boiler plate* contract clauses. These printed statements are usually an attachment to the main body and should not be dismissed as unimportant by the engineer. Many clauses exist, of course, but only a few are discussed here.

The existence of real or potential competition before selection and award ends when the contract is signed. A competing offeror becomes a sole source contractor or supplier. Contract clauses provide for circumstances that may alter some parts of the terms and conditions of an agreement. Contract changes are actions subsequent to the estimated cost and price and directly affect those documents.

Construction contracts give the owner the right to make changes in the design or work after the contract has been signed. Similarly, the U.S. government has the authority to initiate changes.

A general commercial practice accepts what is known as the change clause. The change clause is applicable to fixed-price and cost-reimbursable contracts. The customer may, without notice, make certain changes in the contract requirements if such changes are declared by the customer as not constituting a change in scope. If the change affects, for instance, cost, delivery, schedule, or performance, then the contractor must serve written notice to initiate a review, negotiation, and an equitable resolution of the claim. If the customer authorizes a change that may increase cost, then it is assumed that there is adequate money to pay for the change. Construction contracts provide that the contractor is not to proceed unless there is written authorization.

When those change orders are issued, a supplement to the contract is prepared. The supplement can be made on the basis of a lump sum or cost-reimbursable arrangement.

A change order is a modification of the contract. The parties understand that modification has considered prior negotiations, terms, designs, and conditions.

A disputes clause of the contract provides procedures where resolution may be initially thwarted for a variety of reasons. The dispute clause provides that the customer will initially decide any dispute concerning a question of fact arising under the contract; for example, failure to agree to a price adjustment resulting from a change. The contractor is expected to proceed with the contract as changed.

Details differ between governmental and commercial activities. Construction contractors will first attempt to settle with the owner. There are levels of appeal from an adverse decision, including arbitration and courts, depending on whether the dispute involves questions of fact or law. In a similar way, the government will arbitrarily send a written notice to the customer. After the decision is rendered by the customer, the contractor may appeal.

The *addendum statement* for cost-reimbursable contracts provides for control and regulation by the customer of costs that are accumulated. Costs caused by the contract are allowed and reimbursed. Criteria for determining allowable costs are defined in various documents that may be referenced by the contract, or the terms of the contract may define the kinds of allowable costs.

Reasonableness and prudence are governing practice in accepting allowable costs. For example, a subcontract material was not competitively bid may not be allowable. Thus, this material may be subject to negotiation to determine value.

The inspection and correction of defects, warranty, and full or conditional protection, for example, are clauses that obligate the contractor, vendor, or seller to correct defects, deficiencies, and inability to meet specification in various ways. Those terms are so broad that we are unable to offer specifics, except that the cost estimate needs to include reasonable costs based on the wording of the clause. Mean time to failure and mean time to repair were concepts originally described in chapter 10. Percent returns, conditional service contracts, customer assistance, field correction, recalls, and bulletins alerting the customer are features of a seller's costs. They are the contractual terms that provide specificity. Costs are generally reimbursed in a cost-reimbursable contract.

Acceptance of the work by the owner or customer and payment may constitute waiver of rights for damages if a claim is not made within a reasonable time. In construction, warranty work is normally covered by a performance bond. Exceptions are available to redress *grievances* by both a buyer and seller.

A *subcontract* is an agreement between a prime contractor and a subcontractor or a contractor and vendor. The subcontractor agrees to perform specialized work at a construction site or provide certain materials. A vendor, which is language more common to manufacturing, agrees to supply parts, subassemblies or assemblies, or subcontract material.

A subcontract binds only the parties to the agreement. Many of the same clauses that are required of the prime contractor are applied similarly to a subcontractor, although the value of the contract will dictate its complexity. A $1 million subcontract has greater specificity than one with a $1000 value. Provisions of the general contract, including changes in work, minimum wage laws, warranty clauses, and other laws may extend to the subcontractor.

In fixed-price agreements, *competitive bidding* is encouraged before subcontracts are awarded. The subcontract clauses require the contractor to meet some stipulations in cost-reimbursable contracts. The contractor may be required to advise the owner or government of the anticipated subcontracts. Approval by the owner may be necessary, depending on the size of the award. Consent by an owner to use a specific subcontractor does not relieve the major contractor of failure to perform.

Contracts are concluded in a variety of ways. The usual way is full and satisfactory performance by both parties. Another way is *breach of contract. Failure of progress payments* and unreasonable delays of the project are the most common by the owner. The contractor is entitled to damages caused by the owner's inability to discharge responsibilities required by the contract. Default or failure to perform as required by the contract is the more common breaches committed by the contractor. Nonperformance, poor quality, failure to show progress, disregard of laws or instructions are actions that may allow the termination clause by an owner.

A convenience termination clause allows the customer an opportunity to decide that the material under procurement is no longer required and that the customer is prepared to assume losses associated with termination. The contractor ceases work, and issues cancelations on purchase orders and subcontracts. Eventually, the contractor provides a termination claim that includes incurred costs of work performed and special expenses associated with the termination effort. In cost-reimbursable contracts, the claimed costs are verified by an audit, and the fee is negotiated. In a fixed-price contract, the legitimate costs and fees are determined both by submittal of evidence and negotiation.

There are many other contract clauses. Patent rights, value engineering, excusable delays, retainage, progress payments, interpretation, excess material, shipping papers, insurance bonding, hiring of women and minorities, sex discrimination, and overtime requirements are typical ones.

12.6 NEGOTIATION, AWARD, AND AUDIT

Negotiation may begin once the price and technical proposal are conveyed to a buyer. Consider the situation where several bidders are responding to an RFQ. If the RFQ is for standard commercial materials and ordinary designs, the lowest bidder may be selected. Contrariwise, if the design is complicated and significant in terms of money, then negotiation may be required. Negotiation involves buyers, contract administrators, design engineers, and cost engineers. The discussion relates to technical areas, costs, schedules, and so on, for which the team approach is useful. Representatives are informed about price, schedule, contract type, and the design before negotiation.

Negotiation is a term used broadly and has come to mean tactics and maneuvers by both parties in an effort to reach a decision on whether to contract together.

Not all procurements are negotiated. For technical and complex designs, the various bidders make exceptions or claims regarding the RFQ and their technical response.

It is an inviolate principle to this author that cost estimates be factual. However, profit and pricing is an opinion area. Pricing refers to the fair and reasonable values and is negotiable.

"Horse trading" in negotiation is a simplistic picture when dealing with complicated designs and estimates. Negotiation requires more complex maneuvers.

In small purchases, negotiation consists of letters or telephone calls. In more significant purchases, negotiation is face to face and lasts many days. A plan or list of discussion points is necessary for effective negotiation. Issues regarding the design, RFQ, contract, terms, schedule, estimate, and performance are open to discussion. Each bidder may have exceptions or additional claims for the RFQ. For instance, two bidders on a high voltage electrical transmission line may have a low cost but high voltage line losses, and vice-versa.

The buyer in negotiation must be prepared to ask questions such as, Are there issues that can be traded off if necessary? How realistic is the delivery schedule? If the delivery can be lengthened, is a lower price possible?

If a price reduction is desired, then a vague statement that the "price is too high" represents a weak approach. A price reduction request must be plausible and businesslike. In certain situations the cost estimate is privileged information. In others, such as government work, it may be open to examination. The engineer should be prepared to defend learning theory slope, wage rate, standards, and so on, on a factual basis. The buyer may perform a technical analysis on the estimate.

A variety of contract terms and contract types exist. Usually, standard contract terms are nonnegotiable, although an able negotiator takes nothing for granted. Terms of special clauses are another matter. Those terms need to be carefully examined, because many of the contract terms and specifications may have financial and serious implications. Specifications mean additional practices, such as a particular type of quality control. It may place a burden on a factory or buyer who is unaware of the consequences.

Negotiation may deal with penalty clauses, retainage amounts, or progress payments or omissions, nebulous requirements, unclear accuracies, inconsistencies, and impossible or very expensive requirements.

A competitive negotiation will provide an opportunity for discussion by the offeror and will conclude with the award of a contract to the offeror whose price and design, for instance, are most beneficial to the buyer, owner, prime contractor, or government.

Once the contract is ongoing, terms of the contract may allow audits to verify allowable costs, possibility of fraud, compliance, and documentation. Auditing is a common occurrence for government prime contractors. Audits may be by the customer or may be internal by the firm or by a consultant hired by the organization. We are not referring to an accounting firm auditor who examines the balance sheet and profit-and-loss statement and issues a public notice. We are referring to an audit that deals with the estimating function, although it may be difficult to uncouple that from engineering, purchasing, accounting, or management. Internal audits imply the monitoring of cash flow, accounting, estimating, contracting, and the general business conduct of its operations. Internal audits of this type are commonplace in business. Prime contractors may also audit subcontractors.

Administrative audits deal with several cost factors for a period of time that include overhead cost rates, labor hour rates, or efficiency factors. Those administrative action audits determine if present or future conditions negate the appropriateness of the cost factors.

Estimating *audits* are concerned with mistakes and omissions, procedures, and contingency assessment. Mistakes are $2 + $1 = $4 and are unavoidable. Despite the popularity of mistakes, avoidance begins with attention to detail and checks by others and faultless arithmetic.

Procedures are a significant concern for audit. Consistency of estimate to the accounting system and to other estimates is important. The audit path of verifiable facts is conducive to reassessing consistency.

Separation of direct and indirect costs is watched closely. In general, direct costs are those identified as having been incurred for a particular product, work order, job, or contract. Indirect costs are composed of items of material, labor, and expenses that affect two or more products, work orders, jobs, or contracts, and the amount of cost charged to a specific one cannot be precisely determined. No universal rule exists that under every estimating and accounting system there is an assurance that items of cost be treated fairly as direct or as indirect cost. But it is essential that within estimating each item of cost be treated consistently. Some material costs may be confused as either direct or indirect, even though it is clearly incurred to the final project work order. Paint, for example, which appears on an end item, is an awkward estimate and may be called an indirect charge. Minor hardware is another optional choice. Engineers use one-half to 5% of total direct material cost and claim that as the cost of minor hardware. Or, engineers may agree that it is charged as overhead. Because of those typical alternative choices, auditors check estimates for inconsistent treatment of direct and indirect costs.

Significant direct material cost estimates are audited carefully. Much of what is included as estimated direct material will have been purchased from outside sources. The auditor examines the principal items within each material cost category and will check, for example, sources, quantities, unit prices, and losses as shown on the direct material and subcontract estimate. Whenever engineering is complete and a bill of material is available, key part numbers are traced. For most engineering design efforts, estimates are prepared from less data than appear on a bill. The WBS provides the skeleton for tracing, or the engineer may prepare a tentative bill of material from preliminary drawings.

Estimating forms may serve as the preliminary bill. A less ideal means is the engineer's project manual, a bound record, where costs for the material and parts are estimated and tied to the design. For developmental work where costs are uncertain, planning quotations obtained from potential suppliers are verifiable.

Other options are open to auditing material costs and range from routine supply problems to uncertain development efforts. Follow-on procurement provides realistic costs by using earlier projects, and data can apply to projects even though they may be developmental. In follow-on procurement, reasonable projection of historical costs account for price reductions caused by removal of original design, tooling, rearrangement, excess spoilage, and other startup costs. Economic factors, normal increase or decreases in price, and changes in production rates and quantities are considerations for follow-up estimating. Auditors are aware of those options.

A priced bill of material allows for scrutiny, especially in the amount of materials used. Once the design is fixed, auditors compare the quantity of material to that specified by the bill. This comparison is random, because an entire bill is seldom checked. Historical citations of scrap, waste, obsolescence, and spare parts percentages are helpful in supporting estimates. A priced work breakdown structure can be audited. If a contract work breakdown structure is developed, then estimates, subcontractor bids, purchase offers, vouchers, and bills of lading can be tracked from paper to physical hardware.

The third concern in auditing projects deals with contingency assessment. Though we consent to the auditor's viewpoint that contingency is negotiable and that the external auditor is compassionate to those risks, we do not agree that contingency can be audited at the time of the estimate for developmental projects. Contingency for supply and production contracts may be questioned as to validity, certainly.

12.7 ETHICS

It is fitting that a book on cost estimating close with a discussion on *ethics*. It is no less important than the very first sections of the book, which deals with the necessity of profit and wise stewardship.

The task facing the engineer is providing a measure of the economic want. During the estimating period it is not uncommon that politics or pressures are applied on the person who is estimating. It is natural that an engineer will believe that the new design is "really cheaper" or the sales staff will promote a product that gives encouragement to marketing.

Though those motives are understandable, engineers need to maintain objectivity in fact finding and their analysis. Subjectivity that may influence a policy of estimating out-of-pocket future costs seems to be unprofessional.

The estimate deals with elements of material, work, and money. Because bargaining is the essence of competitive business, there are occasions in which the propriety of some trade practices is questionable. Revealing quotes to other contractors or vendors with the hope that a new bidder will submit an even lower bid is suggested to be improper. "Bid shopping" is the term applied to this practice. On the other hand, some contracts are required to be public knowledge, and, on those occasions, integrity would require that the same value be disclosed equally to all candidate bidders.

Firms known to be unqualified to perform work or supply the product should not be invited to bid. Unless it is understood as a clause in the contract or is mandated by public law, the price and cost estimates of one competitor should not be made known to another competitor.

QUESTIONS

12.1. Define the following terms:

Fixed-price arrangements   Cost plus incentive fee
Cost reimbursement   Cost plus award
Best effort   Subcontract
Open-purchase contract   Change clause
Time and material contract   Allowable costs
Fixed-price incentive   Negotiation
Quote or price in effect   Audit

12.2. In the sharing arrangements for FPI, what are the pros and cons of a sharing plan such as 90/10 versus 50/50 to the contractor and owner for cost reductions below estimated cost?

12.3. Contrast fixed price versus cost reimbursable types of contracts for (a) very high voltage transmission line over rugged terrain and (b) prototype manufacture of an ultra high vacuum chamber environmental test unit.

12.4. When would an owner prefer an FPI contract? Why would a contractor desire an FPI arrangement?

12.5. List the advantages and disadvantages of CPIF from the owners' and contractors' viewpoint.

12.6. Prepare an outline of a negotiation strategy for the project estimate given in section 9.6.

12.7. Assume that the project estimated in section 9.6 was awarded. Itemize a list of documents that will be useful for an auditor on this project.

PROBLEMS

12.1. A construction contractor provides a unit price contract:

Pile Diameter (in.)	Price per Foot Driver
12	$1250
18	2576
20	2685

The owner's engineer observes that 18 12-in. piles were driven to 60 ft, 14 18-in. piles were driven 52 ft, and 5 20-in. were driven 41 ft. What is the net realized contract value?

12.2. A vendor is considering a fixed-price contract where the productive hour cost rates will be quoted. The vendor uses a markup of sales value method of pricing. Markup rates are varied depending on perception of negotiation stages.
    (a) What are the fixed-price quotations?
    (b) The customer indicates that a job is estimated and suggests the potential number of hours. What fixed-price contract is suggested?

Machine Center	Productive Hour Cost Rate	Markup (%)	Job Estimate (hours)
Light machining	$39.16	10	80
Heavy machining	90.98	20	40
Assembly	32.19	15	20
Finishing	42.33	5	15

12.3. **(a)** Reconsider the example given for fixed-price incentive in section 12.3. What is the cost plus profit for an incentive of 75/25 above target cost where the overrun is $135,000?

(b) Assume that the overrun is $150,000 and 75/25 sharing above target cost is negotiated. Find the cost plus for a fixed-price incentive contract. Use the example in section 12.3.

12.4. **(a)** Reconsider the FPI example in section 12.3, where the cost reduction is $80,000. For an 80/20 incentive contract, what is the cost plus profit?

(b) Now let the cost reduction be $70,000 with a 75/25 incentive contract. Find the reimbursement by the owner.

12.5. A contractor estimates and negotiates the following fixed price incentive contract:

Estimated cost	$10,000,000
Target profit	$ 850,000
Target price	$10,850,000
Price ceiling	$11,500,000

There is 70/30 sharing of costs below estimates, and 0/100 sharing above the price ceiling. Sketch the contractor's FPI profit chart, where profit is the y axis and cost dollars are the x axis.

12.6. Hot-rolled alloy steel bar, $1\frac{1}{2}$ in. O.D. × 20 ft, AISI 4140 oil hardening annealed grade, machine straightened, is quoted on a QPE contract.

Weight (lb)	Price ($/100 lb)
120 (= 1 item)	$115.00
2000	71.00
6000	66.50

Steel is indexed according to a grade of scrap that has shown a 5%, 10%, and 8% increase per period. Find the price for the last period.

12.7. Repeat the example of CPIF in section 12.4 for an 80/20 share. Plot a chart for a 75/25 sharing.

12.8. **(a)** An estimate has an estimated cost of $5 million, target fee of $600,000, and maximum and minimum fee of $800,000 and $450,000. The share proportion is 80/20. Find the contractor's CPIF and the owner's cost for a final cost of 15% above estimate.

(b) Repeat for 10% below estimated cost.

(c) Plot a chart for 60/40 sharing.

12.9. The High Voltage Transmission Line Construction Company is under contract to build the transmission line estimated in section 9.6. After agreeing to the price stipulated by Table 9.9, the owner invokes the right to two changes. (1) He increases the length over which the transmission line will travel by 0.6 mile, and (2) he stipulates that construction roads are to be "pioneer" style, thus not being as obvious and more environmentally pleasing. What plan of action do you advise for this company?

CASE STUDY: A LARGE CONTRACT

The following data are determined:

Manufacturing labor	40,000 hours
Manufacturing hourly rate	$18 per hour
Overhead rate on basis of	
direct labor costs	200%
Material	$50,000
Material overhead rate,	15%
Engineering labor,	1500 hours
Engineering hourly rate	$40 per hour
Overhead rate	150%
Profit rate as a markup of full cost	10%

The contract based on those figures is awarded with a 120% ceiling and a split of 75/25 above estimated costs and 80/20 below estimated costs.

(a) If the contractor actually achieves the figures, then what is the reimbursement to the contractor for costs and profit?

(b) If the incurred costs amount to an overrun of 10% above the estimated costs, then what is the reimbursement?

(c) If the incurred costs amount to 5% under the contract, then find the reimbursement.

Appendices

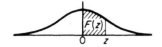

	Areas under the Normal Curve $$F(z) = \int_0^z \frac{1}{\sqrt{2\pi}} e^{-z^2/2} dz$$									
z	0.00	0.01	0.02	0.03	0.04	0.05	0.06	0.07	0.08	0.09
0.0	0.0000	0.0040	0.0080	0.0120	0.0159	0.0199	0.0239	0.0279	0.0319	0.0359
0.1	0.0398	0.0438	0.0478	0.0517	0.0557	0.0596	0.0636	0.0675	0.0714	0.0753
0.2	0.0793	0.0832	0.0871	0.0910	0.0948	0.0987	0.1026	0.1064	0.1103	0.1141
0.3	0.1179	0.1217	0.1255	0.1293	0.1331	0.1368	0.1406	0.1443	0.1480	0.1517
0.4	0.1554	0.1591	0.1628	0.1664	0.1700	0.1736	0.1772	0.1808	0.1844	0.1879
0.5	0.1915	0.1950	0.1985	0.2019	0.2054	0.2088	0.2123	0.2157	0.2190	0.2224
0.6	0.2257	0.2291	0.2324	0.2357	0.2389	0.2422	0.2454	0.2486	0.2518	0.2549
0.7	0.2580	0.2611	0.2642	0.2673	0.2704	0.2734	0.2764	0.2794	0.2823	0.2852
0.8	0.2881	0.2910	0.2939	0.2967	0.2995	0.3023	0.3051	0.3078	0.3106	0.3133
0.9	0.3159	0.3186	0.3212	0.3238	0.3264	0.3289	0.3315	0.3340	0.3365	0.3389
1.0	0.3413	0.3438	0.3461	0.3485	0.3508	0.3531	0.3554	0.3577	0.3599	0.3621
1.1	0.3643	0.3665	0.3686	0.3708	0.3729	0.3749	0.3770	0.3790	0.3810	0.3830
1.2	0.3849	0.3869	0.3888	0.3907	0.3925	0.3944	0.3962	0.3980	0.3997	0.4015
1.3	0.4032	0.4049	0.4066	0.4082	0.4099	0.4115	0.4131	0.4147	0.4162	0.4177
1.4	0.4192	0.4207	0.4222	0.4236	0.4251	0.4265	0.4279	0.4292	0.4306	0.4319
1.5	0.4332	0.4345	0.4357	0.4370	0.4382	0.4394	0.4406	0.4418	0.4430	0.4441
1.6	0.4452	0.4463	0.4474	0.4485	0.4495	0.4505	0.4515	0.4525	0.4535	0.4545
1.7	0.4554	0.4564	0.4573	0.4582	0.4591	0.4599	0.4608	0.4616	0.4625	0.4633
1.8	0.4641	0.4649	0.4656	0.4664	0.4671	0.4678	0.4686	0.4693	0.4699	0.4706
1.9	0.4713	0.4719	0.4726	0.4732	0.4738	0.4744	0.4750	0.4756	0.4762	0.4767
2.0	0.4772	0.4778	0.4783	0.4788	0.4793	0.4798	0.4803	0.4808	0.4812	0.4817
2.1	0.4821	0.4826	0.4830	0.4834	0.4838	0.4842	0.4846	0.4850	0.4854	0.4857
2.2	0.4861	0.4865	0.4868	0.4871	0.4875	0.4878	0.4881	0.4884	0.4887	0.4890
2.3	0.4893	0.4896	0.4898	0.4901	0.4904	0.4906	0.4909	0.4911	0.4913	0.4916
2.4	0.4918	0.4920	0.4922	0.4925	0.4727	0.4929	0.4931	0.4932	0.4934	0.4936
2.5	0.4938	0.4940	0.4941	0.4943	0.4945	0.4946	0.4948	0.4949	0.4951	0.4952
2.6	0.4953	0.4955	0.4956	0.4957	0.4959	0.4960	0.4961	0.4962	0.4963	0.4964
2.7	0.4965	0.4966	0.4967	0.4968	0.4969	0.4970	0.4971	0.4972	0.4973	0.4974
2.8	0.4974	0.4975	0.4976	0.4977	0.4977	0.4978	0.4979	0.4980	0.4980	0.4981
2.9	0.4981	0.4982	0.4983	0.4983	0.4984	0.4984	0.4985	0.4985	0.4986	0.4986
3.0	0.4987	0.4987	0.4987	0.4988	0.4988	0.4989	0.4989	0.4989	0.4990	0.4990
3.1	0.4990	0.4991	0.4991	0.4991	0.4992	0.4992	0.4992	0.4992	0.4993	0.4993

[a]This table gives the probability of a random value of a normal variate falling in the range $z = 0$ to $z = z$ (in the *shaded area in figure*). The probability of the same variate having a deviation greater than z is given by $0.5 -$ probability from the table for the given z. The table refers to a single tail of the distribution; therefore the probability of a variate falling in the range is $\pm z = 2 \times$ probability from the table for the given z. The probability of a variate falling outside the range $\pm z$ is $1 - 2 \times$ probability from the table for the given z.

The values in this table were obtained by permission of author and publishers from C. E. Weatherburn, *Mathematical Statistics*, Cambridge University Press, London, 1946.

APPENDIX 2.   VALUES OF THE STUDENT t DISTRIBUTION

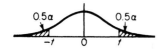

Degrees of Freedom ν	Probability α			
	0.10	0.05	0.01	0.001
1	6.314	12.706	63.657	636.619
2	2.920	4.303	9.925	31.598
3	2.353	3.182	5.841	12.941
4	2.132	2.776	4.604	8.610
5	2.015	2.571	4.032	6.859
6	1.943	2.447	3.707	5.959
7	1.895	2.365	3.499	5.405
8	1.860	2.306	3.355	5.041
9	1.833	2.262	3.250	4.781
10	1.812	2.228	3.169	4.587
11	1.796	2.201	3.106	4.437
12	1.782	2.179	3.055	4.318
13	1.771	2.160	3.012	4.221
14	1.761	2.145	2.977	4.140
15	1.753	2.131	2.947	4.073
16	1.746	2.120	2.921	4.015
17	1.740	2.110	2.898	3.965
18	1.734	2.101	2.878	3.922
19	1.729	2.093	2.861	3.883
20	1.725	2.086	2.845	3.850
21	1.721	2.080	2.831	3.819
22	1.717	2.074	2.819	3.792
23	1.714	2.069	2.807	3.767
24	1.711	2.064	2.797	3.745
25	1.708	2.060	2.787	3.725
26	1.706	2.056	2.779	3.707
27	1.703	2.052	2.771	3.690
28	1.701	2.048	2.763	3.674
29	1.699	2.045	2.756	3.659
30	1.697	2.042	2.750	3.646
40	1.684	2.021	2.704	3.551
60	1.671	2.000	2.660	3.460
120	1.658	1.980	2.617	3.373
∞	1.645	1.960	2.576	3.291

[a]This table gives the values of t corresponding to various values of the probability α (level of significance) of a random variable falling inside the shaded area in the figure, for a given number of degrees of freedom ν available for the estimation of error. For a one-sided test the confidence limits are obtained for $\alpha/2$.

This table is taken from Table III of Fisher and Yates, *Statistical Tables for Biological, Agricultural, and Medical Research*, Oliver & Boyd Ltd., Edinburgh, 1963.

APPENDIX 3. LEARNING TABLES

N	T_u or T_a'	T_c	T_a	T_c'	T_u'
		Learning Table $\phi = 75\%$			
1	1.0000	1.0000	1.0000	1.0000	1.0000
2	0.7500	1.7500	0.8750	1.5000	0.5000
3	0.6338	2.3838	0.7946	1.9015	0.4015
4	0.5625	2.9463	0.7366	2.2500	0.3485
5	0.5127	3.4591	0.6918	2.5637	0.3137
6	0.4754	3.9345	0.6557	2.8523	0.2885
7	0.4459	4.3804	0.6258	3.1214	0.2691
8	0.4219	4.8022	0.6003	3.3750	0.2536
9	0.4017	5.2040	0.5782	3.6157	0.2407
10	0.3846	5.5886	0.5589	3.8456	0.2299
11	0.3696	5.9582	0.5417	4.0661	0.2205
12	0.3565	6.3147	0.5262	4.2784	0.2123
13	0.3449	6.6596	0.5123	4.4835	0.2051
14	0.3344	6.9941	0.4996	4.6821	0.1986
15	0.3250	7.3190	0.4879	4.8749	0.1928
16	0.3164	7.6355	0.4772	5.0625	0.1876
17	0.3085	7.9440	0.4673	5.2453	0.1828
18	0.3013	8.2453	0.4581	5.4236	0.1783
19	0.2946	8.5399	0.4495	5.5979	0.1743
20	0.2884	8.8284	0.4414	5.7684	0.1705
21	0.2826	9.1110	0.4339	5.9354	0.1670
22	0.2772	9.3882	0.4267	6.0991	0.1637
23	0.2722	9.6604	0.4200	6.2598	0.1607
24	0.2674	9.9278	0.4137	6.4176	0.1578
25	0.2629	10.1907	0.4076	6.5727	0.1551
30	0.2437	11.4458	0.3815	7.3124	0.1436
40	0.2163	13.7232	0.3431	8.6526	0.1272
50	0.1972	15.7761	0.3155	9.8590	0.1158
100	0.1479	24.1786	0.2418	14.7885	0.0867
500	0.0758	63.5897	0.1272	37.9137	0.0444

N	T_u or T_a'	Learning Table $\phi = 80\%$ T_c	T_a	T_c'	T_u'
1	1.0000	1.0000	1.0000	1.0000	1.0000
2	0.8000	1.8000	0.9000	1.6000	0.6000
3	0.7021	2.5021	0.8340	2.1063	0.5063
4	0.6400	3.1421	0.7855	2.5600	0.4537
5	0.5956	3.7377	0.7475	2.9782	0.4182
6	0.5617	4.2994	0.7166	3.3701	0.3919
7	0.5345	4.8339	0.6906	3.7414	0.3713
8	0.5120	5.3459	0.6682	4.0960	0.3546
9	0.4929	5.8389	0.6488	4.4365	0.3405
10	0.4765	6.3154	0.6315	4.7651	0.3286
11	0.4621	6.7775	0.6161	5.0832	0.3181
12	0.4493	7.2268	0.6022	5.3922	0.3089
13	0.4379	7.6647	0.5896	5.6929	0.3007
14	0.4276	8.0923	0.5780	5.9863	0.2934
15	0.4182	8.5105	0.5674	6.2730	0.2867
16	0.4096	8.9201	0.5575	6.5536	0.2806
17	0.4017	9.3218	0.5483	6.8286	0.2750
18	0.3944	9.7162	0.5398	7.0985	0.2699
19	0.3876	10.1037	0.5318	7.3635	0.2651
20	0.3812	10.4849	0.5242	7.6242	0.2606
21	0.3753	10.8602	0.5172	5.8806	0.2565
22	0.3697	11.2299	0.5104	8.1332	0.2525
23	0.3644	11.5943	0.5041	8.3820	0.2489
24	0.3595	11.9538	0.4981	8.6274	0.2454
25	0.3548	12.3086	0.4923	8.8696	0.2421
30	0.3346	14.0199	0.4673	10.0368	0.2281
40	0.3050	17.1935	0.4298	12.1986	0.2076
50	0.2838	20.1217	0.4024	14.1913	0.1931
100	0.2271	32.6508	0.3265	22.7062	0.1542
500	0.1352	98.8472	0.1977	67.6232	0.0917

N	T_u or T_a'	T_c	T_a	T_c'	T_u'
		Learning Table $\phi = 85\%$			
1	1.0000	1.0000	1.0000	1.0000	1.0000
2	0.8500	1.8500	0.9250	1.7000	0.7000
3	0.7729	2.6229	0.8743	2.3187	0.6187
4	0.7225	3.3454	0.8364	2.8900	0.5713
5	0.6857	4.0311	0.8062	3.4284	0.5384
6	0.6570	4.6881	0.7813	3.9419	0.5135
7	0.6337	5.3217	0.7602	4.4356	0.4937
8	0.6141	5.9358	0.7420	4.9130	0.4774
9	0.5974	6.5332	0.7259	5.3766	0.4636
10	0.5828	7.1161	0.7116	5.8282	0.4516
11	0.5699	7.6860	0.6987	6.2693	0.4411
12	0.5584	8.2444	0.6870	6.7012	0.4318
13	0.5480	8.7925	0.6763	7.1246	0.4235
14	0.5386	9.3311	0.6665	7.5405	0.4159
15	0.5300	9.8611	0.6574	7.9495	0.4090
16	0.5220	10.3831	0.6489	8.3521	0.4026
17	0.5146	10.8977	0.6410	8.7489	0.3968
18	0.5078	11.4055	0.6336	9.1402	0.3913
19	0.5014	11.9069	0.6267	9.5264	0.3863
20	0.4954	12.4023	0.6201	9.9079	0.3815
21	0.4898	12.8920	0.6139	10.2850	0.3771
22	0.4844	13.3765	0.6080	10.6579	0.3729
23	0.4794	13.8559	0.6024	11.0268	0.3689
24	0.4747	14.3306	0.5971	11.3920	0.3652
25	0.4701	14.8007	0.5920	11.7536	0.3616
30	0.4505	17.0907	0.5697	13.5141	0.3462
40	0.4211	21.4252	0.5356	16.8435	0.3233
50	0.3996	25.5131	0.5103	19.9811	0.3066
100	0.3397	43.7539	0.4375	33.9680	0.2603
500	0.2329	151.4504	0.3029	116.4542	0.1783

N	T_u or T_a'	Learning Table $\phi = 90\%$ T_c	T_a	T_c'	T_u'
1	1.0000	1.0000	1.0000	1.0000	1.0000
2	0.9000	1.9000	0.9500	1.8000	0.8000
3	0.8462	2.7462	0.9154	2.5386	0.7386
4	0.8100	3.5562	0.8891	3.2400	0.7014
5	0.7830	4.3392	0.8678	3.9149	0.6749
6	0.7616	5.1008	0.8501	4.5695	0.6546
7	0.7439	5.8447	0.8350	5.2076	0.6381
8	0.7290	6.5737	0.8217	5.8320	0.6244
9	0.7161	7.2898	0.8100	6.4446	0.6126
10	0.7047	7.9945	0.7994	7.0469	0.6023
11	0.6946	8.6890	0.7899	7.6401	0.5932
12	0.6854	9.3745	0.7812	8.2251	0.5850
13	0.6771	10.0516	0.7732	8.8028	0.5777
14	0.6696	10.7211	0.7658	9.3737	0.5709
15	0.6626	11.3837	0.7589	9.9385	0.5648
16	0.6561	12.0398	0.7525	10.4976	0.5591
17	0.6501	12.6899	0.7465	11.0514	0.5538
18	0.6445	13.3344	0.7408	11.6002	0.5489
19	0.6392	13.9735	0.7354	12.1445	0.5442
20	0.6342	14.6078	0.7304	12.6844	0.5399
21	0.6295	15.2373	0.7256	13.2202	0.5358
22	0.6251	15.8624	0.7210	13.7521	0.5319
23	0.6209	16.4833	0.7167	14.2804	0.5283
24	0.6169	17.1002	0.7125	14.8052	0.5248
25	0.6131	17.7132	0.7085	15.3267	0.5215
30	0.5963	20.7269	0.6909	17.8893	0.5070
40	0.5708	26.5427	0.6636	22.8319	0.4850
50	0.5518	32.1420	0.6438	27.5881	0.4686
100	0.4966	58.1410	0.5814	49.6585	0.4214
500	0.3888	228.7851	0.4576	194.4098	0.3298

N	T_u or T_a'	T_c	T_a	T_c'	T_u'
		Learning Table $\phi = 95\%$			
1	1.0000	1.0000	1.0000	1.0000	1.0000
2	0.9500	1.9500	0.9750	1.9000	0.9000
3	0.9219	2.8719	0.9573	2.7658	0.8658
4	0.9025	3.7744	0.9436	3.6100	0.8442
5	0.8877	4.6621	0.9324	4.4386	0.8286
6	0.8758	5.5380	0.9230	5.2549	0.8163
7	0.8659	6.4039	0.9148	6.0612	0.8063
8	0.8574	7.2610	0.9077	6.8590	0.7978
9	0.8499	8.1112	0.9012	7.6494	0.7904
10	0.8433	8.9545	0.8954	8.4333	0.7839
11	0.8374	9.7919	0.8902	9.2115	0.7781
12	0.8320	10.6239	0.8853	9.9844	0.7729
13	0.8271	11.4511	0.8808	10.7525	0.7682
14	0.8226	12.2736	0.8767	11.5163	0.7638
15	0.8184	13.0921	0.8728	12.2761	0.7598
16	0.8145	13.9066	0.8692	13.0321	0.7560
17	0.8109	14.7174	0.8657	13.7846	0.7525
18	0.8074	15.5249	0.8625	14.5339	0.7493
19	0.8042	16.3291	0.8594	15.2801	0.7462
20	0.8012	17.1302	0.8565	16.0233	0.7433
21	0.7983	17.9285	0.8537	16.7639	0.7405
22	0.7955	18.7241	0.8511	17.5018	0.7379
23	0.7929	19.5170	0.8486	18.2372	0.7354
24	0.7904	20.3074	0.8461	18.9703	0.7331
25	0.7880	21.0955	0.8438	19.7012	0.7308
30	0.7775	25.0032	0.8334	23.3246	0.7208
40	0.7611	32.6838	0.8171	30.4443	0.7054
50	0.7486	40.2339	0.8045	37.4322	0.6938
100	0.7112	76.5864	0.7659	71.1212	0.6588
500	0.6314	340.6472	0.6813	315.6782	0.5847

Selected Answers

CHAPTER 1

1.2. (a) 295. 4 MJ; **(b)** 2.7×10^7 MJ, 11.24 lbm/ft^3 **1.4. (a)** 5.5 m, 0.61 m; **(b)** 254 mm, 0.25 mm, 2540 mm, 0.0038 mm **1.6. (a)** 240.29 kg/m^3. **(b)** 11.24 lbm/ft^3 **1.8.** 2732°F, 392°F, 1832°F; **(b)** 93.3°C, 537.8°C **1.10. (a)** 0.01 m^3, 3.54 m^3, 19.82 m^3; **(b)** 163.8 mm^3, 196,560 mm^3, 2,457,000 mm^3; **(c)** 764.5 m^3; **(d)** 1.5×10^{-2} in.^{3}, 4.88×10^{-2} in^3, 4.88×10^{-3} in.^{3}, 9.16×10^{-2} in.^{3} **1.12. (a)** 26.62 DM/unit; **(b)** 21.91 SF/unit, **(c)** 19.66 C\$/unit **1.14. (a)** \$98.34; **(b)** 0.004718 \$/Yen; **(c)** Reduce **1.16. (a)** Unit billing = \$142.46, lot billing = 5,093,338 renolas **1.18. (a)** Germany cheapest at the time of estimate, England cheapest at time of arrival.

CHAPTER 2

2.2. (a) 90.9091 hr/1000 units; **(b)** 3.390 hr/100 units, 33.8983 hr/1000 units, 338.98305 hr/10,000 units; **(c)** 6.5, 9.20 **2.4.** 0.267 normal minutes, 0.314 standard minutes/unit, 191 pc/hr, 0.523 hr/100 units **2.6. (a)** Total standard minutes = 0.544; **(b)** 110 units/hr, 0.907 hr/100 units; **(c)** 53% increase in output and 35% savings in time **2.8. (a)** Total normal time = 0.031 min, standard min/unit 0.039, pieces per hour = 1538, standard hours/100 units = 0.065; **(b)** Standard min per unit = 0.035 **2.10.** Item 1 23%, 46 hours, . . . , Item 12, 8%, 16 hours **2.12. (a)** 75 man days, 4800 observations; **(d)** Item A, 1.67% occurrence, 10 hours **(d)** Item A interval = 0.0061 and relative accuracy = plus and minus 18.3% **2.14. (a)** For 95% N = 323 **(b)** For 90%, I = 3.9%; **(c)** For 90%, N = 2010 **2.16. (a)** A = 11.6 hour, B = 23.2 hour, C = 46.5 hour **2.18. (a)** About \$33.32/hour; **(b)** \$370.22 **2.20.** \$39.50 **2.22. (a)** Weekly earnings =

$924; **(b)** Actual wage = $16.50 before incentive, with incentive = $23.10/hr. Hourly fringe cost = $3.87. Unit cost = $2.77   **2.24. (a)** Earnings = $240, Hourly wage after incentive = $20/hr; **(b)** Pay = $256, Efficiency = 93.8%   **2.26. (a)** Based on number of machines, A = $0.0046/unit, B = $0.0055/unit; **(b)** Based on product output A = $0.005/unit, B = $0.005/unit; **(c)** Based on value, A = $0.005/unit and B = $0.0049/unit   **2.28. (a)** Dejoint cost per unit = $0.005/unit; **(b)** 0.033 hr/100 units **(c)** $0.490/100 units   **2.30. (b)** Normal time = 1.614 min.; **(c)** Standard time = 2.017 min/fire   **2.32. (a)** Gross hourly cost = $23.71; **(b)** Job cost = $590.50; **(c)** For 85% efficiency, loss = $104.20; for 115% efficiency, gain = $77

CHAPTER 3

3.2. (a) $17.94; **(b)** 48.1%   **3.4. (a)** $13.78 without waste; **(b)** $8.44; **(c)** Yield = 32%   **3.6.** Left design = $0.116/unit, right design = $0.04/unit; **(b)** Left design shape yield = 58%, right design shape yield = 65%   **3.8. (a)** LTR = $12.95/unit, D = 10.01, MOOP = 10.53; **(b)** 12.95, 10.01, 9.81; **(c)** 12.95, 10.01, 10.95; **(d)** 12.95, 10.01, and 11.14   **3.10. (a)** Yield = 87%; **(b)** $1.65/unit; **(c)** $0.223/unit; **(d)** $1.87/unit   **3.12.** Poured metal cost = $12.60/casting, $10.38/casting, $2.08/lb   **3.14.** Original = $8.95/unit, last = 9.00, current = 12.00, LTR = 10.01, D = 11.84, MOOP = 9.78   **Case Study:** Grams in design A, 7.09, B = 9.07, C = 42.06. Shape efficiency = 85%

CHAPTER 4

4.2. Assets = $5350, liabilities and net worth = $5350   **4.4.** Assets = $485,000, liabilities and net worth = $485,000   **4.6.** Net profit = $300,000, Assets = $900,000 and liabilities and net worth = $900,000   **4.8.** Net profit = $190,000   **4.10.** Net income = $105,000, Assets = $540,000   **4.12.** Net income = $72, Assets = $154, Liabilities and net worth = $154   **4.14.** Company A ACR first year depreciation = $50,000 and company B straight line depreciation = $47,000   **4.16. (a)** $0.175/mile; **(b)** $5421.20/year; **(c)** $0.54/mile   **4.18.** At 80, 100, and 125% direct costing overhead rate = 60%, and absorption costing overhead = 1.8, 1.56, and 1.37   **4.20.** Typical overhead rate on the basis of direct labor hours = $24.023   **4.22.** 180%   **4.24.** $27.51   **4.28. (a)** Total variable costs = $1,217,930; **(b)** Full cost = $1,953,930

CHAPTER 5

5.4. Answer will vary with graph; $y = 173.8x + 681.3$. When $x = 11$, $y = 2592.50×10^6 **5.10. (a)** 123.10; **(b)** 1.59; **(c)** Confidence limits = (120.39, 125.80)   **5.12.** $y = 96.6 + 4.736x$. At $x = 10$, $y = 109.83$, P.I. = (110.7, 109.0), C.I. = (1.27, 1.61)   **5.14. (b)** log $a = 2.622$, $b = -0.364$, $T = 419N^{-0.364}$ **5.16. (a)** $60.9e^{0.54Hp}$; **(b)** $60.9(1.72)^{Hp}$ **5.18.** $r = -0.13$, very low or no correlation   **5.20. (a)** $y = 4(1.00022)^x$ and $y = 5.8 - 8.7 \times 10^{-4}x + 4.5 \times 10^{-7}x^2$; **(b)** Polynomial is better   **5.26.** 1959 (20.4), 1969 (23.3), 1979 (52.3), and 1989 (100)   **5.28. (a)** $664,064; **(b)** $601,000   **5.30.** $107,060   **5.34. (a)** Approx equation = log $V = 2.9 - 0.45$ log T **5.36.** Log log plot is more suitable **(b)** $y = 1.4 \times 10^4 + 0.05x$, $r = 0.88$

CHAPTER 6

6.4. $73/ft^2, $4.83/ft^3$ **6.6. (a)** 68 hr; **(b)** At $N = 50$, $114.94 and at $N = 200$, $87.00;
(c) $s = -0.193$, $T_a' = 317$ hr, $T_u' = 257$ hr, $T_c' = 3170$ hr; **(d)** 93.75%. At $N = 5$,
$T_a' = 853$ hr, $T_c' = 4265$, $T_u' = 781$, and at $N = 40$, $T_a' = 703$ hr, $T_c' = 28118$, and $T_u' = 638$ hr; **(e)** 99.8 hr, 11,478 hr, and 114 hr   **6.8. (a)** $s = -0.2615$, average hours $= 162.4$,
and cumulative hours $= 1625$ hr; **(b)** $s = -0.4674$, At $N = 50$, unit $= 391$, cumulative $=$
36,526. At $N = 100$, 282 and 52,835 hr

6.10.

N	Unit	Cumulative	Average
2	0.760	1.760	0.880
3	0.647	2.407	0.802

6.12. At $N = 10$ A is preferred. At $N = 30$, B is preferred. Learning percentage $= 90\%$
(b) $164,316

6.16.

N	Unit, Ave	T_c	T_u	T_c'	T_u'
2	0.830	1.830	0.915	1.660	0.660
3	0.744	2.574	0.858	2.233	0.573
4	0.689	3.263	0.816	2.756	0.523

6.18. (a) $266,783   **6.20.** Approx $m = 0.8$, cost $= 3.8×10^6 **6.22.** $87,.25; **(b)** $50;
(c) $2950; **(d)** 40%   **6.24.** No change $= $245,000/year, Design A $= $445,000,
Design B $= $495,000, expected value design A $= $195,000, expected design B $= $245,000
6.26. (a) Total cost $= $28.75, Total variance $= 3.007$; **(b)** Probability $= 94\%$
6.30. (a) Ignoring setup, cost $= $1.55/unit, 16 pc/hr; **(b)** $1.03/unit, 24.2 pc/hr
6.32. y(weight) $= 0.048 + 0.234$(weight), $r = 99\%$, y(girth) $= 0.3575$ y(girth) $= 0.3575 +
0.3873x_2$, $r = 83\%$, y(location) $= 6.06 - 1.03x_3$, $r = 0.95$ **6.36. (a)** Total marginal
cost $= $327, unit marginal cost $= $1.635, profit $= $23, minimum break even $327
6.38. (a) $200 - 6 \times 10^{-3}n + 3 \times 10^{-7}n^2$; **(b)** $C_m = 170, $n = 10,000$, $C_a = 190 ave;
(d) Marginal profit $= 0$ at $n = 801$ **Case study:** Maximum profit occurs when marginal
profit $= 0$, and cavities per die $=$ are between 4 and 5. For max profit there should be
4 cavities per die. Full cost $= $11,175

CHAPTER 7

7.2. (a) $1.65; **(b)** 3.19 min; **(c)** 30 min; **(d)** $0.20; **(e)** $0.30   **7.4. (a)** Approx 1200 mm;
(b) Approx 16 in; **(c)** Approx 22 mm   **7.6. (a)** Length $= 11.13$ in., $t_m = 4.27$ min,
$N = 325$, feed rate $= 2.64$ in./min; **(b)** Rough pass, Length of cut $= 20.625$ in., time for
cut $= 1.26$ min, feed $= 16.4$ in./min, RPM $= 325$. Finish pass, Length of cut $= 21.125$ in.,
time for cut $= 1.54$ min, RPM $= 363$ **7.8. (a)** For stainless steel, $L = $ about 1.13 in., t_m
for drill $= 1.85$ min, t_m for tap $= 2.65$ min; **(b)** For medium carbon steel, $L = 1.176$, time
to drill $= 0.29$ min, time to tap $= 0.38$ min   **7.10. (a)** $n = 0.11$, $K = 160$; **(b)** 72 min;
(c) 191 RPM   **7.12.** $V_{max} = 120$, $T_{max} = 36$, $T_u = $ approx 11.7 min   **7.14. (a)** $N = 9555$ units; **(b)** $N = 14,000$ units; **(c)** $6335; **(d)** years $= 0.5$ **7.18.** $T_{max} = 11.3$ min,
$V_{max} = 347$

CHAPTER 8

8.2.

Year	0	1	2	3	4	5
Cum. CF	−$60,000	−62,500	−57,500	−35,125	−16,000	−625

8.4.

Year	0	1	2	3	4	5	6
Cum. CF	−$133	−103	−69	−31	11	57	107

Payback about 4 years.

8.6. (c) $s = -0.15$, $K = 191.6$ hr, Lot time $= 127,026$ hr, Average time for lot $= 51.6$ hr.
8.8. For 80% learning, cumulative total at 15 and 30 units $= 159,000$ and $262,000$ hr. For 90% learning, cumulative total at 15 and 30 units $= 102,000$ and $185,000$ hr   **8.10.** ECO time $=$ approx 1514 hr   **8.12.** Composite model at $N = 1$ is 13725 hr and at $N = 500$ units, there are 6180 hours. $s = -0.128$ and learning $= 91.5\%$. Breakeven about 460 units. At $N = 1000$, profit $= \$600$.

8.14.

	Lot	Unit
Direct labor	$ 500	$2.86
Factory overhead	375	2.14
Material	13	.08
Cost of goods	889	5.07
General and administrative	221	1.27
Full cost	1997	11.42
Profit @ 20%	399	2.28
Price	$2398	$13.74

8.18. (a) $200; (b) $228.57   **8.20.** $33.33

8.22.

Price	4	5	6
Contribution	1	2	3
Contribution %	25	40	50
Revenue	$160,000	150,000	90,000

The $5 price provides the greatest contribution.

8.24. (a) 5.1 year, (b) 3.6 year, margin $= 0.57$ **8.26** 3 years

8.28. (a) Labor cost $= \$33,000$
Material
cost $= 1,000$
Full cost $=$ 33,936
Bid $40,723
Time estimates would have to be off about 20% to consume profit.

8.30.

	Single Plant	Twin Plant
Full cost	$2.25	$1.40
Final price	$3.45	$2.15

The labor will have to increase $1.31 to bring the operation back to a single plant and make the decision indifferent.

CHAPTER 9

9.2. Cost of one bridge = $559,787

9.3. Cost of bags = $10,251

9.6. Four levels

9.7. Cumulative baseline cost = $690

9.8.

Period	Commitment	Expenditure
0	650	—
1	1170	520
2	2340	1625
3	1820	2535
4	520	1365
5	—	455
	6500	6500

Approx midpoint = $2\frac{3}{4}$ period for commitment, $3\frac{3}{4}$ period for expenditure. Difference about $1000.

9.9. Project cost total = $11,350

9.10. R_o = 4.74%, R_j = 2.47%, C_{op} = 10.815×10^6

9.12. Job overhead rate = 30.9%, Office overhead rate = 50%, Direct cost = $256,000

9.13. 25% underrun is about 24×10^6, 25% overrun is about 26×10^6, overrun is about 27×10^6

9.15. Contingent present time cost = $43,000 approx. above real cost of $100,000. Contingent cash flow is about $44,075 when indexed. There is very little to commend this practice.

9.17. C_a = $1,600,000, C_m = $2,288,000, t_{ai} = 0.7 monthly, t_{af} = 3.6 month.

Period	CBV %	Profit	Retainage	Net Cash Flow
1	14.3	288,800	1,144,000	(915,200)
2	42.9	686,400	2,402,400	(1,716,000)
3	71.5	1,144,000	2,631,200	(1,487,200)
4	92.9	1,486,400	2,284,000	(797,600)
5	100.0	1,600,000	1,311,200	(288,800)
6	100.0	1,600,000	800,000	800,000
7	100.0	1,600,000	0	1,600,000

9.18.

Period	Net Cash Flow 10^6	Period	Net Cash Flow 10^6
1	(1.3)	9	(8.2)
2	(3.9)	10	(7.4)
3	(6.4)	11	(7.0)
4	(8.6)	12	(6.2)
5	(10.6)	13	(3.5)
6	(10.2)	14	(1.1)
7	(9.4)	15	(5.4)
8	(8.9)	16	7.5
		17	15.0

9.20. Cost to contractor = $7,272,727; Cost to subcontractor = $6,060,606; Total cost of labor = $3,030,303; Manpower = 101,010 hours; 582.9 man months; Average number of direct labor employees = 117; Maximum number of direct labor employers = 167

9.22. Total interest paid = $2,831, Net profit = $12,169

9.23.

Month	10^5 End of Month Outstanding
1	$1.32
2	3.98
3	6.52
4	8.76

9.24.

Month	Outstanding Loan
1	$1,950,000
2	2,181,000
3	1,949,000
4	1,192,000
5	798,000
6	0

9.25. (a) 10%, 20%; (b) 17.5%

9.26. (a) % return = 16.7%, payback = 6 years; (b) For process A, capital is recovered between 3rd and 4th year, for process B, capital is recovered between 2nd and 3rd year

9.28. (a) $90; (b) 4%; (c) $90,000 simple; $107,946 compound; (d) $565; (e) $8572; (f) $422; (g) 22.5 years; (h) 1 year 6.67%, 2 years 3.28%

9.30. (a) NPW = $18, NFW = $24, Annual eq. = $426.80, Net annual equivalent = $7.35; (b) NPW = $55, NFW = $71, Net annual eq. = 22

9.38. Approximate solution to project bid summary

Project: Transmission Line	
Direct labor	$3,417,000
Direct material	1,575,000
Subcontract items	440,000
Facilities	198,800
Equipment	600,000
Engineering	42,000
Office overhead at 4%	217,000
Job overhead at 5%	272,000
Subtotal	$6,761,800
Contingency, $\frac{1}{2}$%	33,800
Interest with retainage at $i = 1\frac{1}{4}$%	84,500
Total	$6,800,100
Profit, 8%	550,500
Bid	$7,430,600

9.39. Approximate solution to project bid summary

Direct labor	$ 9,240,000
Direct material, sub contract	12,133,000
Facilities and equipment	1,315,000
Engineering	84,000
Direct costs	$14,457,000
Office overhead @ 4% direct cost	578,000
Job overhead @ 5% direct costs	722,000
Subtotal	$15,758,000
Contingency @ $\frac{1}{2}$%	79,000
Interest with retainage, $1\frac{1}{4}$%	79,000
Total	$16,034,000
Markup @ 8%	1,283,000
Bid	$17,316,000

CHAPTER 10

10.2. (a) Maximum value is given each goal column

$G_1:A_1$	$E_{11} = 80 > E_{i1}, i = 2, 3, 4, 5$
$G_2:A_2$	$E_{32} = 70 > E_{i2}, i = 1, 2, 4, 5$
$G_3:A_5$	$E_{53} = 75 > E_{i3}, i = 1, 2, 3, 4$

(b) $(XE_{11} + E_{12} + E_{13}) > (X E_{i1} + E_{i2} + E_{i3})$ $i = 2, 3, 4, 5$

i	E_{i1}	$E_{i2} + E_{i3}$	X_i
1	80	41	—
2	50	83	1.4
3	35	91	1.1
4	39	95	1.3
5	25	100	.9

Thus $X > 1.4$

10.3. The $20,000 is a sunk cost and is irrelevant to the decision, except that it gives a feel for winning of the business. You should be willing to spend $50,000. If you win the business after spending exactly $50,000 more, your prosperity is weakened by $20,000. Obviously, business can't be conducted this way for long.

10.5. (a) The new system should include $10,000 for rent of the building since it can be rented and therefore offers an opportunity for return. The meaningful measure = net equivalent annual income − rent. **(b)** If no renter, the answer is yes, as this affects the opportunity. **(c)** If a renter is available for 1 or 2 years, and a system is available, then the discounted value is used to choose the most effective choice. **(d)** Yes, once the decision is made, the alternative is given up, and consideration is useful only in a past-tense way; i.e., helpful to future choices.

10.6. Before the decision:
System cost$_1$ = $E_1 + P_1 + W_1 + OM_1 + I_1 + R_1 + C_1$. $SC_2 = E_2 + P_2 + W_2 + OM_2 + I_2 + R_2 + C_2$
Marginal difference = $E_1 - E_2 + W_1 - W_2 + OM_1 - OM_2 + I_1 - R_2 + C_1 - C_2$
Nonrecurring = engineering + project + (sometimes product)
Investment = product + project
Opportunity cost for selection option $1 : I_1$
Opportunity cost for selection option $2 : I_2 = O, R_1 = 0$
After the decision: There is no opportunity

10.15. $A = 387.50, $B = 385.00, $C = 325.00

10.16. $B/C_A = 1.26$, $B/C_B = 1.46$, $B/C_C = 1.19$, $B/C_D = 1.32$, Select B

10.18.

Cost Element	Total Cash Flow
Design	$ 40,000
Product	265,000
Installation	30,000
Manpower	327,040
Preventive	3,900
Corrective	5,840
Power	26,864
Spares	1,326
Undiscounted LCC	699,970
Discounted LCC	476,498

10.19. Alternatives:

    I. ABC alone with reconstruction, mandatory

   II. ABC and ADC jointly

      Project costs for plan I = 21.68×10^6

      Project costs for Plan II = 45.22×10^6

      Difference between plans = 23.54×10^6

      Discounted traffic costs for ABC = 125.62×10^6

      Undiscounted traffic costs for ABC = 201.66×10^6

      $B/C = .4$ Advise ABC

10.20. Undiscounted B/C = 7.6 = discounted B/C

10.21. C_1 = \$1.39/hr., C_2 = \$2.21/hr., C_3 = \$1.40; Select contractor 1, but contractor 3 is so close that other factors might be considered.

Case Study: No. of PM actions = 1/yr

            Cost/yr = \$104

            Corrective maintenance = \$41.60/yr

            Power cost = \$13.875/yr

            Total undiscounted LCC = \$202.300

            Total discounted LCC = \$136,600

CHAPTER 11

11.1.

Element	Cost
Direct labor	\$1,906.25
Direct material	1,376.00
Other material	145.00
Overhead	953.13
Profit	1,095.10
Total	\$5,475.48

Final profit = \$225.50.

11.3.

Direct labor at 10th unit =	\$ 570,807
Direct material	16,866
Overhead	570,807
Subtotal	\$1,158,480
Profit @ 15%	173,772
Total	\$1,332,252

If actual cost = \$1,250,000, error = \$91,520 in cost

                = 7.9% of cost or

                = 6.9% of total bid

                = \$82,252

For effective profit margin = \$82,252 is 6.2% of total bid or 6.6% of actual cost, and 7.1% of estimated cost.

11.5. Capture rate = 48.7%

11.6.

Quarter	Error %	Capture %	Productivity
1	1.04	51.9	.93
2	−.50	49.8	1.00
3	−6.01	47.1	1.06

11.7. Hours first: Total variance = −$0.034 favorable; Wages first: Total variance = −$0.034 favorable

11.8. Total variance = $224 U

11.10. Total variance = $2.44 U

11.12. Net variance = $60.7 U, PF = 1.06 approx

11.14. $125 variance unfavorable, PF = 1.045

11.16. Net variance for labor = $4.97. Lot hours estimated = 5.83, material net variance = $244 U

11.17. (a) $0.207 unfavorable

11.18. Net variance on constant overhead percentage

	Estimate	Variance		Actual
Direct labor	$ 5.28	+0.53		$ 5.81
Material	17.38	−0.45		16.93
Overhead	7.92	0.80		8.72
	$30.58	+0.88	unfavorable	$31.46

Net variance on constant overhead amount

	Estimate	Variance		Actual
Direct labor	$ 5.28	+0.53		$ 5.81
Material	17.38	−0.45		16.93
Overhead	7.92	0		7.92
	$30.58	+0.08	unfavorable	$30.66

11.20. (a) Net profit percent of net sales = 19.8% **(b)** Percent of net sales of profit contributions at actual: I, 34.5%, II, 47% (best), III, 8.2%

11.23.

	SV 4		CV 4		SV 5		Approx.	
	$	Periods		$		Periods	Overrun	Slippage
	$1000	0.75	$1500	$750		1	$1300	1

11.24. Percent of net sales = 2.6% final

11.25.

Period	Variance Schedule	Cost	Forecast BCWP	ACWP	Variance
4	$\approx \frac{3}{4}$ mo.	$1,500			
5	≈ 1.1	1,200	5800	6800	1000
6	$\approx .5$	1,100	6400	6900	500

CHAPTER 12

12.1.

Pile, in.	Contract value
12	$1,350,000
18	1,875,328
20	550,425
	$3,775,753

12.2.

Center	Hourly Cost	Job	Markup Hourly Cost	Quote
Light	$39.16	80	$ 43.80	$3,446.08
Heavy	90.88	40	109.06	4,362.24
Assembly	32.19	20	37.02	740.37
Finishing	42.33	15	44.47	66.70

12.4. Cost plus profit = $952,000

12.6.

Weight	$/100 lb
120 lb	$143.45
2000	88.57
6000	82.95

References

ADRIAN, JAMES J., *Construction Estimating, An Accounting and Productivity Approach,* Reston Publishing Company, Inc., a Prentice Hall Company, Reston, VA., 1986.

AHUJA, HIRA and WALTER J. CAMBELL, *Estimating from Concept to Completion,* Prentice Hall, Englewood Cliffs, NJ., 1988.

AHUJA, H.N., MICHAEL A. WALSH, *Successful Methods in Cost Engineering,* Wiley, New York, 1983.

AMSTEAD, B.H., PHILLIP F. OSTWALD and MYRON L. BEGEMAN, *Manufacturing Processes,* 8th ed., Wiley, New York, 1987.

ARMSTRONG, J. SCOTT, *Long-Range Forecasting, From Crystal Ball to Computer,* 2nd edition, Wiley Interscience, NY, 1985.

BLANCHARD, BENJAMIN S., *Design and Manage to Life Cycle Cost,* M/A Press, Portland, OR, 1978.

BLANCHARD, BENJAMIN S., WOLTER J. FABRYKY, *System Engineering and Analysis,* Prentice Hall, Englewood Cliffs, NJ, 1981.

BOEHM, BARRY W., *Software Engineering Economics,* Prentice Hall, Englewood Cliffs, NJ, 1981.

Building Cost File, Construction Publishing Co., McKee, Berger, Mansuet, New York, Annual Volume.

CALDER, GRANVILLE, *The Principles and Techniques of Engineering Estimating,* Pergamon Press Ltd., Oxford, 1978.

CANADA, JOHN R., WILLIAM G. SULLIVAN, *Economic and Multiattribute Evaluation of Advanced Manufacturing Systems,* Prentice Hall, Englewood Cliffs, NJ, 1989.

CLARK, FORREST D., and A.B. LORENZONI, *Applied Cost Engineering,* Marcel Dekker, Inc., New York, 1978.

COLLIER, KEITH, F., *Estimating Construction Costs: A Conceptual Approach,* Reston Publishing Co., Reston, VA, 1988.

Cost Engineers' Notebook, American Association of Cost Engineers, Morgantown, WV, 1988.

DUDICK, THOMAS S., Handbook of Product Cost Estimating and Pricing, Prentice Hall, Englewood Cliffs, NY, 1991.

GALLAGER, PAUL F., *Parametric Estimating for Executives and Estimators,* Van Nostrand Reinhold Company, Inc., New York, 1982.

GODFREY, ROBERT STURGES, Ed., *Building Construction Cost Data,* Robert Snow Means Company, Duxbury, MA, Annual Volume.

GREER, WILLIS R., DANIEL A. NUSSBAUM, Ed., *Cost Analysis and Estimating: Tools and Techniques,* Springer Verlag, New York, 1990.

JELEN, F.C., and J. BLACK, Eds., *Cost and Optimization Engineering,* 2nd ed., McGraw-Hill Book Company, New York, 1983.

Machining Data Handbook, 3rd ed., Machinability Data Center, Metcut Research Associates, Inc., Cincinnati, OH, 1980.

MALSTROM, ERIC M., *What Every Engineer Should Know About Manufacturing Cost Estimating,* Marcel Dekker, Inc., New York, 1981.

MATTHEWS, LAWRENCE, M., *Estimating Manufacturing Costs,* McGraw-Hill Book Company, New York, 1983.

MICHAELS, JACK V., WILLIAM P. WOOD, *Design to Cost,* John Wiley & Sons, New York, 1989.

O'NEIL, JAMES N., *Construction Cost Estimating for Project Control,* Prentice Hall, New York, 1982.

OSSENBRUGGEN, P.J., *Systems Analysis for Civil Engineers,* John Wiley & Sons (Chap. 5: Economic Considerations for Resource Allocation).

OSTWALD, PHILLIP F., Ed. *Manufacturing Cost Estimating,* Society of Manufacturing Engineers, Dearborn, MI, 1980.

OSTWALD, PHILLIP F., *AM Cost Estimator,* 4th edition, Penton Publishing Co., Cleveland, Ohio, 1988.

PARK, WILLIAM R., DALE E. JACKSON, *Cost Engineering Analysis: A Guide to Economic Evaluation of Engineering Projects,* 2nd ed., John Wiley & Sons, New York, 1984.

PETERS, MAX S. and KLAUS D. TIMMERHAUS, *Plant Design and Economics for Chemical Engineers,* 4th ed., McGraw-Hill Book Company, New York, 1990.

STEWART, RODNEY D., *Cost Estimating,* 2nd ed., John Wiley & Sons, New York, 1991.

STEWART, RODNEY D., RICHARD M. WYSKIDA, *Cost Estimator's Reference Manual,* John Wiley & Sons, 1987.

TAYLOR, T., *Handbook of Electronics Cost Estimating,* John Wiley & Sons, 1985.

THEUSEN, G.J., W.J. FABRYCKY, *Engineering Economy,* 7th ed., Prentice Hall, Inc., Englewood Cliffs, NJ, 1989.

TYRAN, MICHAEL R., *Product Cost Estimating and Pricing: A Computerized Approach,* Prentice Hall, Englewood Cliffs, NJ, 1982.

Index

A

Absolute budget, 418
Absolute effectiveness, 398
Account, T, 94
  escrow, 358
Accountant, types, 93
Accrual basis, 97
Accumulated depreciation, 117
Accuracy, 16, 191, 216, 219,
  311, 428
ACRS (accelerated cost
  recovery system), 112
Activity based costing, 119
Actual cost, 428
Actual cost of work
  performed, 445
Actual costs, 430, 458
Actual material cost, 436
Actual purchase cost, 70
Actual time, 24
Actual timing, 27
Actual value, 428
ACWP (see actual cost of
  work performed), 445
Ad valorem, 331
Addendum statement,
  contract, 466
Adjusted estimate, 432
Allocators, 76

Allowable, cost reimbursement
  contract, 459
Allowance, 24
  capital estimate, 335
  delay, 24
  depreciation, 276
  fatigue, 24
  interest, 276
  maintenance, 276
  personal, 24
  taxes, 276
Allowance multiplier, 31, 222
Allowances, 219, 222, 253, 309
  delay, 30
  depreciation, 111
  fatigue, 30
  job, 30–31
  personal, 30
  PF&D, 255
Allowances for depreciation,
  111
Allowed time, 24, 34
Allowed time units, project,
  337
All-day time study, 33
Alternatives, 396
Amortization, 111, 273
Amortization, system, 418
Amortizing, 272
  engineering costs, 293

Analogy estimating, 195
Analysis, accounting, 93–128
  Bayesian, 209
  break even, 206, 235, 305
  B/C, 405
  contractural for material, 70
  cost, 141, 395
  cost and bid for projects,
    352–367
  engineering cost, 3
  graphic, 141–146
  investment, 289
  labor, 22–51
  material, 61–81
  micro, 251
  of estimates, 428–433
  project cost and bid, 339
  rate of return, 299–300
  regression, 216
  sensitivity, 397, 404, 449
  trade off for system, 402
  variance, 436
Analysis of estimates,
  428–433
Analytical aids, 394
Annual system cost, 412
Approach length for
  machining, 269
Appropriation estimate, 16
Arithmetic graph paper, 199

Assets, current, 98
  fixed, 111
  intangible, 98
Assets accounts, 97
Assurance, estimate, 428–449
  learning, 442
  operation, 435–440
  product estimate, 441–444
  project estimate, 351
  system estimate, 448–449
Attendance, 42
Audit, estimating, 434
Auditing, 457
Audits, 468
  administrative, 467
  estimating, 467
Available funds, 347
Average, 142, 209
  estimating method, 195
  moving, 164
Average annual rate of return, 368
Average baseline dollars per period, 359
Average concept, 209
Average cost, 231
Average deviation percentage error, 219
Average index rate, 173
Average investment, 369
Average periodic change, 172
Average time error, 219
Average time value, standard data, 218
Averages, moving, 163–170
Avoidable delay, 24
Avoided cost, 400

B

Backcasting, 173
Balance sheet statement, 104
Balance sheets, 93
Balances, trial, 103
Ball park estimate, 17
Bank loans, 98
Baseline dollars per period, 358
Baseline estimate, 342
Basic contract types, 458–460
Basic item costs, 227
Basic items, 226
Battery limit methods, 192

Bayesian analysis, 209
BCWP (*see* budgeted cost of work performed), 445
BCWS (*see* budgeted cost of work scheduled), 445
Behavioral considerations, 433–435
Bench mark, 172
  cost, 227
  year, 226
Benefit, negative, 417
Benefit cost, 404–412
  marginal ratios, 412
Benefits, fringe, 41
  secondary, 408
Beta probability distribution, 213
Bias, in estimates, 37
Bid, 335, 352
  low ball, 431
  project formula, 366
  unsuccessful, 429
Bid contracts, 458
Bid market, 312
Bid shopping, ethics, 470
Bidding, 458
Bill of material, 321
  costed table, 297
  graphical, 295
  priced for audits, 470
  project, 340
  tree, 296
Bill of material explosion, 295
Bill of material tree, 296
Bill of materials, 293
Bills, 94
Bills of exchange, 15
Bills of material, 81
Binomial expression, 36
Binomial variance, 36
BLS (*see* Bureau of Labor Statistics), 11
Blue collar–management, 22
Boeing concept of learning, 204
Boiler plate contract clauses, 465
BOM (*see* bill of material), 297
  project, 340
Book value, 113
Bookkeeping, 94, 401
Bottoms up estimating, 321

Breach of contract, 467
Break even, 276, 305–308
  analysis, 206, 235
Budget, 107, 448
  system, 396
Budgeted cost for work performed, 445
Budgeted cost for work scheduled, 445
Budgeting, 107–110
Budgeting and funding estimate, 16
Budgets, 417–418
Budgets, appropriation, 107
  fixed, 107
  system, 417–418
  variable, 107
Burden, actual rate, 121
Bureau of Labor Statistics, 11
Business entity, 96
Business forecasting, 93–95, 141
Business transactions, 93–95
Buy, 278, 297
Buyer, negotiation, 468
B/C analysis, 405

C

Capacity ratio, 208
Capital, 290
  cost, 334, 405
  fixed, 336
  investment, 299
  net working, 335
  reinvestment, 1
  working, 117, 299
Capital stock, 98
Capture percentage, 138, 251, 432
Cash, working, 336
Cash basis, 97
Cash flow, 288, 307
  gain, 307
  loss, 307
  statement, 298–299
Cash flows, system, 399
Causality, 162, 197, 416
Ceiling price, 460
Central limit theorem, 213
  project, 356
Cents, 13
CER, multivariable, 209

CERs, 197, 317, 448, 465
Change clauses, 465
Change order, 466
Chart of accounts, 99, 109
Checks, 94
Circular cutting velocity for
    machining, 258
Claims, contract, 457
Closed account, 102
Commitments, project, 342
Committed dollars, 347
Commodities, 64
Companion estimates, 308
Comparative economics, 276
Comparison, 314
Comparison estimating, 194
Competition, reason for
    estimating, 2
Competitive bidding, 467
Composite index, 173
Composite learning model,
    306
Compound interest, 371
Compounding periods, 371
Comprehensive public
    viewpoint, 408
Computer in regression,
    161–162
Concepts, design, 4
    engineering, 4
Conceptual methods, 192
Concurrent engineering, 288
Conference, 314
Conference method of
    estimating, 193
Confidence bands, 151
Confidence interval, 37, 152
Confidence limits, 148–153
Conservatism convention, 96
Consignment material, 295
Consistency, 96, 215, 432, 469
Constant dollar cost, 440
Constant dollars, 207, 314
Constant element, 24, 216
Constraints, 403
Construction estimate, 360
    financing, 363
Construction materials, 69
Contingent maximum cost,
    357
Contingency, 338, 354–358,
    399, 430
    auditing, 470

cost, 310, 338
    cost estimating, 354
    risk graph, 386
Continuous timing, 24
Contract, 458
Contract, boiler contract
    clauses, 465
Contract, breach of, 467
    cost plus a fixed fee, 465
    cost plus award fee, 465
    cost plus incentive fee, 463
    firm fixed price, 460
    fixed price incentive, 460
    fixed price with
        redetermination, 463
    open purchase, 459
    repair work, 460
    standard clauses, 465
    termination, 467
    time and material, 460
Contract administrators, 467
Contract clauses, 465–467
Contract considerations, 457–
    470
Contract definition, 341
Contractor's engineer, 335
Contra-depreciation, 111
Contribution, 316
Contribution, for profit, 442
Contribution pricing, 126
Control of cost, 93, 430
Control variable, 133
Conversion cost, 314
    pricing, 315
Conversion factors, 14
Converting type of joint costs,
    76
    termination, 467
    contract, 467
Copyrights, 98
Correction of defects,
    contract, 466
Correlation, 154, 157–159, 225
    coefficient, 159
Cost, 17, 96, 251
    accounting data, 93
    actual, 428, 458
    actual for work performed,
        445
    actual purchase, 70
    analysis, 141
    annual system, 412
    average, 231

avoided, 400
basic item, 226
bench mark, 227
benefit, 404–412
bid analysis for projects,
    352–367
budgeted for work
    performed, 445
budgeted for work
    scheduled, 445
capacity ratio for total, 208
capital, 334, 405
contingent maximum for
    pricing, 357
contingency, 310
contractural for material, 70
control, 93, 430
conversion, 314
converting joint, material,
    76
corrective maintenance, 415
criterion, 212
cumulative baseline, 361
current material, 73
delivery method, material,
    73
design to, 443
differential, 230
direct materials, 64
distributing joint, material,
    76
effective gross hour, 47
engineering, 417
engineering design, 226
engineering speciality, 3
facility and equipment, 338
first, 334
fixed, 399
full, 295
full variable, 316, 320
goods manufactured, of,
    126, 311
gross hourly, 44–46, 123,
    277
gross hourly for labor, 296
handling, 255
highest cost for
    contingency, 355
incremental, 230
indexes, 170
inherited, 399
inherited and marginal, 418
investment, 406

Cost *(cont.)*
 job order, 127
 joint labor, 46–49
 joint material, 75–81
 labor, 459
 labor assurance, 436
 labor for joint material, 78
 labor for product, 295
 last, material, 73
 lead time replacement,
  material, 73
 lowest for contingency, 355
 lump sum, 334
 machining, 256
 marginal, 279, 401
 markup on, 315
 material, 67, 278
 material for product, 295
 material policies, 69–75
 material yield, 67
 minimum average, 237
 modal, 213
 money out of pocket,
  material, 73
 most likely for contingency,
  355
 nonrecurring, 399
 operation, 277–279
 operation unit, 253
 operation, tooling of, 278
 opportunity, 400, 418
 optimistic, 213
 original, material, 73
 overhead, 119
 overrun, 445
 overrun for contingency,
  356
 pessimistic, 213
 plus, 315
 power consumption, 415
 preparation of estimate, 192
 prime, 123, 251
 process cost, 127
 productive hour, 119, 277
 project, 399
 project estimate, 334
 quotation, 70
 quote or price in effect, 70
 reasonable, 459
 recurring, 399
 residual, 401
 salvage, 67, 401

 standard, 431
 standard unit, 436
 sunk, 117, 401
 system fixed, 397
 system, marginal of, 399
 target, 460
 time phased system, 402
 tool, 259
 tool changing, 259
 tooling, 273
 total for operation, 259
 true, system for, 418
 underrun for contingency,
  356
 unit, 259
 unit maufacturing
  operation, 278
 upper limit, 214
 upper limit for project, 356
 variable, 314
 warranty, 414
Cost accounting cycle, 93, 108
Cost analysis, 141, 395
 steps, 197
Cost and bid analysis, 339,
  352–367
Cost and time estimating
  relationships, 197–209
Cost codes, engineering, 109
Cost control, 430
Cost drivers, 197
Cost effectiveness evaluation,
  394
Cost elements, 335
Cost engineers, 467
Cost estimate, 457
Cost estimate, priveleged
  information, 468
Cost estimating, 3, 457
 estimating relationships,
  197, 448
 introduction, 1–20
 performance, 432
Cost for corrective
  maintenance, 415
Cost fundamental, 96
Cost goal, design to cost, 321
Cost indexes, 170
Cost of capital, 316
Cost of direct material, 64
Cost of goods manufactured,
  126, 311

Cost of material, 67
Cost of preparation of
  estimate, 430
Cost plus, 315
 contract for project, 366
Cost plus a fixed fee, 294
Cost plus a negotiated fee, 294
Cost plus award fee, contract,
  465
Cost plus incentive fee, 463
Cost reimbursable contracts,
  457
Cost schedule performance,
  447
Cost scheduling, project, 348
Cost variance, 445
Costed bill of material, 297
Costing, activity based, 119
 direct, 121, 126
 product, 289
 variable machine hour, 137
Costs, actual, 430
 audit for direct and
  indirect, 469
 contingency, 338
 engineering, 98
 engineering estimates,
  292–294
 fixed, 272, 314
 fixed overhead, 125
 fringe, 44
 hidden, 401
 initial, 112, 272
 interest, 338
 investmant, 299
 joint, 47, 410
 joint of overhead, 120
 life cycle, 402
 lump-sum for engineering,
  294
 marginal, 230, 260
 nonrecurring initial fixed,
  272–277, 289
 operation, 400
 operation and maintenance,
  400
 recurring, 308
 reported, 428
 setup, 272
 standard, 431
 variable, 125
 warranty, 414

Costs of engineering, 293
Costs of goods manufactured, 289
Cost-estimating form, 308
Cost/schedule control, 345
CPFF (*see* cost plus a fixed fee), 465
CPIF (*see* cost plus incentive fee), 463
CPM (*see* critical path methods), 341
Credit, 94, 100
Credits, project, 337
Creep, index, 173
Crew time, 48
Crew work, 48
Critical path methods, 341
Critical production rate, 237
Cumulative average curve, 203
Cumulative average learning, 30, 442
Cumulative average number of labor hours, 203
Cumulative baseline value, 361
Cumulative frequency, 141
Cumulative probability of occurrence, 141
Current assets, 98
Current cost, 73
Curve, cumulative average, 203
    density distribution, 429
    learning, 199
    manpower, 363
    normal, 147
    ogive, 342
    S, 360, 444
Curvilinear regression, 207
Cutting speed for machining, 256
Cycle, 24, 253, 309
    accounting, 108
    floor to floor, 51
    standard time data, 254
Cycle time, 253
    direct labor, 49
Cycles, 166

D

Data, 141
    detail standard time, 223
    historical, 224

measured, 224
plotting of, 143
policy, 224
predetermined motion time, 215
project, 340
sample, 146
single valued determined, 403
Day work, 43, 57
Debit, 100
Debits, project, 337
Debts, long term, 98
Debts, short term, 98
Definition, project, 340
Dejoints, 46
Delay, avoidable, 24
Delay allowance, 24, 30
Delays, 30
Density distribution curve, statistics, 429
Dependent variable, 133
Dependent variable time, 217
Depletion, 118
Depreciable property, 113
Depreciation, 107, 111–118, 289, 298, 353, 397
    accumulated, 117
    allowance, 111, 276
    contra, 111
    machine hour rate, 125
    productive hour rate, 123
    reserves for, 111
    straight-line, 115
    Sum of the years', 115
    unit of production, 115
Descriptive statistics, 141
Design, 3, 7
    engineering cost, 226
    for cost, 321, 443
    operation, 8, 334
    problem, 4
    product, 8, 272
    project, 9, 334
    system, 9, 394
Design and evaluation, 181–192
Design engineering, 457
Design engineers, 467

Design feasibility estimate, 16
Design length for cutting, 269
Design to cost, 320–322, 443
Designer, 7
Designs, purpose, 250
Detail estimate, 192, 251, 334, 432
    estimates, 251
Detail methods, 191–238
Detail product estimate, 288
Detail standard time data, 223
Deterioration, 416
Differential cost, 230
Direct costing, 121, 126
Direct costs, audits, 469
Direct hire labor, 348
Direct labor, 22–23, 41, 252, 279, 337
    learning, 300, 308
    man hours for learning, 198
    operation assurance, 435
    system, 417
Direct labor cycle time, 49
Direct labor estimating, 253
Direct labor learning, 196, 300
Direct labor man hours, 198
Direct material, 279, 289, 439
    operation assurance, 435
Direct material and subcontract, 339
Direct material and subcontract estimate, 349
Direct material cost analysis, 70
Direct materials, 63, 417
    project, 337
Disbenefits, 405
Discount factor, 416
Discounted value, system, 403
Discounting, 402, 406
Diseconomy of scale, 208
Dispute clause, 466
Disputes, contract, 458
Distributing type of joint cost, 76
Distribution, beta probability, 213
    standard normal, 213
    Student *t*, 150
Dividends, 1

Dollars, constant, 207
Double entry, 96
  bookkeeping, 94
Drawing change notices, 42
Duality, 94
Duties, 331
Dynamic curve, learning, 198

E

Earned hours, 43
ECO, 303–305
Economic want of design, 9
Economy, engineering,
    367–377
Economy of, principle, 206
  quantity, 206
  scale, 197, 206
  size, 206
  size or scale, 206
Effective gross hour cost, 47
Effectiveness, 109, 395–398,
    448
  fixed, 397
  relative, 398
  system, 399
Efficiency, 109, 395
  labor, 44
Efficiency, material, 67
  workers, 43–44
Efficiency of conversion, 67
Elastic demand, 233, 314
Element, 24, 253, 262
  constant, 24, 216
  foreign, 24
  regular, 25
  variable, 25
Element normal time, 30
Element rating factor, 30
Elemental breakdown, 24
Elemental designs, 395
Elements, noncyclic, 30
  project cost, 335
  variable, 216
Emotional estimating, 140
Employee suggestion, 251
End of year convention, 372,
    406
Engineer, 435
  contractor's, 335
  owner's, 335
  system, 395

Engineering, design, 457
  profit, 3
  project, 334
  value, 467
Engineering bill of materials,
    62
  change notices, 442
  change order, 206, 230,
    303–305, 417
Engineering cost codes, 109
Engineering cost per unit, 293
Engineering costs, 292–294,
    338
Engineering drawings, 252
Engineering economy,
    367–377
Engineering estimate, 351,
    413
Engineering materials, 64
Engineering reserve, 354
Engineering speciality, cost, 3
Engineering work packages,
    446
Engineering-economic
    methods, 370
Engineer's bill of materials, 62
Envelope curves, 292
Equation, accounting, 96
  financial and operating, 100
Equities, 96
Equivalent annual worth, 372
Error, 146, 428
  average deviation percent,
    219
  average time, 219
  gross, 222
  gross overall percentage,
    218
  policy, 197, 431
  risk, 431
  standard, 154
Error estimates, 429
Errors, 37
  probable, 213
Escalation, contract provision,
    463
Escrow account, 358
Estimate, 3, 428
  accuracy of, 16
  adjusted, 432
  appropriation, 17
  assurance, 428–449

ball park, 17
baseline, 342
construction, 360
cost of preparation, 430
design feasibility, 16
detail, 192, 334, 432
detail product, 288
direct material and
    subcontract, 349
document of evidence, 458
engineering, 413
engineering for project, 351
ethics, 470
evaluation study, 17
feasibility, 16
grass roots, 335
greenfield, 335
initial, 17
low ball, 364, 431
operation, 251
order of magnitude, 17
percentage error of, 428
preliminary, 288–322, 334,
    432
product, 288–322
product assurance, 441–444
quality, 428
realized, 432
request for, 251
single, 213
system assurance, 448–449
tolerance, 430
value, 431
verfication, 16
Estimate assurance, 351,
    428–449
  primary goal, 433
Estimate classification, 16
Estimate value, 431
Estimated cost at completion,
    445
Estimated costs, contract, 458
Estimates, analysis, 428–433
  companion, 308
  detail, 251
  engineering, 351
  error, 429
  facilities and equipment,
    351
  high, 429
  low, 429
  operation, 288, 308, 336

perfect, 432
product, 399
project, 399
should be, 435
unit, 195
Estimate-talk-estimate, 193
Estimating, 3, 321
analogy, 195
average, 195
comparison, 193
contingency costs, 354
cost and time, 197–208
cost performance, 432
direct labor, 253
emotional, 140
engineering costs, 292–294
engineering materials, of,
64
factor, 224–228, 273
follow-on, 206
general control model, 434
hidden card, 193
labor, 23, 273
machining, 253
marginal, 234
numerical for system, 396
operation, 250–279
open systems, 434
opinion, 192
order of magnitude, 195
percentage, 224
plant equipment, 228
policies, 429
preliminary, 396
product, 288–322
project, 334–381, 397
range, 213–215, 356, 397
ratio, 224
repair, 320
round table, 193
round table for project, 356
self method for engineering,
293
similarity, 195
system, 394–420
unit, 224
Estimating engineering costs,
292–294
Estimating methods, 191–238
Estimating policies, 429
Estimation by formula, 140
Ethics, 457, 470

Evaluation design, 5
Evaluation study estimate, 16
Exaggeration of precision, 32
Excess material, 467
Expected cost, project range
model, 356
Expected values, 209
Expenditure, project, 362
Expenditure dollars, 347
Expenditure scheduling, 343
Expense, 17, 99
indirect, 436
operating, 106
Experience curve, learning,
198
Explosion, bill of material,
295
Exponential function, 153,
165
Exporters, 15
E-T-E (see
Estimate-talk-estimate),
193

F

Facilities and equipment,
project direct materials,
350
Facility and equipment cost,
338
Factor estimating, 224–228,
273
Factory overhead, 23
Failure, 2
Failure of progress payments,
467
Fatigue, allowance, 24, 30
Favorable variance, 436
Federal Income Contribution
Act, 44
Fee, contract allowable, 464
Feed mode, 256
FFP (see fixed firm price),
460
FICA (see Federal Income
Contribution Act), 44,
353
FIFO (see First In First Out),
73
Financial documents, product
decision, 298

Financial spillovers, 408
Firm fixed price, 460
Firm price, contract for
project, 366
First cost, 334, 399
First In First Out, 73
Fixed assets, 111
Fixed budget, 398
Fixed capital, 336
Fixed capital investment, 335
Fixed cost, system, 397
Fixed costs, 272, 314
make-vs-buy, 278
nonrecurring initial,
272-277
Fixed effectiveness, 397
Fixed lead time, 72
Fixed overhead costs, 125
Fixed price contracts, 457
Fixed price incentive contract,
460
Fixed price with
redetermination, 463
Fixed pricing, 459
Fixed resource, 398
Flat rates, 279
Floor to floor time, 51
Flowchart, 225, 337
FOB (free on board), 331
Folio, 94
Follow-on estimating, 206
Follow-on procurement,
audits, 469
learning, 301–303
Footing, 94
Forecasting, 93, 140–176
Foreign elements, 26
Foreign exchange transaction,
15
Foreign exchanges rates, 15
Foreman's report, 34
Foreman's request, 251
FPR (see fixed price with
redetermination), 463
Frequency distribution, 141
Frequency of occurrence, 24
Fringe, 109
Fringe benefits, 41
Fringe costs, 44
Full cost, 60, 295, 308
variable, 320
Full cost markup, 315

Full cost plus markup, 313
Full variable cost, 316, 320
Fundamentals of time study, 25–33
Funds available, 347
Future amount, 372

G

Gain cash flow, 307
Gang nonrepetitive time study, 59
General accounts, 108
General office overhead, 353
Gilbreth, Frank and Lillian, 25
Going concern, 96
Graphic analysis of data, 141–146
Graphical bill of material, 295
Grass roots estimate, 335
Greenfield estimate, 335
Gross benefits and costs, system, 406
Gross errors, 222
Gross hourly cost, 44–46, 123, 277
  project, 354
Gross hourly direct labor cost, 310
Gross hourly labor cost, 296
Gross income, 99
Gross overall error, 218
Gross product income, 298
Guesstimate, 11, 23, 140, 312

H

Handling cost, 255
Handling time, 254
Handshake, 457
Hidden card technique of estimating, 193
Hidden costs, 401
High estimates, 429
Highest cost, contingency, 355
Histograms, 141
Historical data, 224
Historical information, 10
Historical information, materials, 62
Historical time, 22

Horizontally integrated company, 81
Horse trading, 468
Hour, effective gross cost, 47
  man, 24
  person, 24, 34
Hourly, gross cost, 44–46
Hourly labor, 41
Hours, earned, 43
  lot, 277, 309
  man, 336
Hyperbolic function, 153

I

Idle time, 24
Importers, 15
Incentive, 24
Incentive plans, 43
Income, 17, 99
  gross, 99
Income and expense, 99
Incremental capital, 299
Incremental cost, 230
Independent variable, 133
Index, 357
  bench mark, 172
  composite, 173
  cost, dimensionless, 171
  labor, 171
  market basket, 190
  profitability, 375
  system, 398
Index creep, 173
Index numbers, 170
Index rate, 173
Indexes, 170, 226
  kinds, 176
Indexes of Output Per Man Hour, 176
Indirect costs, audits, 469
Indirect expenses, 436
Indirect labor, 22–23, 337
Indirect materials, 64
Indirect materials, project, 337
Inflation, 3, 81, 207, 399, 448
Information, 10, 394
  historical, 10
  historical, materials, 62
  measured, 10
  measured materials, 62
  policy, 10
  product estimating, 295–297

Information flow structure, 394
Inheritance, 400
Inherited and marginal cost, 418
Inherited cost, 399
Initial costs, 112, 272
Initial estimate, 17
Insurance, life, 45
  bonding, 467
  medical and dental, 45
  unemployment, 45, 279, 353
  workmen's compensation, 353
Intangible property, 113
Intangibles, 396
Intangible assets, 98
Integrity, estimates, 470
Intercept, 133
Interdivisional transfer materials, 63
Interest, 338, 358–364, 371
  allowance, 276
  compound, 371
  short term construction, 358
  simple, 371
Interest charges, 338
Internal rate of return, 375
Interval, 141
Introduction, cost estimating, 1–20
Inventory, 70, 334
Investment, 334
  fixed capital, 335
  return on, 316
  system, 400
Investment analysis, 289
Investment capital, 299
Investment costs, 299, 406
Investors, 288
Invoices, 94
IRS, (Internal Revenue Service), 113, 369
Item, one-of-a-kind, 334

J

JIT (*see* just in time), 72
Job allowances, 30
Job description, 41–42
Job lot production, 252
Job order, 127

Job overhead, 353
Job shop, 34, 69
Job speed, 49
Job standard, 31
Job tickets, 34
Joint cost, distributing type,
    77, 120, 410
    accounting, 120
Joint labor cost, 46–49
Joint material, 75
Joint material cost, 75–81
Journal folio, 94
Journalizing, 94
Journals, 94
Just In Time, 72

L

Labor, 22–23
    cost index, 171
    direct, 22, 41, 252, 279, 289
    direct for projects, 337
    direct for system, 417
    direct hire for projects, 348
    direct labor learning, 196
    direct learning, 300
    direct man hours for
        learning, 198
    efficiency, 44
    gross hourly, 296
    hourly, 41
    index, 173
    indirect, 22–23, 41
    indirect for projects, 337
    joint cost, 46–49
    learning, 49
    net variance, 436
    nondesignated, 22
    nonrecurring, 22
    preventive maintenance, 415
    project, 339
    project gross hourly cost, 354
    project subcontract, 348
    recurring for project, 349
Labor analysis, 22–51
Labor cost, product, 295
Labor costs, 459
    operation assurance, 438
Labor efficiency, 44
Labor estimate, 339
    project, 348
    scheduling for project, 348

Labor estimating, 23, 273
Last cost, 73
Last In First Out, 73
Last inventory withdrawal, 72
Law, wage hour, 44
Laws of Probability, 36
LCC (*see* life cycle cost), 412
Le Systeme International
    d'Unites, 12
Lead time replacement, 73
Learning, 49–50, 154–155,
    179, 198–206, 299,
    300–308,
    average model, 204
    break-even, 305
    composite, 306
    direct labor, 308
    dynamic curve, 198
    experience curve, 198
    pro forma, 311
    product assurance, 442
    unit model, 204
    unlearning, 439
Learning curve, 199
Learning improvement, 49
Learning model assumptions,
    198
Learning pro forma, 311
Least squares, 146–162
Ledger, 94
Legal requirements, 457
Length of cut for machining,
    256, 267
Letter contract, 344
Levels, work breakdown
    structure, 340
Liabilities, 96
Life, 112
    service, 413
Life cycle, 448
    cost, 402, 412–417
Life insurance, 45
LIFO (*see* Last In Last Out),
    73
Likelihood of occurrence, 210
Liquidity, 369
Litigation, contracts, 458
Logarithmic graph paper, 199
Long cycle, 33
Long-lead time components,
    342
Long term debt, 363
Long-term uncertainty, 397

Lookup rules of thumb, 216
Loss cash flow, 307
Loss of material, 65
Lot, 70
    hours, 277, 309
    production, 252
    time, 303
Low ball bid, 431
Low ball estimate, 364
Low estimates, 429
Lowest cost, contingency, 355
Lump-sum contract, 459
Lump-sum cost, 334
Lump-sum turn key,
        engineering costs, 294

M

Machine hour cost, 311
Machine hour, variable cost,
    137
Machine hour overhead rate,
    125
Machine hour rate, 277
Machine hours, 121
Machine interference, 24, 48
Machine time, 24 26
Machining cost, 256
Machining estimating, 253
Machining speeds and feeds,
    256
Machining time, 256
Make, 278, 297
Make-versus-buy, 278, 300
Man days, standard, 336
Man hour reports, 33–36
Man hours, 336, 436
Man hours, preparation of
        estimate, 192
    standard, 36
Man-minutes, 33
Man-month, 33
Manpower curve, 363
Manual operator time, 26
Manufacturing materials, 69
Man-weeks, 33
Man-year, 33
Margin, opportunity, 317
Marginal benefit cost ratios,
    412
Marginal cost, 279, 399
    system, 399

Marginal costs, 230, 233, 260, 401
Marginal effectiveness, 233
Marginal estimating, 234
Marginal profit, 236
Marginal rate of return, 233
Marginal return, 233
Marginal revenues, 260
Marginal savings, 233
Marginal yield, 233
Market basket index, 190
Market value, 97
Marketing, 75, 288
Marketing request, 251
Markup, 313, 366
Markup for project bid, 366
Markup on cost, 315
Mass production, 252
Material, 61–64
  analysis, 61–81
  consignment, 295
  cost, 67
  cost of direct, 64
  current cost, 73
  delivery cost, 73
  direct, 279, 289, 417
  excess, 467
  index, 173
  inventory effects, 70
  joint cost, 75–81
  last cost, 73
  lead time replacement, 73
  loss of, 65
  money out of pocket, 73
  original cost, 73
  raw, 297
  shape yield, 67
  singular raw, 76
  takeoff, 65
  yield, 67
Material cost, product, 295
Material cost policies, 69–75
Material cost yield, 67
Material takeoff, 65
Materials, 337
  commodities, 64
  construction, 69
  contractural, 70
  direct, 63
  direct costs, 70
  engineering, 64
  indirect, 64
  interdivisional, 63

manufacturing, 69
minor raw, project, 344
MRP, 72
normative, 64
project classification, 337
project indirect, 337
raw, 63, 290, 337, 343
raw, splitting, 77
semiengineering, 64
standard commercial, 63, 297, 343
standard commercial items, 337
subcontract, 342
subcontract for project, 349
subcontract items, 63
transfer, 63
Materials requirements planning, 72
Maximum gross profit, 237
Maximum on the project manpower, 363
Maximum production rate, 260
Maximum profit per unit, 237
Mean, system, 404
Mean estimated value, 151
Mean time between failure, 414
Mean time to failure, 466
Mean time to repair, 414, 466
Mean value of beta distribution, 213
Mean-time-between failure, 318
Measure of central tendency, 142
Measure of dispersion, 142
Measured data, 224
Measured information, 10
Measured information, materials, 62
Measured time, 23–41
Measurement, work, 25
Measurement of labor, 22, 142
Medical and dental insurance, 45
Method of estimating, battery limit, 192
  conceptual, 192
  comparison, 193
  conference, 193
  estimating, 191

opinion, 192
order of magnitude, 192
preliminary and detail, 191–238
Methods, product estimating, 292
  product estimating, 308–312
  round table, 193
  Schematic, 192
  unit, 195
Metric units, 11
Microanalysis, 251
Midpoint, 141
Milestones, 342, 434
Minimum average unit cost, 237
Minor raw materials, project, 344
Mistakes, 469
Modal value of cost, 213
Mode, 142
Model, design, 5
  engineering, 5
  power and sizing, 206–208
  regression, 146, 203
  sixth-tenth, 207
Moderate production, 252
Money, 11–16, 334
Money out of pocket, 73, 401, 435
Monopolist's product, 2
Monte Carlo analysis, 404
Mortgage payable, 98
Most likely cost, contingency, 355
Movements within cycles, 163
Moving average value, 164
Moving averages and smoothing, 163–178
MRP (see materials requirements planning), 72
MTBF (see mean time between failure), 414
MTTR (see mean time to repair), 414
Multiple correlation, 158
Multiple linear regression, 159–161
Multiple machine operations, 48
Multivariable CER, 209

N

Negative benefit, 417
Negotiation, 352, 458, 467–470
Net annual benefit, 405
Net cash flow, 362
Net equivalent annual worth, 374
Net future worth, 372
Net labor variance, 436
Net present value, 373
Net present worth, 372
Net return, 316
Net working capital, 335
Net worth, 98
Next In First Out, 73
NIFO (*see* Next In First Out), 73
Noncyclic elements, 30
Nondiscounted basis, cash flow, 299
Nonfundamental units, system, 396
Nonnegativity, 403
Nonrecurring costs, 399
Nonrecurring initial fixed costs, 272–277, 289
Nonreimbursable cost allocations, 409
Nonrepetitive time study, 32–33
Normal curve, 147, 210, 213
Normal cycle time, 30
Normal distribution, 37
Normal time, 30, 219, 222
Normative materials, 64
Numerical estimating, 396

O

Observed time, 24
Obsolescence, 397, 416
Offal, 65
Offfice overhead, 352
Ogive curve, 342, 347
One-of-a-kind product, estimate, 312
One-of-a-kind-end item, 334
Open account, 102
Open market, 312, 364
Open purchase contract, 459
Operating departments, 11, 24
Operating design, 8

Operation and maintenace costs, 400
Operation cost, 277–279
Operation cost for tooling, 278
Operation costs, 400
Operation design, 8, 334
Operation estimate, 251
    assurance, 435–440
Operation estimates, 250–279, 288, 308, 336
Operation process sheet, 310
Operation unit cost, 253
Operational systems, 394
Operations, 251, 459
Operations, multiple machine, 48
Operations, repetitive, 27
Operations sheet, 261, 277
Operator's effort, 27
Opinion, 314, 358
Opinion method of estimating, 192–193
Opinion probabilities, 210
Opportunity cost, 400, 418
Opportunity margin, 317
Optimistic cost, 213
Order market, 312
Order of magnitude, estimating method, 195
Order of magnitude estimate, 17
Order of magnitude methods, 192
Ordinary monetary units, 396
Original cost, 73
Out of pocket, cash flows, 399
Out of pocket cash flow, project, 342
Output, production values, 78
Overhead, 118–126, 272, 352–354
    fixed costs, 125
    for engineering costs, 293
    job for project, 353
    joint cost, 120
    nonvolume variables, 119
    office and field for project, 352
    office for project, 352
    plant rate, 311
    predetermined rates, 121
    primary distribution, 119

    project, 352–354
    secondary overhead, 119
    variable, 279
    variable costs, 125
    volume related variables, 119
Overhead cost, 119
    variable, 125
Overhead methods, 121–127
Overrun, contract, 464
Overrun cost, 356, 445
Overtime, 45–47
Overtime, contract, 467
Overtravel length, 269
Owners, 288
Owner's cash flow, 342
Owner's engineer, 335
O&M (*see* operation and maintenance), 400

P

Paid clock hours, 47
Parabusiness partner, 457
Parameter in estimating, 197
Pareto's law, 435
Partial payments, project, 358
Patent rights, 467
Patents, 98, 458
Pay, sick, 46
    vacation, 45
Payback period, 369
Payback time, 299
Payments, progress, 227
Payrolls, 41
Percent return, engineering economy, 368
Percentage error of estimate, 428
Percentage estimating, 224
Perfect estimates, 432
Performance, 42
Performance bond, 466
Performance measurement, 446
Performance plan, 43
Performance subsidy, 47
Period, project, 346
Perishable tools, 272
Permanent tools, 272
PERs (*see* price estimating relationships), 317
Person hour, 34

Personal allowance, 24, 30
PERT (performance evaluation and review techniques), 213, 341
Pessimistic cost, 213
PF&D (personal, fatigue and delay allowances), 31, 34, 255
PHC (*see* productive hour cost), 277, 295
Physical systems, 395
Piece rate plans, 43
Plant equipment estimating, 228
Plant load, 278
Plant overhead rate, 311
Plotting of data, 143
Plow back, 1
Point of indifference, 235
Policies, estimating, 429
systems, 397
Policy data, 224
Policy error, 197, 431
Policy information, 10
Polynomial function, 153
Pooling, 121
Post estimate analysis, 228–238
Posting, 94
Power consumption cost, 415
Power function, 153
Power law and sizing model, 206–208
Predetermine motion time data, 33, 215
Prediction interval, 152
Prediction limits, 148–153
Preliminary and detail methods, 191–238
Preliminary engineering plan, 334
Preliminary estimate, 191–238, 288, 334, 432
Preliminary bill, audits, 469
Preoperating expenses, 299
Present value, 403
system, 403
Present worth, 402
Price, 288–292, 399, 457
ceiling, 317, 460
definition, 312
estimating relationships, 317
floor, 317

setting, 312
target, 460
Price ceiling, 317
Price estimating relationships, 317
Price floor, 317
Price setting, 312
Price strategy, 289
Priced bill of material, auditing, 470
Price-versus-make, 279
Pricing, 278
contribution, 126
fixed contract, 459
product, 93
Pricing and bid, 364–367
Pricing methods, 312–318
Primary distribution, overhead, 119
Primary products, 81
Prime cost, 123, 251
Principal, 371
Probabilistic techniques, 397
Probability, 150, 403
Problem, design, 4
Process cost, 127
Process time, 24
Product, design, 8
Product, break-even learning, 305
costing, 289
design, 8
estimate, assurance, 441–444
estimates, 399
estimating, 288–322
estimating methods, 292
pricing, 93
strategy, 288
Production planner, 251
Production rate, 260
Production study, 33
Production values of output, 78
Productive hour, product estimating, 309
Productive hour cost, 119, 277, 295
Productive hour cost rate, 123, 305, 459
Productive time, 31
Productivity, 349
Productivity factor, 337, 432

Products, primary, 81
Products, secondary, 81
Profit, 1, 17, 105, 289
contribution of, 442
definition, 312
engineering, 3
marginal, 236
maximum gross, 237
maximum per unit, 237
maximum point, 260
sharing, 460
system, 399
taget, 460
unit, 235
Profit and loss statement, 300
Profit contribution, 442
Profit limit point, 237
Profit maximum point, 260
Profit sharing, 460
Profitability, 289, 299
Profitability index, 375
Pro forma 308, 352, 377, 431
project, 358
system, 409
Program Review and Evaluation Techniques, 213, 341
Progress payment time, 228
Progress payments, 227, 358
failure of, 467
Project bid, formula for, 366
Project cash flow, 362
Project cost, 399
Project cost estimate, 334, 339
Project definition, 340
Project design, 9, 334
Project engineering, 334
Project estimate, 348
Project estimating, 334–381
Project estimating, 397
Project index, 357
Project manual, audits, 469
Project overhead, 352–354
Project savings, 370
Project start time, 228
Proposal plan, 339
Prototypes, 69
Purchased item, 295
Purchasing, 63
Purchasing departments, 11
P&L (*see* profit and loss), 106, 300

Q

QPE (*see* quote or price in
    effect), 463
Quality, index, 173
Quantity survey, 65
Quantity surveyor, 65
Quantity takeoff sheets, 62
Quotation cost, 70
Quote or price in effect, 70.
    173, 463

R

Random variable, 37, 142,
    147, 213, 397
Range, 142
Range estimating, 213–215,
    356, 397
Rate of return, 288, 372, 375
Rate of return analysis,
    299–300
Rates, fringe, 41–46
    wage, 41–46
Rating, 27
    element, 30
Rating factor, 25
Ratio estimating, 224
Raw material, 63, 113, 290,
    297, 337, 343
Real property, 113
Realized estimate, 432
Reasonable costs, 459
Receipts, 94
Recession, 3
Reciprocal function, 153
Recovery property, 113
Recurring costs, 308, 399
Recurring labor, 22
    project, 349
Regression, 146–162, 197,
    207, 231
    analysis, 216
    models, 146, 203, 230
    plane, 160
    polynomial, 231
Regular element, 25
Reimbursable cost allocations,
    409
Relative effectiveness, 398
Relative frequency, 141
Reliability, estimate, 428
Repair estimating, 320

Repair work, 460
Repayment, B/C, 410
Repetitive operations, 27
Replacement, 97
Report costs, 428
Request for estimate, 251,
    291, 335
Request for proposal, 336, 398
Request for quotation, 337
Request for quote, 251
Reserves for depreciation, 111
Residual cost, 399, 401
Retainage, 358, 467
Retained earnings, 98
Retirement, 111
Return, 367, 375
    marginal, 233
    total, 233
Return on investment, 316,
    367, 375
Revenue, 99, 289
    marginal, 260
Reverse engineering, 321
Revolving purchase order, 178
RFE (*see* request for
    estimate), 251, 295,
    335, 398
RFP (*see* request for
    proposal), 398
RFQ (*see* request for
    quotation), 337, 467
Risk, 210, 241
    contracts, 458
    system, 403
Risk errors, 431
Risk graph, contingency, 386
ROI (*see* return on
    investment), 375
Round table estimating, 193,
    356
Rules of thumb, 197
    engineering costs, 294
Run time, 253
R&D (*see* research and
    development), system,
    413

S

S curve, 360, 444
Safety, 290
Safety length for machining,
    269

Salaried employees, 41
Sales, 288
Sales forecasts, 1
Sales income, 1
Sales tickets, 94
Salvage, 64
Salvage cost, 67, 113, 401, 416
Sample data, 146
Sampling error, 40
Sampling interval, 38
Scale of measurement,
    systems, 396
Scatter diagram, 158
Schedule commitment, 445
Schedule variance, 445
Scheduled expenditure, 445
Scheduled operating hours,
    415
Schedules, project, 341
Schematic methods, 192
Scrap, 65, 261, 278, 318, 337
    auditing, 470
    project, 350
    spare parts, 318
Secondary benefits, 408
Secondary distribution,
    overhead, 119
Secondary products, 81
Self estimating, 293
Self-funding, contractor
    project, 345
Selling, general &
    administrative, 126
Semiengineering materials, 64
Semilog function, 153
Sensitivity, 218, 404
    analysis, 397, 449
Service life, 413
Setup, 253, 309
    standard time data, 253
Setup costs, 272
Setup standard time data, 253
SG&A (*see* selling, general,
    and administrative), 126
Shape, 64–69, 436
Shape yield, 67, 261
Share line, 462
Shareholders, 288
Shares, 98
    contract types, 461
Short cycle production, 33
Short term construction
    interest, 358

Should be estimates, 435
Shrinkage, 65, 261, 297, 337
  project, 350
SI, (*see* Le Systeme
    International d'Unites),
    12
Sick leave, 47
Sick pay, 46
Side by side comparison, 434
Signal to noise ratio, 432
Similarity estimating, 195
Simple correlation, 158
Simple interest, 371
Single end item, 377, 399
Single estimate, 213
Single payment compound
    amount factor, 372
Single value estimating, 213
Single valued determined
    data, 403
Single-point tool model, 253
Singular raw material, 76
Sinking fund, depreciation,
    117
Sixth-tenth model, 207
Skewed distributions, 148
Skewed right distributions,
    213
Skill, 49
Slippage, project planning,
    445
Slope, 133
Smoothing, 163–170
  constant, 165
Snap back, 25
Snap observation, 25
Social Security Act, 44
Spare parts, 206, 318–320, 415
  auditing, 470
Specifications, 62
Split point, 76
Split time, 48
Splitting, 48
Splitting, material, 76
Splitting of raw materials, 76
Spot exchange, 15
Squared correlation
    coefficient, 154
Standard, 31, 435
  job, 31
Standard commercial
    materials, 63, 297, 337,
    343

Standard contract clauses, 465
Standard costs, 431
Standard data, 215
Standard deviation, 142
  of the proportion, 37
Standard error, 154
Standard man hours, 36
Standard normal distribution,
    213, 356
Standard parts, 63
Standard purchased parts, 290
Standard time, 25, 49, 309
Standard time data, 215–224,
    253, 271, 310
  average, 218
  tool, 273
Standard time units, project,
    337
Standard unit cost, 436
Statement, balance, 104
Statement, income and
    expense, 104
  profit and loss, 104
Statements of income, 93
Statistical Abstract of the
    United States, 176
Statistical decision theory, 209
Statistical uncertainty, 397
Statistics, 141
Stewardship, 2
Stock purchase plan, 46
Stopwatch procedure, 25
Straight-line method,
    depreciation, 115
Structure of accounts,
    100–104
Student's *t* distribution, 150
Subcontract, 466
Subcontract items, 63
Subcontract labor, project, 348
Subcontract materials, 342
  project, 349
Sum of the years' digits
    method, depreciation,
    115
Sunk cost, 117, 275, 401
Supplemental pension, 46
System, 394
  estimate assurance, 448–449
  present worth, analytical,
    402
System design, 9, 394
System effectiveness, 399

System engineer, 395
System estimate, elements,
    399
System estimate assurance,
    448–449
System estimating, 394–420
System investment, 400
System pro forma, 409
System risk, 403
Systems, operational, 394
Systems, physical, 395

T

T account, 94, 100
Takeoff, material, 65
Tall poles, system estimate
    assurance, 448
Tangible property, 113
Target cost, 460
Target fee, 464
Target price, 460
Target profit, 460
Targets, Design to cost, 321
Tax rate, 298
Taxes, 1, 99, 235
  allowance, 276
Taylor, Frederick, 25
Taylor tool life model, 187,
    258, 281
Technology creep, 174
TERs, 197, 217
Time, allowed, 34
  crew, 48
  cycle, 253
  dependent variable, 217
  direct labor cycle, 49
  fixed lead, 72
  floor to floor, 51
  idle, 24
  lot, 303
  machine, 24
  machining, 256
  manual operator, 26
  normal, 30, 219, 222
  observed, 24
  payback, 299
  process, 24
  productive, 31
  run, 253
  setup, 253
  split, 48

standard, 49, 309
standard data, 310
tool changing, 259
unit, 203
units of labor, 251
Units per, 32
Time and material contract, 460
Time driver, 253
Time estimating relationships, 197
Time horizon, 397
Time phased system cost, 402
Time series, 146, 163
Time series period number, 164
Time study, 25, 215
   all-day, 33
   gang nonrepetitive, 59
   nonrepetitive, 32–33
Time trend, 160
Time units, project, 337
Time value of money, 371, 397
Tolerable maximum sampling error, 37
Tolerance, estimate, 430
Tool changing cost, 259
Tool changing time, 259
Tool estimating data, 273
Tool life equation, 259
Tool materials, 256
Tools, 251
Total cost, capacity ratio, 208
Total cost per operation, 259
Total return, 78, 233
Traceability, 78
Trade off, 275
   analysis, 402
Transaction, 93
   business, 93–95
Tree, bill of material, 296
Trend line, 146
Trial balance, 103
True cost, 418
True rate of return, 375
Truth tellers, 192
T. P. Wright, 204

U

UM BOM (unit material, bill of material), 296

Uncertainty, long term, 397
   statistical, 397
Underrun, contract, 464
Underrun cost, 356
Undesignated work, 33
Unemployment insurance, 45, 279, 353
Unfavorable variance, 436
Union labor, 348
Union–nonunion, 22
Unit, operation cost, 253
Unit cost, 259
   standard, 436
Unit cost estimating method, 224
Unit cost of manufacturing operations, 278
Unit cost of material, 278
Unit estimates, 195
Unit estimating, 224
Unit estimating model, 224
Unit learning, 301
Unit method of estimating, 195
Unit of time, 23
Unit profit, 235
Unit time, 203
Unitizes, 46
Units, 11–16
Units for the wage, 23
Units of labor time per month, 251
Units of labor time per piece, 251
Units of productions method, depreciation, 116
Units of reference, 77
Units per time, 32
Unlearning, 439
Unsuccessful bids, 429
Upper break even point, 237
Upper limit cost, 214
   contingency for project, 356
U.S. Customary units, 11

V

Vacation pay, 45
Value, actual, 428
   book, 113
   estimate, 431
   salvage, 113, 416

Value added, 315
Value engineering, 467
Value foregone, 417
Variable cost, 125, 314
Variable element, 25, 216
Variable machine hour costing, 137
Variable overhead, 279
Variance, 356
   analysis, 436
   cost, 445
   favorable, 436
   net labor, 436
   schedule, 445
   system, 404
   unfavorable, 436
Variance of beta distribution, 213
Vendors, project, 341
Venture worth, 373
Verification-of-vendor-quotation estimate, 16
Vertically integrated company, 81
Voucher, 94, 344

W

Wage, 23, 41, 109
Wage and fringe rates, 41–46, 109
Wage hour law, 44
Wage only method, 42
Wage rates, 41–46
Wages, 22
   attendance, 22
   performance, 22
Wage-salary, 22
Want, 9, 457
Warranty, contract, 466
   claims, 458
   costs, 414
Waste, 65, 261, 278, 297, 318, 337
   auditing, 470
   project, 350
WBS (*see* work breakdown structure), 340, 444
   audits, 469
   package, 444
Wear out, 416
Work, 23
   crew, 48

Work *(cont.)*
  day, 43
  undesignated, 33
Work breakdown structure,
    340, 443
Work in progress, 344
Work measurement, 25, 215
Work order, 251

Work package, 339, 446
  definition, 339
Work packages, 446
Work sampling, 215
Work simplification saving,
  251
Workday, 31
Worker's efficiency, 43–44

Working capital, 117, 299, 336
Workmen's compensation, 45
Worksheet, 103
Workstations, 252

Y

Yield, 67
  shape, 261